G. Meyer, D. Naumann,
L. Wesemann (Eds.)
Inorganic Chemistry in Focus II

Further Titles of Interest

M. Driess, H. Nöth (Eds.)

Molecular Clusters of the Main Group Elements

2004
ISBN 3-527-30654-4

G. Meyer, D. Naumann, L. Wesemann (Eds.)

Inorganic Chemistry Highlights

2002
ISBN 3-527-30265-4

V. Balzani (Ed.)

Electron Transfer in Chemistry

5 Volumes
2001
ISBN 3-527-29912-2

Gerd Meyer, Dieter Naumann, Lars Wesemann (Eds.)

Inorganic Chemistry in Focus II

WILEY-VCH Verlag GmbH & Co. KGaA

Prof. Dr. Gerd Meyer
Department of Inorganic Chemistry
University of Cologne
Greinstraße 6
50939 Köln
Germany

Prof. Dr. Dieter Naumann
Department of Inorganic Chemistry
University of Cologne
Greinstraße 6
50939 Köln
Germany

Prof. Dr. Lars Wesemann
Institute of Inorganic Chemistry
Eberhard-Karls-Universität Tübingen
Auf der Morgenstelle 18
72076 Tübingen
Germany

■ All books published by Wiley-VCH are carefully produced. Nevertheless, authors, editors, and publisher do not warrant the information contained in these books, including this book, to be free of errors. Readers are advised to keep in mind that statements, data, illustrations, procedural details or other items may inadvertently be inaccurate.

Library of Congress Card No.:
applied for

British Library Cataloguing-in-Publication Data:
A catalogue record for this book is available from the British Library.

**Die Deutsche Bibliothek –
CIP Cataloguing-in-Publication-Data:**
A catalogue record for this publication is available from Die Deutsche Bibliothek.

© 2005 WILEY-VCH Verlag GmbH & Co. KGaA, Weinheim

Printed on acid-free paper

All rights reserved (including those of translation in other languages). No part of this book may be reproduced in any form – by photoprinting, microfilm, or any other means – nor transmitted or translated into machine language without written permission from the publishers. Registered names, trademarks, etc. used in this book, even when not specifically marked as such, are not to be considered unprotected by law.

Composition   K+V Fotosatz GmbH, Beerfelden
Printing   betz-druck GmbH, Darmstadt
Bookbinding   Litges & Dopf Buchbinderei GmbH, Heppenheim

Printed in the Federal Republic of Germany.

ISBN-13   978-3-527-30811-8

ISBN-10   3-527-30811-3

Contents

Preface *XIII*

List of Contributors *XV*

1	**On the Track of Reaction Mechanisms:**	
	Characterization and Reactivity of Metal Atom Dimers *1*	
	Hans-Jörg Himmel	
1.1	Introduction *1*	
1.2	Principle and Realization of the Matrix-Isolation Experiment *2*	
1.3	Characterization of Metal Atom Dimers *3*	
1.4	Reactivity of Metal Atom Dimers and Comparison with the Reactivity of Single Metal Atoms *7*	
1.4.1	The Reaction Between Ga_2 and H_2 *7*	
1.4.2	The Reaction Between Ti_2 and N_2 *10*	
1.5	Concluding Remarks *12*	
	References *12*	
2	**Noble Gas Hydride Compounds** *15*	
	Mika Pettersson, Leonid Khriachtchev, Jan Lundell, and Markku Räsänen	
2.1	Introduction *15*	
2.2	Nature of Bonding *16*	
2.3	Computational Properties of HNgY *17*	
2.3.1	Molecular Structure *17*	
2.3.2	Stability *21*	
2.4	Preparative Methods *21*	
2.5	Spectroscopic Properties *23*	
2.5.1	Experimental Spectroscopic Observations *23*	
2.5.2	Computational Vibrational Spectra of HNgY *24*	
2.6	Selected Examples of HNgY Compounds *25*	
2.6.1	Xenon Compounds *25*	
2.6.2	Compounds of Lighter Noble Gases *26*	
2.7	HNgY Complexes *27*	

Inorganic Chemistry in Focus II. Edited by G. Meyer, D. Naumann, and L. Wesemann
Copyright © 2005 WILEY-VCH Verlag GmbH & Co. KGaA, Weinheim
ISBN: 3-527-30811-3

2.8	Reactions of HNgY Molecules with Hydrogen Atoms 29
2.9	Future Directions 31
	References 32

3	**Polycationic Clusters of the Heavier Group 15 and 16 Elements** 35
	Johannes Beck
3.1	Introduction 35
3.2	Principles of Stability for Polycationic Clusters 36
3.3	Synthetic Routes to Polyatomic Cations 37
3.4	Homopolyatomic Cations of the Chalcogens 39
3.4.1	Square-Planar Cations E_4^{2+} 39
3.4.2	Prism-Shaped Cations Te_6^{4+} and Te_6^{2+} 40
3.4.3	Eight-Atomic Molecular Ions E_8^{2+} and Te_8^{4+} 41
3.4.4	Towards Larger Clusters – Chalcogen Polycations with More than Eight Atoms 42
3.5	Homopolyatomic Cations of Bismuth 42
3.6	Heteronuclear Polycationic Clusters of the Heavier Group 15 and 16 Elements 44
3.7	Interactions Between Polycationic Clusters – From Weak Association to One-Dimensional Polymers 46
3.8	Summary and Outlook 49
	References 50

4	**Metal-Catalyzed Dehydrocoupling Routes to Rings, Chains, and Macromolecules based on Elements from Groups 13 and 15** 53
	Cory A. Jaska and Ian Manners
4.1	Introduction 53
4.2	Thermal Dehydrocoupling of Phosphine-Borane Adducts 54
4.3	Catalytic Dehydrocoupling of Secondary Phosphine-Borane Adducts 54
4.4	Catalytic Dehydrocoupling of Primary Phosphine-Borane Adducts 56
4.5	Thermal Dehydrocoupling of Amine–Borane Adducts 57
4.6	Catalytic Dehydrocoupling of Secondary Amine-Borane Adducts 58
4.7	Catalytic Dehydrocoupling of Primary Amine–Borane Adducts and $NH_3 \cdot BH_3$ 59
4.8	Mechanism of Catalytic Dehydrocoupling of Amine-Borane and Phosphine-Borane Adducts 60
4.9	Application of the Catalytic Dehydrocoupling of Amine-Borane Adducts 62
4.10	Summary 63
	References 63

5	**Chemistry with Poly- and Perfluorinated Alkoxyaluminates: Gas-Phase Cations in the Condensed Phase?** *65*
	Ingo Krossing and Andreas Reisinger
5.1	Introduction *65*
5.2	Available Starting Materials *66*
5.3	Weakly Bound Lewis Acid-Base Complexes of the Ag^+ Cation *67*
5.3.1	$Ag(S_8)$ Complexes *68*
5.3.2	Complexes with P_4S_3 *69*
5.3.3	Complexes with P_4 *70*
5.3.4	Complexes with C_2H_4 *72*
5.3.5	Anion Effects *73*
5.4	Highly Electrophilic Cations *74*
5.4.1	Binary Phosphorus–Halogen Cations *74*
5.4.2	Simple Carbenium Ions *78*
5.5	A Rationalization of the Special Properties of Salts of WCAs Based on Thermodynamic Considerations *80*
5.5.1	Stabilization of Weakly Bound Complexes *81*
5.5.2	Relative Anion Stabilities Based on Thermodynamic Considerations *82*
5.6	Conclusions and Outlook *84*
	References *85*
6	**Aluminum(I) Chemistry** *89*
	Herbert W. Roesky
6.1	Introduction *89*
6.2	Preparation of Aluminum(I) Halides *89*
6.3	Reactions of Aluminum(I) Halides *90*
6.4	Chemistry of $(Cp^*Al)_4$ (8) *96*
6.5	Reactions of $[HC(CMeNAr)_2]Al$ (17); $Ar=2,6$-i-$Pr_2C_6H_3$ *98*
6.6	Conclusions *101*
	References *101*
7	**Divalent Scandium** *105*
	Gerd Meyer, Liesbet Jongen, Anja-Verena Mudring, and Angela Möller
7.1	Introduction *105*
7.2	Synthesis *106*
7.3	Scandium Diiodide, Sc_xI_2 *106*
7.4	Other Scandium Dihalides? *110*
7.5	The Ternary Scandium Halides ASc_xX_3 (A=Alkali Metal, X=Cl, Br, I) *110*
7.6	Conclusions *118*
	References *119*

8		**Rare-Earth Metal-Rich Tellurides – A Spectrum of Solid-State Chemistry, Metal–Metal Bonding, and Principles** *121*
		John D. Corbett
8.1		Introduction *121*
8.2		Novel Binary Tellurides *124*
8.3		The Conceptual Polymerization of Sc_2Te by Transition Metals *130*
8.4		Versatility of the Fe_2P-Type Structure and Some Amazing Substitutions *132*
8.5		Diverse, Contrasting, and Yet Related Cluster Chain Motifs *136*
		References *141*
9		**Zintl Phases of Tetrelides – Old Problems and Their Solution** *143*
		Reinhard Nesper
9.1		Introduction and Concept *143*
9.2		$Li_{13}Si_4$ – One-Dimensional Metallicity vs. Cage Orbitals *144*
9.3		$Li_{13}Si_4$ vs. $Li_{13}Ge_4$ Structure Type *149*
9.4		$Li_{12}Si_7$ – Revisited *150*
9.5		The Lanthanide Disilicide Problem *157*
9.6		$YbSi_{1.4}$ – A Beautiful Solution to a "Disorder Problem" *161*
		References *165*
10		**Nonclassical Sb–Sb Bonding in Transition Metal Antimonides** *167*
		Holger Kleinke
10.1		Introduction: Violations of the Lewis Octet Rule? *167*
10.2		Sb–Sb Interactions in Zintl Phases *170*
10.2.1		General Remarks about Zintl Phases *170*
10.2.2		Molecular Antimony Units in Zintl Phases *171*
10.2.3		Extended Antimony Units in Zintl Phases *172*
10.2.4		Beyond Classical Sb–Sb Single Bonds in Zintl Phases *173*
10.3		Antimony Atom Substructures in Transition Metal Antimonides *177*
10.3.1		Valence-Electron-Rich Transition Metal Antimonides *177*
10.3.2		Valence-Electron-Poor Transition Metal Antimonides *180*
10.4		Concluding Remarks *187*
		References *188*
11		**Transition Metal Organosulfur Coordination Polymers** *193*
		Maochun Hong, Rong Cao, Weiping Su, and Xintiao Wu
11.1		Introduction *193*
11.2		Chain-Like Polymers of Silver with Metal–Metal Interactions *194*
11.3		Lamellar Polymers with Thiolato Bridges *197*
11.4		Organosulfur Coordination Polymers with Tubular Structures *202*
11.5		Conclusions *205*
		References *205*

12	**Molybdenum(Tungsten)-Copper(Silver) Thiolates: Rationally Designed Syntheses from "Reactive" Building Blocks** *207*
	Ling Chen and Xin-Tao Wu
12.1	Introduction *207*
12.2	One-Dimensional Thiometalate Polymers *209*
12.2.1	Linear Chains *209*
12.2.2	Wave-Like Chains *211*
12.2.3	Zigzag Chains *211*
12.2.4	Helical Chains *212*
12.2.5	One-Dimensional Chains with Square Units *212*
12.2.6	Double Chains *213*
12.2.7	Hanging Ladder Chains *213*
12.2.8	Semiconductivity and Theoretical Calculations *214*
12.3	Cage-Like Clusters *215*
12.4	Conclusions *217*
	References *218*
13	**Reactivity of Unsaturated Organic Compounds at Ruthenium(II) Centers – The Relevance of Metallacyclopentatriene Intermediates** *221*
	Roland Schmid and Karl Kirchner
13.1	Introduction *221*
13.2	The Cyclotrimerization of Alkynes Mediated by the Cp'RuCl Fragment *223*
13.3	The Cyclocotrimerization of Alkynes with Unsaturated Organic Molecules Mediated by the Cp'RuCl Fragment *227*
13.4	Dramatic Changes Seen Upon Replacement of Cl in Cp'RuCl with ER_3 (E=P, As, Sb) or CO *228*
13.5	What Happens if the Phosphine Ligand is Tethered? *235*
13.6	Outlook *238*
	References *239*
14	**Osmium(VIII) Oxide and Oxide Fluoride Chemistry** *243*
	Michael Gerken and Gary J. Schrobilgen
14.1	Introduction: Oxidation State +8 *243*
14.2	Osmium Tetroxide and Perosmates *243*
14.2.1	Osmium Tetroxide Adducts *245*
14.3	Nitrido Osmates and Organo-Imido Osmates *249*
14.4	Osmium Oxide Fluoride Chemistry *249*
14.4.1	Osmium Trioxide Difluoride *250*
14.4.2	Osmium Dioxide Tetrafluoride *255*
14.4.3	The *trans* Influence *259*
14.4.4	Computational Results and Bonding *261*
14.4.5	NMR Spectroscopy *262*
14.5	Outlook *262*
	References *264*

15	**Liquid-Crystalline Lanthanide Complexes** 267
	Koen Binnemans
15.1	Introduction 267
15.2	Mesophases 268
15.3	Lanthanide Complexes with Schiff's Base Ligands 271
15.4	Complexes of β-Diketonates 275
15.5	Bis(benzimidazolyl)pyridines 277
15.6	Mixed f-d Metallomesogens 279
15.7	Phthalocyanines 280
15.8	Porphyrins 281
15.9	Crown Ether Complexes 283
15.10	Magnetic Anisotropy and Alignment in a Magnetic Field 284
15.11	Luminescent Liquid Crystals 287
15.12	Conclusions 289
	References 289
16	**Rare-Earth Borates: An Overview from the Structural Chemistry Viewpoint** 293
	Jianhua Lin, Y. X. Wang, and L. Y. Li
16.1	Introduction 293
16.2	Rare-Earth Oxyborates 294
16.3	Rare-Earth Orthoborates 296
16.3.1	Aragonite and Calcite-Type Orthoborates 296
16.3.2	Vaterite-Type Orthoborates 297
16.4	Anhydrous Rare-Earth Polyborates 304
16.4.1	$Ln_4B_6O_{15}$ 305
16.4.2	$Ln_2B_4O_9$ 305
16.4.3	Rare-Earth Metaborates 307
16.4.4	Rare-Earth Pentaborates 307
16.5	Hydrated Rare Earth Polyborates 311
16.6	Conclusions and Further Studies 315
	References 316
17	**Ordered Siliceous Mesostructured Materials: Synthesis and Morphology Control** 319
	Pegie Cool, E. F. Vansant, and O. Collart
17.1	Introduction 319
17.2	Mesoporous Siliceous Materials: State-of-the-Art 320
17.2.1	Mesoporous Structures 320
17.2.1.1	Hexagonal Mesoporous Structures MCM-41, SBA-15 322
17.2.1.2	Cubic Mesoporous Structure MCM-48 323
17.2.2	Synthesis of Siliceous Mesoporous Materials 323
17.2.2.1	Key Steps in the Synthesis 329
17.2.2.2	Key Parameters in the Syntheses: Structure-Directing Agents 330

17.3	Case Study: Control of the Morphology and the Structural Ordering of MCM Materials by the Addition of Alcohols *334*	
17.3.1	Structure Determination *334*	
17.3.2	Effect of the Alcohol Concentration *339*	
17.4	Conclusions *344*	
	References *344*	

18 Local Crystal Chemistry, Structured Diffuse Scattering, and Inherently Flexible Framework Structures *347*

Ray L. Withers and Yun Liu

18.1	Introduction *347*
18.2	Recent Methodological Developments *349*
18.3	Applications *352*
18.3.1	Microporous Molecular Sieve Materials *352*
18.3.2	The $A_2BM_2O_8$ (A=Sr, Ba; B=Ti, V; M=Si, Ge, ...) Family of Layered Fresnoites *354*
18.3.3	The Silica Polymorphs *357*
18.4	Future Directions and Conclusions *360*
	References *361*

19 Non-Oxide Optical Glasses: Properties, Structures, and Applications *365*

Bruno Bureau and Jean Luc Adam

19.1	Introduction *365*
19.2	Glass Compositions and Properties *365*
19.2.1	Fluoride Glasses *365*
19.2.2	Chalcogenide Glasses *368*
19.3	Structural Aspects *370*
19.3.1	Fluoride Glasses *370*
19.3.1.1	Fluorozirconate Glass: ZBLAN *370*
19.3.1.2	Transition Metal Fluoride Glasses: PZG *370*
19.3.2	Chalcogenide Glasses *374*
19.4	Optical Properties and Applications *377*
19.4.1	Fiber Lasers *377*
19.4.2	Optical Amplifiers *380*
19.4.3	Lenses for Infrared Cameras *382*
19.4.4	Optical Sensors *384*
19.4.4.1	Optical Fibers *384*
19.4.4.2	Principle and Design *384*
19.4.4.3	Application in the Environment, Biology, and Medicine *386*
	References *389*

Index *393*

Preface

The first volume of *Inorganic Chemistry Highlights* appeared in 2001 and has covered a wide variety of new developments in inorganic chemistry, from areas such as theoretical, physical, solid state, molecular, organometallic chemistry. The anthology was well received. As the expression *Highlight* may be of suggestive character we have changed the name of this second volume to *Inorganic Chemistry in Focus* but have kept the series character by naming it volume II.

We have again drawn together short reviews from all areas of Inorganic Chemistry, beginning with small molecules like HGaGaH and HArF that do only exist under very special conditions, in a solid argon matrix or in the gas phase or only in the computer. We then shed light on polycationic clusters of group 15 and 16 elements and see how the metal-catalyzed dehydrocoupling works to form rings, chains and macromolecules of group 13 to 15 elements. Then it is shown how very weakly coordinating anions like poly- and perfluorinated alkoxyaluminates serve to stabilize very unusual cations like $[CI_3]^+$ or $[CSBr_3]^+$, to name only two. Next the chemistry of monovalent aluminum, for example (Cp^*Al_4), and of divalent scandium, truly stable only at low temperatures for example in $Sc_{0.9}I_2$ lay a bridge to the solid-state chemistry of rare-earth metal-rich tellurides in which not only Sc_2Te but also Lu_8Te and Lu_7Te play an important role, from all viewpoints, preparative, structural, physical and theoretical. "The desire to discover new chemistry" that may be grasped from all these articles is carried further to Zintl phases of tetrelides (in which by no means are only solutions presented to old problems) and nonclassical Sb-Sb bonding in transition metal antimonides. From this chemistry at the edge of intermetallics we travel further to the lands of coordination chemistry, first to transition metal organosulfur coordination polymers and to the rational design of reactive building blocks of group 6/group 11 thiolates. Then the Cp'RuCl fragment plays an important role for the cyclotrimerization of alkynes. The oxide and oxide fluoride chemistry of ruthenium's higher homologue, osmium, is highlighted in the following article. Liquid-crystalline lanthanide complexes lead us from basic research to materials research to which the last four chapters are devoted. Here basic and applied (better applicable) chemistry are always intertwined, in rare-earth borates as well as in ordered siliceous mesostructured materials, in inherently flexible framework structures, for which the local crystal chemistry "seen" through structured diffuse scattering plays an important role, and finally, in non-oxide optical glasses for which structures, properties and applications mix unavoidably together.

With such a great variety of compounds, their synthesis, structure(s) and properties, and (prospective or real) applications we believe that we truly put the focus on Inorganic Chemistry, and we are confident that this second volume with unaltered character but new title will be a success. The editors are grateful that the authors, all of whom have more and other work to do, took the burden and the joy to write chapters for this anthology.

Köln, December 2004

Gerd Meyer
Dieter Naumann
Lars Wesemann

List of Contributors

Jean Luc Adam
Laboratoir Verres et Céramiques
UMR-CNRS 6512
Institut de Chimie de Rennes
Université de Rennes 1
35042 Rennes Cedex
France

Johannes Beck
Institut für Anorganische Chemie
Universität Bonn
Gerhard-Domagk-Str. 1
53121 Bonn
Germany

Koen Binnemans
Katholieke Universiteit Leuven
Department of Chemistry
Coordination Chemistry
Celestijnenlaan 200F
3001 Heverden
Belgium

Bruno Bureau
Laboratoir Verres et Céramiques
UMR-CNRS 6512
Institut de Chimie de Rennes
Université de Rennes 1
35042 Rennes Cedex
France

Ling Chen
State Key Laboratory
of Structural Chemistry
Fujian Institute of Research
on the Structure of Matter
Fuzhou
Fujian, 350002
P.R. China

Rong Coa
State Key Laboratory
of Structural Chemistry
Fujian Institute of Research
on the Structure of Matter
Fuzhou
Fujian, 350002
P.R. China

O. Collart
University of Antwerpen (UA)
Campus Drie Eiken
Department of Chemistry
Laboratory of Adsorption
and Catalysis
Universiteitsplein 1
2610 Wilrijk
Belgium

Pegie Cool
University of Antwerpen (UA)
Campus Drie Eiken
Department of Chemistry
Laboratory of Adsorption
and Catalysis
Universiteitsplein 1
2610 Wilrijk
Belgium

John Corbett
Department of Chemistry
Iowa State University
Ames, Iowa 50011
USA

Inorganic Chemistry in Focus II. Edited by G. Meyer, D. Naumann, and L. Wesemann
Copyright © 2005 WILEY-VCH Verlag GmbH & Co. KGaA, Weinheim
ISBN: 3-527-30811-3

Michael Gerken
FRSC
Department of Chemistry
McMaster University
Hamilton
Ontario L8S 4M1
Canada

Hans-Jörg Himmel
Institut für Anorganische Chemie
Universität Karlsruhe (TH)
Geb. 30.45
Engesser Str.
76128 Karlsruhe
Germany

Maochun Hong
State Key Laboratory
of Structural Chemistry
Fujian Institute of Research
on the Structure of Matter
Fuzhou
Fujian, 350002
P.R. China

Cory A. Jaska
Department of Chemistry
University of Toronto
80 St. George Street
Toronto
Ontario M5S 3H6
Canada

Liesbeth Jongen
Institut für Anorganische Chemie
der Universität zu Köln
Greinstr. 6
50939 Köln
Germany

Leonid Khriachtchev
Department of Chemistry
P.O. Box 35
University of Jyväskylä
40014
Finland

Karl Kirchner
Institut für Angewandte
Synthesechemie
Fakultät für Technische Chemie
Technische Universität Wien
Getreidemarkt 9
1060 Wien
Austria

Holger Kleinke
Department of Chemistry
University of Waterloo
Waterloo
Ontario N2L 3G1
Canada

Ingo Krossing
École Polytechnique Fédérale
De Lausanne
Institut de Sciences
et Ingénierie Chimique (ISIC)
EPFL-BCH
1015 Lausanne
Switzerland

L. Y. Li
State Key Laboratory of Rare Earth
Materials Chemistry and Applications
College of Chemistry and Molecular
Engineering
Peking University
Beijing 100871
P.R. China

Jianhua Lin
State Key Laboratory of Rare Earth
Materials Chemistry and Applications
College of Chemistry and Molecular
Engineering
Peking University
Beijing 100871
P.R. China

Yun Liu
Australian National University
Research School of Chemistry
Canberra, ACT, 0200
Australia

List of Contributors

Jan Lundell
Department of Chemistry
P.O. Box 35
University of Jyväskylä
40014
Finland

Ian Manners
Department of Chemistry
University of Toronto
80 St. George Street
Toronto
Ontario M5S 3H6
Canada

Gerd Meyer
Institut für Anorganische Chemie
der Universität zu Köln
Greinstr. 6
50939 Köln
Germany

Angela Möller
Institut für Anorganische Chemie
der Universität zu Köln
Greinstr. 6
50939 Köln
Germany

Anja-Verena Mudring
Institut für Anorganische Chemie
der Universität zu Köln
Greinstr. 6
50939 Köln
Germany

Reinhard Nesper
ETH-Zürich
Laboratory of Inorganic Chemistry
Wolfgang-Pauli-Str. 10
8093 Zürich
Switzerland

Mika Pettersson
Department of Chemistry
P.O. Box 35
University of Jyväskylä
40014
Finland

Markku Räsänen
Department of Chemistry
P.O. Box 35
University of Jyväskylä
40014
Finland

Andreas Reisinger
École Polytechnique Fédérale
De Lausanne
Institut de Sciences
et Ingénierie Chimique (ISIC)
EPFL-BCH
1015 Lausanne
Switzerland

Herbert W. Roesky
Institut für Anorganische Chemie
der Universität Göttingen
Taumannstr. 4
37077 Göttingen
Germany

Roland Schmid
Institut für Angewandte
Synthesechemie
Fakultät für Technische Chemie
Technische Universität Wien
Getreidemarkt 9
1060 Wien
Austria

Gary J. Schrobilgen
FRSC
Department of Chemistry
McMaster University
Hamilton
Ontario L8S 4M1
Canada

Weiping Su
State Key Laboratory
of Structural Chemistry
Fujian Institute of Research
on the Structure of Matter
Fuzhou
Fujian, 350002
P.R. China

Y. X. Wang
State Key Laboratory of Rare Earth
Materials Chemistry and Applications
College of Chemistry and Molecular
Engineering
Peking University
Beijing 100871
P.R. China

E. F. Vasant
University of Antwerpen (UA)
Campus Drie Eiken
Department of Chemistry
Laboratory of Adsorption
and Catalysis
Universiteitsplein 1
2610 Wilrijk
Belgium

Ray Withers
Australian National University
Research School of Chemistry
Canberra, ACT, 0200
Australia

Xin-Tao Wu
State Key Laboratory
of Structural Chemistry
Fujian Institute of Research
on the Structure of Matter
Fuzhou
Fujian, 350002
P.R. China

1
On the Track of Reaction Mechanisms: Characterization and Reactivity of Metal Atom Dimers

Hans-Jörg Himmel

1.1
Introduction

The matrix-isolation technique is now well established as a valuable method to retain and characterize reaction intermediates [1]. In this chapter, it will be shown how this method can be used to characterize metal atom dimers and to shed light on their special reactivity. An understanding of the bond properties and reactivities of these metal atom dimers is a first step toward an understanding of the physical and chemical properties of metal atom clusters. Because of their high reactivity, metal atom clusters are widely used for catalytic processes. In addition, larger clusters with a diameter in the nm range (ideally 1–2 nm) exhibit quantum-size effects, which make their use in single-electron devices attractive. In spite of these wide-ranging applications, detailed information about the structures and electronic properties of metal atom clusters is sparse. There are not only experimental difficulties. Quantum chemical calculations become extremely difficult and expensive, even in the case of the metal atom dimers. Multi-reference methods have to be applied and inner-core correlation has to be taken into account. Often DFT methods fail or are not really reliable.

Although clusters are generally more reactive than metal atoms or ideal defect-free surfaces, first results show that there are large differences between clusters that consist of different numbers of atoms. In some cases, the maximum reactivity of an M_n particle (where M denotes a metal atom and $n \geq 1$) seems to be reached at the stage of the metal atom dimer. Thus, gas-phase studies revealed that Rh_2^+ brings about a spontaneous dehydrogenation reaction with CH_4, while Rh^+ ions and Rh_n^+ clusters ($n \geq 3$) do not induce spontaneous reaction [2]. In the same vein, Pt_2^+, but not Pt^+ ions or Pt_n^+ clusters, were found to react with NH_3 to give the dehydrogenation product $[Pt_2NH]^+$ [3].

After a brief description of the matrix-isolation technique, the results which have led to a detailed characterization of Ga_2 and Ti_2 will be reported. Thereafter, the reactions between Ga_2 and H_2 and between Ti_2 and N_2 are discussed. These two model reactions underline impressively the high reactivity of metal atom dimers.

Inorganic Chemistry in Focus II. Edited by G. Meyer, D. Naumann, and L. Wesemann
Copyright © 2005 WILEY-VCH Verlag GmbH & Co. KGaA, Weinheim
ISBN: 3-527-30811-3

1.2
Principle and Realization of the Matrix-Isolation Experiment

In the matrix-isolation experiment, the two reactants are isolated in a host material, which generally consists of a frozen inert gas (e.g. argon) to minimize chemical interactions between the host and the isolated reactants (Fig. 1.1). The temperature is kept very low (at a few Kelvin). At these temperatures, even small reaction barriers (of a few kJ mol^{-1}) cannot easily be surmounted ($kT=0.08$ kJ mol^{-1} at 10 K), if tunnelling processes can be neglected. Therefore, reaction intermediates that cannot survive under other conditions can be generated, trapped, and observed. Indeed, one of the great advantages of the matrix-isolation method is that the reaction intermediates can be retained for several hours or even days and therefore can be identified and characterized at leisure by applying standard laboratory techniques.

The identification and characterization of possible intermediates and reaction products usually relies on spectroscopic methods. Vibrational spectroscopy is certainly the most widely applied method, combining as it does the advantages of high sensitivity and the provision of information that can not only be used to identify the species, but also to determine some important properties such as symmetry and more detailed structural information, force constants, and, in some cases, dissociation energies (see below). To allow for a better analysis of the data, the experiments have to be repeated with as many isotopomers as possible. These isotopomers usually have to be synthesized especially for this purpose. The experimental results are often accompanied by quantum chemical calculations, which allow for a further characterization. Absorption spectroscopy with radiation energy sufficiently high to excite electrons within the species isolated in the matrix provides valuable information about possible photochemistry and may be used to analyze the properties of excited states. Thus, if the spectra are vibrationally resolved, the frequency and important structural information can be obtained for some excited states of the species under investigation. Fluorescence spectroscopy can also be extremely informative, although this technique can only be applied in certain cases. Finally, radicals might be studied by means of EPR spectroscopy (see ref. [1] for more information on possible methods of interrogation).

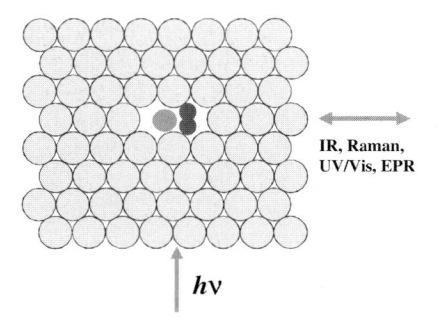

Fig. 1.1 Two reactants are isolated in a defect of the polycrystalline matrix host material. The spontaneous and photolytically induced reactions can be followed by means of spectroscopic methods, such as absorption (vibrational or electronic excitations), emission, or EPR spectroscopy.

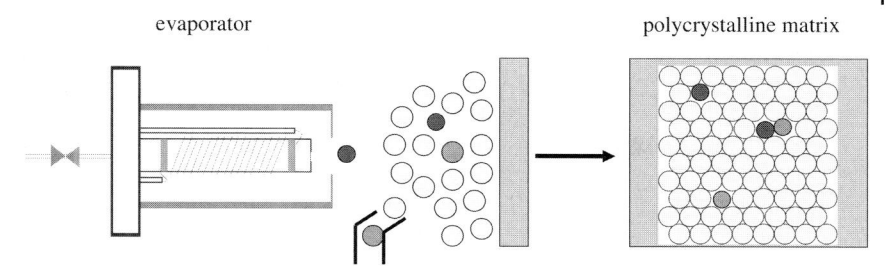

Fig. 1.2 Preparation of a matrix which contains two reactants. One of them is produced in an evaporator, the second is admixed to the matrix gas. The result is a statistical distribution of the two reactants in defect sites of the matrix host.

Fig. 1.2 illustrates the preparation of a matrix in a typical matrix experiment. In the studies discussed herein, one of the reactants is emitted from a metal evaporator. For example, gallium vapor can be generated by resistively heating the metal in a BN cell, or an alumina tube containing a carbon cell, to a temperature of 900–1000 °C. An element like titanium, for which higher temperatures are necessary, may be evaporated by directly heating a pure metal wire (to 1600–1700 °C in the case of titanium). The amount of deposited metal can be monitored with the aid of a microbalance. The other reactant is admixed to the matrix gas. The matrix is deposited onto a metal block (e.g., Cu or Rh-plated Cu) kept at a low temperature (10 K in general), generally by means of a closed-cycle refrigerator. UV/Vis spectra give useful information about the metal atom to metal atom dimer ratio in the matrix.

1.3
Characterization of Metal Atom Dimers

Metal atom dimers in a matrix can be characterized by means of absorption (e.g. UV/Vis), resonance-Raman, and/or fluorescence spectroscopy. In some cases, the resonance-Raman spectra reveal not only the $v(M–M)$ stretching fundamental, but also several overtones. These overtones can be used to estimate the dissociation energy, which is an important parameter in describing the metal–metal bond. It is also of importance for the understanding of reaction mechanisms, since the metal–metal bond is often ruptured in the course of the reaction. Thus, the energy required for the rupture of the metal–metal bond has to be compensated by the formation of other bonds if the reaction is to proceed spontaneously. In the following, the results obtained for Ti_2 and Ga_2 from matrix-isolation experiments and quantum chemical calculations are discussed. These two metal atom dimers were chosen exemplarily because, as detailed below, they show remarkably high reactivities.

Ti_2 Fig. 1.3 shows the resonance-Raman spectra obtained for Ti_2 measured with the $\lambda = 514.532$ nm line of an Ar^+ ion laser [4]. Some regions of the spectrum are

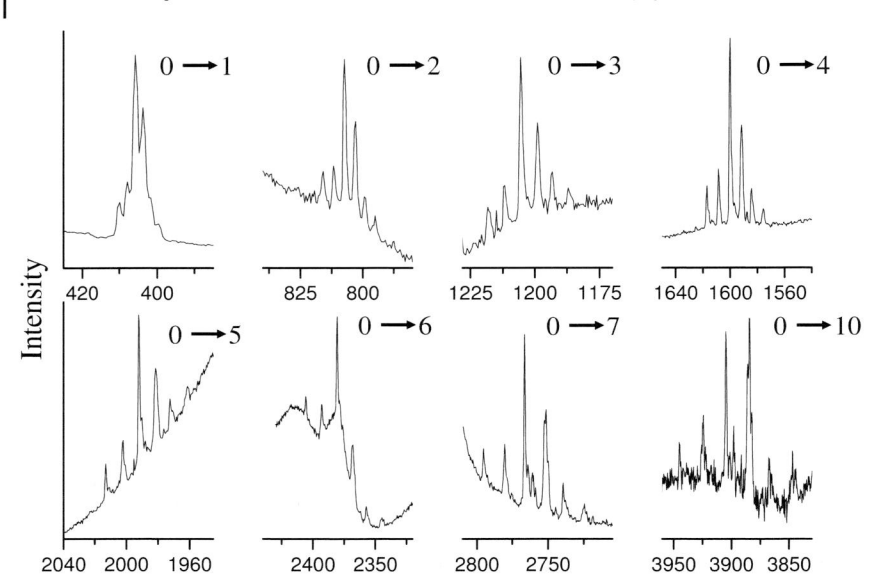

Fig. 1.3 Resonance Raman spectra of Ti_2 isolated in an Ar matrix (measured with the 514.532 nm line of an Ar^+ ion laser).

covered by fluorescence signals, which for the most part belong to Ti atoms. Nevertheless, in the regions free from fluorescence signals, a series of overtones can be measured, which exhibit an isotopic splitting. The results can be used to estimate the dissociation energy on the basis of a LeRoy–Bernstein–Lam analysis [5]. In this analysis, the potential near the dissociation limit is assumed to be a quadrupole-quadrupole-type interaction between two Ti atoms in their ^3F electronic ground state. One then interpolates between the formulas derived for the vibrational level energies near the dissociation limit (resulting from a WKB treatment) and those that are closer to the bottom of the potential-energy curve. The analysis yields a dissociation energy (D_e value) of 113.9 kJ mol^{-1} [4].

Additional valuable information is provided by absorption measurements. Fig. 1.4 shows the absorption spectrum of Ti_2 in the region between 4000 and 6500 cm^{-1} [6]. Two series of bands are visible, which can be assigned to excitations into different vibrational levels of the $1\,^3\Pi_u$ and the $1\,^3\Phi_u$ states. The relative intensities of the bands in each series can be used to estimate the difference Δr_e in the bond distances between the excited state and the ground state on the basis of a Franck–Condon analysis. To this end, Morse-type functions were assumed for each state. The analysis resulted in Δr_e values of 9 pm for the $1\,^3\Pi_u \leftarrow\,^3\Delta_g$ and of 10 pm for the $1\,^3\Phi_u \leftarrow\,^3\Delta_g$ transition. Thus, the Ti–Ti bond length is found to increase in both electronically excited states relative to the ground state.

Experimental information was also obtained for the next higher $2\,^3\Pi_u$ and $2\,^3\Phi_u$ states. In these states, the Ti–Ti bond is even longer. Tab. 1.1 includes Δr_e,

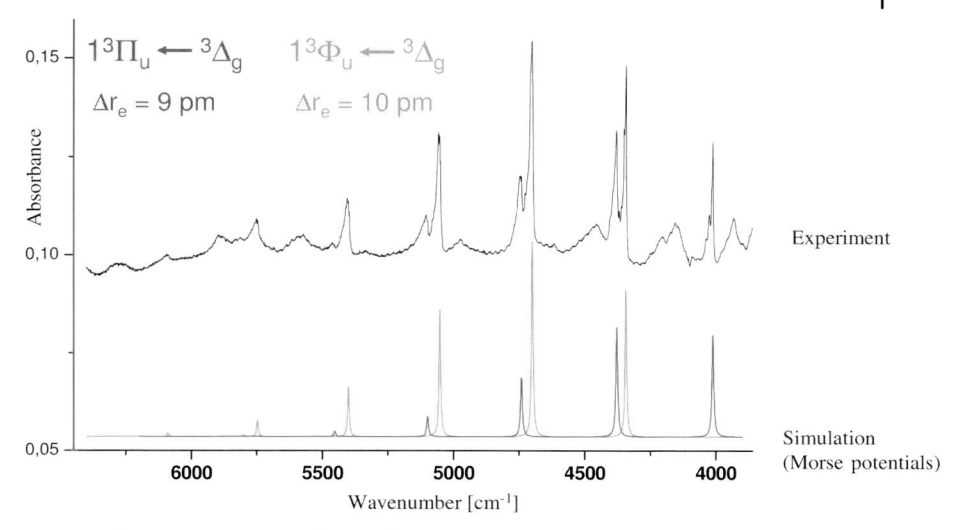

Fig. 1.4 Absorption spectrum of Ti$_2$ in the region between 4000 and 6500 cm^{-1}. The bands are due to two vibrationally resolved electronic excitations.

Spectrum labels: $1\,^3\Pi_u \leftarrow {}^3\Delta_g$, $\Delta r_e = 9$ pm; $1\,^3\Phi_u \leftarrow {}^3\Delta_g$, $\Delta r_e = 10$ pm. Experiment (top); Simulation (Morse potentials) (bottom).

Tab. 1.1 Comparison between the experimentally determined and calculated [MRCI (ANO 76432)] relative bond distances Δr_e (in pm), the harmonically corrected frequencies, and the excitation energies T_e (in kJ mol^{-1})

Electronic state	Δr_e		ω_e		T_e	
	exp.	calcd.	exp.	calcd.	exp.	calcd.
$^3\Delta_g$	0	0	407.0	369	0	0
$1\,^3\Pi_u$	9±2	11.3	371.5	330	48.2	40.5
$1\,^3\Phi_u$	10±2	12.2	359.5	317	52.1	42.5
$2\,^3\Pi_u$	13±2	17.4	282.0	259	83.9	71.4
$2\,^3\Phi_u$	13±4	17.0	288.2	268	99.4	91.7

the harmonic frequencies ω_e, and the excitation energies T_e for all states for which detailed experimental information is available. Tab. 1.1 also compares the experimentally derived values with those predicted by high-level quantum chemical calculations (MRCI method).

Ga$_2$ Again, resonance-Raman spectroscopy proves to be extremely useful to obtain information about the ground state of the dimer. The resonance-Raman spectrum of Ga$_2$ is displayed in Fig. 1.5 [7]. The ν(Ga–Ga) stretching fundamental occurs at 176.5 cm^{-1}. Three additional signals in the spectrum can be assigned to the first, second and third overtones. The signals show an isotopic

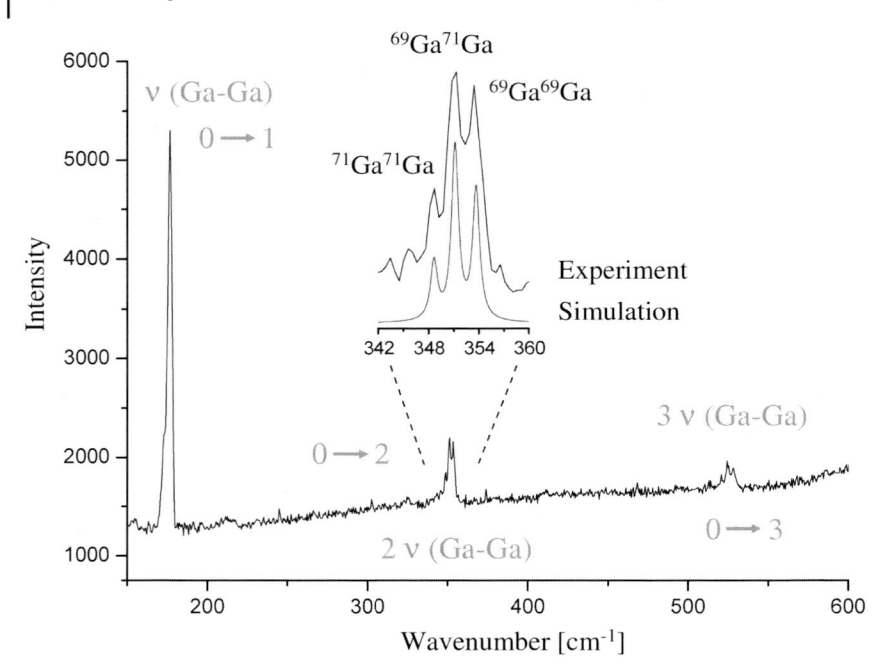

Fig. 1.5 Resonance Raman spectrum of Ga_2, obtained with the 514.532 nm line of an Ar^+ ion laser.

Tab. 1.2 Calculated dimensions (in pm), harmonic vibrational frequencies (in cm^{-1}), and relative energies (in kJ mol^{-1}) of Ga_2 in various electronic states

Ga_2		CASSCF/SVP	MP2/TZVPP	Exp.
$^3\Pi_u$	d(Ga–Ga)	276.3	271.4	
	ω(Ga–Ga)	161	178	175
$^3\Sigma_g^-$	ΔE	7.1	7.8	
	d(Ga–Ga)	251.0	247.1	
	ω(Ga–Ga)	204	222	
$^1\Sigma_g^+$	ΔE	19.0	41.5	
	d(Ga–Ga)	309.3	300.3	
	ω(Ga–Ga)	121	146	
$^1\Pi_u$	ΔE	46.3		
	d(Ga–Ga)	278.1		
	ω(Ga–Ga)	154		
$^1\Delta_g$	ΔE	56.4		
	d(Ga–Ga)	258.7		
	ω(Ga–Ga)	170		

splitting that is in pleasing agreement with the pattern expected for Ga_2 (see inset in Fig. 1.5) in its three isotopic forms $^{69}Ga^{69}Ga$, $^{69}Ga^{71}Ga$, and $^{71}Ga^{71}Ga$. The dissociation energy of Ga_2 can be estimated to be ca. 130 kJ mol^{-1}.

Unfortunately, it has hitherto not been possible to obtain detailed information about excited states of Ga_2 with energies close to the ground state. Calculations indicate that a $^3\Sigma_g^-$ state has an energy which is only ca. 7 kJ mol^{-1} higher than that of the $^3\Pi_u$ ground state. Several singlet states are also predicted to be close by in energy (see Tab. 1.2) [10]. It will be shown below that these excited states play a significant role in relation to the reactivity of Ga_2.

1.4
Reactivity of Metal Atom Dimers and Comparison with the Reactivity of Single Metal Atoms

In the following, two examples are discussed which should underline the strikingly high reactivity of metal atom dimers, namely the reaction of Ga_2 with H_2 and the reaction of Ti_2 with N_2. Both reactions proceed spontaneously with the metal atom dimers in their ground electronic states.

1.4.1
The Reaction Between Ga_2 and H_2

Ga atoms in their electronic ground state (^2P) do not react with dihydrogen. A reaction can only be brought about by photoactivation of the Ga atoms ($^2S \leftarrow {}^2P$ excitation), thereby leading to the bent radical GaH_2 [8, 9]. This result is at first glance surprising, since transition metal atoms are known to react spontaneously with H_2 only after one of the d electrons has been excited into an empty p orbital. On this basis, it has been argued that the attractive interaction between the half-filled p orbital and the σ^* orbitals at the dihydrogen initiates the reaction. A correlation diagram shows, however, that the ground state of the Ga$\cdots H_2$ system (with a large separation between Ga and H_2) correlates with an excited state of the product GaH_2. Therefore, the thermal reaction is subject to a massive reconfiguration barrier [10]. In fact, a radical mechanism is favored, which leads first to GaH and H atoms, and these combine in the second step to give the GaH_2 radical. Thus, although the overall reaction is slightly endothermic (by 16 kJ mol^{-1} according to MP2/TZVPP), [10] photolysis is needed for the reaction to take place.

For the reaction between Ga_2 and H_2 one might also assume at first glance a significant barrier, since the reaction is formally spin-forbidden (Ga_2 exhibits a triplet ground electronic state and Ga_2H_2 a singlet one). However, spin-orbit coupling is significant and provides a means by which the system can change its multiplicity from triplet to singlet. The experiments show that Ga_2 reacts spontaneously with H_2 to give the cyclic, D_{2h} symmetric Ga(μ-H)$_2$Ga molecule (see Fig. 1.6) [10]. Calculations indicate that the reaction proceeds via excited states of Ga_2. Thus, in the early stage of the approach between the two reactants, the $^3\Pi_u$ and $^3\Sigma_g^-$ type states

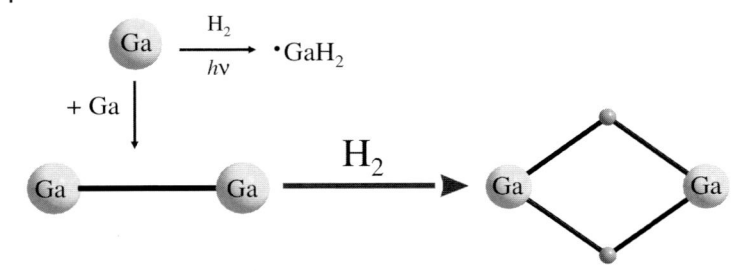

Fig. 1.6 Ga_2 reacts spontaneously with H_2 to give the cyclic, D_{2h} symmetric $Ga(\mu\text{-}H)_2Ga$ ring. Ga atoms react with H_2 only upon photoactivation, the product being the bent radical GaH_2.

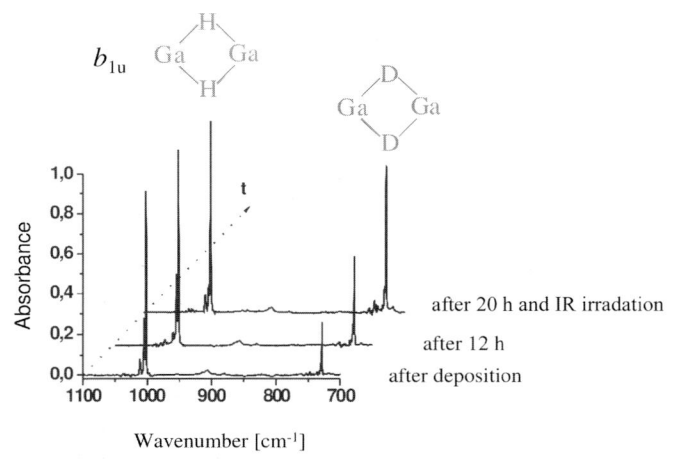

Fig. 1.7 IR spectra taken for an Ar matrix containing Ga_2 and a 1:1 mixture of H_2/D_2.

of Ga_2 mix. At the point of intersystem crossing from triplet to singlet state, a $^1\Delta_g$ type state is adopted. As a consequence, the Ga–Ga distance first shortens from 276 to 255 pm. The relevant excited states have energies which are relatively close to that of the ground state. Therefore, the barrier to reaction is relatively low. The experiments give additional information about the reaction mechanism. Thus, the reaction proceeds spontaneously with H_2, but not with D_2. In the case of D_2, the matrix has to be kept for several hours in the dark or irradiated with IR light to complete the reaction (see Fig. 1.7). This isotopic effect indicates that the barrier to reaction is of the order of the zero-point energy difference between H_2 and D_2, viz. ca. 30 kJ mol^{-1}. This value is slightly lower than the calculated estimate (ca. 50 kJ mol^{-1}). Nevertheless, both experiment and theory agree in that Ga_2 is much more reactive toward H_2 than a single Ga atom.

It is worth mentioning that $Ga(\mu\text{-}H)_2Ga$ can be converted into two other isomeric forms when the molecule is selectively photolyzed (see Fig. 1.8). In both

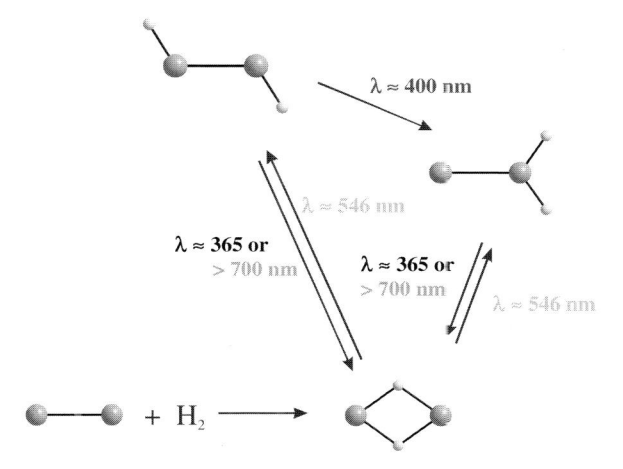

Fig. 1.8 Photoconversion of Ga(μ-H)$_2$Ga into the two isomers HGaGaH and GaGaH$_2$.

of these isomers, direct Ga–Ga bonds exist. One of the isomers is the *trans*-bent species HGa–GaH. Analysis of the spectra obtained for this molecule in combination with quantum chemical calculations clearly shows that the Ga–Ga bond is most adequately described as a relatively weak donor–acceptor interaction between two GaH diatomics. According to quantum chemical calculations, the energy for fragmentation of HGa–GaH into two non-relaxed GaH units with singlet electronic state amounts to no more than 57 kJ mol^{-1} [11]. This again shows that the Ga–Ga bond is weak. At 262.0 pm, the calculated Ga–Ga bond distance is relatively long. Meanwhile, there are structural data (from X-ray diffraction analyses) available for some derivatives Ar'GaGaAr' (e.g., with Ar' being 2,6-Dipp$_2$C$_6$H$_3$, Dipp = 2,6-iPr$_2$C$_6$H$_3$) [12, 13]. These data confirm the analysis made for HGaGaH. Like HGaGaH, the derivatives exhibit a *trans*-bent structure. The Ga–Ga bond distance in Ar'GaGaAr' amounts to 262.68(7) pm. The results also indicate that the Ga–Ga bond in the dianion [HGaGaH]$^{2-}$ cannot be adequately described as a triple bond and thus the properties differ to a large extent from those found for HCCH. This is also reflected in the different structures (*trans*-bent in the case of [HGaGaH]$^{2-}$ vs. linear for HCCH). The crystal structures of derivatives [RGaGaR]$^{2-}$ [e.g., R being 2,6-(2,6-iPr$_2$C$_6$H$_3$)$_2$C$_6$H$_3$] [14] were determined and the Ga–Ga distance was found to be short (231.9 pm). However, this short distance does not necessarily imply the presence of a triple bond. An analysis indicates that the Na$^+$ cations are engaged in the bonding [15, 16]. At the same time, the Na$^+$ ion interacts with the aromatic rings on the ligands. That the Ga–Ga bond length has to be taken with caution as a criterion for the bond order also becomes evident if the values determined for typical Ga–Ga single bonds are compared. Thus, in Ga$_2$[Ga$_2$I$_6$], the Ga–Ga bond length is 238.7(5) pm [17]. At the other extreme, a value of 254.1(1) pm was determined in the case of Ga$_2$(Trip)$_4$ (Trip = 2,4,6-iPr$_3$C$_6$H$_2$) [18] and also for Ga$_2$[CH(SiMe$_3$)$_2$]$_4$ [19].

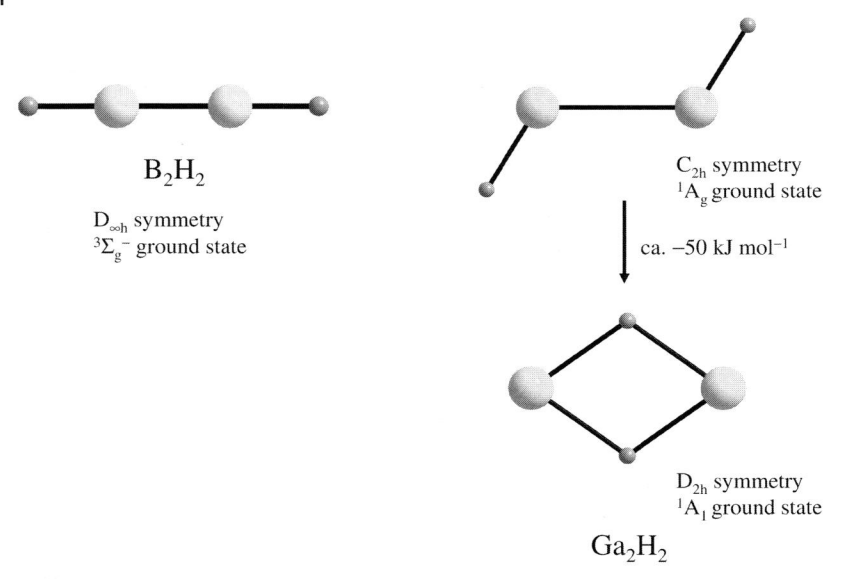

Fig. 1.9 Comparison between the structures of B_2H_2 and its homologue Ga_2H_2.

In contrast to HGaGaH, the lighter homologue HBBH is a linear molecule with a triplet electronic ground state (see Fig. 1.9). The molecule has been characterized in matrix by IR [20] and EPR [21] spectroscopies, and has also been the subject of theoretical work [22]. The results show that, as anticipated, the triplet state arises from the presence of two degenerate π-orbitals, which are each occupied by one electron. The B–B bond is strong (in line with a formal bond order of 2) and the calculated B–B distance amounts to 150.7 pm. Fragmentation of HBBH into two geometrically non-relaxed HB units with singlet electronic states requires ca. 450 kJ mol^{-1} [CCSD(T) estimate] [11]. Nevertheless, the dimerization of the molecule to give B_4H_4 in its tetrahedral global energy minimum structure is predicted to be highly exothermic [ca. –482 kJ mol^{-1} according to CCSD(T) calculations] [23].

Recently, it has been shown that Ga_2 reacts spontaneously not only with H_2, but also with SiH_4 [24]. Interestingly, the product formed in this reaction is HGa(μ-SiH$_3$)Ga, featuring a terminal Ga–H bond and the SiH$_3$ group in a bridging position.

1.4.2
The Reaction Between Ti$_2$ and N$_2$

According to matrix experiments, Ti atoms in their electronic ground state do not engage in a complex with dinitrogen. Ti$_2$ dimers, however, undergo spontaneous reaction with N$_2$. In the course of this reaction, which proceeds without

1.4 Reactivity of Metal Atom Dimers and Comparison with the Reactivity of Single Metal Atoms

$$Ti_2 + N_2 \longrightarrow Ti_2N_2$$

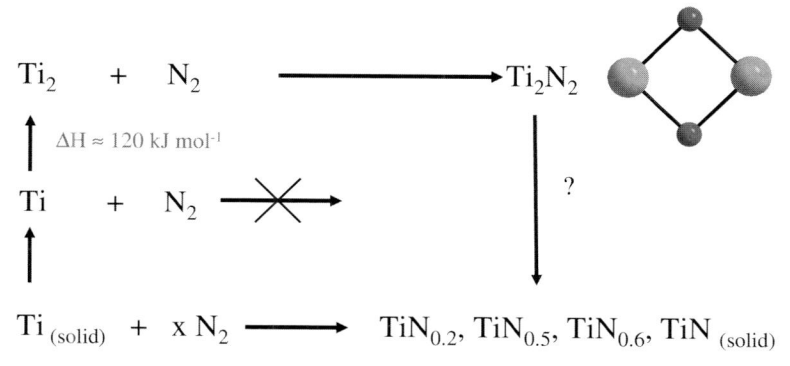

$\uparrow \Delta H \approx 120 \text{ kJ mol}^{-1}$

$Ti + N_2 \not\longrightarrow$

$\uparrow$?

$Ti_{(solid)} + xN_2 \longrightarrow TiN_{0.2}, TiN_{0.5}, TiN_{0.6}, TiN_{(solid)}$

Fig. 1.10 The reaction between Ti_2 and N_2 leads to the cyclic $Ti(\mu\text{-}N)_2Ti$ molecule, which might be an intermediate on the way to solid nitrides of titanium.

a significant barrier, four Ti–N bonds are formed at the expense of the strong NN triple bond, leading to a cyclic $Ti(\mu\text{-}N)_2Ti$ molecule (see Fig. 1.10) [25]. The spectra indicate that the product features no direct N–N interaction. The ground state of $Ti(\mu\text{-}N)_2Ti$ is a singlet state, but a triplet state is of very similar energy. It was possible to detect some vibrationally resolved excitations around 9000 cm^{-1} attributable to this species [25]. This reaction is especially interesting since solid Ti has been shown to react at higher temperatures with molecular nitrogen to first give compounds which contain N atoms dissolved in solid titanium ($TiN_{0.20}$ intercalation compound), with the h.c.p. structure of α-Ti. Finally, with increasing concentration of nitrogen and at ca. 900 °C, a defect NaCl structure is adopted. Thus, in the course of this reaction, the NN bond of dinitrogen has to be ruptured. Ti_2N_2 might be a possible intermediate on the way to these solid phases. A goal of future studies should be the estimation of the reaction enthalpy of Ti_2N_2 formation. This value could then be used to calculate the enthalpy difference between Ti_2N_2 and solid titanium nitride. Solid titanium nitride coatings are of interest as protection layers and as semiconductors.

The isodesmic reaction $2\,TiCl_4 + Si_2N_2 \rightarrow 2\,SiCl_4 + Ti_2N_2$ can be used to estimate the standard enthalpy of formation for Ti_2N_2. This reaction was calculated to be exothermic by -194 kJ mol^{-1}. First, the enthalpy of formation for Si_2N_2 has to be estimated. The enthalpy for the reaction $2Si(g) + N_2(g) \rightarrow Si_2N_2(g)$ was calculated to be ca. -650 kJ mol^{-1} [26]. With values of $+450$ kJ mol^{-1} for the standard enthalpy of formation for Si(g) [27], the standard enthalpy of formation of Si_2N_2 amounts to ca. $+250$ kJ mol^{-1}. With this value, the standard enthalpy of formation of $Ti_2N_2(g)$ can be estimated to be of the order of -49 kJ mol^{-1} (with values of -763.2 and -662.8 kJ mol^{-1} for the standard enthalpies of formation of $TiCl_4$ and $SiCl_4$, respectively [27]). Considering a value of 945.4 kJ mol^{-1} for the standard enthalpy of formation of two single N atoms, this value demonstrates the high affinity of titanium for nitrogen.

It also implies that the reaction between $Ti_2(g)$ (for which the enthalpy of formation is ca. +827 kJ mol^{-1}) and N_2 has to be highly exothermic (standard reaction enthalpy ca. –876 kJ mol^{-1}). The enthalpy of formation of solid TiN was determined to be –337.7 kJ mol^{-1}. Thus, the enthalpy for the reaction $2\,Ti(s)+N_2 \rightarrow 2\,TiN(s)$ amounts to –675.4 kJ mol^{-1}. This value is smaller than the –876 kJ mol^{-1} estimated for the enthalpy of formation of $Ti_2N_2(g)$ from $Ti_2(g)$ and N_2. These considerations do not prove, but support the view that the barrier for the formation of solid TiN from solid Ti and N_2 is caused by the thermal energy required to form "Ti_2" or other small clusters from solid Ti. According to these calculations, Ti_2N_2 could very well be an intermediate on the way to solid titanium nitride.

1.5
Concluding Remarks

The examples discussed herein demonstrate impressively how reactive metal atom dimers are. Electronically excited states with energies close to the ground state are often responsible for these high reactivities (as shown explicitly for the reaction between Ga_2 and H_2) [10]. Therefore, an understanding of the reaction mechanisms requires knowledge of the properties of the ground state and the excited states of these species. A detailed characterization can only be achieved through a combination of experimental and quantum chemical results. However, calculations are extremely difficult and multi-reference methods have to be applied. Inner-core correlation effects have to be rigorously included. As regards experiments, absorption and Raman spectroscopies have been shown to provide useful information on the matrix-isolated species.

The aims of future studies include the characterization of metal atom trimers and other small clusters and the analysis of their reactivity. The clusters can be generated by diffusion of metal atoms or dimers into the matrix upon annealing. Many new fascinating results are expected to emerge from these studies. They might prove to be valuable for possible applications in materials science and catalytic processes.

References

1 See, for example: H.-J. HIMMEL, A.J. DOWNS, T.M. GREENE, *Chem. Rev.* **2002**, *102*, 4191.

2 ALBERT, C. BERG, M. BEYER, U. ACHATZ, S. JOOS, G. NIEDNER-SCHATTEBURG, V.E. BONDYBEY, *Chem. Phys. Lett.* **1997**, *268*, 235.

3 K. KOSZINOWKSI, D. SCHRÖDER, H. SCHWARZ, *J. Phys. Chem. A* **2003**, *107*, 4999.

4 H.-J. HIMMEL, A. BIHLMEIER, *Chem. Eur. J.* **2004**, *10*, 627.

5 See, for example: R.J. LE ROY, W.-H. LAM, *Chem. Phys. Lett.* **1980**, *71*, 544.

References

6 O. Hübner, H.-J. Himmel, W. Klopper, L. Manceron, *J. Chem. Phys.*, **2004**, *121*, 7195.
7 H.-J. Himmel, B. Gaertner, *Chem. Eur. J.* **2004**, *10*, 5936. See also a previous report: F. W. Froben, W. Schulze, U. Kloss, *Chem. Phys. Lett.* **1983**, *99*, 500.
8 P. Pullumbi, C. Mijoule, L. Manceron, Y. Bouteiller, *Chem. Phys.* **1994**, *185*, 13.
9 L. B. Knight, Jr., J. J. Banisaukas III, R. Babb, E. R. Davidson, *J. Chem. Phys.* **1996**, *105*, 6607.
10 a) A. Köhn, H.-J. Himmel, B. Gaertner, *Chem. Eur. J.* **2003**, *9*, 3909. b) H.-J. Himmel, L. Manceron, A. J. Downs, P. Pullumbi, *Angew. Chem.* **2002**, *111*, 829; *Angew. Chem. Int. Ed.* **2002**, *41*, 796. c) H.-J. Himmel, C. Manceron, A. J. Downs, P. Pullumbi, *J. Am. Chem. Soc.* **2002**, *124*, 4448.
11 H.-J. Himmel, H. Schnöckel, *Chem. Eur. J.* **2003**, *9*, 748.
12 N. J. Hardman, R. J. Wright, A. D. Phillips, P. P. Power, *Angew. Chem.* **2002**, *114*, 2966.
13 N. J. Hardman, R. J. Wright, A. D. Phillips, P. P. Power, *J. Am. Chem. Soc.* **2003**, *125*, 2667.
14 J. Su, X. W. Li, R. C. Crittendon, G. H. Robinson, *J. Am. Chem. Soc.* **1997**, *119*, 5471.
15 N. Takagi, M. W. Schmidt, S. Nagase, *Organometallics* **2001**, *20*, 1646.
16 H.-J. Himmel, H. Schnöckel, *Chem. Eur. J.* **2002**, *8*, 2397.
17 (a) J. C. Beamish, M. Wilkinson, I. J. Worrall, *Inorg. Chem.* **1978**, *17*, 2026. (b) G. Gerlach, W. Hönle, A. Simon, *Z. Anorg. Allg. Chem.* **1982**, *486*, 7.
18 X. He, R. A. Bartlett, M. M. Olmstead, K. Ruhlandt-Senge, R. W. Sturgeon, P. P. Power, *Angew. Chem.* **1993**, *105*, 761.
19 W. Uhl, M. Layh, T. Hildenbrand, *J. Organomet. Chem.* **1989**, *364*, 289.
20 T. J. Tague, Jr., L. Andrews, *J. Am. Chem. Soc.* **1994**, *116*, 4970.
21 L. B. Knight, Jr., K. Kerr, P. K. Miller, C. A. Arrington, *J. Phys. Chem.* **1995**, *99*, 16842.
22 See, for example: (a) J. D. Watts, R. J. Bartlett, *J. Am. Chem. Soc.* **1995**, *117*, 825. (b) P. R. Schreiner, H. F. Schaefer III, P. v. R. Schleyer, *J. Chem. Phys.* **1994**, *101*, 7625.
23 H.-J. Himmel, *Eur. J. Inorg. Chem.* **2003**, 2153.
24 B. Gaertner, H.-J. Himmel, V. A. Macrae, A. J. Downs, T. M. Greene, *Chem. Eur. J.* **2004**, *10*, 3430.
25 H.-J. Himmel, L. Manceron, publication in preparation. This molecule was previously detected in laser ablation experiments: G. P. Kushto, P. F. Souter, G. V. Chertihin, L. Andrews, *J. Chem. Phys.* **1999**, *110*, 9020.
26 BP/TZVPP calculations. The structure calculated for Si_2N_2 was compared to the one calculated previously (G. Maier, H. P. Reisenauer, J. Glatthaar, *Organometallics* **2000**, *19*, 4775). Si_2N_2 was generated and characterized in matrix-isolation experiments.
27 M. W. Chase, Jr., NIST-JANEF Thermochemical Tab.s, 4th Edition, *J. Phys. Chem. Ref. Data, Monograph* **1998**, *9*, 1–1951.

2
Noble Gas Hydride Compounds

Mika Pettersson, Leonid Khriachtchev, Jan Lundell, and Markku Räsänen

2.1
Introduction

After their discoveries, the noble gas elements were intact from chemical bond formation for quite a long time. In 1962, the situation suddenly changed when Bartlett succeeded in synthesizing an ionic "Xe$^+$[PtF$_6$]$^-$" compound [1]. Almost simultaneously, XeF$_2$ and XeF$_4$ were prepared [2, 3]. After this, several compounds in which xenon is bound to highly electronegative elements, mostly fluorine or oxygen, were discovered [4–9] and the theory of bonding in these compounds was developed [10]. Active research on xenon- and krypton-containing molecules has produced a number of interesting compounds, which have been thoroughly reviewed in several papers [11–23].

Exploration of the chemical reactivity of noble gases has often benefited from studies in solid noble gases at low temperatures. Examples of noble gas containing molecules identified in noble gas matrices include KrF$_2$ [24–26], XeCl$_2$ [27], and XeClF [26, 28]. By using matrix-isolation techniques, a new class of noble gas containing compounds was experimentally introduced in 1995 at the University of Helsinki [29, 30]. These noble gas hydride molecules have a common structure HNgY, where Ng=Ar, Kr or Xe and Y is an electronegative atom or fragment [22, 23]. The first examples included HXeI, HXeBr, HXeCl, HKrCl, and HXeH, and these were followed by a number of analogous species, the total number of synthesized HNgY molecules being 20 by the end of 2004 (Ar: 1, Kr: 5, and Xe: 14). These compounds are listed in Tab. 2.1 and they are truly chemical compounds having both covalent and ionic contributions in bonding. As a demonstration of this, they typically feature an H$\cdots$Ng bond distance that is shorter than the van der Waals distance by a factor of more than two. Their dissociation energies depend on the properties of the noble gas atom and on the electronegativity of the Y fragment, resulting in computational values varying from $\sim$15 to $\sim$200 kJ mol^{-1}. Even though these compounds are neutral, the strongly bound ionic HNg$^+$Y$^-$ structure plays an important role in the bonding. From an experimental point of view, the extremely large infrared intensity of the H–Ng stretching vibration makes the detection of these molecules easy. The

Inorganic Chemistry in Focus II. Edited by G. Meyer, D. Naumann, and L. Wesemann
Copyright © 2005 WILEY-VCH Verlag GmbH & Co. KGaA, Weinheim
ISBN: 3-527-30811-3

Tab. 2.1 Experimentally prepared HNgY compounds known by the end of 2004

Xe compounds	Kr compounds	Ar compounds
HXeH		
HXeI		
HXeBr		
HXeCl	HKrCl	
	HKrF	HArF
HXeCN	HKrCN	
HXeNC		
HXeOH		
HXeSH		
HXeNCO		
HXeCCH	HKrCCH	
HXeCCXeH		
HXeC$_4$H	HKrC$_4$H	
HXeCC		
HXeO		

HNgY molecules are intrinsically stable [31] and the first observations of their formation in gas-phase Xe clusters have also been made [32–34].

The HNgY molecules have peculiar properties due to their strong ionic character and relatively weak bonding. Their large dipole moments result in strong interactions with other molecules, and these interactions can be easily studied experimentally by monitoring the large shifts in the vibrational frequencies of the molecules. Complexation of the HNgY molecules with other species (for example, with N$_2$) leads to interesting blue shifts in the H–Ng stretching frequencies [35–50].

The investigation of these molecules has already produced many interesting findings, including argon chemistry [51, 52], compounds with new bonds (Xe–S, Xe–I, Kr–Cl, Kr–C, etc.) [22], organokrypton chemistry [53, 54], noble gas radical molecules [55, 56], and so on. We expect this area of research to continue to yield similarly interesting findings in the future.

2.2
Nature of Bonding

The nature of the bonding in the HNgY compounds can be understood in terms of a simple ionic model [22, 29, 57]. In this model, the equilibrium electronic structure is best described as HNg$^+$Y$^-$. Accordingly, the molecule may be viewed as an ion pair consisting of HNg$^+$ and Y$^-$. This suggests that the bonding in the HNg fragment is mostly covalent, while that between Ng and Y is mostly of an ionic nature. This electronic structure is energetically favorable due to two factors. The NgH$^+$ ions are strongly bound [58–61], and the Y fragments in the compounds found to date all have high electron affinities. An additional factor that needs to be considered in relation to energetic stability is the size of the Y$^-$ fragment. Small fragments allow a closer approach of the negative and positive ions and thus offer stronger coulombic stabilization. Thus, it is no surprise that fluorine atoms as the Y fragment form the most strongly bound compounds [29, 51, 62, 63]. The model presented so far is supported by numerous ab initio calculations (see later), which indicate a substantial charge-transfer character with a strong positive charge on the Ng and a negative charge on Y [29, 51, 64–68]. It should be noted that the first prediction of the existence of these compounds was obtained from semiempirical DIIS (Diatomics In Ionic Systems) calculations, which also suggested ionic structures [57]. The H–Ng distances in the HNgY compounds are close to the corresponding distances in HNg$^+$ ions, supporting this view [58–61]. It should be emphasized that this model is simplified and that other electronic configurations contribute to the wavefunction. For example, the H·NgY· configuration adds repulsive contributions, and it becomes dominant as the bonds are stretched, finally leading to dissociation into neutral fragments [22, 69]. The H$^-$Ng$^+$Y configuration certainly plays a non-negligible role, and it may be thought of as the source of the covalent contribution to the Ng–Y bond. Experimental evidence for the covalent contribution to the Ng–Y bond is provided by comparing the experimental Kr–Cl stretching frequency in HKrCl with the corresponding frequency in the Kr$^+$Cl$^-$ ion pair [70]. Although it is approximate, the simple HNg$^+$Y$^-$ model offers an intuitive way of understanding why and how such compounds are bound.

2.3
Computational Properties of HNgY

2.3.1
Molecular Structure

The properties of the HNgY molecules have been studied by quantum chemical methods. Electronic structure algorithms such as perturbation theory (MP2, MP3, MP4) and coupled cluster (CCSD, CCSD(T)) have been frequently employed for this purpose. In Tabs. 2.2 and 2.3, we have collected computational information on the properties of HXeY (Tab. 2.2) and HKrY/HArY (Tab. 2.3). The majority of the results were obtained at the MP2 level, which in most cases works well at the equilibrium structure. More detailed studies at higher correlated levels have demonstrated that, in general, MP2 calculations underestimate the H–Ng bond distance, whereas the bond distance between the heavy atoms (Ng–Y) is overestimated. According to the ab initio calculations, the noble gas hydride compounds have a strong ionic nature. For all species studied to date, the inserted noble gas atom possesses a large positive partial charge, while the

Tab. 2.2 Reported computational[a] bond distances and harmonic frequencies of the H–Xe–Y functional group of the noble gas hydrides. Experimental v(Xe–H) frequencies are also given where applicable

		r(H–Xe) [Å]	ω(H–Xe) [cm^{-1}]	v_{exp}(H–Xe) [cm^{-1}]	r(Xe–Y) [Å]	ω(Xe–Y) [cm^{-1}]	Refs.
HXeC$_6$H$_5$	MP2	1.880	1317.7		2.401	178.3	[90]
HXeH	MP2	1.861	1558.5	1166, 1181			[30, 84]
	CCSD(T)	1.918	1384.6				
HXeCCXeH	MP2	1.777	1594.7	1301	2.314	60.3	[56, 90]
HXeSH	MP2	1.774	1520.6	1118.6	2.729	251.4	[65]
	CCSD(T)	1.844	1147.8		2.775	218.3	
HXeCCXeCCXeH	MP2	1.765	1655.9		2.322	50.5	[90]
HXeCCH	MP2	1.750	1735.9	1486	2.322	327.2	[56, 90]
	CCSD(T)	1.767	1620.8		2.351	313.2	
HXeI	MP2	1.747	1514.8	1193	3.095	143	[22, 23, 29, 85]
HXeC$_4$H	MP2	1.741	1759.2	1532	2.329	235	[53]
HXeCC[b]	CCSD(T)	1.725	1754	1478	2.419		[56]
HXeOH	MP2	1.718	1822.7	1577.6	2.208	436.4	[66]
	CCSD(T)	1.740	1677.9		2.218	419.2	
HXeCN	MP2	1.707	1874.9	1623.8	2.392	297.3	[64]
HXeOC$_6$H$_5$	MP2	1.697	1785.3		2.253	307.1	[90]
HXeO[b]	CCSD(T)	1.690	1681	1466	2.153	382	[55]
HXeBr	MP2	1.694	1830.6	1504	2.837	182	[22, 23, 29]
HXeCl	MP2	1.685	1740	1648	2.663	267	[22, 23, 29]
HXeNCO	MP2	1.675	1986.2	1788.1	2.326	325.5	[80]
HXeOOCH	MP2	1.666			2.322		[89]
HXeNC	MP2	1.659	2010.5	1851.0	2.342	305.2	[64]

Tab. 2.2 (continued)

		r(H–Xe) [Å]	ω(H–Xe) [cm^{-1}]	ν_{exp}(H–Xe) [cm^{-1}]	r(Xe–Y) [Å]	ω(Xe–Y) [cm^{-1}]	Refs.
HXeF	MP2	1.666	2068.9		2.146	438.9	[63]
	CCSD(T)	1.681	1954.7		2.150	433.2	
HXeOCN	MP2	1.644	2118.3		2.361	321.9	[80]

a) The LJ18 ECP was used for Xe, employing 18-valence electrons in the valence space. For all other atoms, the 6-311++G(2d,2p) basis set was used.

b) Radical species.

negative partial charge is located on the electronegative fragment Y. This polarization of the HNgY molecules is reproduced with various electron population analysis schemes (Mulliken, NBO, AIM) in similar proportions. A relative positive partial charge ranging from +0.6 to +0.8 is typically found for Xe in xenon-containing molecules. The electronic structures of HNgY molecules have been further analyzed by various topological surveys of their computed electron densities, such as the so-called atoms-in-molecules method and topological analyses of the electron localization function (ELF). The electron density Laplacian plot of HArF is shown in Fig. 2.1, along with the computed atomic partial charges. The hydrogen atom is clearly covalently bound to the noble gas atom, which is seen as a shared electron density between the atoms. The interaction between the noble gas atom and the electronegative fragment Y is mainly of unshared-electron interaction type, which indicates primarily electrostatic interactions, as expected based on the computed partial charges of the atoms. However, the Ng–Y bond is not purely ionic in nature since there exists a certain covalent component in the interactions that manifests itself in rather notable values of computed electron densities in the bond [68, 71, 72]. The HNgY molecules can be described in terms of ionic and neutral resonance structures, with the most important contributions coming from HNg$^+$Y$^-$. In the case of HArF, this reso-

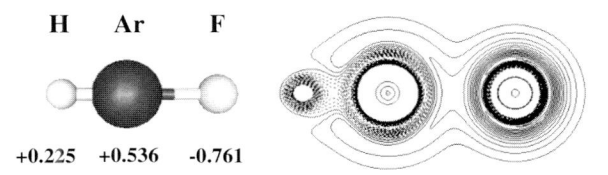

Fig. 2.1 Computed atomic partial charges (from NBO analysis) and electron density Laplacian plot for HArF. The solid lines denote positive Laplacian values, the dotted ones negative values. The contours are spaced at 0.02e a_0^{-5} intervals. Adapted from Fig. 3 of ref. [71].

Tab. 2.3 Reported computational[a] bond distances and harmonic frequencies of the H–Ng–Y functional group of hydrides containing lighter noble gas atoms. Experimental ν(Ng–H) frequencies are also given where applicable

		r(H–Ng) [Å]	ω(H–Ng) [cm^{-1}]	ν_{exp}(H–Ng) [cm^{-1}]	r(Ng–Y) [Å]	ω(Ng–Y) [cm^{-1}]	ν_{exp}(Ng–Y) [cm^{-1}]	Refs.
Krypton								
HKrCCH	MP2[b]	1.59	1575	1241	2.25	306		[54]
HKrC$_4$H	MP2	1.581	1517.2	1290	2.255	238.0		[53]
HKrCl	MP2[b]	1.500	1943.1	1476.1	2.563	275.7	253.1	[22, 23, 29]
HKrCN	MP2	1.468	2010.6	1497.4	2.342	247.8		[64]
	CCSD(T)	1.496	1780.6		2.357	251.7		
HKrO[c]	CCSD(T)	1.490	1728.8		2.087	372.1		[55]
HKrF	MP2	1.423	2315.9	1951.6	2.134	396.1	417.0	[29, 62, 63, 94]
	CCSD(T)	1.439	2173.8		2.138	396.8		
HKrNC	MP2	1.413	2016.6		2.339	266.8		[64]
	CCSD(T)	1.418	2065.8		2.354	269.1		
Argon								
HArCl	MP2	1.377	1592.1		2.460	304.2		[29, 46, 47]
HArF	MP2	1.326	2148.9	1969.4	1.996	481.2	435.7	[51, 52, 63, 67, 94]
	CCSD(T)	1.355	1793.1		2.005	478.9		
HArO[c]	CCSD(T)	1.343	1639.0		2.021	396.4		[55]
Helium								
HHeF	MP2	0.793	2701.6		1.413	1006.3		[63, 94, 97]
	CCSD(T)	0.825	2110.1		1.422	963.6		

a) For all atoms the 6-311++G(2d,2p) basis set was used.
b) The aug-cc-pVDZ basis set was used.
c) A radical species.

nance structure amounts to about 80% of the total interaction [72]. These computational results also support the simple model described above for the nature of the HNgY molecules.

2.3.2
Stability

All HNgY molecules are metastable local minimum structures on their potential energy surfaces and the barriers to their decomposition into H+Ng+Y and HY+Ng should be considered. The involvement of several electronic configurations in the H+Ng+Y dissociation path naturally lends itself to a computational problem that is inherently multiconfigurational in nature, and therefore the barrier defining the kinetic stability of the HNgY molecules towards this dissociation channel is extremely difficult to compute. Simple DFT or MP2 calculations are not sufficient, and an MRCI approach is needed [67, 69, 73–76]. However, up to now, all experimentally observed noble gas hydrides have been found computationally to lie *below* the neutral fragment H+Ng+Y dissociation limit, and the existence of the barrier is not a determining factor for the stability. A possible exception is HArF, for which the height of the barrier is calculated to be of similar magnitude as the well depth [67, 73].

The stability of the HNgY molecules is also determined by another dissociation process, whereby the molecule decomposes via its H–Ng–Y bending coordinate. This process (HNgY → Ng+HY) leads to the global minimum energy structure corresponding to a van der Waals bound complex between the hydrogen-containing species and the noble gas atom. This process has been found to be single-determinant in nature and so DFT and MP2 calculations are also able to give reasonable estimates for the barrier height besides the more sophisticated computational methods [73]. This dissociation process is usually less important compared with the three-body dissociation process since the bending barrier is higher than the three-body dissociation barrier. This is evident, for example, in the case of HArF, for which the bending barrier has been reported to be almost 100 kJ mol^{-1}, about three times higher than the barrier for the H+Ar+F dissociation process [67, 73]. Similar barriers have also been computed for other HNgY compounds [39, 69, 73].

2.4
Preparative Methods

Almost all of the experimental data on HNgY compounds have been obtained from matrix-isolation studies. Only very recently have HXeI and HXeCl been prepared in the gas phase in elegant experiments involving the photolysis of HI- and HCl-doped Xe clusters [32–34]. There are basically two preparative methods, both of which involve the generation of atomic hydrogen and neutral Y fragments in solid Ng matrices (or clusters). Usually, the most suitable source of these species is HY, which is dissociated in the matrix by UV or VUV radiation, although separate sources of H and Y can also be used [22, 23]. In the first, mostly used, method, the H and Y fragments are stored in an Ng matrix at low temperatures (< 30 K for Xe) until a suitable degree of dissociation of HY is achieved by photolysis.

When the temperature is raised up to the point at which H atoms become mobile (~40 K in xenon), HNgY species are formed in a H+Ng+Y reaction [22, 23].

$$H + Ng + Y \rightarrow HNgY \tag{1}$$

The kinetics of this reaction is usually limited by the diffusion of hydrogen atoms, indicating that intrinsic barriers for the formation are rather low [77]. The yield of HNgY is usually rather high, some tens of percent of the dissociated HY, typical values being in the range 20–40% [77–79].

As the other possibility, HNgY is formed directly in the photolysis of HY without the need for additional mobilization of H atoms by raising the temperature of the sample. This process involves dissociation of HY and some motion of the photofragments after dissociation, and finally the formation of HNgY.

$$HY + Ng \xrightarrow{h\nu} H + Y + Ng \rightarrow HNgY \tag{2}$$

The difference between this and the first mechanism lies in the timescale of the formation of HNgY. In the first (delayed) mechanism, there is a macroscopic period of time between the photolysis of HY and the formation of HNgY. In the second process [Eq. (2)], the whole dissociation-formation process probably takes place within a few picoseconds, as has been found in molecular dynamics simulations [76]. The second preparative method has been shown to work at least for HXeNCO, HKrCl, and HArF [51, 80, 81]. It is most probably operative for other molecules as well, but in most cases the yield of HNgY remains very low due to its secondary photolysis by the same radiation that is used to photodissociate HY.

$$HY + Ng \xrightarrow{h\nu} HNgY \xrightarrow{h\nu} H + Ng + Y \tag{3}$$

HNgY species typically have photodissociation cross-sections that are orders of magnitude larger than that for HY and therefore the amount of HNgY observed after the photolysis step is often below the detection limit [78, 82]. While the preparation in Ng matrices is very general and successful, it would be highly desirable to prepare these compounds by more traditional methods of chemical synthesis as well. Although no such breakthrough has yet been reported, there are theoretical estimates that predict the crystals of some of these compounds to be stable. At least XeH_2 would seem to be a possible candidate in this respect [41].

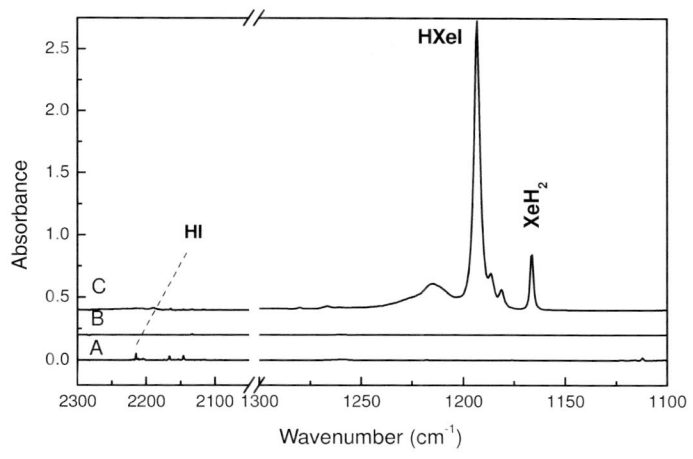

Fig. 2.2 Production of HXeI and XeH$_2$ from HI in a xenon matrix. Trace A shows the situation after the deposition of an HI/Xe = 1/1000 matrix. The absorption of monomeric HI is marked in the figure. Trace B shows the situation after the photolysis of HI with 248 nm radiation. It is evident that HI has been practically totally decomposed. Trace C shows the situation after annealing the photolyzed matrix at 50 K after the photolysis of HI. Note the tremendous intensity of HXeI and XeH$_2$. All spectra were recorded at 7.5 K.

2.5 Spectroscopic Properties

2.5.1 Experimental Spectroscopic Observations

Experimentally, the information on HNgY has been obtained mostly from IR spectroscopic measurements and, to a lesser extent, from UV spectroscopy. The first experimentally observed molecules, HXeI, HXeBr, HXeCl, and HKrCl, were detected by IR spectroscopy in UV-photolyzed hydrogen halide-doped Ng matrices [29]. The most characteristic feature in the IR spectrum is the very strong H–Ng stretching absorption. There are two features of this fingerprint mode that make it very useful: very high intensity and characteristic spectral position for each molecule. The intensities, according to ab initio calculations, are usually several thousands of km mol^{-1}, placing these compounds among the strongest IR absorbers known. The tremendous intensity of the H–Ng stretching absorption of HXeI is demonstrated in Fig. 2.2. The position of this vibration depends strongly on the nature of the Y fragment. In Tabs. 2.2 and 2.3, experimental data for the H–Ng stretching vibrations of various HNgY compounds are collected. The other absorptions are usually much weaker. In many cases, H–Ng–Y bending modes have been detected, and these are typically about two orders of magnitude weaker than the stretching modes. The Ng–Y stretching modes are interesting because they are directly probing the Ng–Y bonds, but they are usually weak and lie below 400 cm^{-1}, which is technically a more difficult region to measure than the mid-IR region. However, for HArF, HKrF, and HKrCl, the Ar–F, Kr–F, and Kr–Cl stretching absorptions have been directly observed [51, 62, 70].

UV absorption spectra of HXeI, HXeBr, HXeCl, XeH$_2$, HXeSH, HXeCN, and HXeOH have been measured [78, 82, 83]. They show strong absorptions in the region between 200 and 400 nm. The peak absorption cross-sections for these broad transitions can be estimated to be of the order of 10^{-16} cm^2 [78, 82], indicating very high transition moments. This suggests that UV spectroscopy could be equally well used to detect these compounds in some experimental situations. These observations have been supported by ab initio calculations on the excited states, which predict large transition moments of 7–8 Debye [78, 82]. In addition, photodissociation spectra have been measured for XeH$_2$, HXeOH, HKrCl, HKrF, and HArF, and they are in agreement with the UV absorption measurements [62, 81, 83].

2.5.2
Computational Vibrational Spectra of HNgY

Computational predictions of the vibrational spectra of HNgY have always played an important role in the experimental search for these compounds. Of special interest is the intense Ng–H stretching vibration, which is the strongest band in the computed spectrum. This is also the vibrational mode that is mainly affected by the computational level used. Depending on the level of theory and its balance with the basis set used, the position of the Ng–H band can vary by several hundred cm^{-1}. The Ng–H stretching mode is known to be rather anharmonic in nature. This is evidenced by the experimental detection of overtones of this vibration in various Xe compounds [23]. The H–Ng–Y bending and Ng–Y stretching modes have been found to be more harmonic in nature, and harmonic vibrational calculations even at the MP2 level are able to produce qualitatively good results.

Anharmonic vibrational calculations of HNgY molecules are important both because of their usefulness for experimental identification of these molecules, and because of their relevance to dynamic processes in these systems. Anharmonic vibrational calculations based on MP2 computed potential energy surfaces have hitherto been carried out for HXeH, HXeI, HXeCl, HXeBr, HXeOH, HArF, HKrF, and HKrCl [51, 63, 67, 70, 84–87]. In the case of HXeI [85], several weak spectral features were assigned to combination bands by a comparison with anharmonic calculations. Based on this assignment, the Xe–I vibrational frequency was suggested to lie at around 146 cm^{-1}, in good agreement with ab initio calculations (see Tab. 2.2).

2.6
Selected Examples of HNgY Compounds

2.6.1
Xenon Compounds

The first HXeY compounds reported in 1995 were triatomic HXeI, HXeBr, HXeCl, and XeH$_2$ [29, 30]. Since then, several other compounds have been prepared, including more complex molecules. The early stages of development have been described in two reviews in 1999 and 2000 [22, 23]. In this section, we highlight some specific examples of more recent progress.

HXeOH This compound is remarkable in the sense that it is made from water and xenon and thus, together with XeH$_2$, it represents an example of a noble gas compound made in a photoreaction with a naturally occurring abundant and inert molecule [66]. Therefore, it was suggested that, in principle, this molecule could be formed, for example, in water ice under suitable conditions. In order to predict the properties of HXeOH in a water environment, a molecular modeling study was carried out [39]. The results showed that upon complexation with three or four water molecules, HXeOH becomes unstable with respect to the bending coordinate and decomposes to H$_2$O and Xe (see later). Thus, the computational results indicate that this species would be only transiently stable if formed in a water environment.

HXeO This compound is interesting because it is an open-shell molecule ($^2\Sigma$). It can be made from water in solid xenon by sequential photolysis ultimately producing hydrogen and oxygen atoms [55]. Alternatively, different sources of hydrogen and oxygen atoms may be used. It has been suggested that HXeO is formed from singlet oxygen atoms, and this hypothesis is currently under investigation [55, 88]. Being a radical, this molecule would be an interesting target for EPR measurements, which so far are lacking. Interestingly, according to computations, HKrO and even HArO could also be stable species, but they have yet to be observed experimentally.

Organoxenon Compounds In 2000, it was predicted by ab initio calculations that Xe could be inserted into carboxylic acids, and chemical binding of xenon to proteins was hypothesized [89]. However, no experimental verification of these results has yet been reported. Another computational prediction, made in 2002, was of the existence of stable insertion compounds of xenon into acetylene, benzene, and phenol [90]. Soon after this prediction, the experimental preparation of HXeCCH was reported independently by two groups [56, 91]. Additionally, two other interesting species were reported: an open-shell species, HXeCC, and a compound incorporating two xenon atoms, HXeCCXeH [56]. Soon thereafter, a new member was added to the series, HXeC$_4$H [53]. These examples show that there is a promising new field of fluorine-free organoxenon chemistry connected with xenon hydride compounds. An interesting aspect of these compounds is that the energetic stability seems to increase upon lengthening of the carbon chain [53]. It has been suggested that similar compounds could be promising candidates for stable bulk noble gas hydride materials [53].

2.6.2
Compounds of Lighter Noble Gases

The chemistry of lighter noble gases is naturally more challenging than that of xenon due to the higher ionization potentials and thus more stable electronic configurations of the lighter noble gases. Several krypton hydride compounds have nevertheless been made since the first report of HKrCl in 1995 [29]. These include HKrCN, HKrF, and as the latest developments, HKrCCH and HKrC$_4$H, which are the first examples of organokrypton compounds [53, 54, 62, 64]. Similar to the corresponding organoxenon species, it is anticipated that these compounds represent the first examples of a whole new field of interesting chemistry of krypton.

A breakthrough in argon chemistry took place in 2000, when the experimental preparation of HArF was reported [51]. This was accomplished by VUV photolysis of HF isolated in an argon matrix and subsequent thermal mobilization of hydrogen and fluorine atoms. An unambiguous assignment was achieved by observing all three fundamental IR absorptions and using isotopic substitution of hydrogen and argon. The first prediction of the existence of HArF was made five years before its experimental realization [29]. Since then, several theoretical papers on the properties of HArF have been published and they have confirmed the experimental findings [63, 67, 72, 73, 92–96]. According to ab initio calculations, HArF has a linear structure with bond lengths H–Ar=1.33 Å and Ar–F=1.97 Å [67] (see Fig. 2.1). The molecule is stable with respect to dissociation into atomic fragments by 14.5 kJ mol^{-1} and is additionally stabilized by a barrier of 17.4 kJ mol^{-1}. The dissociation into HF+Ar is exothermic by 551.8 kJ mol^{-1}, but this channel is prevented by a substantial barrier of $\sim$100 kJ mol^{-1} [67, 73].

Obviously, it is interesting to investigate whether the concept of noble gas hydrogen chemistry can be extended to Ne and He. There are several theoretical investigations that predict HHeF to be a metastable compound [74, 94, 97]. This molecule is metastable with respect to both channels H+He+F and He+HF and its lifetime would be limited by tunneling. Predictions for its lifetime are in the femtosecond to picosecond range [74, 97]. In any case, it is clear that the experimental detection of this molecule is very demanding. It has been suggested that high-pressure (>23 GPa) solid helium could be used as a stabilizing medium that would extend the lifetime to macroscopic timescales [98]. Nevertheless, the experimental effort needed would still be considerable. Another useful stabilizing strategy might be the complexation of HHeF with other species [99]. According to our knowledge, no other serious candidates for stable He hydride compounds exist. It is curious that the corresponding Ne compound, HNeF, is not even computationally stable, and no other candidates have been found to date [63, 94]. The chemistry of neon, like that of helium, remains a challenge for both experimentalists and theoreticians.

2.7
HNgY Complexes

HNgY molecules should be very reactive, which makes the question of the stability of their complexes of fundamental interest. In 1999, it was found computationally that HXeH forms a stable complex with H_2O [35] and this seems to be the first work on the HNgY complexes. A number of studies on various HNgY complexes have since been performed [36–50, 99–101]. These works have had at least a threefold purpose. First, it was intriguing to study shifts of the H–Ng stretching frequencies upon complexation that had to be intuitively large (if not record-breaking) because of the charge-transfer nature of these weakly bound species. Second, it was practically important to distinguish absorption bands of the monomers and complexes for reasons of reliable assignments in matrix-isolation experiments. Third, the studies of complexes could make a link to the species in clusters and in bulk and introduce a sensitive probe of local solid-state morphology (see later).

The experimental research on HNgY complexes in our laboratory was initiated by the studies on HArF in solid Ar. This argon compound was made by means of vacuum-UV photolysis of HF in solid Ar at 7 K and annealing at ~20 K, leading to the characteristic H–Ar stretching absorption bands at 1965.7, 1969.4, and 1972.3 cm^{-1} [51]. This triplet absorption was found to decrease and disappear upon annealing above 28 K, which was tentatively attributed to chemical reactions of HArF with thermally mobile fragments, for instance, with fluorine atoms. However, it was subsequently found that HArF exhibits two additional blue-shifted H–Ar stretching absorption bands at 2016.3 and 2020.8 cm^{-1}, with thermal stability limited only by the degradation of the matrix structure above 40 K [52]. This doublet absorption was experimentally assigned to HArF in a different, thermally relaxed, solid-state configuration (matrix site), referred to as "stable" HArF in order to distinguish it from the "unstable" HArF absorbing at ~1970 cm^{-1}. Thermal conversion of the unstable HArF geometry to the stable HArF geometry was experimentally demonstrated, and this relaxation process formed the basis of the proposed experimental model [52]. This seemed to be the first experimental observation demonstrating the high sensitivity of the HNgY molecules to specific interactions with the local surroundings. Qualitatively similar behavior of the matrix-site structure upon annealing was found for HKrF [62], and tentatively also for HXeD [83]. Theoretical modeling confirmed high sensitivity of the H–Ng stretching frequencies to fine adjustment of the local morphology [92, 93]. In particular, Jolkkonen et al. attributed the experimental blue-shift of the stable HArF absorption bands to the formation of the tight Ar···HArF complex [93].

Complexes of HArF, HKrF, and HKrCl with dinitrogen were studied both experimentally in rare-gas matrices and computationally by Lignell *et al* [38,40].

For all the complexes studied, two configurations with bent and linear organization of the $N_2\cdots HNgY$ structure were found by the *ab initio* methods and assigned experimentally. The computational interaction energies of the $N_2\cdots HNgY$ complexes before zero-point-energy correction were found to be between −4.8 and −9.6 kJ mol^{-1}. Most remarkably, large blue-shifts of the H–Ng stretching frequencies were obtained as compared with the corresponding monomers. The shifts of the H–Ng stretching frequencies were found to be larger for the linear N_2 complexes, and the largest shift observed experimentally was 113 cm^{-1} for the $HKrCl\cdots N_2$ complex. The calculated shifts were somewhat larger, being, for instance, 194 cm^{-1} for the $HKrCl\cdots N_2$ complex at the MP2/aug-cc-pVTZ level of theory. This numerical discrepancy may have been caused by inaccuracy of the calculations as well as by the matrix effect. The latter phenomenon is fundamental and interesting. The theory compares vibrations of the complexed molecule and the isolated molecule, whereas in the experiment the "uncomplexed" molecule is isolated in a solid matrix, and its interaction with the matrix atoms can shift the vibrational frequencies from the position in vacuum. Studies on thermal matrix-site modifications of HArF and HKrF have highlighted this effect [52, 62]. The complexation-induced blue-shifts were attributed to the enhancement of the $(HNg)^+Y^-$ charge-transfer character, strengthening the HNg covalent bonding [38, 40]. The computations indicated that the H–Ng stretching absorption intensity should decrease upon such blue-shifting complexation, actually leading to complication of the experimental studies.

The $HXeOH\cdots(H_2O)_n$ structures were studied by Nemukhin *et al.* both computationally and experimentally [39]. The $HXeOH\cdots H_2O$ ($n=1$) structure is essentially nonlinear, and a hydrogen atom of H_2O is bonded to the oxygen atom of HXeOH. The theoretical binding energy of this complex is quite large, ~ 45 kJ mol^{-1}. The computations show that complexation with one water molecule enhances charge separation in the HXeOH molecule, shifts the H–Xe stretching frequency to higher energy by 107 cm^{-1}, and decreases the absorption intensity by a factor of 1.30. Complexation with two and three water molecules continues the trend of stabilization of HXeOH along the stretching coordinate, and the H–Xe stretching frequency is blue-shifted by 182 and 276 cm^{-1}, respectively. However, it was found that the stabilization with respect to the stretching coordinate is accompanied by destabilization along the bending coordinate, i.e., recombination to the global $Xe+(H_2O)_{n+1}$ minimum is catalyzed by water molecules. The barrier along the bending coordinate was found to be 165.5 kJ mol^{-1} for isolated HXeOH, and it decreases to 111.2, 46.8, and 1.7 kJ mol^{-1} for $n=1$, 2, and 3, respectively. In the same work, experimental evidence of $HXeOH\cdots(H_2O)$ and $HXeOH\cdots(H_2O)_2$ complexes in an Xe matrix was presented, the blue-shifts of the H–Xe stretching frequency being 103 and 164 cm^{-1}, respectively.

As has already been mentioned, a number of purely computational works on various HNgY complexes have been published. In the complex of HXeH with water, the interaction occurs via a so-called dihydrogen bond, with the interaction energy of ca. −10 kJ mol^{-1} arising mainly from electrostatic forces [35–37]. The $(HXeH)_n$ ($n=2, 3, 4$) system was found to be weakly bound but still stable, demonstrating the potential of HNgY molecules to constitute high-energy bulk materials [41]. However, the practical procedure to make these structures is unclear. The results for the complexes of HArF, HKrF, and HKrCl with dinitrogen, reported by McDowell [42, 43, 45, 47, 49], agree well with the computational data of Lignell *et al* [38, 40]. McDowell has also performed calculations on the complexes of HArF with CO, P_2, and HF, of HKrCl with CO and HF, and of HArCl with N_2 [43–46, 48]. The complexes of HXeF, HKrF, and HArF with HF were considered by Yen *et al* [100]. The predicted HHeF molecule complexed with N_2 and CO (binding energies up to 8.2 kJ mol^{-1}) was computed by Wang *et al* [50]. Interestingly, the HHeF$\cdots$Xe complex appears to be about six times stronger [99]. Alkorta and Elguero considered the hydrogen-bonding properties of various krypton derivatives [101]. With the exception of the $P_2\cdots$HArF complex [43] all studied complexes were found to exhibit contraction of the H–Ng bond and elongation of the Ng–Y bond, with the H–Ng stretching frequency increasing accordingly. A very large binding energy of ~ 75 kJ mol^{-1} (MP2) has been found for the HArF$\cdots$HF complex with the bent geometry, in which the H atom of HF is bonded to the F atom of HArF [44]. The H–Ar stretching frequency of this complex (MP2) is shifted by 420 cm^{-1} to higher energy with respect to the HArF monomer. The exception of the $P_2\cdots$HArF complex, which shows a large (116 cm^{-1}) red shift (MP2) of the H–Ar stretching frequency, is remarkable [43].

2.8
Reactions of HNgY Molecules with Hydrogen Atoms

As proposed earlier, the HNgY molecules can react efficiently with various atomic and molecular fragments. In studies of HArF, it was suggested that this molecule could be destroyed in reactions with thermally mobilized fragments such as fluorine atoms [51]. In another study, the reaction between HXeO radicals and H atoms was suggested [55]. However, the established experimental evidence is currently limited to the reactions of HXeH and HKrCl with thermally mobilized D atoms [102, 103]. In solid Kr and Xe, extensive motion of H atoms starts at ~ 28 and 40 K, and the activation energies associated with this motion are ~ 6 and 12 kJ mol^{-1}, respectively [104]. It is reasonable to expect that mobile H atoms will be capable of reacting with the already formed HNgY molecules, and that this reaction constitutes a loss channel for H atoms and a limiting factor for the resulting HNgY concentration [23].

$$HNgY + H \rightarrow H_2 + Ng + Y \tag{4}$$

$$HNgY + H \rightarrow HY + Ng + H \tag{5}$$

To the best of our knowledge, no theoretical studies on chemical reactions of the HNgY molecules are available.

A series of experiments has been performed on the HXeI and HXeH molecules [102]. The photostabilities of these two noble gas molecules are very different

(HXeI is much less stable), which allowed their selective decomposition by light. After selective decomposition of HXeI, additional annealing restores ~50% of the decomposed HXeI molecules, but the HXeH concentration remains practically unchanged. These decomposition-annealing cycles can be repeated several times, but the result is qualitatively always the same, i.e., the annealing-induced HXeI concentration gradually decreases, while the HXeH concentration does not change significantly. At first glance, this observed stability of the HXeH concentration looks surprising because one might expect an annealing-induced increase of the HXeH concentration due to reactions of H atoms originating from decomposed HXeI molecules. This "surprising" experimental observation could be satisfactorily explained by taking into account reactions between mobile H atoms and HXeH molecules. According to this model, the HXeH+H reaction decreases the HXeH concentration and leads to H_2 molecules. Under this assumption, the stability of the HXeH concentration means that the HXeH destruction rate is numerically similar to the HXeH formation rate.

In the next group of experiments [102], a Xe matrix was doped with H_2O and D_2O ($H_2O/D_2O/Xe$ ~ 1/3/6000). After extensive irradiation at 193 nm and annealing at 39 K, various isotopologues of HXeH and HXeOH had formed. The formation of HXeH and HXeD is faster than that of DXeD, and this observation was ascribed to the higher thermal mobility of H atoms compared with D atoms [105]. Additional annealing at 45 K (after 100 min at 39 K) led to qualitatively different changes of the HXeH and DXeD concentrations. In fact, the HXeH concentration decreased by ~10% at 45 K, whereas the DXeD concentration was somewhat increased. The thermal decomposition of HXeH at this temperature was ruled out on the basis of the existing experimental data [83]. To explain this opposite change of the HXeH and DXeD concentrations upon annealing at 45 K, it was proposed that HXeH molecules could be destroyed in reactions with D atoms.

These experimental results were modeled on the basis of kinetic equations taking into account the reactions of HNgY molecules with mobile H and D atoms. In particular, the modeling shows that the H+HNgY reactions are basically diffusion-controlled, i.e. no additional barrier for their formation is indicated by the experiments. The proposed reaction scheme forms a buffering mechanism stabilizing the HNgY concentration in solid Xe, and it most probably leads to the production of H_2 molecules, hence constituting an efficient loss channel for hydrogen atoms.

Similar conclusions were extracted on the basis of experiments with HCl(DCl)/Kr matrices [103]. According to the developed kinetic model, the formed HKrCl molecules can further react with hydrogen atoms producing H_2 and HCl molecules. This process was tested using the slower thermal mobility of D atoms as compared with H atoms. HKrCl molecules are formed quickly with the mobility of H atoms. The motion of D atoms extends into a longer time scale, after formation of the main part of HKrCl molecules. If HKrCl molecules react with D atoms, these kinetic distinctions should lead to a decrease of the HKrCl concentration upon selective mobilization of D atoms, and this has been demonstrated experimentally. The decrease in the HKrCl concentration upon reaction with D atoms occurs on the same time scale as the increase in the DKrCl concentration, which is a fingerprint of the D+HKrCl diffusion-controlled reaction.

The example of the first "constructive" reaction involving HNgY molecules is the HXeCC+Xe+H reaction leading to the HXeCCXeH molecule. Evidence for this reaction has recently been obtained based on formation kinetics of various noble gas species upon annealing of photolyzed H_2C_2/Xe matrices. These studies strongly indicate that the HXeCCXeH molecules are formed from the HXeCC precursor but not from CC molecules in a concerted reaction with two H atoms in a Xe matrix [106].

2.9
Future Directions

The chemistry of noble gas hydride compounds will remain an active field of research in the near future. Prediction and preparation of new compounds will continue with emphasis on new developments. Hopefully, these will include chemistry of neon and helium, new bonds between noble gases and other elements, and the extension of organoxenon and organokrypton chemistry. A special challenge is to make more stable compounds since these may offer a route for the preparation of these molecules in pure form by traditional synthetic methods. This is clearly desirable, since it would extend the field to experimental determination of structures and characterization of chemical and physical properties. Related to this, the production of these molecules in the gas phase is needed to open new possibilities for experimental determination of their properties. Progress in this direction has already been made [32–34]. Studies of the interaction of HNgY compounds with other molecules will continue. A field that is only in its infancy is the dynamics of the formation mechanisms of these compounds. So far, the experimental studies have been conducted on macroscopic timescales, giving highly averaged information. Some theoretical effort has been devoted to the study of dynamics of formation of these compounds [76]. Experimental efforts in this field would be highly desirable, but they are clearly very complicated. This direction seems to be an attractive field of research for the future.

Acknowledgements

We would like to thank the past and present group members who have contributed to this research: Antti Lignell, Tiina Kiviniemi, Hanna Tanskanen, Santtu Jolkkonen, Mia Saarelainen, Esa Isoniemi, and Robbert-Jan Roozeman. Additionally, we are grateful to many colleagues for fruitful cooperation and many discussions: Nino Runeberg, Alexander Nemukhin, Bella Grigorenko, Vladimir Feldman, Benny Gerber, Galina Chaban, Zsolt Bihary, Udo Buck, and Henrik Kunttu. This work was supported by the Academy of Finland, the Finnish Cultural Foundation, and Helsinki University research funds.

References

1 (a) N. BARTLETT, *Proc. Chem. Soc.* **1962**, 218; (b) F.O. SLADKY, P.A. BULLINER, N. BARTLETT, *J. Chem. Soc. A* **1969**, 2179.
2 R. HOPPE, W. DÄHNE, H. MATTAUCH, K.H. RÖDDER, *Angew. Chem.* **1962**, *74*, 903.
3 H.H. CLAASSEN, H. SELIG, J.G. MALM, *J. Am. Chem. Soc.* **1962**, *84*, 3593.
4 F.O. SLADKY, P.A. BULLINER, N. BARTLETT, D.G. DEBOER, A. ZANKIN, *Chem. Commun.* **1968**, 1048.
5 V.M. MCRAE, R.D. PEACOCK, D.R. RUSSELL, *Chem. Commun.* **1969**, 62.
6 J.H. HOLLOWAY, J.G. KNOWLES, *J. Chem. Soc. A* **1969**, 756.
7 N. BARTLETT, M. GENNIS, D.D. GIBLER, B.K. MORRELL, A. ZALKIN, *Inorg. Chem.* **1973**, *12*, 1717.
8 D.E. MCKEE, C.J. ADAMS, N. BARTLETT, *Inorg. Chem.* **1973**, *12*, 1722.
9 J.R. MORTEN, W.E. FALCONER, *J. Chem. Phys.* **1963**, *39*, 427.
10 C.A. COULSON, *J. Chem. Soc.* **1964**, 1442.
11 J.G. MALM, H. SELIG, J. JORTNER, S.A. RICE, *Chem. Rev.* **1965**, *65*, 199.
12 N. BARTLETT, F.O. SLADKY, in *Comprehensive Inorganic Chemistry*, Pergamon Press, New York, 1973, vol. 3, pp. 213–330.
13 E. TOMMILA, *Suomen Kemistilehti, A* **1963**, *36*, 209.
14 K. SEPPELT, D. LENTZ, *Prog. Inorg. Chem.* **1982**, *29*, 167.
15 J.H. HOLLOWAY, *J. Fluorine Chem.* **1986**, *33*, 149.
16 D. CREMER, G. FRENKING, K. HILPERT, C.K. JØRGENSEN, G.B. KAUFFMAN, *Structure and Bonding* **1990**, *73*, 2–95.
17 M. GERKEN, G.J. SCHROBILGEN, *Coord. Chem. Rev.* **2000**, *197*, 335–395.
18 H.-J. FROHN, V.A. BARDIN, *Organometallics* **2001**, *20*, 4750–4762.
19 V.K. BREL, N. SH. PIRKULIEV, N.S. ZEFIROV, *Russ. Chem. Rev.* **2001**, *70*, 231–264.
20 W. TYRRA, D. NAUMANN, in *Inorg. Chem. Highlights*, Wiley-VCH, 2002, pp. 297–316.
21 R.B. GERBER, *Ann. Rev. Phys. Chem.* **2004**, *55*, 55.
22 M. PETTERSSON, J. LUNDELL, M. RÄSÄNEN, *Eur. J. Inorg. Chem.* **1999**, 729.
23 J. LUNDELL, L. KHRIACHTCHEV, M. PETTERSSON, M. RÄSÄNEN, *Low-Temp. Phys.* **2000**, *26*, 680.
24 J.J. TURNER, G.C. PIMENTEL, *Science* **1963**, *140*, 975.
25 D.R. MACKENZIE, *Science* **1963**, *141*, 1171.
26 W.F. HOWARD, L. ANDREWS, *J. Am. Chem. Soc.* **1974**, *96*, 7864.
27 L.Y. NELSON, G.C. PIMENTEL, *Inorg. Chem.* **1967**, *6*, 1758.
28 V.E. BONDYBEY, *Doctoral Thesis*, Univ. of California, 1971.
29 M. PETTERSSON, J. LUNDELL, M. RÄSÄNEN, *J. Chem. Phys.* **1995**, *102*, 6423.
30 M. PETTERSSON, J. LUNDELL, M. RÄSÄNEN, *J. Chem. Phys.* **1995**, *103*, 205.
31 M. LORENZ, M. RÄSÄNEN, V.E. BONDYBEY, *J. Phys. Chem. A* **2000**, *104*, 3770.
32 R. BAUMFALK, N.H. NAHLER, U. BUCK, *J. Chem. Phys.* **2001**, *114*, 4755.
33 N.H. NAHLER, R. BAUMFALK, U. BUCK, Z. BIHARY, R.B. GERBER, B. FRIEDRICH, *J. Chem. Phys.* **2003**, *119*, 224.
34 N.H. NAHLER, M. FARNIK, U. BUCK, *Chem. Phys.* **2004**, *301*, 173.
35 J. LUNDELL, M. PETTERSSON, *Phys. Chem. Chem. Phys.* **1999**, *1*, 1691.
36 J. LUNDELL, S. BERSKI, Z. LATAJKA, *Phys. Chem. Chem. Phys.* **2000**, *2*, 5521.
37 S. BERSKI, J. LUNDELL, Z. LATAJKA, *J. Mol. Struct.* **2000**, *552*, 223.
38 A. LIGNELL, L. KHRIACHTCHEV, M. PETTERSSON, M. RÄSÄNEN. *J. Chem. Phys.* **2002**, *117*, 961.
39 A.V. NEMUKHIN, B.L. GRIGORENKO, L. KHRIACHTCHEV, H. TANSKANEN, M. PET-

tersson, M. Räsänen. *J. Am. Chem. Soc.* **2002**, *124*, 10706.
40 A. Lignell, L. Khriachtchev, M. Pettersson, M. Räsänen. *J. Chem. Phys.* **2003**, *118*, 11120.
41 J. Lundell, S. Berski, Z. Latajka, *Chem. Phys. Lett.* **2003**, *371*, 295.
42 S. A. C. McDowell, *J. Chem. Phys.* **2003**, *118*, 4066.
43 S. A. C. McDowell, *Phys. Chem. Chem. Phys.* **2003**, *5*, 808.
44 S. A. C. McDowell, *Chem. Phys. Lett.* **2003**, *377*, 143.
45 S. A. C. McDowell, *J. Chem. Phys.* **2003**, *119*, 3711.
46 S. A. C. McDowell, *J. Mol. Struct. (Theochem.)* **2003**, *625*, 243.
47 S. A. C. McDowell, *Mol. Phys.* **2003**, *101*, 2261.
48 S. A. C. McDowell, *Chem. Phys. Lett.* **2003**, *368*, 649.
49 S. A. C. McDowell, *J. Chem. Phys.* **2003**, *118*, 7283.
50 J.-T. Wang, Y. Feng, L. Liu, X.-S. Li, Q.-X. Guo, *Chem. Lett.* **2003**, *32*, 746.
51 L. Khriachtchev, M. Pettersson, N. Runeberg, J. Lundell, M. Räsänen, *Nature (London)* **2000**, *406*, 874.
52 L. Khriachtchev, M. Pettersson, A. Lignell, M. Räsänen, *J. Am. Chem. Soc.* **2001**, *123*, 8610.
53 H. Tanskanen, L. Khriachtchev, J. Lundell, H. Kiljunen, M. Räsänen, *J. Am. Chem. Soc.* **2003**, *125*, 16361.
54 L. Khriachtchev, H. Tanskanen, A. Cohen, R. B. Gerber, J. Lundell, M. Pettersson, H. Kiljunen, M. Räsänen, *J. Am. Chem. Soc.* **2003**, *125*, 6876.
55 L. Khriachtchev, M. Pettersson, J. Lundell, H. Tanskanen, T. Kiviniemi, N. Runeberg, M. Räsänen, *J. Am. Chem. Soc.* **2003**, *125*, 1454.
56 L. Khriachtchev, H. Tanskanen, J. Lundell, M. Pettersson, H. Kiljunen, M. Räsänen, *J. Am. Chem. Soc.* **2003**, *125*, 4696.
57 I. Last, T. F. George, *J. Chem. Phys.* **1988**, *89*, 3071.
58 J. W. C. Johns, *J. Mol. Spectrosc.* **1984**, *106*, 124.
59 H. E. Warner, W. T. Connor, R. C. Woods, *J. Chem. Phys.* **1984**, *81*, 5413.
60 K. A. Peterson, R. H. Petrmichl, R. L. McClain, R. C. Woods, *J. Chem. Phys.* **1991**, *95*, 2352.
61 R. Klein, P. Rosmus, *Z. Naturforsch.* **1984**, *39a*, 349.
62 M. Pettersson, L. Khriachtchev, A. Lignell, M. Räsänen, Z. Bihary, R. B. Gerber, *J. Chem. Phys.* **2002**, *116*, 2508.
63 J. Lundell, G. Chaban, R. B. Gerber, *Chem. Phys. Lett.* **2000**, *331*, 308.
64 M. Pettersson, J. Lundell, L. Khriachtchev, M. Räsänen, *J. Chem. Phys.* **1998**, *109*, 618.
65 M. Pettersson, J. Lundell, L. Khriachtchev, E. Isoniemi, M. Räsänen, *J. Am. Chem. Soc.* **1998**, *120*, 7979.
66 M. Pettersson, L. Khriachtchev, J. Lundell, M. Räsänen, *J. Am. Chem. Soc.* **1999**, *121*, 11904.
67 N. Runeberg, M. Pettersson, L. Khriachtchev, J. Lundell, M. Räsänen, *J. Chem. Phys.* **2001**, *114*, 836.
68 S. Berski, Z. Latajka, B. Silvi, J. Lundell, *J. Chem. Phys.* **2001**, *114*, 4349.
69 M. Johansson, M. Hotokka, M. Pettersson, M. Räsänen, *Chem. Phys.* **1999**, *244*, 25.
70 A. Lignell, J. Lundell, M. Pettersson, L. Khriachtchev, M. Räsänen, *Low-Temp. Phys.* **2003**, *29*, 844.
71 J. Panek, Z. Latajka, J. Lundell, *Phys. Chem. Chem. Phys.* **2002**, *4*, 2504.
72 S. Berski, B. Silvi, J. Lundell, S. Noury, Z. Latajka, *The nature of binding in HRgY compounds (Rg= Ar, Kr, Xe; Y= F, Cl) based on the topological analysis of the electron localization function (ELF)*, in *New Trends in Quantum Chemistry and Physics* (Eds.: J. Maruani, C. Minot, R. McWeeny, Y. G. Smeyes, S. Wilson), Vol. 1, Kluwer, Dordrecht, 2001, pp. 259–279.
73 G. M. Chaban, J. Lundell, R. B. Gerber, *Chem. Phys. Lett.* **2002**, *364*, 628.
74 T. Takayanagi, A. Waga, *Chem. Phys. Lett.* **2002**, *352*, 91.
75 T. Takayanagi, *Chem. Phys. Lett.* **2003**, *371*, 675.
76 A. Cohen, M. Y. Niv, R. B. Gerber, *Faraday Discuss. Chem. Soc.* **2001**, *118*, 269.

77 M. Pettersson, J. Nieminen, L. Khriachtchev, M. Räsänen, *J. Chem. Phys.* **1997**, *107*, 8423.
78 J. Ahokas, K. Vaskonen, J. Eloranta, H. Kunttu, *J. Phys. Chem. A* **2000**, *104*, 9506.
79 V. I. Feldman, F. F. Sukhov, A. Yu. Orlov, *Chem. Phys. Lett.* **1997**, *280*, 507.
80 M. Pettersson, L. Khriachtchev, J. Lundell, S. Jolkkonen, M. Räsänen, *J. Phys. Chem. A* **2000**, *104*, 3579.
81 L. Khriachtchev, M. Pettersson, J. Lundell, M. Räsänen, *J. Chem. Phys.* **2001**, *114*, 7727.
82 J. Ahokas, H. Kunttu, L. Khriachtchev, M. Pettersson, M. Räsänen, *J. Phys. Chem. A* **2002**, *106*, 7743.
83 L. Khriachtchev, H. Tanskanen, M. Pettersson, M. Räsänen, J. Ahokas, H. Kunttu, V. Feldman, *J. Chem. Phys.* **2002**, *116*, 5649.
84 J. Lundell, G. M. Chaban, R. B. Gerber, *J. Phys. Chem. A* **2000**, *104*, 7944.
85 J. Lundell, M. Pettersson, L. Khriachtchev, M. Räsänen, G. M. Chaban, R. B. Gerber, *Chem. Phys. Lett.* **2000**, *322*, 389.
86 L. Khriachtchev, J. Lundell, M. Pettersson, H. Tanskanen, M. Räsänen, *J. Chem. Phys.* **2002**, *116*, 4758.
87 H.-G. Yu, J. T. Muckerman, *J. Theor. Comput. Chem.* **2003**, *2*, 573.
88 T. Kiviniemi, M. Pettersson, L. Khriachtchev, M. Räsänen, N. Runeberg, *J. Chem. Phys.* **2004**, *121*, 1839.
89 J. Lundell, M. Pettersson, M. Räsänen, *Comput. Chem.* **2000**, *24*, 325.
90 J. Lundell, A. Cohen, R. B. Gerber, *J. Phys. Chem. A* **2002**, *106*, 11950.
91 V. I. Feldman, F. F. Sukhov, A. Yu. Orlov, I. V. Tyulpina, *J. Am. Chem. Soc.* **2003**, *125*, 4698.
92 Z. Bihary, G. M. Chaban, R. B. Gerber, *J. Chem. Phys.* **2002**, *116*, 5521.
93 S. Jolkkonen, M. Pettersson, J. Lundell, *J. Chem. Phys.* **2003**, *119*, 7356.
94 M. W. Wong, *J. Am. Chem. Soc.* **2000**, *122*, 6289.
95 Z. Bihary, G. M. Chaban, R. B. Gerber, *J. Chem. Phys.* **2003**, *119*, 11278.
96 M. Lein, J. Frunzke, G. Frenking, *Structure and Bonding* **2003**, *106*, 181.
97 G. M. Chaban, J. Lundell, R. B. Gerber, *J. Chem. Phys.* **2001**, *115*, 7341.
98 Z. Bihary, G. M. Chaban, R. B. Gerber, *J. Chem. Phys.* **2002**, *117*, 5105.
99 A. Lignell, L. Khriachtchev, M. Räsänen, M. Pettersson, *Chem. Phys. Lett.* **2004**, *390*, 256.
100 S.-Y. Yen, C.-H. Mou, W. P. Hu, *Chem. Phys. Lett.* **2004**, *383*, 606.
101 I. Alkorta, J. Elguero, *Chem. Phys. Lett.* **2003**, *381*, 505.
102 L. Khriachtchev, M. Pettersson, H. Tanskanen, M. Räsänen, *Chem. Phys. Lett.* **2002**, *359*, 135.
103 L. Khriachtchev, M. Saarelainen, M. Pettersson, M. Räsänen, *J. Chem. Phys.* **2003**, *118*, 6403.
104 J. Eberlein, M. Creuzburg, *J. Chem. Phys.* **1997**, *106*, 2188.
105 L. Khriachtchev, H. Tanskanen, M. Pettersson, M. Räsänen, V. Feldman, F. Sukhov, A. Orlov, A. F. Shestakov, *J. Chem. Phys.* **2002**, *116*, 5708.
106 H. Tanskanen, L. Khriachtchev, J. Lundell, M. Räsänen, *J. Chem. Phys.* **2004**, *121*, 8291.

3
Polycationic Clusters of the Heavier Group 15 and 16 Elements

Johannes Beck

3.1
Introduction

Main group element clusters have a long history, reaching back to the beginning of modern chemistry. It was some years after the discovery of the element tellurium by von Reichenstein in 1782, that Klaproth reported in 1798 the red coloration that occurs when Te is dissolved in sulfuric acid [1]. In 1804, Buchholz reported a similar behavior for sulfur. Brown, green, or blue solutions are formed depending on the SO_3 concentration in the sulfuric acid [2]. Magnus, in 1827, found an analogous behavior for selenium [3]. Low-valent bismuth chlorides of the tentative formula BiCl were assumed to exist in the binary system Bi/$BiCl_3$ since the studies of Heintz in 1844 [4]. For more than a century, the nature and compositions of the colored solutions obtained from elemental chalcogens and H_2SO_4 remained unknown, despite the fact that these reactions were used by generations of chemists as specific analytical tests for the presence of the respective elemental chalcogens.

A breakthrough was achieved in 1961 with the crystal structure determination of Bi_6Cl_7 by J. Corbett and co-workers, which revealed the presence of the unprecedented Bi_9^{5+} cluster embedded in the structure. This ended the long discussion about the constitution of "BiCl" [5]. Pioneering work on the chalcogen polycations was carried out in the 1960s by N. Bjerrum, who found by vibrational and optical spectroscopies that the colored species present in sulfuric acid are identical to those in NaCl/$AlCl_3$ melts [6]. In 1968, R. J. Gillespie and co-workers showed the intensely red colored species to be the square-planar cationic cluster Te_4^{2+} [7].

This initial work was mainly supported by applying the method of crystal structure determination by X-ray diffraction. The number of isolated and characterized compounds containing polycationic clusters grew rapidly during the 1970s and 1980s, in parallel with the number of related polyanionic Zintl-type clusters and of compounds containing transition metal clusters. Research efforts concerning new synthetic methods have made polycationic main group element cluster chemistry a challenging and still growing field of modern inorganic chemistry.

Inorganic Chemistry in Focus II. Edited by G. Meyer, D. Naumann, and L. Wesemann
Copyright © 2005 WILEY-VCH Verlag GmbH & Co. KGaA, Weinheim
ISBN: 3-527-30811-3

3.2
Principles of Stability for Polycationic Clusters

The Lewis acid-base concept plays a basic role in the understanding of the syntheses and stabilization of polycationic clusters. As an example, the following reactions describe the equilibria of the decomposition of compounds containing polycationic clusters, as exemplified by Se_{10}^{2+} and Bi_8^{2+}, and anions X^- into the normal-valent compounds SeX_4 and BiX_3 and the respective element.

$$2\,Se_{10}^{2+}\,(X^-)_2 \leftrightarrows 19\,Se + SeX_4$$

$$3\,Bi_8^{2+}\,(X^-)_2 \leftrightarrows 22\,Bi + 2\,BiX_3$$

The existence of a cluster-containing phase in the systems Se/X and Bi/X depends on the thermodynamics of these reactions. A base X^- which forms strong covalent bonds with Se^{4+} or Bi^{3+}, e.g. Cl^-, shifts the equilibria to the right side and favors disproportionation into elemental Se or Bi and the respective normal-valent compound containing Se(IV) or Bi(III). A weak base X^-, e.g. $[AlCl_4]^-$, forming only weak bonds to Se and Bi, opens the possibility of shifting the equilibrium to the left side and to thereby stabilize the polycationic clusters. These clusters, therefore, do not exhibit covalent bonds to external ligand atoms and are called "bare" or "naked" clusters. The bonding between cations and anions in the structures of such compounds is mainly ionic. The weakly coordinating anions are well separated from the atoms of the clusters by distances normally not significantly less than the sum of the van der Waals radii. This principle is termed "acid stabilization" and was introduced by the early work of J. Corbett [8]. Using the simple rules of the acid-base concept, one can understand which anions are suitable for this purpose. HX/X^- is a pair of corresponding acid and base. The base strength of X^- corresponds inversely to the acid strength of HX. So, for the stabilization of naked polycationic clusters only anions deriving from strong acids are suitable. The anions of compounds containing polycationic clusters characterized to date all fall in this category. Examples are $[AsF_6]^-$, $[SbF_6]^-$, $[Sb_2F_{11}]^-$, $[SO_3F]^-$, $[HS_2O_7]^-$, $[AlCl_4]^-$, $[NbCl_6]^-$, $[WOCl_4]^-$, and $[Bi_2Br_8]^{2-}$.

The predominantly ionic nature of the compounds containing polycationic clusters and weakly coordinating anions permits the estimation of thermodynamic properties through calculated lattice energies and Born-Haber cycles. This method was implemented by J. Passmore in 1989 [9], and has since been refined and expanded to the majority of the known polycations of the halogens and the chalcogens [10]. Through these calculations, a rationalization of formation reactions and the non-existence of specific ion combinations is possible. For example, the observation that polycations of Se and Te can be obtained with $[AlCl_4]^-$ as counterion whereas $S_4[AlCl_4]_2$ is unknown can be ascribed to the enthalpy of the formation reaction of the latter, which is calculated to be positive (endothermic reaction).

$$S_2Cl_2 + 2\,AlCl_3 + 1/4\,S_8 \rightarrow S_4^{2+}\,[AlCl_4^-]_2 \quad \Delta H_{298} = +97\,kJ\,mol^{-1}$$

Another example is the explanation of the experimental fact that for E = S, Se, Te only E_4^{2+} cations are known, whereas for oxygen only the radical diatomic cation O_2^+, as in $O_2[PtF_6]$ [11], is observed. Ab initio calculations have shown for all chalcogen elements that the formation of diatomic E_2^+ radical cations is favored over the formation of square-planar E_4^{2+} cations by energies ranging from about 100 kJ mol^{-1} (E = Te) up to 1000 kJ mol^{-1} (E = O) [12]. The estimation of the solid-state energetics for the formation of $E_4^{2+}[AsF_6^-]_2$ salts showed that $E_4[AsF_6]_2$ is clearly favored over $E_2[AsF_6]$ by around 300 kJ mol^{-1} for E = S, Se, Te. The lattice energy, which is three times higher for $E_4[AsF_6]_2$ than for $E_2[AsF_6]$, overcompensates the exothermic dissociation $E_4^{2+} \rightarrow 2\,E_2^+$ for E = S, Se, Te. This is not the case for O_4^{2+}, which has the highest exothermic dissociation energy of all the chalcogens. The main reason for this different behavior is the large difference in σ and π bond strengths of O–O bonds in comparison to E–E bonds of the heavier chalcogens.

3.3
Synthetic Routes to Polyatomic Cations

The successfully employed preparative routes for obtaining polycationic clusters of main group elements all involve the oxidation of the element accompanied by the formation of a weakly basic and weakly coordinating counter ion. All of these routes imply the use of highly electrophilic and acidic reaction media and the need for a mild and specific oxidation of the main group element to low positive oxidation states.

The first route, historically speaking, proceeds in superacidic media such as H_2SO_4, $H_2S_2O_7$, HSO_3F, $S_2O_6F_2$, and HSO_3F/SbF_5 [13]. The isolation of well-defined, crystalline compounds from such media is generally difficult. The high viscosity and low volatility of these solvents makes the process of distilling off the solvent almost impossible. Increasing the concentration of dissolved material in order to initiate the crystallization process has been achieved only in some cases, e.g. with $Se_4[HS_2O_7]_2$ [14] and $Se_{10}[SO_3F]_2$ [15].

The use of AsF_5 and SbF_5 as oxidants for elemental chalcogens is more convenient, especially if liquid SO_2 is used as a solvent. A large number of polycationic clusters have been obtained from this reaction system by the groups of R. Gillespie and J. Passmore [9]. Both AsF_5 and SbF_5 act as two-electron oxidants under the formation of AsF_3 and SbF_3 and of the desired counterions $[AsF_6]^-$ and $[SbF_6]^-$. The use of SbF_5 has the disadvantage that it forms adducts with SbF_3 and $[SbF_6]^-$ which occasionally leads to anions with complicated composition and structure. At ambient temperatures, liquid SO_2 has a moderate vapor pressure of about 3 bar and can be handled in a thick-walled glass apparatus using Teflon-coated screw valves. The disadvantage lies in the generally observed incorporation of SO_2 into crystalline compounds. SO_2 often weakly coordinates the atoms of the cationic clusters via an oxygen atom. The vapor pressure of SO_2 leads to a high susceptibility of the crystalline products to loss of SO_2, which is notoriously accompanied by the destruction of the crystal lattice. Handling the products at low temperature is therefore a necessity.

Molten tetrachloroaluminates are well-known media for the preparation of polycationic clusters of the chalcogens and of bismuth [16]. An advantage is the possibility of adjusting the acidity of the melt by selecting the appropriate alkali halide/aluminum trihalide ratio. To keep the melt acidic, an excess of aluminum halide is normally used. Care has to be taken to ensure that the appropriate quantity of melt is used with respect to the amount of the polycationic species. The melts solidify below about 100 °C and include all precipitated crystalline material. The separation of the desired product from the solidified melt is always a problem that has to be overcome. The use of the minimum amount of melt and the separation of the crystalline product by decanting the hot melt has turned out to be a good method for the separation of crystalline products.

Some chalcogen polycation-containing compounds can be transported by chemical vapor transport. This was observed for $Te_4[AlCl_4]_2$ and $Se_8[AlCl_4]_2$ [16] and for $Te_4[Nb_2OCl_{10}]$ [17]. The basic principle is the transformation of a solid to purely gaseous products by reaction with gaseous reagents. The reaction must be close to equilibrium since the gaseous compounds must yield solid compounds by vapor deposition under conditions differing only slightly from those of their formation, e.g. a temperature differential of the order of 10 °C. Elemental selenium and tellurium react in this way with volatile transition metal halides in which the metal is in a high oxidation state. An example is the reaction of tellurium with WCl_6. In closed evacuated glass ampoules at a temperature of 170 °C and with a temperature gradient of 20 °C, $Te_4[WCl_6]_2$ is deposited in two modifications [18, 19] in the colder part of the ampoule. With a higher Te load, $Te_8[WCl_6]_2$ [20] is formed under similar conditions.

$$4\,Te + 2\,WCl_6 \rightarrow Te_4\,[WCl_6]_2$$

$$8\,Te + 2\,WCl_6 \rightarrow Te_8\,[WCl_6]_2$$

Investigation of the gas-phase composition either by mass spectrometry or by thermodynamic calculations showed that Te is transported mainly as $TeCl_2$ and W mainly as WCl_4. This principle could be extended to many other transition metal halides and halide oxides such as $WOCl_4$, $WOBr_4$ or $MoOCl_4$. The oxidation state of Mo and W in these halides is +VI. In the resulting ionic compounds containing the anions $[MOX_4]^-$ (M = Mo, W; X = Cl, Br), the oxidation state is reduced to +V. Many other metal halides without strong oxidizing properties can be used, e.g. $NbBr_5$, $TaCl_5$, $ZrBr_4$, $ReCl_4$, $BiCl_3$, and these yield a broad variety of polycationic clusters of the chalcogens [21]. In these cases, a stoichiometric amount of the respective main group element halide is added as the oxidant, as exemplified by the following equations.

$$7\,Te + TeBr_4 + 4\,ZrBr_4 \rightarrow 2\,Te_4\,[Zr_2Br_{10}]$$

$$13\,Te + TeCl_4 + 4\,BiCl_3 \rightarrow (Te_4)(Te_{10})[Bi_4Cl_{14}]$$

$$9\,Bi + BiCl_3 + 3\,NbCl_5 \rightarrow Bi(Bi_9)\,[NbCl_6]_3$$

In some cases, chemical vapor transport does not work. An example is a reaction involving uranium pentabromide. UBr$_5$ is volatile, but UBr$_4$ has a very low vapor pressure. If UBr$_5$ acts as an oxidant towards Te, the resulting ionic compound consisting of a tellurium polycationic cluster and a bromouranate(IV) anion is not amenable to chemical vapor transport. In this case, the use of SiBr$_4$ as a solvent proved helpful. Under solvothermal conditions, Te$_8$[U$_2$Br$_{10}$] could be isolated in crystalline form [22].

3.4
Homopolyatomic Cations of the Chalcogens

3.4.1
Square-Planar Cations E_4^{2+}

The smallest discrete cationic chalcogen clusters are the tetraatomic square-planar E_4^{2+} (E = S, Se, Te) ions. Among this series, compounds containing the colorless S_4^{2+} are rare. It was not until 1980 that (S$_4$)(S$_7$Br)$_4$[AsF$_6$]$_6$ was structurally characterized as the first compound containing an S_4^{2+} ion [23]. Se_4^{2+} and Te_4^{2+} are more stable and are easily formed, so that more than 25 compounds containing these ions have been prepared and studied. This number is still increasing, the latest reported compounds being Se$_4$[Bi$_4$Cl$_{14}$] [24] with a two-dimensional polymeric anion [Bi$_4$Cl$_{14}$]$^{2-}$, Se$_4$[ReCl$_6$], [25] Te$_4$[AsF$_6$]$_2 \cdot$SO$_2$, [26] and (Se$_4$)$_2$[Mo$_2$O$_2$Cl$_8$][MCl$_6$], the latter containing as a unique feature two different anions [Mo$_2$O$_2$Cl$_8$]$^{2-}$ and [MCl$_6$]$^{2-}$ (M = Zr, Hf) [27]. The square-planar ions (Fig. 3.1a) can be regarded as 6π electron aromatics. The molecular orbitals deriving from the p orbitals of the four chalcogen atoms are arranged according to the point symmetry group D_{4h} in four MOs of A_{2u}, E_g, and B_{2u} symmetry, and are occupied by six electrons. Since the doubly-degenerate electron-filled E_g orbital set is non-bonding, essentially only one pair of electrons (A_{2u}) contributes to the aromatic system. Ab initio calculations have shown that for E_4^{2+} (E = S, Se, Te), the square-planar D_{4h} molecular geometry is energetically favorable by about 100 kJ mol^{-1} over the alternative C_{2v} butterfly structure with a transannular bond [28]. This structure is observed for the isoelectronic 22 valence electron polysilicide anion Si$_4^{6-}$ present in the crystal structure of Ba$_3$Si$_4$ [29] (Fig. 3.1b).

In the E_4^{2+} ions, the E–E bonds are substantially shorter than the length of an E–E single bond, indicating partial π bonding. There is an obvious relationship between the coordination of the cations and the E–E bond lengths. In the crystal structures, the E_4^{2+} squares are always coordinated by four halogen or oxygen atoms, each bridging two chalcogen atoms in the plane of the square. Electron density is transferred from the lone pairs of the halogen/oxygen into antibonding σ^* orbitals of E_4^{2+}. The shorter the halogen/oxygen–E coordinating bond, the longer are the E–E bonds, underlining the proposed interaction mechanism [29].

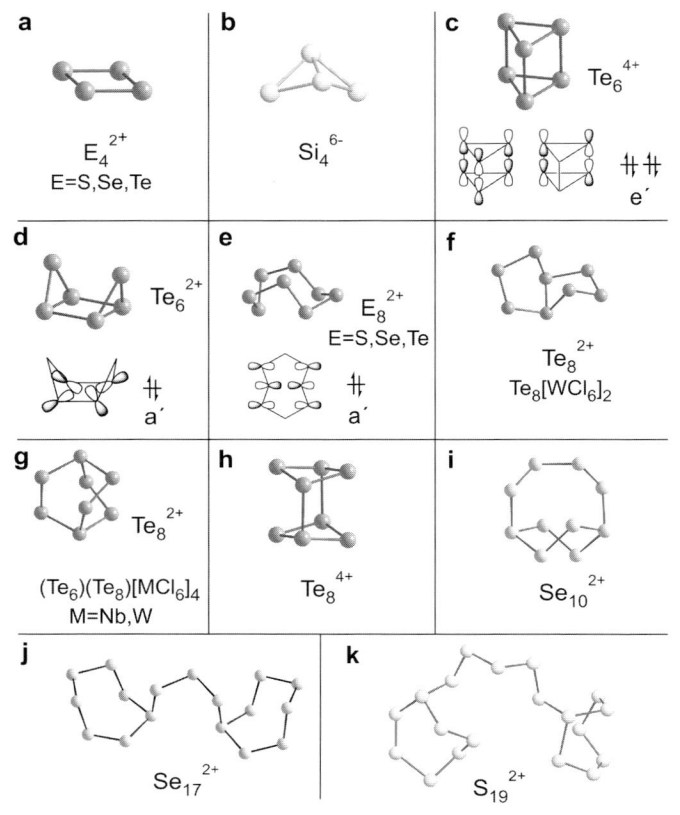

Fig. 3.1 The structures of the known homopolyatomic cations of the chalcogens. For Te_6^{4+} (c), Te_6^{2+} (d), and E_8^{2+} (e), the orbital overlap of selected molecular orbitals, i.e. those responsible for multi-center bonding, is also shown.

3.4.2
Prism-Shaped Cations Te_6^{4+} and Te_6^{2+}

Te_6^{4+} was first prepared in 1979 as $Te_6[AsF_6]_4 \cdot 2AsF_3$ and $Te_6[AsF_6]_4 \cdot 2SO_2$ [31, 32] and occurred again in 1988 in the structure of $(Te_6)(Se_8)[AsF_6]_6 \cdot SO_2$, a unique compound containing two different cationic clusters [33]. Te_6^{4+} has the structure of a trigonal prism with two parallel triangular faces connected via three long Te–Te bonds of average length 312 pm (Fig. 3.1c). The triangular monomer Te_3^{2+} is unknown as a discrete ion, but was found as a π-bonded ligand at a transition metal center in $[W(CO)_4(Te_3)][SbF_6]_2$ [34]. A theoretical investigation based on density functional theory showed that the elongated prism is the ground-state structure of Te_6^{4+} [35]. The bonding can be viewed in terms of a π^*-π^* six-center-four-electron bond between the two Te_3^{2+} moieties [9], leading to a bond order of 2/3 along one prism edge. Fig. 3.1c shows the doubly-degenerate molecular orbital obtained by combining $5p_z$ orbitals of the Te atoms. These orbitals are antibonding within the Te_3 subunits but bonding along the prism edges.

Formal addition of two electrons to Te_6^{4+} leads to Te_6^{2+}. The addition of one electron pair leads one to expect the loss of one bond by occupation of antibonding orbitals. This ion has been found in the structures of $Te_6[MOCl_4]_2$ (M = Nb [36], W [37]). The actual structure is a boat-shaped six-membered ring with two weak transannular interactions (Fig. 3.1d). The two positive charges are apparently delocalized over the four Te atoms of the rectangular basal plane. A detailed analysis of the bonding situation in Te_6^{2+} has recently been performed [10]. Each Te atom of the rectangular plane contributes a p orbital that is directed towards the center of the rectangle. A set of four MOs is formed, three of which are occupied by electrons. Since the HOMO and the HOMO-1 compensate themselves in their bonding effect, only HOMO-2 has a substantial bonding contribution (see Fig. 3.1d) and represents a $5p_\pi$-$5p_\pi$ bond along the edges and a four-center–two-electron transannular π^*-π^* bond.

3.4.3
Eight-Atomic Molecular Ions E_8^{2+} and Te_8^{4+}

Following their discovery, the structure and bonding situation of the eight-atomic cations S_8^{2+}, first prepared as $S_8[AsF_6]_2$, [38] and Se_8^{2+}, first prepared as $Se_8[AlCl_4]_2$, [39] was a matter of discussion. The heavy homologue Te_8^{2+} could not be obtained, neither by oxidation of Te with AsF_5 in SO_2 nor from $Te/TeCl_4$ in chloroaluminate melts. By chemical vapor transport starting from Te_3Cl_2 and $ReCl_4$ at 200 °C, $Te_8[ReCl_6]$, containing a structurally equivalent ion Te_8^{2+}, was formed [40]. The use of $BiCl_3$, $ZrCl_4$, and $HfCl_4$ expanded the series of characterized E_8^{2+} ions to $Se_8[Bi_4Cl_{14}]$, mixed species such as $(Te_{6.5}Se_{1.5})[ZrCl_6]$, and $Te_8[HfCl_6]$ [41]. The structure of these eight-atomic ions can be derived from the crown-shaped rings of the neutral molecules S_8, Se_8, and Te_8. The latter is still unknown as a polymorph of elemental tellurium but occurs embedded in the structure of the polytelluride Cs_3Te_{22} [42]. Deformation of the ring to an oval and flipping one atom from an *exo* to an *endo* position, accompanied by the removal of one electron pair, leads to a substantially new structure with three weak transannular E–E interactions (Fig. 3.1e). Therefore, a six-center–two-electron π^*-π^* bond is the best description of the bonding situation. Up to 1997, numerous attempts to obtain fully optimized structures had failed. Only in 2000 were successful ab initio optimizations of the structures of the gas-phase ions accomplished [43].

For Te_8^{2+}, further structural possibilities have been found. In $Te_8^{2+}[WCl_6^-]_2$, a bicyclic ion composed of two five-membered rings, each in an envelope conformation, is present [44] (Fig. 3.1f). The transannular bond of 295 pm is significantly shorter than in the respective isomers in the structures of $Te_8[HfCl_6]$ and $Te_8[ReCl_6]$ (306 and 315 pm). In $(Te_6)(Te_8)[MCl_6]_4$ (M = Nb, W), two different tellurium polycations Te_6^{2+} and Te_8^{2+} are present. The synthesis of this compound is another example of a reaction that only proceeds under solvothermal conditions. $SnCl_4$ was used as the solvent at 150 °C [21]. The Te_8^{2+} cations in this compound exhibit a bicyclo[2.2.2]octane structure with two threefold-coordinated Te atoms (Fig. 3.1 g).

The reaction of Te_3Cl_2 with $VOCl_2$ yields $Te_8^{4+}[VOCl_4^{2-}]_2$ [45]. This 44 valence electron cluster has a unique structure of a cube with two open edges (Fig. 3.1 h). Four Te atoms are threefold coordinated and formally bear the four positive charges. Te_8^{4+} can be viewed as two Te_4^{2+} ions that have dimerized under the formation of two localized Te–Te bonds with simultaneous loss of electron delocalization and planarity of the Te_4^{2+} ions.

3.4.4
Towards Larger Clusters – Chalcogen Polycations with More than Eight Atoms

Only a few examples of chalcogen polycations composed of more than eight atoms are known. Se_{10}^{2+} was first obtained as $Se_{10}[SbF_6]_2$, [46] later as $Se_{10}[SO_3F]_2$, [47] and recently as $Se_{10}[Bi_5Cl_{17}]$, the latter containing a complex, three-dimensional chlorobismuthate anion [48]. The ions exhibit a bicyclo[4.2.2]decane structure with two threefold-coordinated Se atoms (Fig. 3.1 i). In solution, however, ^{77}Se NMR studies have provided evidence for a non-rigid structure [49]. The largest known discrete molecular polycationic chalcogen clusters are Se_{17}^{2+} and S_{19}^{2+}. Se_{17}^{2+} can be obtained by vapor-phase transport from selenium and WCl_6 as $Se_{17}[WCl_6]_2$ [50] or from selenium and $NbCl_5$ or $TaBr_5$ in $SnCl_4$ or $SiBr_4$, respectively, by the solvothermal route as $Se_{17}[NbCl_6]_2$ or $Se_{17}[TaBr_6]_2$ [51]. S_{19}^{2+} was obtained as the $[MF_6]^-$ salts (M = As, Sb) [52, 53]. The structures of the two ions are closely related. They consist of two seven-membered rings in a chair conformation connected by an Se_3 chain or by an S_5 chain, respectively (Figs. 3.1 j and 3.1 k). Both ions contain two threefold coordinated chalcogen atoms, which formally bear the two positive charges.

3.5
Homopolyatomic Cations of Bismuth

For polycationic clusters of bismuth, reaction media and terms of stabilization are coincident with those for the polycationic clusters of the chalcogens. Bi cluster cations are highly electrophilic and require an extremely weakly basic environment to be stable. Thus, the anions of compounds containing Bi cluster cations are the same as those of compounds containing chalcogen cluster cations.

Bismuth was one of the first elements for which subvalent species in oxidation states between 0 and +III were studied systematically. The occurrence of a series of discrete, homonuclear cations Bi^+, Bi_2^{4+}, Bi_5^+, Bi_5^{3+}, Bi_6^{2+}, Bi_8^{2+}, and Bi_9^{5+} has made bismuth the classical example of the existence of naked clusters among the main group elements. Because of these exceptional properties, bismuth was called "The Wonder Metal" [16].

3.5 Homopolyatomic Cations of Bismuth

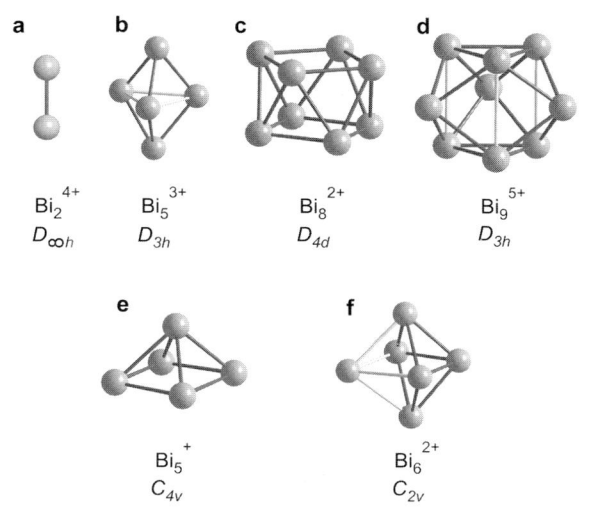

Fig. 3.2 The structures of polybismuth cations. Light-grey bond sticks represent long interatomic separations of more than 330 pm. Structures have been idealized for the point group assignments.

a) Bi_2^{4+}, $D_{\infty h}$
b) Bi_5^{3+}, D_{3h}
c) Bi_8^{2+}, D_{4d}
d) Bi_9^{5+}, D_{3h}
e) Bi_5^{+}, C_{4v}
f) Bi_6^{2+}, C_{2v}

The monoatomic cation Bi^+ was found in dilute solutions of Bi in $BiCl_3$ and in the more acidic solvent systems $NaCl/AlCl_3$ and $KCl/ZnCl_2$ [54]. In the solid state, this ion appears in the structure type $Bi(Bi_9)[MX_6]_3$. The diamagnetism of this compound with $[MX_6]^{2-}=[HfCl_6]^{2-}$ shows that Bi^+ has a diamagnetic $^3P_0$ ground state [55]. Bi_2^{4+} dumbbells were discovered in the structures of the ternary bismuth(II) chalcogenometallates(III) $Bi_2M_4E_8$ (M = Al, Ga; E = S, Se) [56] (Fig. 3.2a). Calculations of the electron localization function (ELF) led to the interpretation of the Bi–Bi bond length of 314 pm as a covalent single bond, sp hybridization of the Bi atoms, and one lone pair on each Bi atom sticking out along the Bi–Bi axis. Bi_5^{3+} is obtained from acidic melts as $Bi_5[AlCl_4]_3$ [57, 58] and $Bi_5[GaCl_4]_3$ [59]. The analogous antimony cation Sb_5^{3+} was also reported as the tetrachlorogallate [59]. These ions exhibit a trigonal-bipyramidal structure (Fig. 3.2b). Bi_5^+ and Bi_6^{2+} have been discovered in the complex structure of $Bi_{34}Ir_3Br_{37}$ [60], which, besides these two Bi cations, also contains two independent cluster anions $[IrBi_6Br_{12}]^-$ and $[IrBi_6Br_{13}]^{2-}$. Bi_5^+ has the structure of a distorted square pyramid (Fig. 3.2e), thus exhibiting a fundamental difference to Bi_5^{3+}. Bi_6^{2+} has the structure of a strongly distorted octahedron (Fig. 3.2f). One of the six Bi atoms is only loosely bound, with four Bi–Bi distances of more than 330 pm. Bi_8^{2+} has been found in the compounds $Bi_8[AlCl_4]_2$ [57, 58, 61], $Bi_8[InBr_4]_2$ [62], and $Bi_8[Ta_2O_2Br_7]$ [63]. Its structure is close to a regular square antiprism with Bi–Bi bond lengths of around 310 pm (Fig. 3.2c). Bi_9^{5+} is the polycationic main group element cluster with the highest charge per atom ratio (Fig. 3.2d). It was discovered in the bismuth subchloride Bi_6Cl_7 [5, 61] and thereafter in the structures of $Bi(Bi_9)[MX_6]_3$ (M = Hf, X = Cl [55]; M = Zr, X = Cl

[64]; M=Zr, X=Br [64]; M=Hf, X=Br [64]; M=Nb, X=Cl [63, 65]), in the bismuth subbromide Bi_6Br_7 [66], and in the complex bromo bismuthate indate $Bi_9[InBr_5][Bi_3Br_{13}][Bi_7Br_{30}]$ [67].

Numerous articles have appeared dealing with theoretical approaches to explain the bonding situation in these predominantly deltahedral clusters. The positive charges are delocalized almost evenly over all atoms of the ions. The structures of Bi_5^{3+} and Bi_8^{2+} can be understood as *closo* and *arachno* clusters that fulfil Wade's rules [68]. Three valence electrons (VE) are counted for each Bi atom, while the $6s^2$ electron pair is assumed to be inert. Thus, Bi_5^{3+} has 12 VE and is derived from a *closo* polyhedron with 5 corners that is a trigonal bipyramid and has 5+1 skeletal electron pairs (SEP). Bi_8^{2+} has 22 VE and is derived from an *arachno* polyhedron with 10−2 corners that is a square antiprism with 8+3 SEP. Bi_9^{5+} also has 22 VE and a *nido* structure with 9+2 SEP. Thus, a monocapped square antiprism with C_{4v} symmetry is expected. In the crystal structures of the respective compounds, the structure of the cluster ion is generally close to a three-capped trigonal prism with D_{3h} symmetry approximately fulfilled (Fig. 3.2d). Recently, it was shown by an ab initio theoretical study that the energy difference between these two isomers is very small, amounting to less than 9 kJ mol^{-1} [69]. The two youngest members of this cluster family, Bi_5^+ and Bi_6^{2+}, can be treated in an analogous manner. Bi_5^+ with 14 VE represents a *nido* polyhedron with 5 corners and 5+2=7 SEP that is a tetragonal pyramid with C_{4v} symmetry. For Bi_6^{2+} with 16 VE a *nido* structure with 6+2=8 SEP is expected. The actual structure, though reminiscent of a *closo*-type octahedron, is strongly distorted and can be derived from a pentagonal bipyramid with one atom removed from the equatorial plane, leaving a C_{2v} *nido* cluster behind. The high degree of electron delocalization in the polyhedral clusters has recently been termed as "three-dimensional" or "spherical" aromaticity [70].

3.6
Heteronuclear Polycationic Clusters of the Heavier Group 15 and 16 Elements

Besides the homonuclear polycationic clusters of the chalcogens and of bismuth, there is also a series of heteronuclear clusters. Among them, cationic clusters made up of two different kinds of chalcogen atoms represent the largest group. For the tetraatomic E_4^{2+} cations, various compositions are possible. $(Te_2Se_2)^{2+}$, $(Te_3Se)^{2+}$ [71], and $(S_3Se)^{2+}$ [72] have all been obtained from mixtures of chalcogens by oxidation with AsF_5/SbF_5. In the structures, the D_{4h} symmetry of the cations is not significantly distorted, indicating disorder of the different atom types over the four atom positions. In the six-membered, boat-shaped rings of both $(Te_3S_3)^{2+}$ and $(Te_2Se_4)^{2+}$, one definite transannular bond is observed [73] (Figs. 3.3a and 3.3b). Typically, the heavier Te atoms occupy the positions with the higher coordination number three and thus formally carry the positive charges. The different electronegativities of the two atom types cause a charge and bond localization which is not observed for the corresponding

3.6 Heteronuclear Polycationic Clusters of the Heavier Group 15 and 16 Elements

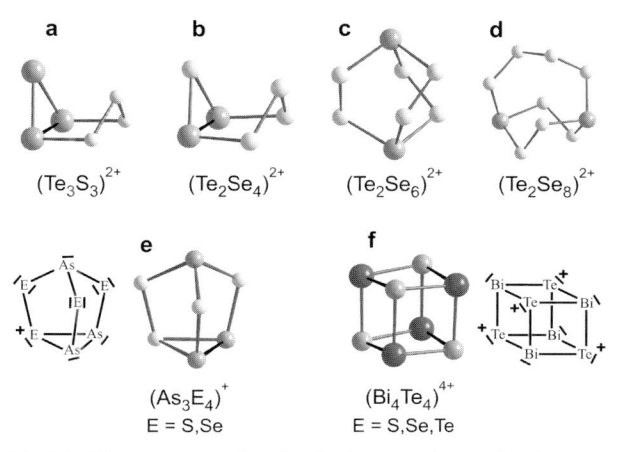

Fig. 3.3 The structures of molecular heteronuclear polycationic clusters composed of chalcogen and heavier group 15 element atoms. For $(As_3E_4)^+$ (e) and $(Bi_4E_4)^{4+}$ (f) a Lewis formula is also given.

homonuclear cluster Te_6^{2+}. Both the mixed eight- and ten-atomic polycations $(Te_2Se_6)^{2+}$ and $(Te_2Se_8)^{2+}$ [74] have their homonuclear congeners (Figs. 3.3c and 3.3d). The bicyclo[2.2.2]octane isomer of Te_8^{2+} and $(Te_2Se_6)^{2+}$ on the one hand and Se_{10}^{2+} and $(Te_2Se_8)^{2+}$ on the other have analogous structures. In these two heteronuclear ions, a clear separation into threefold and twofold coordinated chalcogen atoms is present. In these mixed ions the principle of the occupation of the threefold coordinated positions by the heavier Te atoms is again observed.

The only known arsenic/chalcogen polycations are $(As_3S_4)^+$ and $(As_3Se_4)^+$ [75]. Both are 38 valence electron species and adopt the structure often observed for seven-atom cage molecules, as found, for example, for P_7^{3-} and P_4S_3 (Fig. 3.3e). In these ions of C_s symmetry, the As atoms occupy the three-coordinate positions and are formally neutral. One of the chalcogen atoms also occupies a three-coordinate position and formally bears the positive charge. Recently, the first bismuth/chalcogen polycationic clusters were characterized [76, 77]. The complete series of $(Bi_4E_4)^{4+}$ (E = S, Se, Te) ions could be prepared in acidic $NaCl/AlCl_3$ melts and isolated as the respective tetrachloroaluminates. The structures of the cations are unusual since cube-shaped polycations were hitherto unknown (Fig. 3.3f). All atoms at the corners of the cubes are threefold coordinated, so the Bi atoms count as neutral whereas the chalcogen atoms are formally positive. Since the structures can be described with only one valence dash formula, delocalization of charges and bonds does not have to be taken into consideration, and the structures are classical. An ab initio study, performed in parallel for $(Bi_4Te_4)^{4+}$ and Bi_8^{2+}, clearly showed that the two additional valence electrons in $(Bi_4E_4)^{4+}$ are responsible for the structural transition of the square-antiprismatic 38 VE (including $6s^2$ electrons) species Bi_8^{2+} to the cuboidal

$(Bi_4E_4)^{4+}$ structure with 40 VE [76]. Though Bi_8^{2+} cannot be described in terms of two-center–two-electron bonds and therefore has a non-classical electronic structure, both clusters have a large HOMO-LUMO gap and must be interpreted as electron-precise molecules.

3.7
Interactions Between Polycationic Clusters – From Weak Association to One-Dimensional Polymers

It is a striking feature of the heavy elements bismuth and tellurium that they form a larger number of polycationic clusters than their lighter homologues. Both hypervalent bonding and the special role of $np^2 \to n\sigma^*$ interactions are responsible for the rich bismuth and tellurium polycation chemistry. The energy difference between occupied np^2 orbitals with lone pair character and antibonding $n\sigma^*$ orbitals of E–E bonds in the clusters decreases with increasing n and becomes important for $n > 3$. So, for Te $5p^2 \to 5\sigma^*$ intra- and intermolecular bonding is a generally observed structural feature. Besides the covalent Te–Te bonds with a length of 275–285 pm, secondary Te $\cdots$ Te bonds between 300 and 350 pm are generated by this mechanism.

A special feature in the structures of $Te_6^{2+}[MOCl_4^-]_2$ (M = Nb [36], W [37]) and $Te_8^{4+}[VOCl_4^{2-}]_2$ [45] is the arrangement of the cationic clusters in linear arrays (Figs. 3.4a and 3.4b). The one-dimensional arrangement of the anions, which are associated via linear M=O $\cdots$ M bridges, surely favors this cation arrangement. The interionic distances are long in the case of Te_6^{2+} (Te $\cdots$ Te > 460 pm), but are quite short in the case of Te_8^{4+} (Te $\cdots$ Te > 359 pm). Another example is the occurrence of dimers $(Te_8^{2+})_2$ in the structure of $Te_8[HfCl_6]$ (Fig. 3.4c). Two short interionic contacts of 347 pm are observed between two neighboring cations leading to a rectangular Te_4 ring as the connecting link [71]. This kind of intermolecular interaction is found in an extended way in the structure of $Te_8[WCl_6]_2$ [20]. Here, the polycations are linked to an ondulated band by secondary Te $\cdots$ Te bonds of 342 pm (Fig. 3.4d).

The second group of compounds contains one-dimensional polymeric cationic strands connected via Te–Te covalent bonds all along the chain. This represents a relatively new structural feature in structures of compounds containing chalcogen polycations. The coordination number of the chalcogen atoms in the strands is either two or three. Since the charge of the chain corresponds exactly to the number of three-coordinated atoms, the bonding situation can be understood in terms of classical two-center–two-electron bonds. The first example is $(Te_4)(Te_{10})[Bi_4Cl_{16}]$, which is made up of two different one-dimensional cations [78]. In contrast to the only hitherto known molecular square-planar Te_4^{2+} ions, the polymeric $(Te_4^{2+})_n$ consists of a chain of linked rectangular four-membered rings (Fig. 3.5a). In the chain, there is a balanced number of two- and three-coordinated Te atoms and formal charge localization on the Te atoms with coordination number three. A study of the bonding situation showed that the details

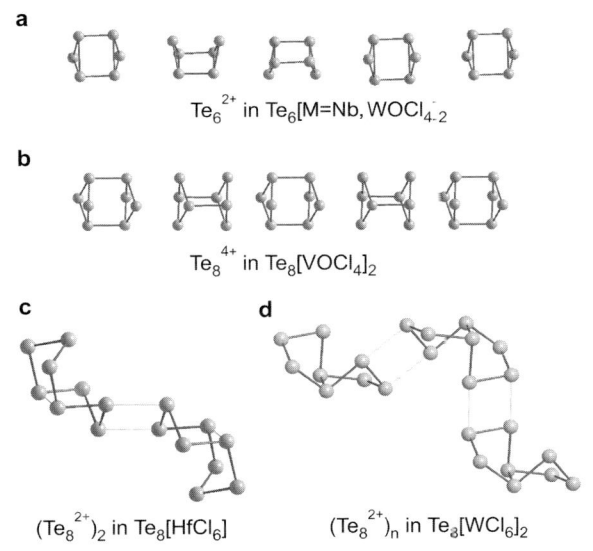

Fig. 3.4 The arrangement of the polycations Te_6^{2+} in the structures of $Te_6[MOCl_4]_2$ (M = Nb, W) (a) and Te_8^{4+} in the structure of $Te_8[VOCl_4]_2$ (b); the arrangement of Te_8^{2+} ions in the structures of $Te_8[HfCl_6]$ (c) and $Te_8[WCl_6]_2$ (d). In $Te_6[MOCl_4]_2$ and $Te_8[VOCl_4]_2$ the ions form linear strands, in $Te_8[HfCl_6]$ the cations form dimers, and in $Te_8[WCl_6]_2$ the cations are associated and form an undulated chain.

of this structure can be understood in terms of donation of electron density from the $5p^2$ lone pairs of the two-coordinated Te atoms into the empty $5\sigma^*$ orbital of the Te–Te bond connecting the four-membered rings [10]. By this mechanism, the positive charge is delocalized over all atoms of the chain and the length of the ring-connecting bond (297 pm) is increased with respect to the bonds in the ring (275 and 281 pm).

Meanwhile, a whole series of one-dimensional chain-like chalcogen polycations is known. All have the same structural principle and are built-up of four- to six-membered rings which are directly connected as in $(Te_4^{2+})_n$, by one chalcogen atom or by short chains of two or three atoms (Fig. 3.5). Four-membered rings connected by three-atom bridges are present in $(Te_3Se_4^{2+})_n[WOCl_4]_2$ [79]. As in the heteronuclear molecular clusters, Te atoms occupy the threefold-coordinated positions in this mixed polycationic chain (Fig. 3.5b). The structure of $Te_6[HfCl_6]$ contains a polymeric isomer $(Te_6^{2+})_n$ of the molecular Te_6^{2+} [80]. Here, five-membered rings are connected by single-atom bridges (Fig. 3.5c). Five-membered rings, but connected differently via –Se–Te–Se– bridges, are also present in the structure of $(Te_{3.15}Se_{4.85}^{2+})[WOCl_4]_2$ [79] (Fig. 3.5d). Six-membered rings connected by bridging Te atoms make up the structure of $(Te_7^{2+})_n[AsF_6]_2$ obtained from $Te_4[AsF_6]$ and $[Fe(CO)_5]$ in liquid SO_2 [81] (Fig. 3.5e). Finally, polymeric $(Te_8^{2+})_n$ is present in the structures of $(Te_8^{2+})_n[U_2Br_{10}^{2-}]$ [82] and $(Te_8^{2+})_n[Bi_4Cl_{14}^{2-}]$ [83]. Both compounds contain a polycationic chain of six-membered

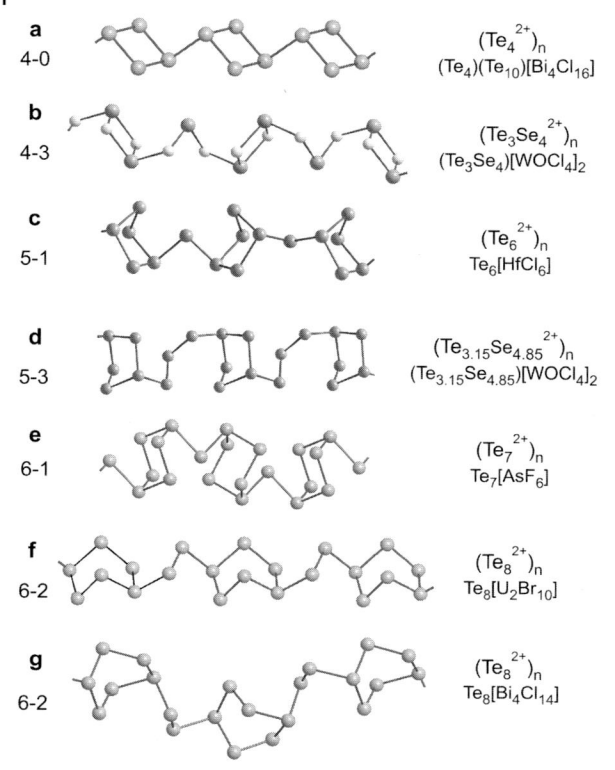

Fig. 3.5 The structures of one-dimensional polymeric chalcogen cations. The numbers on the left refer to a simple structure designator code. The first digit gives the number of atoms in the respective rings, the second digit gives the number of atoms in the bridges that connect the rings. $(Te_8^{2+})_n$ is made up of six-membered rings connected via –Te–Te– groups and thus obtains the designator 6-2.

rings connected via –Te–Te– groups. This arrangement is very flexible and so two different spatial arrangements for the two chains are observed (Figs. 3.5f and 3.5g).

"Hypervalent" bonding is observed in the last group of polymeric tellurium cations. In these ions, Te atoms are present with a coordination number greater than three. A common feature is the occurrence of Te in a square-planar coordination environment forming TeTe$_4$ groups. These groups are commonly observed in tellurium-rich polytellurides, e.g. in the structure of Ga$_2$Te$_5$ as $(TeTe_4)^{6-}$ units [84] or as butterfly-shaped $(Te_7)^{2-}$ in (Re$_6$Te$_8$)Te$_7$ [85]. In the $(Te_7^{2+})_n$ polycation of Te$_7$[Be$_2$Cl$_6$], synthesized in an Na$_2$[BeCl$_4$] melt, Te$_7$ groups are connected to a folded band [78] (Fig. 3.6a). The structure is markedly different from the polymeric cation in $(Te_7^{2+})_n$[AsF$_6$]$_2$ of identical formula. If one tries to explain the bonding situation in classical terms, i.e. only in terms of two-electron–two-center bonds, a hypervalent central Te atom carrying twelve electrons

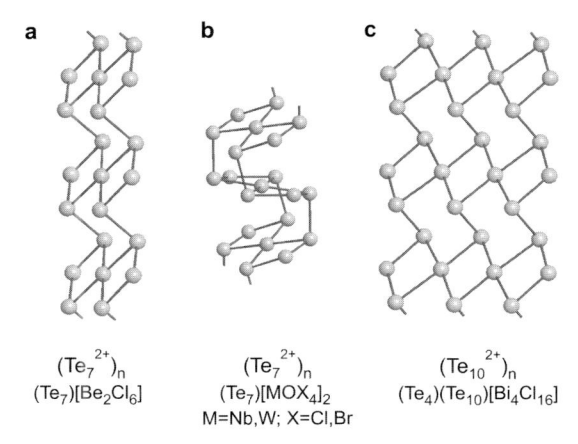

Fig. 3.6 The structures of polymeric tellurium cations with "hypervalent" bonding.

$(Te_7^{2+})_n$
$(Te_7)[Be_2Cl_6]$

$(Te_7^{2+})_n$
$(Te_7)[MOX_4]_2$
M=Nb,W; X=Cl,Br

$(Te_{10}^{2+})_n$
$(Te_4)(Te_{10})[Bi_4Cl_{16}]$

and two formal negative charges would be necessary. The relatively long bonds around the central four-coordinate Te atom (300 pm) are better described as two orthogonal three-center–four-electron bonds. Since antibonding orbitals are occupied in this model, the bond order of each of the Te–Te bonds is only 0.5. Another isomer of a polymeric $(Te_7^{2+})_n$ cation is present in the structures of $Te_7[MOX_4]X$ (M = Nb, W; X = Cl, Br) [86–88]. The structural difference is caused by a *cis/trans* isomerism of the folding of the Te_7 groups (Fig. 3.6b). While in $Te_7[Be_2Cl_6]$ all Te_7 groups are translationally equivalent, in $Te_7[MOX_4]X$ this is the case only for every second group. $(Te_{10}^{2+})_n$ as a part of the structure of $(Te_4)(Te_{10})[Bi_4Cl_6]$ represents a higher extended system of the same type. Two four-coordinated Te atoms in a planar environment are present in each Te_{10} group (Fig. 3.6c). The linkage isomerism is of the $Te_7[Be_2Cl_6]$ type.

3.8
Summary and Outlook

The chemistry of main group element clusters has historical roots dating back to the 18th century. However, it is only in the last 40 years that knowledge about this class of ions has been dramatically improved through the application of modern spectroscopic and diffraction methods for structure determination. Their solubility in super acidic media, in $AlCl_3$ melts, and in liquid SO_2 made the molecular, discrete clusters a well-known class of compounds, of which many examples are now known. Relatively new developments include synthesis by chemical vapor transport and solvothermal synthesis in liquid halides such as $SnCl_4$ or $SiBr_4$. Novel structural features established in the last decade are chain-like polymeric polycations, of which a variety of isomers are also now known. Thus, the chemistry of polycationic main group element clusters remains an active and developing field in modern inorganic chemistry. This is underlined by recent reviews concerning homoatomic cages and clusters, with special emphasis on the clusters of phosphorus [89] and polychalcogenide cations and anions [90]. Particular progress has recently been achieved through the theoretical work of I. Krossing and J. Passmore. For example, the blue color of solutions containing sulfur polycations was usually attributed to S_8^{2+} and the radical cation S_5^{+}. Ab initio calculations and reaction enthalpy estimates revealed that the hitherto unknown $10\,\pi$ electron S_6^{2+} dication is in fact responsible for the coloration [91]. The synthetic possibilities for obtaining salt-like compounds containing heavy main group polycationic clusters are by no means exhausted. The use of extremely weakly coordinating anions such as perfluorinated aluminum tetraalkoxides $[Al(OR^F)_4]^-$ [92] could help to stabilize new and more highly charged cations. Novel, unusual cations that have recently been obtained, such as N_5^+ [93], $P_5Br_2^+$ [94], Sb_8^{2+} [95] and metal-centered clusters like $(Pd@Bi_{10})^{4+}$ [96], are illustrative of how research in this area is likely to proceed.

Acknowledgements

The author has to thank all of his coworkers who have contributed to the results established in our group and the *Deutsche Forschungsgemeinschaft* and the *Fonds der Chemischen Industrie* for their long-term financial support.

References

1 M. H. Klaproth, *Phil. Mag.* **1798**, *1*, 78.
2 C. F. Buchholz, *Gehlens Neues J. Chem.* **1804**, *3*, 7.
3 G. Magnus, *Ann. Phys. Leipzig* **1827**, *10*, 491.
4 A. Heintz, *Ann. Phys.* **1844**, *63*, 59.
5 A. Hershaft, J. D. Corbett, *J. Chem. Phys.* **1961**, *36*, 551; *Inorg. Chem.* **1963**, *2*, 979.
6 N. Bjerrum, G. P. Smith, *J. Am. Chem. Soc.* **1968**, *90*, 4472.
7 J. Barr, R. J. Gillespie, R. Kapoor, G. P. Pez, *J. Am. Chem. Soc.* **1968**, *90*, 6855.
8 J. D. Corbett, R. K. McMullan, *J. Am. Chem. Soc.* **1956**, *78*, 2906.
9 N. Burford, J. Passmore, J. C. P. Sanders, in *From Atoms to Polymers, Isoelectronic Analogies* (Eds.: J. F. Liebman, A. Greenberg), 1989, p. 53.
10 S. Brownridge, I. Krossing, J. Passmore, H. D. B. Jenkins, H. K. Roobottom, *Coord. Chem. Rev.* **2000**, *197*, 397.
11 N. Bartlett, D. H. Lohman, *J. Chem. Soc.* **1962**, 5253.
12 H. D. B. Jenkins, L. C. Jitariu, I. Krossing, J. Passmore, R. Suontamo, *J. Comput. Chem.* **2000**, *21*, 218.
13 T. A. O'Donnell, *Superacids and Acidic Melts as Inorganic Chemical Reaction Media*, VCH Publishers, New York, 1993.
14 I. D. Brown, D. B. Crump, R. J. Gillespie, *Inorg. Chem.* **1971**, *10*, 2319.
15 M. J. Collins, R. J. Gillespie, J. F. Sawyer, G. J. Schrobilgen, *Acta Crystallogr.* **1986**, *C42*, 13.
16 J. D. Corbett, *Progr. Inorg. Chem.* **1976**, *21*, 129.

17 M. J. Collins, R. J. Gillespie, J. W. Kolis, J. F. Sawyer, *Acta Crystallogr.* **1987**, *C43*, 2033.
18 J. Beck, *Z. Naturforsch.* **1990**, *B45*, 413.
19 J. Beck, *Z. Naturforsch.* **1990**, *B45*, 1159.
20 J. Beck, *Angew. Chem.* **1990**, *102*, 301; *Angew. Chem. Int. Ed. Engl.* **1990**, *29*, 293.
21 J. Beck, *Coord. Chem. Rev.* **1997**, *163*, 55.
22 J. Beck, A. Fischer, *Z. Anorg. Allg. Chem.* **2002**, *628*, 369.
23 J. Passmore, G. Sutherland, P. S. White, *J. Chem. Soc., Chem. Commun.* **1980**, 330.
24 J. Beck, T. Hilbert, *Z. Anorg. Allg. Chem.* **2000**, *626*, 837.
25 J. Beck, A. Desgrosseillers, K. Müller-Buschbaum, K.-J. Schlitt, *Z. Anorg. Allg. Chem.* **2002**, *628*, 1145.
26 J. Beck, F. Steden, A. Reich, H. Fölsing, *Z. Anorg. Allg. Chem.* **2003**, *629*, 1073.
27 J. Beck, M. Kellner, M. Kreuzinger, *Z. Anorg. Allg. Chem.* **2002**, *628*, 2656.
28 I. Krossing, J. Passmore, *Inorg. Chem.* **1999**, *38*, 5203.
29 B. Eisenmann, K. H. Janzon, H. Schäfer, A. Weiss, *Z. Naturforsch.* **1969**, *B24*, 457.
30 J. Beck, G. Bock, *Z. Naturforsch.* **1996**, *B51*, 119.
31 R. C. Burns, R. J. Gillespie, W. C. Luk, D. R. Slim, *Inorg. Chem.* **1979**, *18*, 3086.
32 The structure of Te$_6$[AsF$_6$]$_4$ · 2SO$_2$ was recently redetermined: J. Beck, F. Steden, *Acta Crystallogr.* **2003**, *E59*, i158.
33 M. J. Collins, R. J. Gillespie, J. F. Sawyer, *Acta Crystallogr.* **1988**, *C44*, 405.
34 R. Faggiani, R. J. Gillespie, C. Campana, J. W. Kolis, *J. Chem. Soc., Chem. Commun.* **1987**, 485.
35 P. D. Lyne, D. M. P. Mingos, T. Ziegler, *J. Chem. Soc., Dalton Trans.* **1992**, 2743.
36 J. Beck, G. Bock, *Z. Anorg. Allg. Chem.* **1996**, *622*, 823.
37 J. Beck, *Chem. Ber.* **1994**, *128*, 23.
38 C. G. Davies, R. J. Gillespie, J. J. Park, J. Passmore, *Inorg. Chem.* **1971**, *10*, 2781.
39 R. K. McMullan, D. J. Prince, J. D. Corbett, *Inorg. Chem.* **1971**, *10*, 1749.
40 J. Beck, K. Müller-Buschbaum, *Z. Anorg. Allg. Chem.* **1997**, *623*, 409.
41 A. Baumann, J. Beck, T. Hilbert, *Z. Naturforsch.* **1999**, *B54*, 1253.
42 W. S. Sheldrick, M. Wachhold, *Angew. Chem.* **1995**, *107*, 490; *Angew. Chem. Int. Ed. Engl.* **1995**, *34*, 450.
43 T. S. Cameron, R. J. Deeth, I. Dionne, H. Du, H. D. B. Jenkins, I. Krossing, J. Passmore, H. K. Roobottom, *Inorg. Chem.* **2000**, *39*, 5614.
44 J. Beck, *Angew. Chem.* **1996**, *102*, 301; *Angew. Chem. Int. Ed. Engl.* **1990**, *29*, 293.
45 J. Beck, G. Bock, *Angew. Chem.* **1995**, *107*, 2739; *Angew. Chem. Int. Ed. Engl.* **1995**, *34*, 2559.
46 R. C. Burns, W.-L. Chan, R. J. Gillespie, W.-C. Luk, J. F. Sawyer, D. R. Slim, *Inorg. Chem.* **1980**, *19*, 1432.
47 M. J. Collins, R. J. Gillespie, J. F. Sawyer, G. J. Schrobilgen, *Acta Crystallogr.* **1986**, *C42*, 13.
48 J. Beck, T. Hilbert, *Z. Anorg. Allg. Chem.* **2000**, *626*, 837.
49 R. C. Burns, M. J. Collins, R. J. Gillespie, G. J. Schrobilgen, *Inorg. Chem.* **1986**, *25*, 4465.
50 J. Beck, J. Wetterau, *Inorg. Chem.* **1995**, *34*, 6206.
51 J. Beck, A. Fischer, *Z. Anorg. Allg. Chem.* **1997**, *623*, 780.
52 R. C. Burns, R. J. Gillespie, J. F. Sawyer, *Inorg. Chem.* **1980**, *19*, 1423.
53 R. Faggiani, R. J. Gillespie, J. F. Sawyer, J. E. Vekris, *Acta Crystallogr.* **1989**, *C45*, 1847.
54 N. J. Bjerrum, C. R. Boston, G. P. Smith, *Inorg. Chem.* **1967**, *6*, 1162; ibid. **1967**, *6*, 1968.
55 R. M. Friedman, J. Corbett, *Inorg. Chem.* **1973**, *12*, 1134.
56 H. Kalpen, W. Hönle, M. Somer, U. Schwarz, K. Peters, H. G. v. Schnering, R. Blachnick, *Z. Anorg. Allg. Chem.* **1998**, *624*, 1137.
57 J. D. Corbett, *Inorg. Chem.* **1968**, *7*, 198.
58 B. Krebs, M. Mummert, C. Brendel, *J. Less-Common Met.* **1986**, *116*, 159.
59 S. Ulvenlund, K. Stahl, L. Bengtsson-Kloo, *Inorg. Chem.* **1996**, *35*, 223.
60 M. Ruck, *Z. Anorg. Allg. Chem.* **1998**, *624*, 521; M. Ruck, S. Hampel, *Polyhedron* **2002**, *21*, 651.

61 J. Beck, C. J. Brendel, L. Bengtsson-Kloo, B. Krebs, M. Mummert, A. Stankowski, S. Ulvenlund, *Chem. Ber.* **1996**, *129*, 1279.
62 A. N. Kuznetsov, B. P. Popovkin, *Z. Anorg. Allg. Chem.* **2002**, *628*, 2178.
63 J. Beck, T. Hilbert, *Eur. J. Inorg. Chem.* **2004**, 2019.
64 A. N. Kuznetsov, A. V. Shevel'kov, S. I. Troyanov, B. A. Popovkin, *Russ. J. Inorg. Chem.* **1996**, *41*, 920; A. N. Kuznetsov, A. V. Shevel'kov, B. A. Popovkin, *Russ. J. Coord. Chem.* **1998**, *24*, 861.
65 A. N. Kuznetsov, P. I. Naumenko, B. A. Popovkin, L. Kloo, *Russ. Chem. Bull.* **2003**, *52*, 2100.
66 H. v. Benda, A. Simon, W. Bauhofer, *Z. Anorg. Allg. Chem.* **1978**, *438*, 37.
67 V. Dubenskyy, M. Ruck, *Z. Anorg. Allg. Chem.* **2003**, *629*, 375.
68 K. Wade, *Adv. Inorg. Chem. Radiochem.* **1976**, *18*, 1.
69 A. N. Kuznetsov, L. Kloo, M. Lindsjö, J. Rosdahl, H. Stoll, *Chem. Eur. J.* **2001**, *7*, 2821.
70 A. Hirsch, Z. Chen, H. Jias, *Angew. Chem.* **2001**, *113*, 2916; *Angew. Chem. Int. Ed.* **2001**, *40*, 2834.
71 P. Boldrini, I. D. Brown, M. J. Collins, R. J. Gillespie, E. Maharajh, D. R. Slim, J. F. Sawyer, *Inorg. Chem.* **1985**, *24*, 4302.
72 M. J. Collins, R. J. Gillespie, J. F. Sawyer, G. J. Schrobilgen, *Inorg. Chem.* **1986**, *25*, 2053.
73 R. J. Gillespie, W. Luk, E. Maharajh, D. R. Slim, *Inorg. Chem.* **1977**, *16*, 852.
74 M. J. Collins, R. J. Gillespie, J. F. Sawyer, *Inorg. Chem.* **1987**, *26*, 1476.
75 B. H. Christian, R. J. Gillespie, J. F. Sawyer, *Inorg. Chem.* **1981**, *20*, 3410.
76 J. Beck, M. Dolg, S. Schlüter, *Angew. Chem.* **2001**, *113*, 2347; *Angew. Chem. Int. Ed.* **2001**, *40*, 2287.
77 J. Beck, S. Schlüter, N. Zotov, *Z. Anorg. Allg. Chem.* **2004**, *630*, 2512.
78 J. Beck, A. Fischer, A. Stankowski, *Z. Anorg. Allg. Chem.* **2002**, *628*, 2542.
79 J. Beck, Th. Schlörb, *Phosphorus, Sulfur, Silicon, Relat. Elem.* **1997**, *124*, 125.
80 A. Baumann, J. Beck, *Z. Anorg. Allg. Chem.* **2004**, *630*, 2078.
81 G. W. Drake, G. L. Schimek, J. W. Kolis, *Inorg. Chem.* **1996**, *35*, 1740.
82 J. Beck, A. Fischer, *Z. Anorg. Allg. Chem.* **2002**, *628*, 369.
83 J. Beck, A. Stankowski, *Z. Naturforsch.* **2001**, *B56*, 453.
84 M. Joulien-Pouzol, S. Jaulmes, F. Alapini, *Acta Crystallogr.* **1977**, *B33*, 2270.
85 F. Kaliber, W. Petter, F. Hulliger, *J. Solid State Chem.* **1983**, *46*, 112.
86 J. Beck, *Angew. Chem.* **1991**, *103*, 1149; *Angew. Chem. Int. Ed. Engl.* **1991**, *30*, 1128.
87 J. Beck, *Z. Anorg. Allg. Chem.* **1993**, *619*, 237.
88 J. Beck, G. Bock, *Z. Anorg. Allg. Chem.* **1994**, *620*, 1971.
89 I. Krossing, in *Molecular Clusters of the Main Group Elements* (Eds.: M. Driess, N. Nöth), Wiley-VCH, Weinheim, 2004, p. 209.
90 W. S. Sheldrick, in *Molecular Clusters of the Main Group Elements* (Eds.: M. Driess, N. Nöth), Wiley-VCH, Weinheim, 2004, p. 230.
91 I. Krossing, J. Passmore, *Inorg. Chem.* **2004**, *43*, 1000.
92 I. Krossing, *Angew. Chem.* **2004**, *116*, 2116; *Angew. Chem. Int. Ed. Engl.* **2004**, *43*, 2066.
93 A. Vij, W. W. Wilson, V. Vij, F. S. Tham, J. A. Sheehy, K. O. Christe, *J. Am. Chem. Soc.* **2001**, *123*, 6308.
94 I. Krossing, *J. Chem. Soc., Dalton Trans.* **2002**, 500.
95 M. Lindsjö, A. Fischer, L. Kloo, *Angew. Chem.* **2004**, *116*, 2594; *Angew. Chem. Int. Ed. Engl.* **2004**, *43*, 2540.
96 M. Ruck, V. Dubenskyy, T. Soehnel, *Angew. Chem.* **2003**, *115*, 3086; *Angew. Chem. Int. Ed.* **2003**, *42*, 2978.

4
Metal-Catalyzed Dehydrocoupling Routes to Rings, Chains, and Macromolecules based on Elements from Groups 13 and 15

Cory A. Jaska and Ian Manners

4.1
Introduction

Transition metal catalysis plays a profound role in chemical synthesis, and is widely utilized in the manufacturing of both bulk and specialty chemicals. Industrial scale preparations of commodity chemicals rely heavily upon metal-catalyzed processes to increase efficiency, reduce waste generation, and decrease energy consumption. The production of fine and specialty organic chemicals also extensively utilizes metal-catalyzed routes, allowing for precise control over factors such as enantioselectivity and regioselectivity in the formation of new C–C, C–N, C–O, and C–H bonds. In contrast, the use of metal-catalyzed routes to form new homonuclear or heteronuclear bonds between inorganic elements is virtually unexplored. The extension of catalytic methods to inorganic chemistry is particularly desirable, as current synthetic strategies are often limited to routes that are haphazard, of narrow scope or restricted to salt elimination reactions. In addition, detailed mechanistic studies of any catalytic cycles may permit the development of generalized reactions, which should also help to advance this promising field [1].

One route that has emerged as a promising method for the synthesis of new bonds between inorganic elements is catalytic dehydrocoupling (or dehydrogenative coupling). Eq. (1) shows an example of homodehydrocoupling, in which bonds are made between similar elements; Eq. (2) demonstrates heterodehydrocoupling, in which bonds are made between different elements. Dehydrocoupling routes have the advantage that the only by-product is hydrogen gas, which can be efficiently removed from the reaction medium. This field originated in the mid-1980s through studies on the catalytic dehydropolymerization of primary and secondary group 14 hydrides such as silanes [2], germanes, [3], and stannanes [4]. This field is now well developed and has benefited from detailed mechanistic studies by multiple research groups [5]. More recently, homodehydrocoupling chemistry has been extended to include P–P bond formation [6], and catalytic heterodehydrocoupling reactions to form, for example, B–C [7], Si–P [8], Si–N [9], and Si–O bonds [10] have also been reported. The use of metal complexes to catalyze ring-opening polymerization (ROP) reactions has also

Inorganic Chemistry in Focus II. Edited by G. Meyer, D. Naumann, and L. Wesemann
Copyright © 2005 WILEY-VCH Verlag GmbH & Co. KGaA, Weinheim
ISBN: 3-527-30811-3

been developed to prepare macrocycles and polymers containing inorganic elements. The ROP of silaheterocycles such as silacyclobutanes [11] and [1] silaferrocenophanes [12] has been shown to afford high molecular weight polycarbosilanes and polyferrocenylsilanes under relatively mild conditions. In addition, cyclic thionylphosphazenes undergo Lewis acid (e.g. $GaCl_3$) catalyzed ROP to give polythionylphosphazenes [13].

$$2\ R_3EH \xrightarrow{\Delta\ \text{or Catalyst}} R_3E\text{---}ER_3 + H_2 \quad (1)$$

$$R_3XH + R_3YH \xrightarrow{\Delta\ \text{or Catalyst}} R_3X\text{---}YR_3 + H_2 \quad (2)$$

In this article, we highlight some of our recent work directed at the development of new routes to inorganic rings, chains, and macromolecules via the catalytic dehydrocoupling of group 13–group 15, Lewis acid–Lewis base adducts.

4.2
Thermal Dehydrocoupling of Phosphine-Borane Adducts

The preparation of polymers with backbones of alternating phosphorus and boron atoms attracted significant attention in the 1950s and early 1960s as a consequence of their anticipated high thermal stability and resistance to oxidation and hydrolysis. The main synthetic route explored at that time involved thermally-induced dehydrocoupling of phosphine-borane adducts at 180–200 °C, to afford predominantly six-membered rings. Only negligible yields of low molecular weight, partially characterized polymers were claimed, mainly in patents [14]. Since this initial period of activity, little else has been investigated with regard to the thermal dehydrocoupling of phosphine-borane adducts to afford cyclic or polymeric species. However, we have recently re-investigated this dehydrocoupling process utilizing late transition metal complexes as catalysts [15]. This has permitted the formation of six- and eight-membered phosphinoborane rings under more facile conditions, novel linear oligomeric chains, and high molecular weight polyphosphinoboranes.

4.3
Catalytic Dehydrocoupling of Secondary Phosphine-Borane Adducts

The uncatalyzed dehydrocoupling of the secondary phosphine-borane adduct $Ph_2PH \cdot BH_3$ at 170 °C gave a mixture of the cyclic trimer $[Ph_2P-BH_2]_3$ (**1**) and tetramer $[Ph_2P-BH_2]_4$ (**2**) in an 8:1 ratio. However, upon heating $Ph_2PH \cdot BH_3$ in the presence of $[Rh(1,5\text{-cod})(\mu\text{-Cl})]_2$ or $[Rh(1,5\text{-cod})_2]OTf$ (0.5–1 mol% Rh) at 120 °C, **1** and **2** were formed in a 2:1 ratio (Eq. (3), [Rh]=rhodium catalyst) [16].

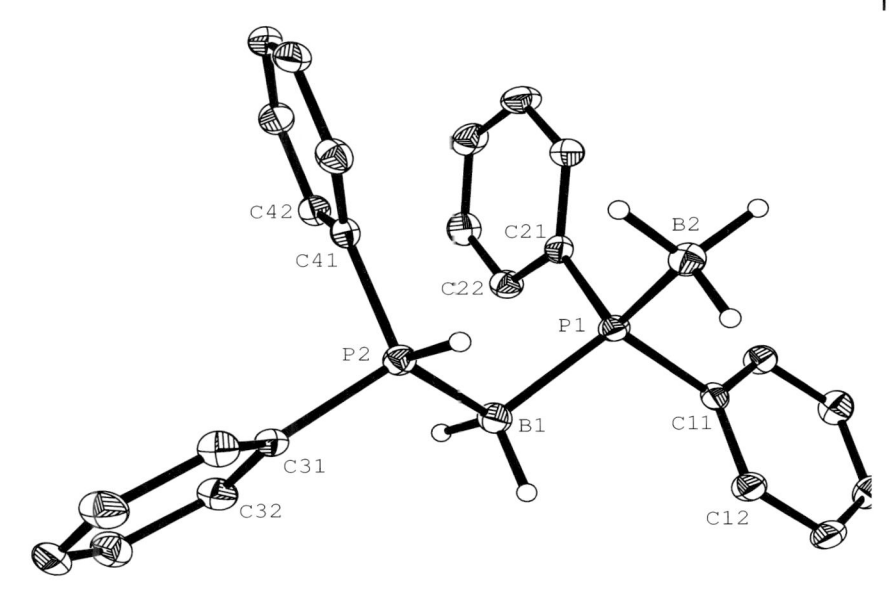

Fig. 4.1 Molecular structure of 3.

Upon lowering the temperature to 90 °C the novel linear compound $Ph_2PH-BH_2-PPh_2-BH_3$ (3) was formed as the exclusive product (Eq. (4)) [15, 16]. In addition, the molecular structure of 3 was determined (Fig. 4.1). In the absence of catalyst at 90 °C, no conversion of $Ph_2PH \cdot BH_3$ was observed; at 120 °C, only a minor amount (<5%) of 3 and no cyclic products were observed. The catalytic activity of other transition metal complexes (e.g. of Ti, Ru, Rh, Ir, Pd, Pt) for the formation of 3 was also demonstrated. In general, Rh(I) and Rh(III) complexes were found to possess the highest catalytic dehydrocoupling activities.

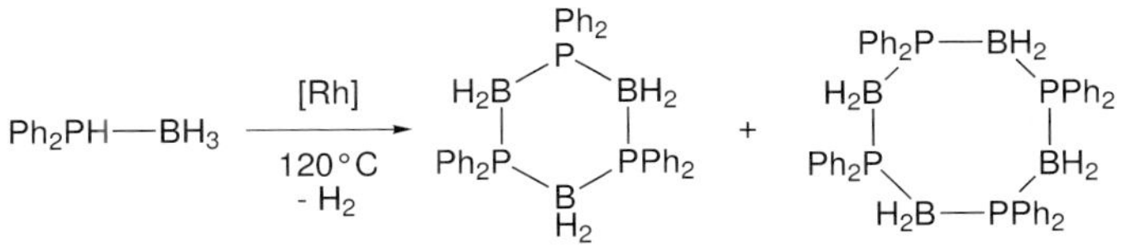

$$Ph_2PH-BH_3 \xrightarrow[-H_2]{[Rh], 120°C} \mathbf{1} + \mathbf{2} \qquad (3)$$

$$R_2PH-BH_3 \xrightarrow[-H_2]{[Rh], 90°C} Ph_2PH-BH_2-PPh_2-BH_3 \qquad (4)$$
$$\mathbf{3}$$

Heating of the sterically encumbered adduct $^tBu_2PH \cdot BH_3$ in the presence of [Rh(1,5-cod)$_2$]OTf or $Rh_6(CO)_{16}$ at 140 °C also resulted in dehydrocoupling to form the linear compound $^tBu_2PH-BH_2-P^tBu_2-BH_3$ (4) (Eq. (5)). The use of the chlorinated catalysts [Rh(1,5-cod)(μ-Cl)]$_2$ or RhCl$_3 \cdot$ 3H$_2$O at 160 °C was found to give an inseparable mixture of **4** and the chlorinated compound ^{t}Bu$_2$PH–BH$_2$–P^{t}Bu$_2$–BH$_2$Cl (**5**) (Eq. (6)) [17].

$$\text{tBu}_2\text{PH}-\text{BH}_3 \xrightarrow[\substack{140\,°C \\ -\,H_2}]{\substack{[\text{Rh}(1,5\text{-cod})_2]\text{OTf} \\ \text{or Rh}_6(\text{CO})_{16}}} \text{tBu}_2\text{PH}-\text{BH}_2-\text{PtBu}_2-\text{BH}_3 \qquad (5)$$
$$\hspace{10cm} \mathbf{4}$$

$$\text{tBu}_2\text{PH}-\text{BH}_3 \xrightarrow[\substack{160\,°C \\ -\,H_2}]{\substack{[\text{Rh}(1,5\text{-cod})(\mu\text{-Cl})]_2 \\ \text{or RhCl}_3 \cdot 3\,H_2O}} \text{tBu}_2\text{PH}-\text{BH}_2-\text{PtBu}_2-\text{BH}_2\text{Cl} + \mathbf{4} \qquad (6)$$
$$\hspace{10cm} \mathbf{5}$$

4.4
Catalytic Dehydrocoupling of Primary Phosphine-Borane Adducts

In the early 1960s, the uncatalyzed pyrolysis of the primary phosphine-borane adduct PhPH$_2 \cdot$ BH$_3$ at 100–150 °C for 13 h was reported to give a benzene-soluble polymer [PhPH–BH$_2$]$_n$ with a molecular weight (M_n) of 2150. Prolonged heating at elevated temperatures (250 °C) led to the formation of insoluble material, without significant molecular weight increase of the benzene-soluble fraction (maximum M_n = 2630) [14h]. We found that heating of a solution of PhPH$_2 \cdot$ BH$_3$ in toluene at reflux (110 °C) in the presence of [Rh(1,5-cod)$_2$]OTf (0.5–1 mol% Rh) for 15 h led to dehydrogenative coupling to afford poly(phenylphosphinoborane) [PhPH–BH$_2$]$_n$ (**6**) as an off-white powder (Eq. (7)) [15, 16, 18]. Static light scattering (SLS) of **6** in THF showed the material to be of relatively low molecular weight with M_w = 5600. However, dehydrocoupling in the absence of solvent and modification of the reaction temperatures led to higher molecular weight polymer. Thus, neat PhPH$_2 \cdot$ BH$_3$ in the presence of [Rh(1,5-cod)(μ-Cl)]$_2$ or RhCl$_3 \cdot$ 3H$_2$O (0.6 mol% Rh) was heated at 90 °C for 3 h, and then at 130 °C for 3 h. The reaction mixture gradually became viscous during the heating stage at 90 °C and completely solidified after 3 h at 130 °C. The polymer obtained was spectroscopically identical to the material prepared in toluene, but SLS confirmed a higher molecular weight of M_w = 33300.

$$\text{RPH}_2-\text{BH}_3 \xrightarrow[\substack{90\,-\,130\,°C \\ -\,H_2}]{[\text{Rh}]} \left[\text{PRH}-\text{BH}_2\right]_n \qquad (7)$$

6: R = Ph
7: R = ^{i}Bu
8: R = p-^{n}BuC$_6$H$_4$
9: R = p-dodecylC$_6$H$_4$

Prolonged heating of **6** at 130 °C in the presence of a catalyst afforded insoluble products that were found to swell significantly in THF or CH_2Cl_2. As branching positions could not be detected in the ^{31}P and ^{11}B NMR spectra, the polymers may either become likely cross-linked through additional inter-chain P–B coupling, or may increase in molecular weight above the solubility limit. We confirmed that in the absence of a transition metal catalyst, the thermally-induced dehydrocoupling of $PhPH_2 \cdot BH_3$ proceeds very slowly and forms only low molecular weight materials [16].

Several other primary phosphine-borane adducts have been shown to undergo metal-catalyzed dehydrocoupling to form polyphosphinoboranes (**7**: R = ^{i}Bu; **8**: p-^{n}BuC$_6$H$_4$; **9**: p-dodecyl-C$_6$H$_4$; Eq. (7)). The alkyl-substituted adduct ^{i}BuPH$_2 \cdot$BH$_3$ was found to require more forcing dehydrocoupling conditions (15 h at 120 °C) in order to form **7** [16, 19]. This is likely due to the decreased polarity of the P–H bond as a result of the strong (+)-inductive effect of the alkyl group. SLS studies of **7** in THF gave molecular weight values of $M_w = 13100$. In contrast to **6** and **7**, molecular weight analysis of **8** and **9** by GPC gave M_w values of ca. 19000 and 168000 (relative to polystyrene standards), respectively.

Wide-angle X-ray scattering of solution-cast films of **6** and **8** indicated that the polymers were essentially amorphous [19]. This was expected, as the dehydropolymerization process should not provide any stereochemical control, and thus atactic polymers should result. Glass transition temperatures (T_g) below room temperature were found for polymers **7–9**, which may be attributed to the high degree of torsional flexibility in the polymer main chains. This may be a result of the long B–P bonds (1.9–2.0 Å), which would reduce the steric interference between the side groups, facilitating polymer motion and thereby lowering T_g. Thermogravimetric analysis of **6** indicated the onset of decomposition at 160 °C with a ceramic yield of 75–80% upon heating to 1000 °C [19]. Lower onset temperatures of 120 °C were found for both **7** and **8**. Substantial weight loss was observed upon heating to 1000 °C, resulting in moderate ceramic yields of 40–45% and 35–40% for **7** and **8**, respectively. The high ceramic yield of **6** suggests that it may function as a useful pre-ceramic polymer for BP-based materials [19]. Indeed, preliminary pyrolysis studies of **6** at 1000 °C under nitrogen and subsequent powder X-ray diffraction analysis showed the formation of boron phosphide as the major crystalline component.

4.5
Thermal Dehydrocoupling of Amine–Borane Adducts

The generalization of this catalytic dehydrocoupling method to other group 13–15 systems was of particular interest to our group. It is well known that primary and secondary amine-borane adducts undergo thermally-induced dehydrocoupling at elevated temperatures (>100 °C) to yield cyclic aminoborane $[H_2B–NRR']_x$ ($x = 2$ or 3) and borazine derivatives $[HB–NR]_3$ [20]. For example, $Me_2NH \cdot BH_3$ thermally eliminates hydrogen at or above 130 °C in the condensed phase to form the cyclic aminoborane dimer [Me$_2$N–BH$_2$]$_2$ [21]. MeNH$_2$·BH$_3$ undergoes thermal dehydrocoupling in a two-step process. Initial pyrolysis at 100 °C yields the intermediate trimer [H$_2$B–HNMe]$_3$, which upon heating to 200 °C eliminates a second equivalent of hydrogen to yield the borazine [HB–NMe]$_3$ [22]. The uncatalyzed formation of borazine [HB–NH]$_3$ directly from NH$_3$·BH$_3$ typically requires high temperature forcing conditions (>150 °C) [23]. In addition, the dehydrocoupling of borazine at 70 °C yields polyborazylene [B$_3$N$_3$H$_4$]$_n$, a useful precursor to boron nitride ceramics [24]. We anticipated that the use of a transition metal catalyst might allow the dehydrocoupling to be achieved under milder conditions, thereby improving the pre-existing routes to boron-nitrogen rings and chains.

4.6
Catalytic Dehydrocoupling of Secondary Amine-Borane Adducts

Treatment of the secondary amine-borane adduct Me$_2$NH·BH$_3$ with a catalytic amount of [Rh(1,5-cod)(μCl)]$_2$ (0.5 mol% Rh), either in the melt at 45 °C or in solution at 25 °C, was found to result in the formation of the cyclic dimer [Me$_2$N–BH$_2$]$_2$ (**10**) in high yields (Eq. (8); Fig. 4.2) [25, 26]. This method represents a significant improvement over the uncatalyzed route to **10**, which requires temperatures of up to 130 °C [21]. The catalytic activity of other transition metal complexes (e.g., of Ru, Rh, Ir, Pd) for the formation of **10** was also demonstrated. The catalytic dehydrocoupling of the cyclic amine-borane (1,4-C$_4$H$_8$)NH·BH$_3$ was found to afford the spirocyclic aminoborane dimer [(1,4-C$_4$H$_8$)N–BH$_2$]$_2$ (**11**), while the unsymmetrically substituted (PhCH$_2$)MeNH·BH$_3$ gave a mixture of cis and trans isomers [(PhCH$_2$)MeN–BH$_2$]$_2$ (**12**) under similarly mild conditions of 25–45 °C (Eq. (8)) [26].

$$R'RNH-BH_3 \xrightarrow[-H_2]{[Rh]} \frac{1}{2} [\text{cyclic dimer}] \quad (8)$$

10: R = R' = Me
11: R = R' = 1,4-C$_4$H$_8$
12: R = PhCH$_2$, R' = Me

The catalytic dehydrocoupling of the adduct ^{i}Pr$_2$NH·BH$_3$ was found to result in the formation of the monomeric aminoborane ^{i}Pr$_2$N=BH$_2$ (**13**), rather than the cyclic dimer (Eq. (9)) [26]. The dimerization of this species is likely inhibited by the steric bulk of the large ^{i}Pr groups.

$$iPr_2NH-BH_3 \xrightarrow[\substack{25\,°C \\ -H_2}]{[Rh]} {}^iPr_2N=BH_2 \quad (9)$$

13

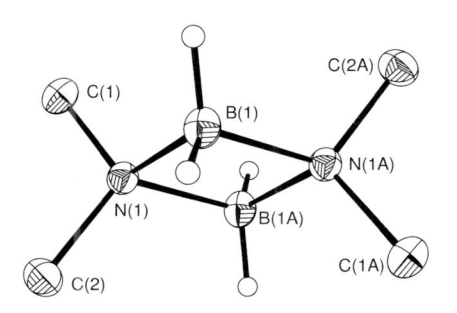

Fig. 4.2 Molecular structure of **10**. Reproduced from Ref. [25] with permission.

One of the possible intermediates in the catalytic cycle is the linear compound R$_2$NH–BH$_2$–NR$_2$–BH$_3$ (**14**: R = Me; **15**: R = 1,4-C$_4$H$_8$), which would result from a single intermolecular dehydrocoupling reaction between two adduct species. Treatment of **14** and **15** with an Rh pre-catalyst was found to result in the formation of **10** and **11** (Eq. (10)), most likely via an intramolecular dehydrocoupling reaction that gives the cyclic dimer [26].

$$R_2NH\text{—}BH_2\text{—}NR_2\text{—}BH_3 \xrightarrow[-H_2]{[Rh]} \begin{array}{c} \text{cyclic dimer} \end{array} \qquad (10)$$

14: R = Me
15: R = 1,4-C$_4$H$_8$

10: R = Me
11: R = 1,4-C$_4$H$_8$

4.7
Catalytic Dehydrocoupling of Primary Amine–Borane Adducts and NH$_3$·BH$_3$

The treatment of NH$_3$·BH$_3$ with a catalytic amount of [Rh(1,5-cod)(μ-Cl)]$_2$ (1.5 mol% Rh) was found to result in the formation of borazine [HB-NH]$_3$ (**16**) after ca. 48–84 h at 45 °C (Eq. (11)). Complete conversion to **16** was confirmed by the ^{11}B NMR spectrum of the reaction mixture. However, isolation by vacuum fractionation routinely yielded only small quantities (ca. 10%) of pure **16**. This was attributed to undesirable intermolecular dehydrocoupling reactions of the intermediates or the product itself to give involatile species. The primary adducts RNH$_2$·BH$_3$ were also found to undergo similar mild catalytic dehydrocoupling reactions to afford the substituted borazines [HB–NR]$_3$ (**17**: R = Me; **18**: R = Ph) (Eq. (11)). Improved isolated yields were also obtained for both **17** (40%) and **18** (56%), most probably due to the reduced possibility of intermolecular coupling reactions of the intermediate species. In the case of the methyl derivative **17**, the intermediate cyclic trimer [H$_2$B–HNMe]$_3$ was initially formed in solution after 4–6 h, and slowly converted to **17** by subsequent dehydrocoupling reactions over 48–72 h [26]. This difference in the dehydrocoupling rates for the stepwise loss of hydrogen from MeNH$_2 \cdot$BH$_3$ is also reflected in the uncatalyzed thermal route to **17**, in which the first equivalent of H$_2$ is removed at 100 °C and the second equivalent at 200 °C [22].

$$\text{RNH}_2\text{—BH}_3 \xrightarrow[-\text{H}_2]{[\text{Rh}]} \tfrac{1}{3} \begin{array}{c} \text{H} \\ | \\ \text{B} \\ \text{RN} \diagup \diagdown \text{NR} \\ | | \\ \text{HB} \diagdown \diagup \text{BH} \\ \text{N} \\ | \\ \text{R} \end{array} \qquad (11)$$

16: R = H
17: R = Me
18: R = Ph

4.8
Mechanism of Catalytic Dehydrocoupling of Amine-Borane and Phosphine-Borane Adducts

An understanding of the dehydrocoupling mechanism for both phosphine-borane and amine-borane adducts is of considerable interest, and could allow for the development of more efficient catalysts that operate at lower temperatures. For the amine-borane systems, some preliminary evidence for the operation of a heterogeneous mechanism involving colloidal Rh(0) metal was observed [26]. In order to rule out catalysis by metal particles for the dehydrocoupling of phosphine-borane adducts, a series of comparative tests using Me$_2$NH$\cdot$BH$_3$ and Ph$_2$PH$\cdot$BH$_3$ with the pre-catalyst [Rh(1,5-cod)(μ-Cl)]$_2$ was performed to determine the exact nature of the active catalyst. Using an approach proposed by Finke and co-workers [27], transmission electron microscopy (TEM), kinetic evaluation, poisoning experiments, filtration experiments, and isolation of the active catalyst were performed and compared for the two systems to determine whether the catalysis is homogeneous or heterogeneous.

Initial TEM experiments on catalytically active solutions for the dehydrocoupling of Me$_2$NH$\cdot$BH$_3$ and Ph$_2$PH$\cdot$BH$_3$ both indicated the presence of metal particles [28]. However, TEM micrographs of the precatalyst alone also showed the presence of metal particles in solution. This suggests the decomposition of the precatalyst under the electron beam, making these results ambiguous. Nevertheless, the visual observation that Me$_2$NH$\cdot$BH$_3$ dehydrocoupling solutions undergo a color change from orange to black, along with the formation of a black precipitate, imply that the metal particles are formed prior to the TEM experiment. This is in contrast to dehydrocoupling solutions of Ph$_2$PH$\cdot$BH$_3$, which only change in color from orange to dark red with no observable black precipitate.

Monitoring the dehydrocoupling reaction by ^{11}B NMR spectroscopy, an induction period was observed in the case of Me$_2$NH$\cdot$BH$_3$ (Fig. 4.3). This was followed by a rapid increase in catalytic activity resulting in a sigmoidal kinetic curve, which is characteristic for the metal particle formation reactions A → B

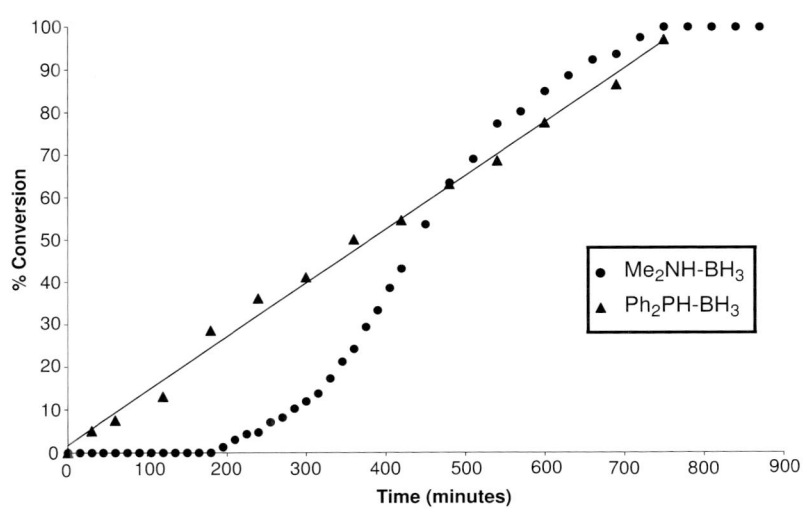

Fig. 4.3 Graph of % conversion vs time for the catalytic dehydrocoupling of $Me_2NH \cdot BH_3$ (●, 2 mol% Rh, toluene, 25 °C) and $Ph_2PH \cdot BH_3$ (▲, 10 mol% Rh, toluene, 90 °C) using [Rh(1,5-cod)(μ-Cl)]$_2$ as a pre-catalyst. Reproduced from Ref. [28a] with permission.

(nucleation) and A+B → 2B (autocatalytic surface growth) [29]. In contrast, the dehydrocoupling of $Ph_2PH \cdot BH_3$ showed no observable induction period, and a sigmoidal kinetic curve was noticeably absent (Fig. 4.3), suggesting the operation of a homogeneous catalyst [28].

The addition of excess mercury to dehydrocoupling trials involving $Me_2NH \cdot BH_3$ showed complete suppression of all catalytic activity, whereas the dehydrocoupling of $Ph_2PH \cdot BH_3$ was unaffected by treatment with Hg [28, 30]. The addition of the strongly coordinating ligand PPh_3 (0.5 equiv.) in fractional poisoning experiments involving $Me_2NH \cdot BH_3$ resulted in a strong reduction in the dehydrocoupling rate [31]. However, for dehydrocoupling trials involving $Ph_2PH \cdot BH_3$, no detectable effect on the catalytic activity was observed [28]. Filtration of a catalytically active $Me_2NH \cdot BH_3$ solution through a 0.5 μm filter resulted in almost complete suppression of the dehydrocoupling activity, suggesting an insoluble heterogeneous catalyst [32]. For the dehydrocoupling of $Ph_2PH \cdot BH_3$, no change in the dehydrocoupling rate was observed upon filtration through a 0.5 μm filter [28]. In the case of $Me_2NH \cdot BH_3$, bulk Rh metal was identified as the active catalyst as it displayed further dehydrocoupling activity without the presence of an induction period. It is thought that Rh colloids may be initially formed in solution, with [Me_2NH_2]Cl acting as an electrostatic stabilizing agent that helps to slow aggregation to bulk metal [28]. The results from all of these experiments suggest that the dehydrocoupling of $Me_2NH \cdot BH_3$ proceeds by a heterogeneous mechanism involving bulk Rh metal, while the dehydrocoupling of $Ph_2PH \cdot BH_3$ involves a soluble, homogeneous catalytic species.

It remains an intriguing problem to explain why the catalytic dehydrocoupling is heterogeneous in the case of amine-borane adducts, but homogeneous in the case of phosphine-borane analogues. Differences in adduct reducing strengths, and also the extent of dissociation, may be two factors that influence this behavior. The enhanced hydridic nature of the B–H protons in amine-borane adducts would be expected to promote reduction of the precatalyst, effectively functioning as a "borohydride-type" reducing agent. In addition, phosphine-borane adducts are expected to have lower dissociation energies than amine-borane adducts, as they possess weaker dative bonds. The presence of free Ph_2PH and BH_3 (as B_2H_6) in the reaction mixture may favor alternative reaction pathways such as oxidative addition or ligation of the phosphine over reduction of the metal center.

Having ruled out metal particle catalysis for the dehydrocoupling of $Ph_2PH \cdot BH_3$, several efforts have been made to synthesize potential intermediate complexes and to attempt the formation of new B–P bonds directly at the metal center. Initial reactions involving insertion of the metal into either P–H or B–H bonds, followed by subsequent σ-bond metathesis and/or oxidative addition/reductive elimination steps may define the catalytic cycle. We have found that the reaction of $Pt(PEt_3)_3$ with $PhRPH \cdot BH_3$ resulted in insertion of the Pt center into the P–H bond of the adduct to afford the *trans* hydride complexes $[PtH(PPhR \cdot BH_3)(PEt_3)_2]$ (**19**: R = H; **20**: R = Ph) [33]. However, subsequent B–P bond formation at the Pt center was not observed. This lack of reactivity may imply that metal-boryl complexes could be key intermediates in the catalytic cycle, or that the bond-forming processes involve more complicated reaction sequences.

```
        PEt3
         |
   H−Pt-PPhR−BH3
         |
        PEt3
```

19: R = H
20: R = Ph

4.9
Application of the Catalytic Dehydrocoupling of Amine-Borane Adducts

We have also found that the active catalyst (either Rh colloids or bulk Rh metal) for $Me_2NH \cdot BH_3$ dehydrocoupling can also be used in the catalytic hydrogenation of simple alkenes. These two reactions can be performed concurrently, with the evolved hydrogen from the dehydrocoupling reaction being used as a stoichiometric hydrogen source for the hydrogenation reaction. Thus, the reaction of $Me_2NH \cdot BH_3$ (1 equiv.) with an appropriate alkene (1 equiv.) and $[Rh(1,5\text{-cod})(\mu\text{-Cl})]_2$ (ca. 2 mol% Rh) in a *closed* vessel led to quantitative dehydrocoupling and hydrogenation conversions after 24 h at 25 °C [34].

4.10
Summary

The transition metal-catalyzed dehydrocoupling of amine-borane and phosphine-borane adducts has been found to be an efficient route to the formation of rings, chains, and polymers based on backbones of alternating four-coordinate boron and either phosphorus or nitrogen atoms. The dehydrocoupling mechanism has been found to be heterogeneous for amine-borane adducts, while phosphine-borane adducts follow a homogeneous mechanism. While the details of the active catalyst species and the catalytic cycle for the dehydrocoupling of phosphine-borane adducts are not explicitly known, it will continue to be an area of interest. The ability to perform the dehydropolymerization reactions at lower temperatures and with more control over molecular weight, and also to further extend the catalytic processes to other substrates, are future goals in our research group.

References

1. For general reviews, see: a) F. GAUVIN, J.F. HARROD, H.G. WOO, *Adv. Organomet. Chem.* **1998**, *42*, 363; b) J.A. REICHL, D.H. BERRY, *Adv. Organomet. Chem.* **1998**, *43*, 197.
2. a) C. AITKEN, J.F. HARROD, E. SAMUEL, *J. Organomet. Chem.* **1985**, *279*, C11; b) C.T. AITKEN, J.F. HARROD, E. SAMUEL, *J. Am. Chem. Soc.* **1986**, *108*, 4059.
3. a) C. AITKEN, J.F. HARROD, A. MALEK, E. SAMUEL, *J. Organomet. Chem.* **1988**, *349*, 285; b) N. CHOI, M. TANAKA, *J. Organomet. Chem.* **1998**, *564*, 81.
4. a) T. IMORI, T.D. TILLEY, *J. Chem. Soc., Chem. Commun.* **1993**, 1607; b) T. IMORI, V. LU, H. CAI, T.D. TILLEY, *J. Am. Chem. Soc.* **1995**, *117*, 9931; c) P. BRAUNSTEIN, X. MORISE, *Chem. Rev.* **2000**, *100*, 3541.
5. a) J.F. HARROD, S.S. YUN, *Organometallics* **1987**, *6*, 1381; b) J.Y. COREY, X.-H. ZHU, T.C. BEDARD, L.D. LANGE, *Organometallics* **1991**, *10*, 924; c) S.P. NOLAN, M. PORCHIA, T.J. MARKS, *Organometallics* **1991**, *10*, 1450; d) T.D. TILLEY, *Acc. Chem. Res.* **1993**, *26*, 22; e) V.K. DIOUMAEV, J.F. HARROD, *J. Organomet. Chem.* **1996**, *521*, 133.
6. a) N. ETKIN, M.C. FERMIN, D.W. STEPHAN, *J. Am. Chem. Soc.* **1997**, *119*, 2954; b) A.J. HOSKIN, D.W. STEPHAN, *Angew. Chem. Int. Ed.* **2001**, *40*, 1865; c) V.P.W. BÖHM, M. BROOKHART, *Angew. Chem. Int. Ed.* **2001**, *40*, 4694.
7. H. CHEN, S. SCHLECHT, T.C. SEMPLE, J.F. HARTWIG, *Science* **2000**, *287*, 1995.
8. R. SHU, L. HAO, J.F. HARROD, H.G. WOO, E. SAMUEL, *J. Am. Chem. Soc.* **1998**, *120*, 12988.
9. a) H.Q. LIU, J.F. HARROD, *Organometallics* **1992**, *11*, 822; b) J. HE, H.Q. LIU, J.F. HARROD, R. HYNES, *Organometallics* **1994**, *13*, 336.
10. a) Y. LI, Y. KAWAKAMI, *Macromolecules* **1999**, *32*, 3540; b) Y. LI, Y. KAWAKAMI, *Macromolecules* **1999**, *32*, 6871; c) R. ZHANG, J.E. MARK, A.R. PINHAS, *Macromolecules* **2000**, *33*, 3508; d) T.E. READY, B.P.S. CHAUHAN, P. BOUDJOUK, *Macromol. Rapid Commun.* **2001**, *22*, 654.
11. a) H. YAMASHITA, M. TANAKA, K. HONDA, *J. Am. Chem. Soc.* **1995**, *117*, 8873; b) X. WU, D.C. NECKERS, *Macromolecules* **1999**, *32*, 6003; c) M. LIENHARD, I. RUSHKIN, G. VERDECIA, C. WIEGAND, T. APPLE, L.V. INTERRANTE, *J. Am. Chem. Soc.* **1997**, *119*, 12020.
12. a) Y. NI, R. RULKENS, J.K. PUDELSKI, I. MANNERS, *Macromol. Rapid Commun.* **1995**, *16*, 637; b) N.P. REDDY, H. YAMASHITA, M. TANAKA, *J. Chem. Soc., Chem. Commun.* **1995**, 2263; c) P. GÓMEZ-ELIPE, P.M. MACDONALD, I. MANNERS, *Angew. Chem. Int. Ed. Engl.* **1997**, *36*, 762; d) P.

Gómez-Elipe, R. Resendes, P. M. Macdonald, I. Manners, *J. Am. Chem. Soc.* **1998**, *120*, 8348; e) K. Temple, F. Jäkle, J. B. Sheridan, I. Manners, *J. Am. Chem. Soc.* **2001**, *123*, 1355.

13 a) D. P. Gates, M. Edwards, L. M. Liable-Sands, A. L. Rheingold, I. Manners, *J. Am. Chem. Soc.* **1998**, *120*, 3249; b) A. R. McWilliams, D. P. Gates, M. Edwards, L. M. Liable-Sands, I. Guzei, A. L. Rheingold, I. Manners, *J. Am. Chem. Soc.* **2000**, *122*, 8848.

14 a) A. B. Burg, R. I. Wagner, *J. Am. Chem. Soc.* **1953**, *75*, 3872; b) A. B. Burg, *J. Inorg. Nucl. Chem.* **1959**, *11*, 258; c) R. I. Wagner, F. F. Caserio, Jr., *J. Inorg. Nucl. Chem.* **1959**, *11*, 259; d) A. B. Burg, P. J. Slota, Jr., *J. Am. Chem. Soc.* **1960**, *82*, 2145; e) A. B. Burg, R. I. Wagner, **1960**, US Patent 2,925,440; f) A. B. Burg, **1963**, US Patent 3,071,552; g) A. B. Burg, R. I. Wagner, **1963**, US Patent 3,071,553; h) V. V. Korshak, V. A. Zamyatina, A. I. Solomatina, *Izv. Akad. Nauk. SSSR, Ser. Khim.* **1964**, *8*, 1541.

15 H. Dorn, R. A. Singh, J. A. Massey, A. J. Lough, I. Manners, *Angew. Chem. Int. Ed.* **1999**, *38*, 3321.

16 H. Dorn, R. A. Singh, J. A. Massey, J. M. Nelson, C. A. Jaska, A. J. Lough, I. Manners, *J. Am. Chem. Soc.* **2000**, *122*, 6669.

17 H. Dorn, E. Vejzovic, A. J. Lough, I. Manners, *Inorg. Chem.* **2001**, *40*, 4327.

18 For a recent report of $B(C_6F_5)_3$-catalyzed formation of P-B bonds through dehydrocoupling, see: J.-M. Denis, H. Forintos, H. Szelke, L. Toupet, T.-N. Pham, P.-J. Madec, A.-C. Gaumont, *Chem. Commun.* **2003**, 54.

19 H. Dorn, J. M. Rodezno, B. Brunnhöfer, E. Rivard, J. A. Massey, I. Manners, *Macromolecules* **2003**, *36*, 291.

20 K. Niedenzu, J. W. Dawson, in *The Chemistry of Boron and Its Compounds* (Ed.: E. L. Muetterties), John Wiley & Sons, New York, **1967**, p. 377.

21 A. B. Burg, C. L. Randolph Jr., *J. Am. Chem. Soc.* **1949**, *71*, 3451.

22 T. C. Bissot, R. W. Parry, *J. Am. Chem. Soc.* **1955**, *77*, 3481.

23 T. Wideman, L. G. Sneddon, *Inorg. Chem.* **1995**, *34*, 1002.

24 a) P. J. Fazen, J. S. Beck, A. T. Lynch, E. E. Remsen, L. G. Sneddon, *Chem. Mater.* **1990**, *2*, 96; b) P. J. Fazen, E. E. Remsen, J. S. Beck, P. J. Carroll, A. R. McGhie, L. G. Sneddon, *Chem. Mater.* **1995**, *7*, 1942; c) T. Wideman, L. G. Sneddon, *Chem. Mater.* **1996**, *8*, 3.

25 C. A. Jaska, K. Temple, A. J. Lough, I. Manners, *Chem. Commun.* **2001**, 962.

26 C. A. Jaska, K. Temple, A. J. Lough, I. Manners, *J. Am. Chem. Soc.* **2003**, *125*, 9424.

27 J. A. Widegren, R. G. Finke, *J. Mol. Catal. A: Chem.* **2003**, *198*, 317.

28 a) C. A. Jaska, I. Manners, *J. Am. Chem. Soc.* **2004**, *126*, 1334; b) C. A. Jaska, I. Manners, *J. Am. Chem. Soc.* **2004**, *126*, 9776.

29 a) M. A. Watzky, R. G. Finke, *J. Am. Chem. Soc.* **1997**, *119*, 10382; b) J. A. Widegren, M. A. Bennett, R. G. Finke, *J. Am. Chem. Soc.* **2003**, *125*, 10301.

30 Mercury is a well-known poison of heterogeneous catalysts through the formation of an amalgam or adsorption onto the catalyst surface. See: D. R. Anton, R. H. Crabtree, *Organometallics* **1983**, *2*, 855.

31 A heterogeneous catalyst can be completely poisoned by < 1 equiv. of ligand (per metal atom) because only a fraction of the metal atoms are on the surface, whereas a homogeneous catalyst would require > 1 equiv. of ligand to completely poison the active site.

32 Filtration using small-pore membrane filters can also distinguish between soluble and insoluble catalysts. If the activity is lowered upon filtration, an insoluble catalyst is assumed.

33 a) H. Dorn, C. A. Jaska, R. A. Singh, A. J. Lough, I. Manners, *Chem. Commun.* **2000**, 1041; b) C. A. Jaska, H. Dorn, A. J. Lough, I. Manners, *Chem. Eur. J.* **2003**, *9*, 271.

34 C. A. Jaska, I. Manners, *J. Am. Chem. Soc.* **2004**, *126*, 2698.

5
Chemistry with Poly- and Perfluorinated Alkoxyaluminates: Gas-Phase Cations in the Condensed Phase?

Ingo Krossing and Andreas Reisinger

5.1
Introduction

Weakly coordinating anions (WCAs) are of great interest for applied as well as fundamental chemistry, i.e. to stabilize strongly acidic gas-phase species, highly electrophilic metal- and nonmetal cations, or weakly bound Lewis acid-base complexes of metal cations. In applied chemistry, WCAs are important for Li^+- or Ag^+-catalyzed organic reactions, electrochemistry, ionic liquids, Li ion batteries, as photoacid generators, and more [1].

Apart from the classical $[BF_4]^-$ and $[MF_6]^-$ anions, larger anions of the type $[M_nF_{5n+1}]^-$ (M=As, Sb; $n=2$–4) are known in superacid solution. Several other types of WCAs are now established: halogenated and (trifluoro)methylated carboranes $[CB_{11}X_nH_{12-n}]^-$ ($n=0$–12, X=F, Cl, Br, I, CH_3, CF_3) [1a, 2], polyfluorinated tetraarylborates $[B(Ar^F)_4]^-$ ($Ar^F=C_6F_5$, $C_6H_3(CF_3)_2$, and others) [3, 4], and tetra- or hexateflatometallates $[M(OTeF_5)_n]^-$ ($n=4$, M=B [5]; $n=6$, M=As [6], Sb [7, 8], Bi [9], Nb [10, 11]).

The two main limiting factors for all WCAs are anion coordination and anion degradation. Thus, it is clear that good WCAs share some common properties:

- With respect to their weakly coordinating nature, they should only possess a low negative charge (ideally –1), which should be delocalized over many atoms so that electrostatic cation–anion interactions are minimized.
- With respect to their inertness towards reactive cations, no sterically accessible basic groups should be present and all chemical bonds within the anions should be very strong so that anion decomposition is unlikely.

A frequently used approach to fulfil these demands is to attach large, sterically demanding fluorinated alkyl or aryl groups to a Lewis acidic central atom (see above). A recent development in WCA chemistry that combines the latter approach with the oxygen-linked but sterically shielded teflate ($OTeF_5$) based anions, is the use of poly- and perfluorinated alkoxy ligands OR^F that produce a smooth and non-adhesive "Teflon surface" of the anion. However, in contrast to teflate anions, these do not liberate anhydrous HF upon hydrolysis and are thus easier to handle in conventional laboratories. Using aluminum as a central

Inorganic Chemistry in Focus II. Edited by G. Meyer, D. Naumann, and L. Wesemann
Copyright © 2005 WILEY-VCH Verlag GmbH & Co. KGaA, Weinheim
ISBN: 3-527-30811-3

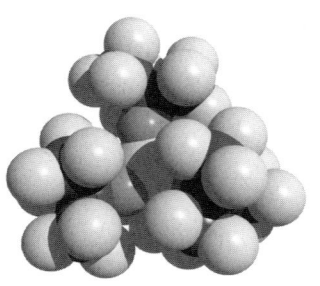

Fig. 5.1 Space-filling model of the anion [Al{OC(CF$_3$)$_3$}$_4$]$^-$.

atom, this leads to alkoxyaluminate anions of the type [Al(OR)$_4$]$^-$, the chemistry of which is highlighted in the present article [R=C(H)(CF$_3$)$_2$, C(CH$_3$)(CF$_3$)$_2$, and C(CF$_3$)$_3$; an example is shown as a space-filling model in Fig. 5.1] [12, 13].

One notes from the space-filling model that the most basic sites of these anions, the oxygen atoms, are not available for coordination to all but the smallest of electrophiles (Li$^+$). With the large number of 36 peripheral C–F bonds, [Al{OC(CF$_3$)$_3$}$_4$]$^-$ is, together with [CB$_{11}$(CF$_3$)$_{12}$]$^-$, presumably one of the least coordinating anions known. However, the carborane anion is explosive. One of the major advantages of these aluminates is that they are easily accessible on a preparative scale. They can be prepared on a 100 g scale in well over 90% yield within two days. Moreover, these anions are remarkably stable and the [Al(OR)$_4$]$^-$ anion with R=C(CF$_3$)$_3$, in contrast to R=C(H)(CF$_3$)$_2$, C(CH$_3$)(CF$_3$)$_2$, is stable in water and even in 6 N HNO$_3$ [13].

In the following, we summarize available starting materials for these aluminates, present some of the obtained electrophilic as well as weakly bound cations, and attempt to rationalize the reasons for the unusual chemistry possible in the environment created by such very weakly coordinating aluminates [Al(OR)$_4$]$^-$.

5.2
Available Starting Materials

Important inputs for the choice of a suitable anion from the multitude of known WCAs are its availability, accessibility, and price, as well as knowledge of the desired degree of "*non coordination*" necessary for a given problem. Hence, the right choice is normally a WCA that is stable and weakly coordinating enough to allow the desired transformation, but not necessarily the ultimate most weakly coordinating anion. In this respect, the poly- and perfluorinated alkoxyaluminates are good WCAs that may be prepared with little synthetic effort in all standard inorganic/organometallic laboratories. Li[Al{OC(CF$_3$)$_3$}$_4$] is also commercially available (Strem).

All syntheses of [Al(OR)$_4$]$^-$ starting materials begin with the preparation of the Li$^+$ salts using LiAlH$_4$ and the appropriate alcohols as shown in Scheme 5.1 [12, 13]. Most other salts can then be easily prepared in high yields through me-

Scheme 5.1

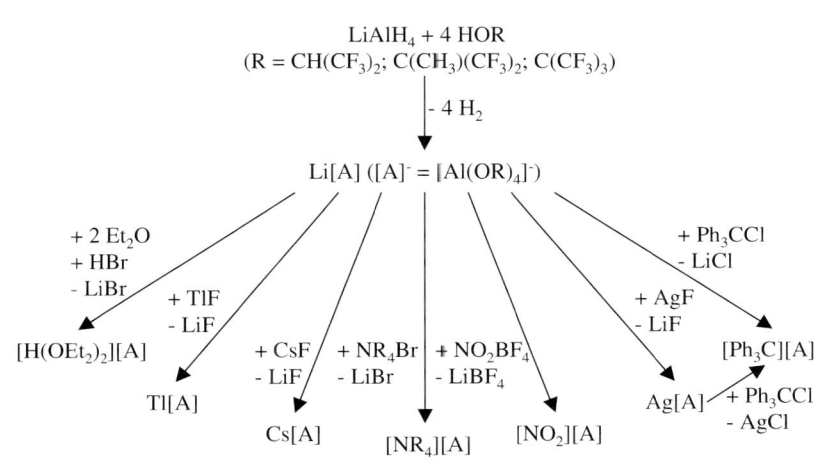

$$\text{LiAlH}_4 + 4\text{ HOR}$$
$$(R = CH(CF_3)_2;\ C(CH_3)(CF_3)_2;\ C(CF_3)_3)$$
$$\downarrow -4\text{ H}_2$$
$$\text{Li[A]}\ ([A]^- = [Al(OR)_4]^-)$$

Products via metathesis:
- $+ 2\text{ Et}_2\text{O}$, $+ \text{HBr}$, $- \text{LiBr}$ → $[\text{H}(\text{OEt}_2)_2][A]$
- $+ \text{TlF}$, $- \text{LiF}$ → $\text{Tl}[A]$
- $+ \text{CsF}$, $- \text{LiF}$ → $\text{Cs}[A]$
- $+ \text{NR}_4\text{Br}$, $- \text{LiBr}$ → $[\text{NR}_4][A]$
- $+ \text{NO}_2\text{BF}_4$, $- \text{LiBF}_4$ → $[\text{NO}_2][A]$
- $+ \text{AgF}$, $- \text{LiF}$ → $\text{Ag}[A]$ $\xrightarrow{+ \text{Ph}_3\text{CCl},\ - \text{AgCl}}$ $[\text{Ph}_3\text{C}][A]$
- $+ \text{Ph}_3\text{CCl}$, $- \text{LiCl}$ → $[\text{Ph}_3\text{C}][A]$

Scheme 5.1 Available starting materials, obtained by metathesis reactions with Li[Al(OR)$_4$]. NO$_2^+$ and H(OEt$_2$)$_2^+$ salts are not available with all R groups.

tathesis reactions with the precipitation of LiX (X = F, Cl, Br, BF$_4$). Scheme 5.1 gives an overview of preparations of Cs$^+$ [14], Ag$^+$ [13], Tl$^+$ [15], CPh$_3^+$ [14], H(OEt$_2$)$_2^+$ [16], NR$_4^+$ [12, 17], NO$^+$ [18], and NO$_2^+$ [19] salts of the WCA [Al(OR)$_4$]$^-$.

The Li aluminates themselves have been used in catalysis [20] and as electrolytes [12, 21]. [Al{OC(Ph)(CF$_3$)$_2$}$_4$]$^-$-based ionic liquids have been investigated as media for transition metal catalysis [22].

5.3 Weakly Bound Lewis Acid-Base Complexes of the Ag$^+$ Cation

Many complexes of the late transition metal Ag$^+$ with weak Lewis bases such as S$_8$, P$_4$ or C$_2$H$_4$ were unknown in condensed phases but have been characterized in the gas phase by one of the advanced mass spectrometric methods. Unfortunately, all MS methods are hampered by the lack of structural information provided by this technique. Thus, although it has been established that the [Ag(P$_4$)$_n$]$^+$, [Ag(S$_8$)$_n$]$^+$, and [Ag(C$_2$H$_4$)$_n$]$^+$ ($n = 1, 2$) cations exist in the gas phase, their geometries remained unknown, except for insights provided by quantum chemical calculations and the structure of a distorted [Ag(S$_8$)$_2$]$^+$ salt with the [AsF$_6$]$^-$ anion [23]. Lewis acid–base complexes of weak (polydentate) Lewis bases reside within shallow potential energy wells on an extended PES (Potential Energy Surface). Thus, it proved very difficult to verify experimentally conclusions drawn from the delicate quantum chemical calculations. However, such complexes are now accessible in the solid state, if provided with a suitable WCA such as an alkoxyaluminate. Some of these gas-phase cations in condensed phases are detailed in the following sections.

5.3.1
Ag(S$_8$) Complexes

It has been shown by mass spectrometry that Ag$^+$ and S$_8$ give complexes of type Ag(S$_8$)$_x^+$ ($x = 1, 2$). The only available structural model for such complexes was (S$_8$)$_2$AgMF$_6$ (M = As, Sb) [23]. However, due to anion coordination it is unlikely that the gaseous Ag(S$_8$)$_2^+$ cation will adopt the same geometry (Fig. 5.2).

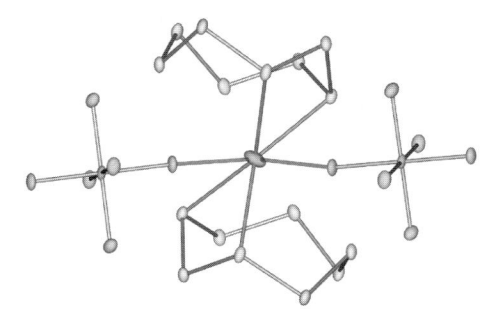

Fig. 5.2 Section of the solid-state structure of (S$_8$)$_2$AgSbF$_6$: coordinated MF$_6^-$ anions that lead to a distortion of the Ag(S$_8$)$_2^+$ structure.

Therefore, the larger [Al(OR)$_4$]$^-$ anions were used to approximate the still unknown geometries of undistorted Ag(S$_8$)$_x^+$. The compounds [Ag(S$_8$)][Al{OC(H)(CF$_3$)$_2$}$_4$] and [Ag(S$_8$)$_2$][Al{OC(CF$_3$)$_3$}$_4$] (Fig. 5.3) were obtained [24] and include the first examples of undistorted homoleptic metal S$_8$ complexes, i.e. the almost C_{4v}-symmetric [Ag(η^4-S$_8$)]$^+$ and the approximately centrosymmetric [Ag(η^4-S$_8$)$_2$]$^+$ cations. As also shown by quantum chemical calculations, they provide the best structural models for the gaseous AgS$_8^+$ and AgS$_{16}^+$ cations.

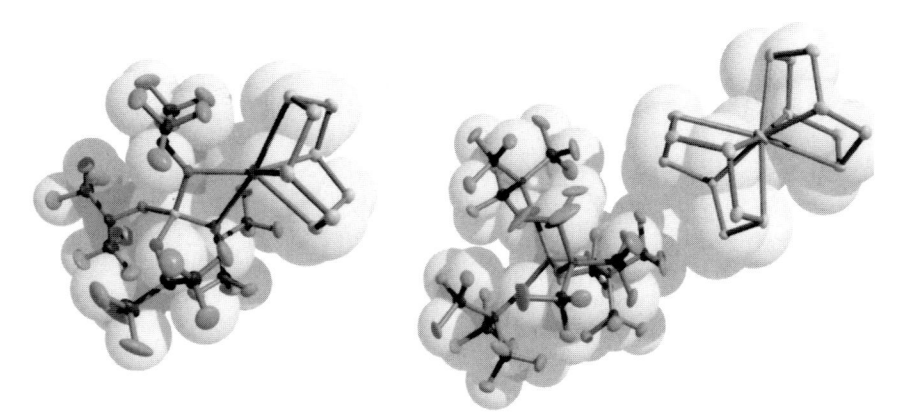

Fig. 5.3 Sections of the solid-state structures of [Ag(S$_8$)$_2$][Al{OC(CF$_3$)$_3$}$_4$] and [Ag(S$_8$)$_2$][Al{OC(CF$_3$)$_3$}$_4$]. Thermal ellipsoids are drawn at the 25% probability level.

It should be noted that quantum chemical calculations of these cations are rather difficult and without the experimental structures it would have been hard to establish the quality of the computation. Weak dispersive Ag–S interactions are structure-determining and, therefore, DFT and HF-DFT theory failed to describe the Ag(S$_8$)$_x^+$ cations. Only MP2 with a large basis set (TZVPP) gave minimum-energy geometries close to the experimental ones.

5.3.2
Complexes with P$_4$S$_3$

Similar to the Ag–S$_8$ complexes, the coordination chemistry of small inorganic cage or ring molecules such as P$_4$S$_3$ is of fundamental interest, while structural knowledge on intact coordinated P$_4$S$_3$ cages is rather limited [25]. Normally, the P$_4$S$_3$ molecule degrades upon coordination to transition metal fragments and phosphide or sulfide groups are incorporated into the complex [26]. Since it was possible to obtain undistorted homoleptic S$_8$ complexes with the weak Lewis acid Ag$^+$ in the presence of the WCAs [Al(OR)$_4$]$^-$, we extended our investigations to the coordination behavior of undistorted P$_4$S$_3$ with the goal to learn more about the most reactive coordination sites of this molecule. Prior to our work, only a few examples of intact P$_4$S$_3$ cages coordinated to transition metal atoms were known [25]. All of them were exclusively bound through the apical P atom. No coordination of an S atom or a P atom of the P$_3$ base was reported. However, S-coordination was postulated as a possible pathway for sulfidic degradation [26a, 27]. The Ag–P$_4$S$_3$ complexes revealed the first polymeric Lewis acid-base adducts of Ag[Al(OR)$_4$] (one-dimensional chains), which previously had always led to isolated molecular or salt structures [28]. With the more basic [Al(OR)$_4$]$^-$ anion (R=C(CH$_3$)(CF$_3$)$_2$), one anion remained coordinated to the sil-

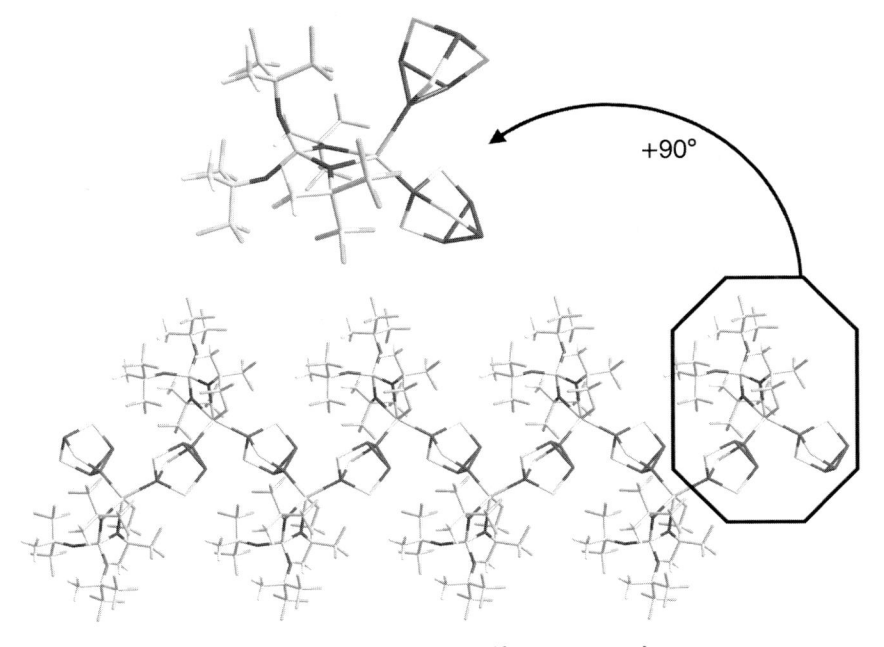

Fig. 5.4 Extended solid-state structure of [(P$_4$S$_3$)AgAl{OC(CH$_3$)(CF$_3$)$_2$}$_4$]$_\infty$.

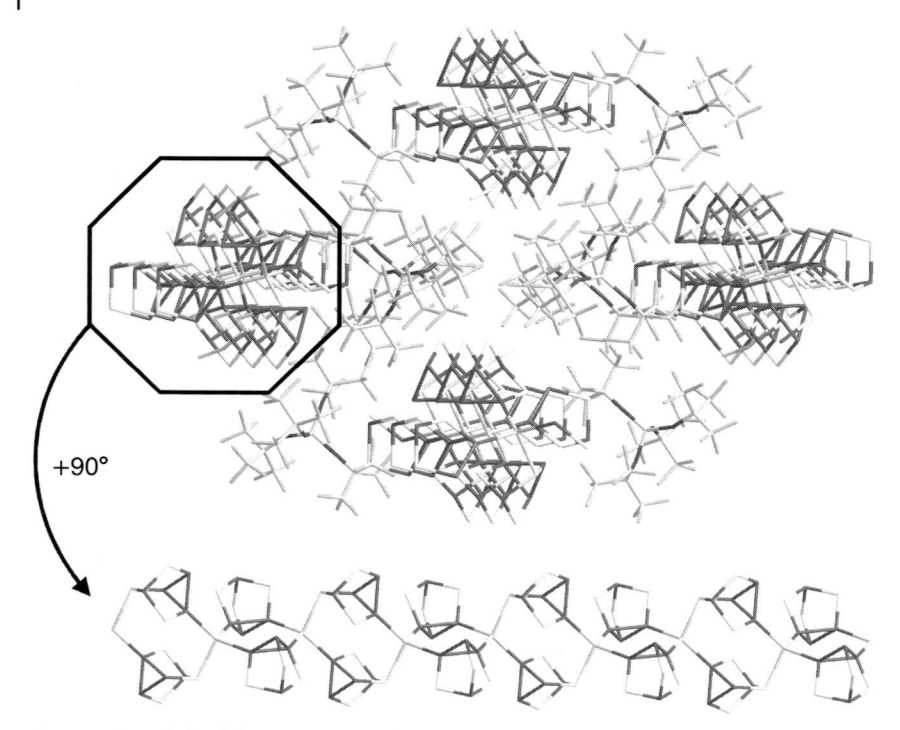

Fig. 5.5 Extended solid-state structure of $\{Ag(P_4S_3)_2^+[Al\{OC(CF_3)_3\}_4]^-\}_\infty$. In the lower part of the image, all anions have been omitted for clarity.

ver atom in addition to one bridging P_4S_3 cage (Fig. 5.4). However, with the least basic perfluorinated aluminate with $R=C(CF_3)_3$, a cationic polymer of the general formula $[Ag(P_4S_3)_2^+]_\infty$ with no anion contacts was formed (Fig. 5.5).

From this study, it was evident that – in contrast to P_4 discussed below – P_4S_3 has sterically active lone-pair orbitals on all of its atoms. Although incorporating a cyclo-P_3 unit similar to P_4, molecular P_4S_3 prefers η^1-coordination. Moreover, the structures of $[Ag(P_4S_3)][Al\{OC(CF_3)_3\}_4]$ and $[Ag(P_4S_3)_2][Al\{OC(CF_3)_3\}_4]$ revealed the first examples of sulfur coordination within P_4S_3. All cage atoms are similar with regard to their coordination behavior. Thus, P_4S_3 can act in a 1,3-, 2,4- or 3,4- fashion as a bridging ligand. This flexible S or P coordination is the starting point for the frequently observed further degradation of the P_4S_3 molecule. After our report that all atoms of the P_4S_3 cage may be coordinated, several reports on such coordination modes appeared [29].

5.3.3
Complexes with P_4

In a pioneering MS study, two binary silver–phosphorus cations were observed: AgP_4^+ and AgP_8^+. On the basis of these mass spectrometric experiments and initial DFT calculations, a structure with two η^1-bound molecules was predicted. Similar-

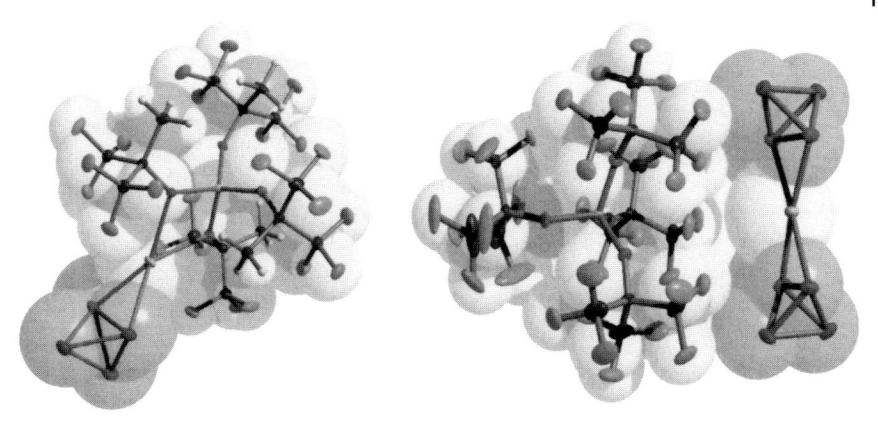

Fig. 5.6 Sections of the solid-state structures of [Ag(P$_4$)] [Al{OC(CH$_3$)(CF$_3$)$_2$}$_4$] and [Ag(P$_4$)$_2$][Al{OC(CF$_3$)$_3$}$_4$]. Thermal ellipsoids are drawn at the 25% probability level.

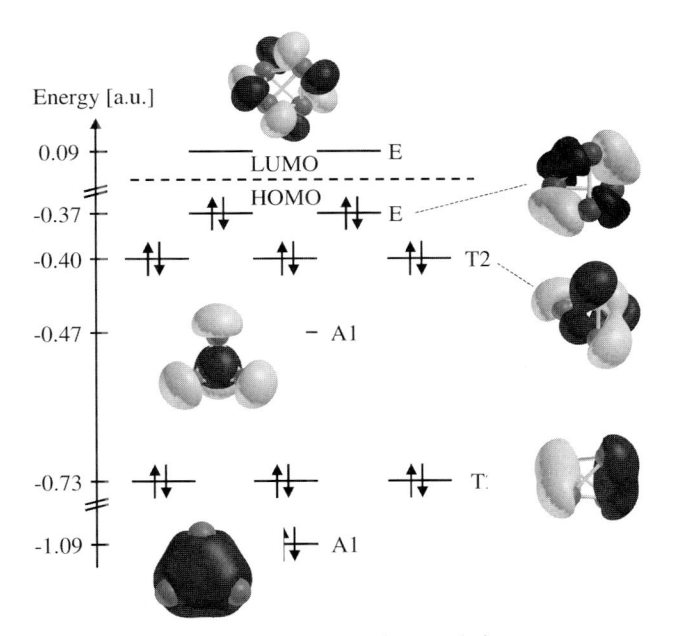

Fig. 5.7 Section of the MO diagram of tetrahedral P$_4$.

ly to the situation in the [Ag(η^4-S$_8$)$_n$]$^+$ cations, unusual and weak bonding interactions in Ag$^+$–P$_4$ complexes can be stabilized by WCAs. In agreement with this, Ag$^+$–P$_4$ complexes containing one or two almost undistorted tetrahedral P$_4$ units can be formed, i.e. [Ag(P$_4$)][Al{OC(CH$_3$)(CF$_3$)$_2$}$_4$] and [Ag(P$_4$)$_2$][Al{OC(CF$_3$)$_3$}$_4$], containing the first homoleptic metal P$_4$ cation (Fig. 5.6) [30]. In contrast to the initial computational prediction, the crystal structure determination revealed a $[Ag(P_4)_2]^+$ cation with D_{2h} symmetry and two coplanar η^2-bound P_4 molecules.

To understand why η^2-coordination – as compared to η^1- or η^3-coordination – is favored by 27 to 76 kJ mol^{-1} [30], an analysis of the MO diagram of tetrahedral P_4 is helpful (Fig. 5.7): s- and p-type orbitals are well separated and the lone-pair orbitals have $3s^2$ character and are not available for coordination. Thus, coordination has to proceed through an interaction with the frontier orbitals (HOMOs). The major contribution to the Ag–P bonding is made by donation of electron density from the HOMOs of P_4 to the empty 5s orbital of Ag^+. Therefore, the P_4 molecules prefer side-on to end-on coordination. For the same reason, P_4 is protonated on an edge and not at a corner [31].

5.3.4
Complexes with C_2H_4

After the investigation of the coordination chemistry of group 15 and group 16 elemental rings and cages with the Ag^+ cation, it was obvious to extend this kind of chemistry. D. Deubel, a theoretician, became interested in the $Ag(P_4)_x^+$ cations and compared their bonding to that in $Ag(C_2H_4)_2^+$ [32] and, therefore, we decided to study these $Ag(C_2H_4)_x^+$ complexes experimentally. Since the discovery of Zeise's salt $K[PtCl_3(C_2H_4)]$ in 1827 [33], the investigation of simple ethene–metal complexes has been an area of continuing interest in many fields of chemistry, fueled by the interest in transformations of the coordinated C_2H_4 molecule. Several ethene–metal coordination modes, including μ and $\eta^2{:}\eta^2$, are known, and ethene–silver complexes with little or no back-bonding served to formulate the Dewar–Chatt–Duncanson model for the binding of alkenes to transition metal atoms [34]. However, there is still much controversy about the nature of the bonding in $[Ag(C_2H_4)_n]^+$ ions ($n=1–3$), with proposals ranging from mainly electrostatic to mainly covalent and mixtures thereof [35]. Although Ag-C_2H_4 complexes are textbook compounds, experimental information on this system is very limited [36] and no solid-state structure of a $[Ag(C_2H_4)_n]^+$ unit has been determined. In the gas phase, $[Ag(C_2H_4)_n]^+$ cations have been characterized by MS for $n=1$ and 2 [37].

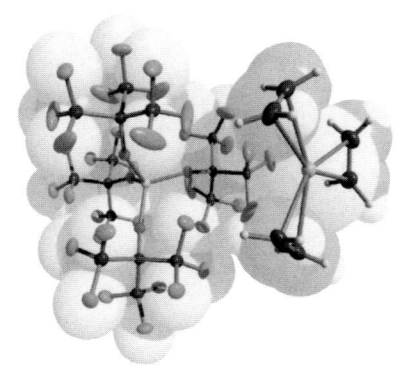

Fig. 5.8 Section of the solid-state structure of $[Ag(C_2H_4)_3]^+[Al\{OC(CF_3)_3\}_4]^-$. Superposition of a space-filling and a 25% probability ellipsoid model.

Using $[Al\{OC(CF_3)_3\}_4]^-$ as a WCA, a cationic $[Ag(C_2H_4)_3]^+$ complex with an ionic lattice in the solid state and an almost planar D_{3h} symmetric cation in a spoke-wheel arrangement has been synthesized (Fig. 5.8) [38].

The synthesis and full characterization of $[Ag(C_2H_4)_3]^+[Al\{OC(CF_3)_3\}_4]^-$ in the condensed phase is rather remarkable, especially since only $Ag(C_2H_4)_n^+$ complexes with $n=1, 2$ have been observed in the gas phase. This is probably due to the low (calculated) binding energy of the third C_2H_4 molecule of $\Delta G° = -55$ kJ mol^{-1} in the gas phase (MP2/TZVPP; see also Section 5.5.1).

5.3.5
Anion Effects

Generally, anion–cation as well as solvent–cation interactions have to be minimized and to be smaller in magnitude than the interaction with the weakly basic ligand to stabilize weak Lewis acid–base adducts of the present type. The influence of the counterion may be illustrated by the series of silver-P_4 and -S_8 complexes highlighted in the present article.

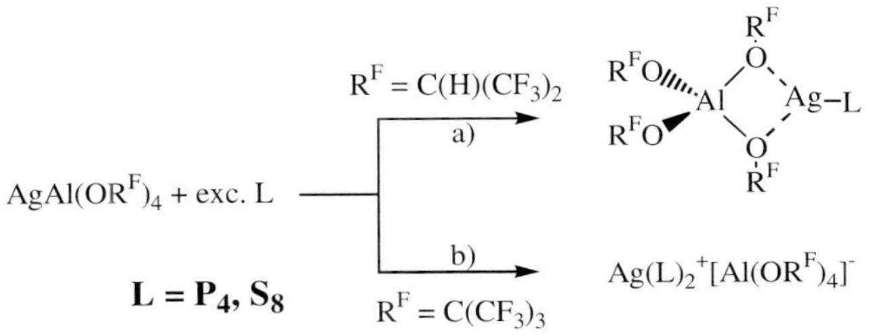

Scheme 5.2 Counterion dependence of the structures of weakly bound Lewis acid–base adducts of the silver cation.

Role of the counterion

From the solid-state structures, it is evident that $[Al(OR)_4]^-$ anions with R=C(H)(CF$_3$)$_2$, C(CH$_3$)(CF$_3$)$_2$ coordinate to the Ag$^+$ cation and give molecular compounds; see (P$_4$)AgAl{OC(H)(CF$_3$)$_2$}$_4$ [30] and (S$_8$)AgAl{OC(H)(CF$_3$)$_2$}$_4$ [24] (Scheme 5.2a). With the least basic anion, $[Al\{OC(CF_3)_3\}_4]^-$, P$_4$ and S$_8$ are stronger Lewis bases towards the cation than the anion. Consequently, in the analogous reactions,

$$Ag(solvent)_n^+[X]^- + 2\,P_4 \underset{Solvent = C_6H_6}{\overset{Solvent = CH_2Cl_2}{\rightleftharpoons}} Ag(P_4)_2^+[X]^- + n\text{ solvent}$$

$$[X]^- = [Al\{OC(CF_3)_3\}_4]^-$$

Scheme 5.3 Solvent dependence of the stability of the structures of weakly bound Lewis acid base adducts of the silver cation.

the Ag$^+$ cation coordinates two ligands and forms [Ag(P$_4$)$_2$]$^+$[Al{OC(CF$_3$)$_3$}$_4$]$^-$ [30] and [Ag(S$_8$)$_2$]$^+$[Al{OC(CF$_3$)$_3$}$_4$]$^-$ [24] with truly ionic structures (Scheme 5.2b).

Role of the solvent

When Ag$^+$[Al{OC(CF$_3$)$_3$}$_4$]$^-$ and 2 P$_4$ are mixed in CH$_2$Cl$_2$, the equilibrium of the reaction shown in Scheme 5.3 lies completely on the right-hand side. However, replacing the weakly basic solvent CH$_2$Cl$_2$ by the more strongly coordinating solvent C$_6$H$_6$ leads to decomposition of the Ag(P$_4$)$_2^+$ cation, replacement of P$_4$ by C$_6$H$_6$, and hence the formation of Ag(C$_6$H$_6$)$_2^+$. Thus, in benzene the equilibrium lies completely on the left-hand side [30].

5.4
Highly Electrophilic Cations

The present section attends to another major WCA chemistry field: highly electrophilic non-metal cations and reactive intermediates, which are otherwise at least very difficult or even impossible to synthesize and stabilize in condensed phases. Again, some additional information may be gained from mass spectrometric experiments as well as quantum chemical calculations. Due to the coordinating ability (see also Section 5.3.5) of [Al(OR)$_4$]$^-$ with R=C(H)(CF$_3$)$_2$, C(CH$_3$)(CF$_3$)$_2$, reactions with P–X cations such as P$_2$X$_5^+$ (X=Br, I) led to immediate decomposition of the anions. Thus, for reactions to prepare highly electrophilic cations the least coordinating and chemically most robust [Al{OC(CF$_3$)$_3$}$_4$]$^-$ anion was exclusively used. Silver-salt metathesis reactions as presented in this section provide an alternative approach to polyatomic non-metal cations that are conventionally prepared by direct oxidation with strong one- or two-electron oxidants such as PtF$_6$, WX$_6$ (X=F, Cl) or MF$_5$ (M=As, Sb, Nb, Au) [39], or strong Lewis acids such as AlX$_3$ (X=Cl, Br, I) [39b] and transition metal halides or oxohalides.

5.4.1
Binary Phosphorus–Halogen Cations

In the course of our investigations directed towards the preparation of the still unknown homopolyatomic phosphorus cations, we fortuitously found a simple

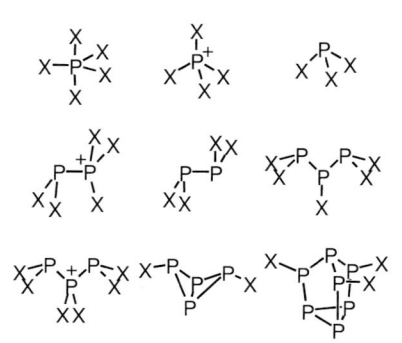

Fig. 5.9 Known neutral or cationic binary P–X species (X=F, Cl, Br, I); note that some compounds only exist with one type of X.

5.4 Highly Electrophilic Cations

approach to generate new binary phosphorus–halogen cations. The knowledge on cationic or neutral binary P–X species, where X may be F, Cl, Br, I, is still very sparse and has been limited to P^VX_5 (X=F, Cl), $P^VX_4^+$ (X=F–I) [40], $P^{III}X_3$ (X=F–I), $P^{III}_2I_5^+$ [41], $P^{II}_2X_4$ (X=F, Cl, I), and P_3X_5 (X=F, I [42]) (Fig. 5.9).

Solution studies showed the additional existence of small amounts of the phosphorus-rich cage molecules P_4Br_2 and P_7I_3 in CS_2 solutions of P_4 and X_2 mixtures (X=Br, I). Structural data are only available for a few of these types, i.e. among the binary P–X cations only for PX_4^+ (X=F–I) [40] and $P_2I_5^+$ [41]. Binary phosphorus–iodine cations are very electrophilic and $P_2I_5^+$ immediately decomposed the EF_6^- counterions when prepared from P_2I_4 and $I_3^+EF_6^-$ (E=As, Sb) at −78°C [41 b,c]. This kind of decomposition can be avoided by the use of the weakly basic $[Al\{OC(CF_3)_3\}_4]^-$ anion. Thus, a silver-salt metathesis of $Ag[Al\{OC(CF_3)_3\}_4]$ with PX_3 resulted in the formation of the highly reactive PX_2^+ intermediates, which were previously only investigated in the gas phase [43]. These intermediates acted as electrophilic carbene analogues [43a]; the electronic structure of this PBr_2^+ intermediate is compared with that of $P(NH_2)_2^+$ in Fig. 5.10.

Fig. 5.10 Calculated HOMOs of PBr_2^+ (left) and $P(NH_2)_2^+$ (right).

From Fig. 5.10 it is evident that the largest coefficient of the HOMO of PBr_2^+ is located at the P atom, while the HOMO of $P(NH_2)_2^+$ mainly consists of contributions from the NH_2 groups but not from P. Thus, as also shown earlier [43a], only PX_2^+ (X=Cl, Br, I) can react as a carbene-analogous reagent. These PX_2^+ intermediates then inserted into the X–X, P–X or P–P bonds of X_2, PX_3, P_4 or P_4S_3 to quantitatively afford stable but highly electrophilic and soluble PX_4^+,

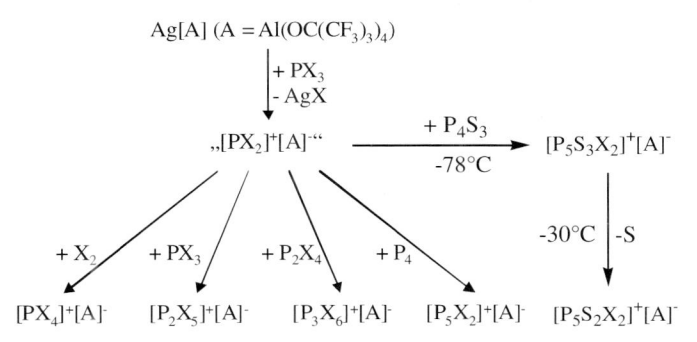

Scheme 5.4 Insertion reactions of the carbene analogue "PX_2^+" intermediate. Note that X may be I for $P_3X_6^+$ but Br and I for the other cations.

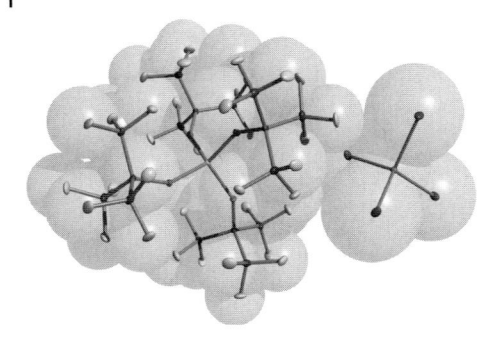

Fig. 5.11 Section of the solid-state structure of $[PI_4]^+[Al\{OC(CF_3)_3\}_4]^-$. Superposition of a space-filling and a 25% probability ellipsoid model.

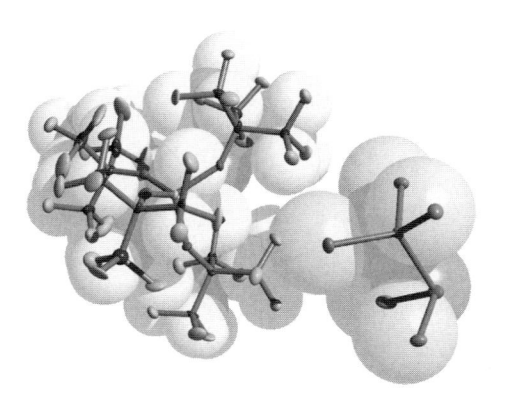

Fig. 5.12 Section of the solid-state structure of $[P_2Br_5]^+[Al\{OC(CF_3)_3\}_4]^-$. Superposition of a space-filling and a 25% probability ellipsoid model.

$P_2X_5^+$ [44a], phosphorus-rich $P_5X_2^+$ [44], $P_5S_3X_2^+$, and $P_5S_2X_2^+$ [48] as well as subvalent $P_3I_6^+$ [45] salts of the weakly basic $[Al\{OC(CF_3)_3\}_4]^-$ anion (Scheme 5.4).

Thus, the preparation of $[PX_4]^+$ salts (X = Br, I) with the $[Al\{OC(CF_3)_3\}_4]^-$ counterion is also possible (Fig. 5.11). Recent results have shown that this reaction can be extended to the $[AsBr_4]^+$ salt [46]. These solid EX_4^+ salts are stable at room temperature for at least several hours. This stability may be contrasted with the absence of experimental evidence for PI_5 and $AsBr_5$ and the fact that $AsCl_5$ decomposes at a low temperature of about –50 °C [47].

Similarly, using two equivalents of PX_3 for the reaction, $P_2X_5^+$ salts (X = Br, I) were prepared (Fig. 5.12). Dissolved $P_2X_5^+$ is fluxional on the time scale of ^{31}P NMR spectroscopy and disproportionates for X = I at a rate dependent on the concentration to give reduced $P_3I_6^+$ and an oxidized species that was assigned as PI_4^+ [44a]. $P_2Br_5^+$ is even more fluxional than $P_2I_5^+$ and may be in equilibrium with $PBr_2^+(CH_2Cl_2)$ and PBr_3. $P_2Br_5^+$ serves as a PBr_2^+ source and reacts with P_4 with the formation of $P_5Br_2^+$ as well as PBr_3 [44b].

Unlike other oxidative methods, the silver-salt metathesis approach also allows the stabilization of reduced subvalent cations. Thus, insertion of the PI_2^+ carbene analogue into the P–P bond of P_2I_4 results in the formation of the subvalent $P_3I_6^+$ cation (average valency: 2.33), a derivative of the poorly characterized P_3I_5 [42]. Fig. 5.13 shows a section of the solid-state structure of $[P_3I_6]^+[(RO)_3Al–F–Al(OR)_3]^-$

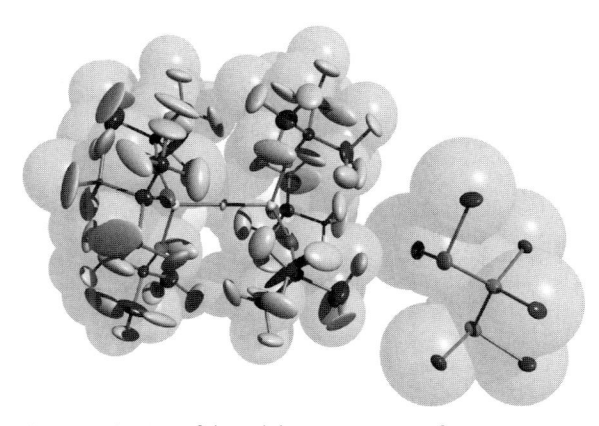

Fig. 5.13 Section of the solid-state structure of [P_3I_6]$^+$[{(CF_3)$_3$CO}$_3$Al–F–Al{OC(CF_3)$_3$}$_3$]$^-$. Superposition of a space-filling and a 25% probability ellipsoid model.

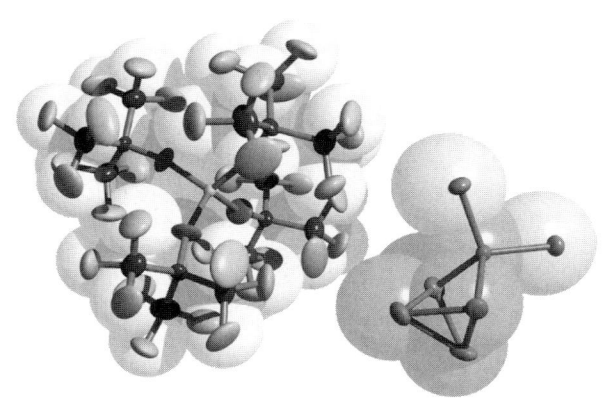

Fig. 5.14 Section of the solid-state structure of [P_5Br_2]$^+$[Al{OC(CF_3)$_3$}$_4$]$^-$. Superposition of a space-filling and a 25% probability ellipsoid model.

(R = C(CF_3)$_3$); in contrast to this salt, [P_3I_6]$^+$[Al{OC(CF_3)$_3$}$_4$]$^-$ is only marginally stable in the solid state [45].

The synthesis of $P_5X_2^+$ (X = Cl, Br, I) [44] salts with [Al{OC(CF_3)$_3$}$_4$]$^-$ revealed that "PX_2^+" insertion into a P cage is also possible. The $P_5X_2^+$ cations are the first phosphorus-rich binary P–X cations and include a hitherto unknown phosphorus cage. They are probably the closest approximation of the, at least on a preparative scale, as yet unknown homopolyatomic phosphorus cations, i.e. "P_5^+". Fig. 5.14 shows a section of the solid-state structure of [P_5Br_2]$^+$[Al{OC(CF_3)$_3$}$_4$]$^-$.

The synthesis of $P_5S_2X_2^+$ (X = Br, I) salts [48] with the [Al{OC(CF_3)$_3$}$_4$]$^-$ counterion represents the last example of the variety of "PX_2^+" insertion reactions. Interestingly, after the initial formation of $P_5S_3X_2^+$, identified by low-temperature NMR studies,

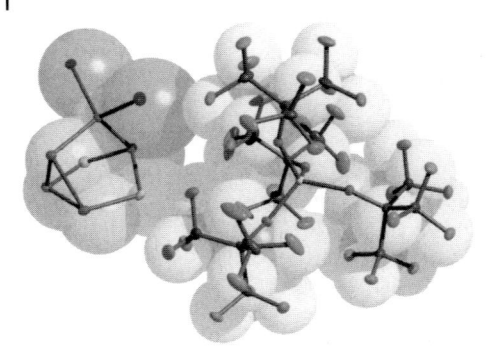

Fig. 5.15 Section of the solid-state structure of $[P_5S_2I_2]^+[Al\{OC(CF_3)_3\}_4]^-$. Superposition of a space-filling and a 25% probability ellipsoid model.

one S was eliminated and the less sulfur containing $[P_5S_2X_2]^+[Al\{OC(CF_3)_3\}_4]^-$ was formed (Fig. 5.15).

By in situ ^{31}P NMR spectroscopy, we observed an initial PX_2^+ insertion into the P_{basal}–S bond, although ab initio calculations predicted that an insertion into the P–P bond of the P_3 base of P_4S_3 would be thermodynamically more favorable by 52 kJ mol^{-1}. This unexpected behavior may be explained by the fact that the LUMO of P_4S_3 is a σ^* orbital in relation to the P_{basal}–S bond. Therefore, the initial reaction of the HOMO of PX_2^+ involves insertion into the P_{basal}–S bond and not, as expected on thermodynamic grounds, into the P–P bond.

5.4.2
Simple Carbenium Ions

In the course of our investigations of new arsenic–halogen cations, we fortuitously prepared a very simple yet electronically very interesting carbenium ion: the $CS_2Br_3^+$ cation [49] (Fig. 5.16), which may be obtained by the quantitative reaction of $AsBr_4^+$ with CS_2 yielding $CS_2Br_3^+$ and a polymeric As^IBr.

The C–Br bond length in $CS_2Br_3^+$ is very close to that expected for the simple CBr_3^+ cation (181.7 vs. 181.3 pm (MP2/TZVPP)), and thus it appeared likely that

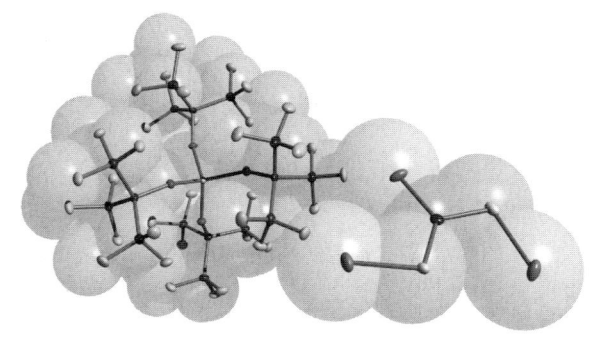

Fig. 5.16 Section of the solid-state structure of $[CS_2Br_3]^+$ $[Al\{OC(CF_3)_3\}_4]^-$. Superposition of a space-filling and a 25% probability ellipsoid model.

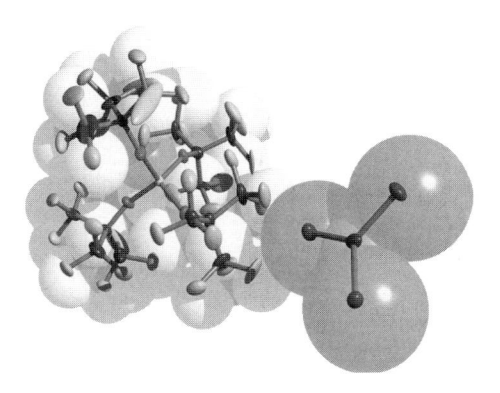

Fig. 5.17 Section of the solid-state structure of $[CI_3]^+[Al\{OC(CF_3)_3\}_4]^-$. Superposition of a space-filling and a 25% probability ellipsoid model.

simple binary carbon–halogen cations such as CX_3^+ may also be accessible by metathesis of CX_4 with Ag[A].

Small, highly reactive carbon-based cations, such as CX_3^+, OCX^+ (X=F–I), and others, are frequently observed in fragmentation reactions of gaseous ions in the mass spectrometer and are thought to be reactive intermediates in organic synthesis. The closest structurally characterized approximations of this kind are Me_2CF^+ [50], $C(OH)_3^+$, and $(H_3CO)C(H)X^+$ salts (X=F, Cl) [51]. However, the heavier CX_3^+ cations with X=Cl, Br, I have only been prepared as long-lived intermediates at –78 °C in SO_2ClF solution [52]. From recent computational work [53], it became clear that, at variance with earlier conclusions [52 a, e], cations bearing the heavier halogens should be thermodynamically more stable and less electrophilic than those with X=F, Cl [52]. Therefore, we decided to examine reactions of CI_4 and $Ag[Al(OR)_4]$ with the goal to prepare $CI_3^+[Al(OR)_4]^-$ (and AgI). For the successful preparation of $CI_3^+[Al(OR)_4]^-$, the complete exclusion of light and the use of carefully purified diiodine-free CI_4 is absolutely necessary; given this, the reaction is quantitative and the structure of $CI_3^+[Al(OR)_4]^-$ is shown in Fig. 5.17 [54].

CI_3^+ was the first structurally characterized binary C–X carbenium ion (X=H, F, Cl, Br, I). One year after our report on the successful preparation of CI_3^+, G. J. Schrobilgen published CCl_3^+ and CBr_3^+ salts of the $[Sb(OTeF_5)_6]^-$ anion [64]. Experimental and computational work finally showed that the Lewis acidity of the CX_3^+ cations decreases on going from F to I, while it increases in the same direction for the isoelectronic BX_3 molecules. To compare the Lewis acidities of BX_3 and CX_3^+, the fluoride ion affinity (FIA) is a well established measure (Tab. 5.1; see also Section 5.2). The higher the FIA, the stronger the Lewis acid.

The data in Tab. 5.1 include solvation energies (COSMO model) and thus one may directly compare the Lewis acidities of the isoelectronic BX_3 and CX_3^+ particles in CH_2Cl_2 solution. CI_3^+ has about the same acidity as BBr_3 and BI_3; however, CF_3^+ is an exceedingly aggressive Lewis acid with little hope for its stabilization in the condensed phase, while BF_3 is a long known and only mild Lewis acid. Reasons for this discrepancy may be found in the more favorable charge delocalization of the heavier CX_3^+ cations, while a similar stabilization for the boron halides involves charge separation and the formation of a formal B^- center as shown in Scheme 5.5 [55].

Tab. 5.1 Fluoride ion affinities (FIAs) for CX_3^+ and BX_3 (X = H, F, Cl, Br, I).

X =	FIA (CX_3^+) [a)] in CH_2Cl_2 [b)]	FIA (BX_3) [a)] in CH_2Cl_2 [b)]
H	501	164
F	497	225
Cl	359	279
Br	343	307
I	302	322

a) MP2/TZVPP [kJ mol^{-1}]
b) COSMO solvation model.

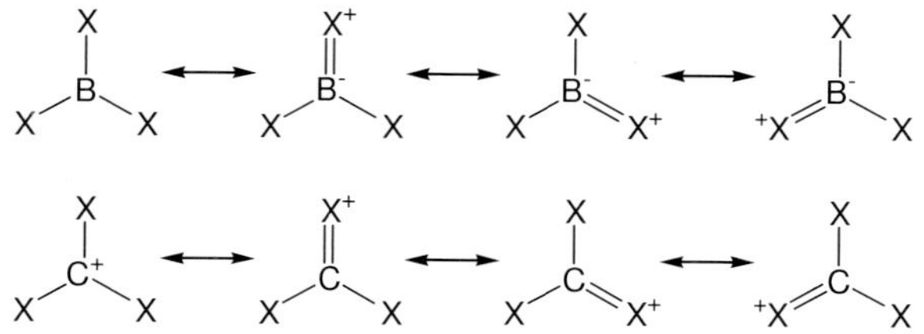

Scheme 5.5 Possible mesomeric structures of CX_3^+ and BX_3.

The driving force for such an interaction is smaller in BX_3 and thus the Lewis acidities of the isoelectronic CX_3^+ and BX_3 particles are opposite and for CX_3^+ follow the trend expected on the basis of simple electronegativity arguments [55].

5.5
A Rationalization of the Special Properties of Salts of WCAs Based on Thermodynamic Considerations

Many unusual and fundamentally important species have already been observed in the gas phase by means of advanced mass spectrometric methods. Often, these cationic species are highly electrophilic or only weakly bound Lewis acid–base complexes that react with the counterions and/or solvents that are necessarily present in condensed phases. Therefore, it has often been impossible to generate stable salts of these species in condensed phases and to analyze their properties by classical physical measurements such as vibrational and NMR spectroscopy or X-ray crystallography. To stabilize a given gas-phase cation in a condensed phase, a counterion and usually also a solvent have to be introduced. In the solid state, the ion–counterion pairs form an ionic lattice in which the gas-phase cation is subjected to an electrostatic field that is absent in the gas phase (= lattice energy). Since lattice energies are inversely proportional to the sum of the ionic radii (or volumes) of the constituent ions, the lattice energies of the WCA salts are so low that they approach the values of sublimation enthalpies of molecular solids of comparable atomic weight, as can clearly be seen by comparing the lattice energy of $[Ag(C_2H_4)_3]^+[Al\{OC(CF_3)_3\}_4]^-$ (347 kJ mol^{-1}) with the lattice energies of typical salts such as Li$^+$F$^-$ (1036 kJ mol^{-1}) and Cs$^+$F$^-$ (740 kJ mol^{-1}). Moreover, large WCAs have diameters on the nanometer scale, e.g. 1.25 nm for $[Al(OR)_4]^-$ or 1.20 nm for $[Sb(OTeF_5)_6]^-$, and thus considerably separate anions and cations, which effectively diminishes coulombic interactions. Due to the highly fluorinated surface of most WCAs, dispersive interactions are weak and are not structure-determining. To account for the special environment of the cations within the framework of an ensemble of large and very weakly coordinating anions such as $Al(OR)_4^-$, $Sb(OTeF_5)_6^-$ or other suitable WCAs, the expression "pseudo gas-phase conditions" in the solid state was coined [24].

"Pseudo gas-phase conditions" are also present in solution: while salts with smaller anions are usually only soluble in polar media with high dielectric constants such as ethanol ($\varepsilon_r=24$), CH$_3$CN ($\varepsilon_r=35$) or even strong acids such as anhydrous HF ($\varepsilon_r=83$), WCA salts are generally very soluble and allow the use of almost non-polar solvents with low dielectric constants ($\varepsilon_r=2$ to 9; toluene to CH$_2$Cl$_2$). This high solubility of WCA salts obviously results from their decreased lattice potential enthalpies.

5.5.1
Stabilization of Weakly Bound Complexes

Apart from being weakly basic and stable to oxidation, another stabilizing term arises for weak Lewis acid-base adducts M(L)$_n^+$[X]$^-$ by using large counterions (M$^+$=univalent cation, L=non-charged weakly bound ligand, [X]$^-$=suitable anion), i.e., the reduced gain in M$^+$[X]$^-$ lattice energy upon dissociation of M(L)$_n^+$[X]$^-$ into M$^+$[X]$^-$ and n L.

Let us explicitly consider two cases for the recently prepared $[Ag(\eta^2\text{-}C_2H_4)_3]^+$ cation (Fig. 5.8): L=C$_2$H$_4$ and X$^-$=BF$_4^-$ or X$^-$=Al(OR)$_4^-$ (R=C(CF$_3$)$_3$). The calculated lattice potential enthalpies (thermochemical volumes) [56] for $[Ag(C_2H_4)_3]^+$ [X]$^-$ are 498 and 347 kJ mol^{-1}; those of Ag$^+$[X]$^-$ are 655 and 361 kJ mol^{-1}. The gas-phase enthalpy Δ_rH(gas) of C$_2$H$_4$ is the same in both cases, and, therefore, the only differences are due to the lattice energies. For [X]$^-$=[BF$_4$]$^-$, the resulting gain in lattice energy upon dissociation is 157 kJ mol^{-1}, while that for [X]$^-$=[Al(OR)$_4$]$^-$ is only 14 kJ mol^{-1}. Therefore, $[Ag(C_2H_4)_3]^+[Al(OR)_4]^-$ is more stable towards dissociation than the [BF$_4$]$^-$ salt by 143 kJ mol^{-1} (Scheme 5.6). From Scheme 5.6, it is not surprising that [Ag(C$_2$H$_4$)$_3$][BF$_4$], with the comparatively small [BF$_4$]$^-$ anion, is only stable up to 0 °C in an atmosphere of C$_2$H$_4$ [36].

Only Ag(η^2-C$_2$H$_4$)$_x^+$ complexes with $x=1, 2$ are known in the gas phase (MS) [37]. The formation and structural characterization of an $[Ag(\eta^2\text{-}C_2H_4)_3]^+[Al(OR)_4]^-$ salt with $x=3$ was therefore surprising and shows that the environment provided by the solid-state arrangement of [Al(OR)$_4$]$^-$ anions is very close to that of the gas phase in thermal equilibrium at low temperature. The gas-phase addition of a third molecule of gaseous C$_2$H$_4$ to gaseous Ag(η^2-C$_2$H$_4$)$_2^+$ was impossible in the

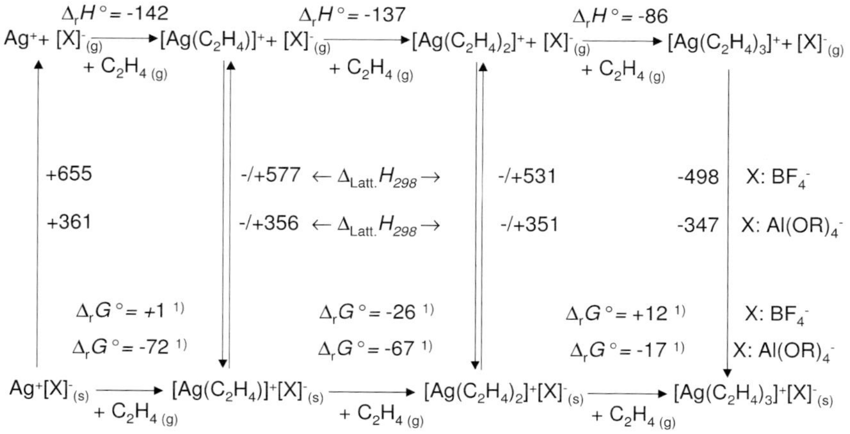

$\Delta_r H° = -142$ $\Delta_r H° = -137$ $\Delta_r H° = -86$

$Ag^+ + [X]^-_{(g)} \longrightarrow [Ag(C_2H_4)]^+ + [X]^-_{(g)} \longrightarrow [Ag(C_2H_4)_2]^+ + [X]^-_{(g)} \longrightarrow [Ag(C_2H_4)_3]^+ + [X]^-_{(g)}$
$+ C_2H_{4(g)}$ $+ C_2H_{4(g)}$ $+ C_2H_{4(g)}$

| +655 | -/+577 ← $\Delta_{Latt.}H_{298}$ → | -/+531 | -498 | X: BF_4^- |
| +361 | -/+356 ← $\Delta_{Latt.}H_{298}$ → | -/+351 | -347 | X: $Al(OR)_4^-$ |

| $\Delta_r G° = +1$ [1)] | $\Delta_r G° = -26$ [1)] | $\Delta_r G° = +12$ [1)] | X: BF_4^- |
| $\Delta_r G° = -72$ [1)] | $\Delta_r G° = -67$ [1)] | $\Delta_r G° = -17$ [1)] | X: $Al(OR)_4^-$ |

$Ag^+[X]^-_{(s)} \longrightarrow [Ag(C_2H_4)]^+[X]^-_{(s)} \longrightarrow [Ag(C_2H_4)_2]^+[X]^-_{(s)} \longrightarrow [Ag(C_2H_4)_3]^+[X]^-_{(s)}$
$+ C_2H_{4(g)}$ $+ C_2H_{4(g)}$ $+ C_2H_{4(g)}$

Scheme 5.6 Born–Fajans–Haber cycle for the formation of $[Ag(\eta^2-C_2H_4)_n]^+$ complexes; all energies given in kJ mol^{-1}; [1)] contains 65 kJ mol^{-1} contribution to the Gibbs energy upon removal of free gaseous C_2H_4.

MS, presumably due to the low binding Gibbs energy of the third ligand ($\Delta G° = -55$ kJ mol^{-1}). This points to a failure to dissipate the energy stored in translational, vibrational, and rotational levels, which in sum can be larger than the binding energy of the third ligand thus preventing coordination. In WCA salts, this internal energy can be removed through intermolecular vibrational coupling, thus allowing equilibrium conditions to be reached at a given temperature. In this respect, the pseudo gas-phase conditions provided by the best WCAs in the solid state are even better than the gas phase inside a mass spectrometer, where thermal equilibrium conditions are difficult to reach.

5.5.2
Relative Anion Stabilities Based on Thermodynamic Considerations

One question that is still open to discussion is a reliable ordering of the relative stabilities and coordinating abilities of all types of WCAs known today. Earlier attempts used the ^{29}Si NMR shift of the $Si(iPr)_3^{\delta+}X^{\delta-}$ silylium ion pair (X=WCA) as a measure, with shifts at lower field indicating a higher cationic character of the $Si(iPr)_3^{\delta+}$ part, this being an indication of a less coordinating anion X$^-$. However, inertness and coordinating ability of WCAs do not always go hand in hand and, therefore, the reactive nature of the $Si(iPr)_3^{\delta+}$ part precludes the investigation of many anions that else proved to be very weakly coordinating due to anion decomposition (for example, all fluorometallates and teflate-based anions). In another very recent contribution to this subject, the N–H stretching vibrations of a series of $(n\text{-oct})_3N–H^+[X]^-$ ammonium salts in CCl$_4$ solution (X=WCA) were evaluated on the basis of the following assumption: the higher the frequency of the stretching vibration, the lesser the interaction be-

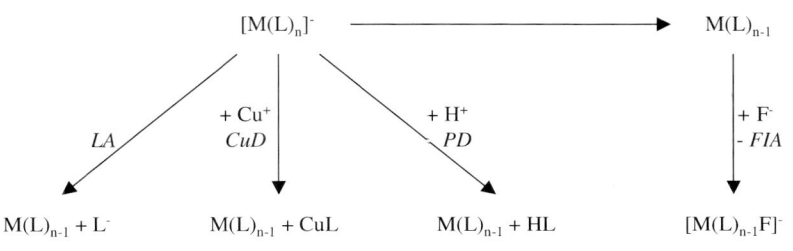

```
                      [M(L)ₙ]⁻  ─────────────────────→  M(L)ₙ₋₁
                     ╱    │    ╲                            │
                    ╱     │     ╲                           │
              LA   ╱  +Cu⁺│  +H⁺ ╲                          │ +F⁻
                  ╱    CuD│   PD  ╲                         │ -FIA
                 ↓        ↓        ↓                        ↓
          M(L)ₙ₋₁+L⁻   M(L)ₙ₋₁+CuL   M(L)ₙ₋₁+HL        [M(L)ₙ₋₁F]⁻
```

Scheme 5.7 Definition of the thermodynamic quantities: ligand affinity (LA), fluoride ion affinity (FIA), proton decomposition (PD), and copper decomposition (CuD). L = F, OTeF$_5$, CF$_3$, C$_6$F$_5$, OC(CF$_3$)$_3$,...; M = B, Al, P, Sb, ...

tween the anion and the cation and the more weakly coordinating is [X]⁻. This scale gives an ordering of the relative coordinating ability of the WCA towards the (n-oct)$_3$N–H⁺ cation in CCl$_4$ solution but still awaits full publication [57]. To investigate the stabilities of fluorometallate anions such as BF$_4^-$, MF$_6^-$ (M = P, As, Sb,...), the fluoride ion affinities (FIAs) of their parent Lewis acids A, i.e. BF$_3$, MF$_5$, etc., were estimated on thermodynamic grounds [58]. The higher the FIA of the parent Lewis acid A of a given WCA, the more stable it is against decomposition on thermodynamic grounds [59]. Starting with the calculation of the FIA, one can obtain a larger relative WCA stability scale by using an entirely computational approach (Scheme 5.7) [60]. Using the approach delineated in Scheme 5.7, one is not limited to experimental data in order to compare the properties of very different types of WCAs such as the fluoroantimonates and the perfluoroarylborates. With the calculated structures and data, the thermodynamic stabilities and coordinating abilities of the WCAs of type [M(L)$_n$]⁻ were established based on the ligand affinity LA of the parent Lewis Acid A = M(L)$_{n-1}$, the FIA of the parent Lewis acid A as presented above, and the decomposition of a given anion in the presence of a hard (H⁺, proton decomposition PD) and a soft electrophile (Cu⁺, copper decomposition CuD) as shown in Scheme 5.7.

The calculations presented above were used for a variety of different WCAs and their values are collected in Tab. 5.2. While these data may not be taken as absolute, since the same methods were used for all computations the relative trends will definitely be correct [60].

From Tab. 5.2 emerges the outstanding capability of the [Sb$_4$F$_{21}$]⁻ and [Sb(OTeF$_5$)$_6$]⁻ WCAs to stabilize highly oxidizing cations in anhydrous HF solution (see FIA, PD). This stability versus oxidation has to be traded in for sensitivity towards reduction and moisture as well as an increased coordinating ability. For the borate-based anions, fluorination greatly increases the thermodynamic stability of all fluorinated borates compared to the non-fluorinated parent [B(C$_6$H$_5$)$_4$]⁻ anion (see FIA, LA, PD, CuD). The differences between the commercially available [B(C$_6$F$_5$)$_4$]⁻ and [B{C$_6$H$_3$(CF$_3$)$_2$}$_4$]⁻ WCAs are small. Of all borates, the novel [B(CF$_3$)$_4$]⁻ anion is the best. However, recent reports have revealed that [B(CF$_3$)$_4$]⁻ decomposes via another pathway. It is not stable in the

Tab. 5.2 Calculated properties of WCAs: The FIA of the parent Lewis acid and the LA, PD, and CuD of the WCA.

Anion	Sy.	FIA	LA	PD	CuD
$[BF_4]^-$	T_d	338	[a]	−1212	−521
$[SbF_6]^-$	D_{4h}	489	[a]	−1061	−371
$[Sb_4F_{21}]^-$ vs. Sb_4F_{20}	C_{2v}	584	[a]	−991	−301
$[B(OTeF_5)_4]^-$	C_1	550	274	−1040	−420
$[Sb(OTeF_5)_6]^-$	C_3	633	341	−973	−353
$[Al(OR)_4]^-$ (R=C(CF$_3$)$_3$)	S_4	537	342	−1081	−395
$[(RO)_3Al-F-Al(OR)_3]^-$ (R=C(CF$_3$)$_3$) [b, c]	C_i	685 [b]	441 (363)	−983 (−1061 [c])	−297 (−375 [c])
$[B\{C_6H_3(CF_3)_2\}_4]^-$	S_4	471	382	−1251	−506
$[B(C_6F_5)_4]^-$	S_4	444	296	−1256	−538
$[B(CF_3)_4]^-$	T	552	490	−1136	−379

a) LA and FIA are identical;
b) FIA vs. 2 Al(OR)$_3$;
c) Values in parentheses correspond to the formation of the Al$_2$(OR)$_5$F isomer I.

presence of AsF$_5$ or SiR$_3^+$ ions [61]. The stability of the perfluoroalkoxyaluminate $[Al(OR)_4]^-$ (R=C(CF$_3$)$_2$) with respect to FIA, PD, and CuD in Tab. 5.2 is remarkable and higher than that of all borates except the $[B(CF_3)_4]^-$ anion, which has comparable values. The $[Al(OR)_4]^-$ anion even comes close to the oxidation resistance and low PDs of the fluoroantimonates and in some respects is even better than the teflate-based anions [60].

5.6
Conclusions and Outlook

This article presents a cross-section through the work with WCAs of type $[Al(OR)_4]^-$ (R=C(H)(CF$_3$)$_2$, C(CH$_3$)(CF$_3$)$_2$, C(CF$_3$)$_3$) and highlights their possibilities as well as limitations. It could be demonstrated that starting with the facile preparation of Li[Al(OR)$_4$] several metal or non-metal cations can be introduced into the system in high yields. This then allowed the synthesis of various starting materials for different challenges in WCA chemistry. Using weakly coordinating fluorinated alkoxyaluminate anions, it is possible to prepare and characterize several very weak Lewis acid–base complexes of the silver cation in the condensed phase, e.g. silver adducts of molecular S$_8$, P$_4$S$_3$, P$_4$ or C$_2$H$_4$. Most of these cations were initially observed in the gas phase by mass spectrometry. Now stable salts of such species are known and, therefore, present a bridge between gas-phase and condensed-phase chemistry. Moreover, it has been possible to stabilize a variety of highly electrophilic cations, such as PX_4^+, $P_2X_5^+$, $P_5X_2^+$, $P_5S_2X_2^+$, $P_5S_3X_2^+$, $P_3I_6^+$, $CS_2Br_3^+$, and CX_3^+. All of the phosphorus-based cations were derived from PX_2^+ and, similarly to CX_3^+ [62], may be viewed as gas-phase cations [43]. Thus, we may conclude by stating that gas-phase cations may be stabilized in the condensed phase, if suitable weakly coordinating and chemically robust counterions are introduced.

Nevertheless, it is absolutely necessary to develop new, easily accessible, chemically robust but very weakly coordinating anions. The fluoride-bridged WCA $[(RO)_3Al–F–Al(OR)_3]^-$ (R = C(CF$_3$)$_3$) provides an example of a compound that could extend the solid-state stability of some reactive cation salts unstable with the homoleptic $[Al(OR)_4]^-$ WCA [44c, 60, 63]. Our long-term goal in this direction is the preparation of the least coordinating and chemically most robust WCAs at the cheapest price. Time alone will tell which application or simple but fundamental cations may prosper from the availability of such special anions.

References

1. For recent reviews, see: a) C. REED, *Acc. Chem. Res.* **1998**, *31*, 133; b) S. H. STRAUSS, *Chem. Rev.* **1993**, *93*, 927; c) I. KROSSING, I. RAABE, *Angew. Chem.* **2004**, in press; E. Y.-X. CHEN, T. J. MARKS, *Chem. Rev.* **2000**, *100*, 1391.
2. a) J. PLESEK, T. JELINEK, S. HERMANEK, B. STIBR, *Collect. Czech. Chem. Commun.* **1986**, *51*, 819; b) C.-W. TSANG, Q. YANG, E. TUNG-PO SZE, T. C. W. MAK, D. T. W. CHAN, Z. XIE, *Inorg. Chem.* **2000**, *39*, 5851; c) Z. XIE, C.-W. TSANG, E. TUNG-PO SZE, Q. YANG, D. T. W. CHAN, T. C. W. MAK, *Inorg. Chem.* **1998**, *37*, 6444; d) Z. XIE, C.-W. TSANG, F. XUE, T. C. W. MAK, *J. Organomet. Chem.* **1999**, *577*, 197; e) B. T. KING, Z. JANOUSEK, B. GRÜNER, M. TRAMELL, B. C. NOLL, J. MICHL, *J. Am. Chem. Soc.* **1996**, *118*, 3313.
3. A. G. MASSEY, A. J. PARK, *J. Organomet. Chem.* **1964**, *2*, 245.
4. a) J. H. GOLDEN, P. F. MUTOLO, E. B. LOBROVSKI, F. J. DISALVO, *Inorg. Chem.* **1994**, *33*, 5374; b) K. FUJIKI, S. IKEDA, H. KOBAYASHI, A. MORI, A. NAGIRA, J. NIE, T. SONODA, Y. YAGUPOLSKII, *Chem. Lett.* **2000**, 66.
5. D. M. VAN SEGGAN, P. K. HURLBURT, M. D. NOIROT, O. P. ANDERSON, S. H. STRAUSS, *Inorg. Chem.* **1992**, *31*, 1423.
6. E(OTeF$_5$)$_6^-$ (E = As, Sb, Bi): H. P. A. MERCIER, J. C. P. SAUNDERS, G. T. SCHROBILGEN, *J. Am. Chem. Soc.* **1994**, *116*, 2921.
7. E(OTeF$_5$)$_6^-$ (E = Sb, Nb): D. M. VAN SEGGEN, P. K. HURLBURT, O. P. ANDERSON, S. H. STRAUSS, *Inorg. Chem.* **1995**, *34*, 3453.
8. Sb(OTeF$_5$)$_6^-$: T. S. CAMERON, I. KROSSING, J. PASSMORE, *Inorg. Chem.* **2001**, *40*, 2001.
9. E(OTeF$_5$)$_6^-$ (E = As, Sb, Bi): H. P. A. MERCIER, J. C. P. SAUNDERS, G. T. SCHROBILGEN, *J. Am. Chem. Soc.* **1994**, *116*, 2921.
10. E(OTeF$_5$)$_6^-$ (E = Sb, Nb): D. M. VAN SEGGEN, P. K. HURLBURT, O. P. ANDERSON, S. H. STRAUSS, *Inorg. Chem.* **1995**, *34*, 3453.
11. K. MOOCK, K. SEPPELT, *Z. Anorg. Allg. Chem.* **1988**, *561*, 132.
12. S. M. IVANOVA, B. G. NOLAN, Y. KOBAYASHI, S. M. MILLER, O. P. ANDERSON, S. H. STRAUSS, *Chem. Eur. J.* **2001**, *7*, 503.
13. I. KROSSING, *Chem. Eur. J.* **2001**, *7*, 490.
14. I. KROSSING, H. BRANDS, R. FEUERHAKE, S. KOENIG, *J. Fluor. Chem.* **2001**, *112*, 83.
15. M. GONSIOR, I. KROSSING, N. MITZEL, *Z. Anorg. Allg. Chem.* **2002**, *628*, 1821.
16. I. KROSSING, A. REISINGER, *Eur. J. Inorg. Chem.*, in press.
17. M. GONSIOR, I. KROSSING, I. RAABE, publication in progress.

18 J. Passmore et al., submitted to Dalton Trans.
19 I. Krossing, I. Raabe, publication in progress.
20 T. J. Barbarich, S. T. Handy, S. M. Miller, O. P. Anderson, P. A. Grieco, S. H. Strauss, Organometallics 1996, 15, 3776.
21 S. H. Strauss, B. G. Nolan, B. P. Fauber, International Patent WO 00/53611 of September 2000.
22 A. Bösmann, G. Francio, E. Janssen, W. Leitner, P. Wasserscheid, Angew. Chem. 2001, 113, 2769; Angew. Chem. Int. Ed. 2001, 40, 2697.
23 a) H. W. Roesky, M. Thomas, J. Schimkowiak, P. G. Jones, W. Pinkert, G. M. Sheldrick, J. Chem. Soc., Chem. Commun. 1982, 895; b) H. W. Roesky, M. Witt, Inorg. Synth. 1986, 24, 72.
24 T. S. Cameron, A. Decken, I. Dionne, M. Fang, I. Krossing, J. Passmore, Chem. Eur. J. 2002, 8, 3386.
25 a) M. di Vaira, M. Peruzzini, P. Stoppioni, Inorg. Chem. 1983, 22, 2196; b) A. W. Cordes, R. D. Joyner, R. D. Shores, Inorg. Chem. 1974, 13, 132; c) C. Aubauer, E. Irran, T. M. Klapötke, W. Schnick, A. Schulz, J. Senker, Inorg. Chem. 2001, 40, 4956.
26 Reviews: a) M. di Vaira, P. Stoppioni, Coord. Chem. Rev. 1992, 120, 259; b) J. Wachter, Angew. Chem. 1998, 110, 8782; Angew. Chem. Int. Ed. 1998, 37, 750; c) J. O. Scherer, Chem. in unserer Zeit 2000, 374; d) K. H. Whitmire, Adv. Organomet. Chem. 1998, 42, 2.
27 a) L. Yoong Goh, W. Chen, R. C. S. Wong, Angew. Chem. 1993, 105, 1838; Angew. Chem. Int. Ed. Engl. 1993, 32, 1728; b) L. Yoong Goh, W. Chen, R. C. S. Wong, K. Karaghiosoff, Organometallics 1995, 14, 3886.
28 A. Adolf, M. Gonsior, I. Krossing, J. Am. Chem. Soc. 2002, 124, 7111.
29 a) M. Di Vaira, I. de los Rios, F. Mani, M. Peruzzini, P. Stoppioni, Eur. J. Inorg. Chem. 2004, 2, 293; b) I. de los Rios, F. Mani, M. Peruzzini, P. Stoppioni, J. Organomet. Chem. 2004, 689(1), 164.
30 a) I. Krossing, J. Am. Chem. Soc. 2001, 123, 4603; I. Krossing, L. v. Wüllen, Chem. Eur. J. 2002, 8, 700.

31 J.-L. M. Abboud, M. Herreros, R. Notario, M. Esseffar, O. Mo, M. Yanez, J. Am. Chem. Soc. 1996, 118(5), 1126.
32 D. V. Deubel, J. Am. Chem. Soc. 2002, 124, 12312.
33 Crystal structure: M. Black, R. H. B. Mais, P. G. Owston, Acta Crystallogr. Sect. B 1969, 25, 1753.
34 a) M. J. S. Dewar, Bull. Soc. Chim. Fr. 1951, 79, 18; b) J. Chatt, L. A. Duncanson, J. Chem. Soc. 1953, 2939.
35 a) T. Ziegler, A. Rauk, Inorg. Chem. 1979, 18, 1558; b) J. C. Ma, D. A. Dougherty, Chem. Rev. 1997, 97, 1303; c) G. W. Gokel, S. L. De Wall, E. S. Meadows, Eur. J. Org. Chem. 2000, 2967; d) J. Kaneti, L. de Smet, R. Boom, H. Zuilhof, E. Sudhölter, J. Phys. Chem. A 2002, 106, 11197.
36 a) H. W. Quinn, J. S. McIntyre, Can. J. Chem. 1965, 43, 2896; b) D. B. Powell, J. G. V. Scott, N. Sheppard, Spectrochim. Acta A 1972, 28, 327; c) H. W. Quinn, D. N. Glew, Can. J. Chem. 1962, 40, 1103.
37 a) B. C. Guo, A. W. Castleman, Jr., Chem. Phys. Lett. 1991, 181, 16; b) D. Schröder, R. Wesendrup, R. H. Hertwig, T. K. Dargel, H. Grauel, W. Koch, B. R. Bender, H. Schwarz, Organometallics 2000, 19, 2608.
38 I. Krossing, A. Reisinger, Angew. Chem. 2003, 115, 5903; Angew. Chem. Int. Ed. 2003, 42, 5725.
39 a) S. Brownridge, H. D. B. Jenkins, I. Krossing, J. Passmore, H. K. Roobottom, Coord. Chem. Rev. 2000, 197, 397; b) I.-C. Hwang, K. Seppelt, Angew. Chem. 2001, 113, 3803; Angew. Chem. Int. Ed. 2001, 40, 3690.
40 a) J. Passmore, J. Chem. Soc., Dalton Trans. 1978, 1251; b) C. Aubauer, M. Kaupp, T. M. Klapötke, H. Nöth, H. Pietrowski, W. Schnick, J. Senker, J. Chem. Soc., Dalton Trans. 2001, 1880, and references therein.
41 a) S. Pohl, Z. Anorg. Allg. Chem. 1983, 498, 20; b) C. Aubauer, G. Engelhardt, T. M. Klapötke, A. Schulz, J. Chem. Soc., Dalton Trans. 1999, 1729; c) C. Aubauer, T. M. Klapötke, P. M. Mayer, Acta Cryst. 2001, E57, i1.
42 K. B. Dillon, B. Y. Xue, Inorg. Chim. Acta 2001, 320(1,2), 172.

43 a) D. Gudat, *Eur. J. Inorg. Chem.* **1998**, 1087; b) O. A. Mazyar, T. Baer, *Chem. Phys. Lett.* **1998**, *288*, 327; c) L. Latifzadeh-Masoudipour, K. Balasubramanian, *Chem. Phys. Lett.* **1997**, *267*, 545; L. Latifzadeh, K. Balasubramanian, *Chem. Phys. Lett.* **1996**, *258*, 393; L. Latifzadeh, K. Balasubramanian, *Chem. Phys. Lett.* **1995**, *241*, 13.
44 a) M. Gonsior, I. Krossing, L. Müller, I. Raabe, M. Jansen, L. v. Wüllen, *Chem. Eur. J.* **2002**, *8*, 4475; b) I. Krossing, I. Raabe, *Angew. Chem.* **2001**, *113*, 4544; *Angew. Chem. Int. Ed.* **2001**, *40*, 4406; c) M. Gonsior, I. Raabe, A. Bihlmeier, N. Trapp, I. Krossing, *Chem. Eur. J.* **2004**, *10*, 5041.
45 I. Krossing, *J. Chem. Soc., Dalton Trans.* **2002**, 500.
46 a) M. Gonsior, I. Krossing, publication in progress; b) M. Gonsior, I. Krossing, Poster presented at the GDCh meeting in Munich, October 2003.
47 S. Haupt, K. Seppelt, *Z. Anorg. Allg. Chem.* **2002**, *628(4)*, 729.
48 M. Gonsior, I. Krossing, publication in progress.
49 M. Gonsior, I. Krossing, *Chem. Eur. J.* **2004**, *10*, 5730.
50 K. O. Christe, X. Zhang, R. Bau, J. Hegge, G. A. Olah, G. K. S. Prakash, J. A. Sheehy, *J. Am. Chem. Soc.* **2000**, *122*, 481.
51 a) R. Minkwitz, S. Schneider, *Angew. Chem.* **1999**, *111*, 749; *Angew. Chem. Int. Ed.* **1999**, *38*, 714; b) R. Minkwitz, S. Reinemann, O. Becher, H. Hartl, I. Brüdgam, *Inorg. Chem.* **1999**, *38*, 844.
52 a) G. A. Olah, G. Rasul, L. Heiliger, G. K. S. Prakash, *J. Am. Chem. Soc.* **1996**, *118*, 3580; b) G. A. Olah, L. Heiliger, G. K. S. Prakash, *J. Am. Chem. Soc.* **1989**, *111*, 8020; c) H. Vancik, K. Percac, D. E. Sunko, *J. Am. Chem. Soc.* **1990**, *112*, 7418; d) J. W. Hudgens, R. D. Johnson III, B. P. Tsai, S. A. Kafafi, *J. Am. Chem. Soc.* **1990**, *112*, 5763;

e) G. A. Olah, Y. K. Mo, E. G. Melby, H. C. Lin, *J. Org. Chem.* **1973**, *38*, 367; f) G. A. Olah, G. Rasul, A. K. Yudin, A. Burrichter, G. K. S. Prakash, A. L. Chistyakov, A. V. Stankevich, I. S. Akhrem, N. P. Gambaryan, M. E. Vol'pin, *J. Am. Chem. Soc.* **1996**, *118*, 1446.
53 G. Frenking, S. Fau, C. M. Marchand, H. Grützmacher, *J. Am. Chem. Soc.* **1997**, *119*, 6648.
54 I. Krossing, A. Bihlmeier, I. Raabe, N. Trapp, *Angew. Chem.* **2003**, *115*, 1569; *Angew. Chem. Int. Ed.* **2003**, *42*, 1531.
55 a) I. Krossing, I. Raabe, *J. Am. Chem. Soc.* **2004**, *126*, 7571; b) I. Krossing, I. Raabe, publication in progress.
56 H. K. Roobottom, H. D. B. Jenkins, J. Passmore, L. Glasser, *Inorg. Chem.* **1999**, *38*, 3609.
57 C. A. Reed, ACS Meeting, New Orleans, 2003.
58 a) T. E. Mallouk, G. L. Rosenthal, G. Muller, R. Brusasco, N. Bartlett, *Inorg. Chem.* **1984**, *23*, 3167; b) H. K. Roobottom, H. D. B. Jenkins, J. Passmore, L. Glasser, *Inorg. Chem.* **1999**, *38*, 3609; H. D. B. Jenkins, H. K. Roobottom, J. Passmore, *Inorg. Chem.* **2003**, *42*, 2886.
59 K. O. Christe, D. A. Dixon, D. McLemore, W. W. Wilson, J. Sheehy, J. A. Bootz, *J. Fluor. Chem.* **2000**, *101*, 151.
60 I. Krossing, I. Raabe, *Chem. Eur. J.* **2004**, *10*, 5017.
61 H. Willner, ACS Meeting, New Orleans, 2003.
62 J.-L. M. Abboud, O. Castafio, J. Elguero, M. Herreros, N. Jagerovic, R. Notario, K. Sak, *Int. J. Mass Spectrom. Ion Proc.* **1998**, *175*, 35.
63 M. Gonsior, I. Krossing, L. O. Müller, German patent application DE10356768.2.
64 H. P. A. Mercier, M. D. Moran, G. J. Schrobilgen, C. Steinberg, R. Suontamo, *J. Am. Chem. Soc.* **2004**, *126*, 5533–5548.

6
Aluminum(I) Chemistry

Herbert W. Roesky

6.1
Introduction

The most abundant elements in the Earth's crust are oxygen (50%), silicon (25%), and aluminum (7.4%). Aluminum is a major constituent of many common minerals, such as feldspars and micas. Aluminum never occurs free in nature; it is mostly bound to oxygen or, in the case of cryolite (Na_3AlF_6), to fluorine. Aluminum metal is produced in large amounts by the Hall-Héroult process through electrolysis of Al_2O_3 dissolved in cryolite [1].

The vast majority of aluminum compounds contain aluminum in the formal oxidation state +III. A comparison of the atomic radii of the group 13 elements indicates that there is a tremendous increase on going from boron (0.98 Å) to aluminum (1.43 Å), whereas that of gallium (1.41 Å) is of the same order of magnitude as that of aluminum. This phenomenon is also reflected in the first ionization energies, for which the values are B 8.30, Al 5.98, and Ga 6.00 eV, respectively. In general, aluminum uses all its valence electrons ($3s^2$, $3p^1$) for its trivalent chemistry. However, the difference between the first (5.98 eV) and second ionization energies (18.82 eV) of aluminum allows the development of the chemistry of aluminum(I).

6.2
Preparation of Aluminum(I) Halides

The pursuit of compounds of low-valent aluminum has a long history of over 50 years. As long ago as 1948, Klemm et al. [2] reported on the reaction of elemental chlorine with aluminum at 1000 °C to obtain AlCl.

An alternative method for the preparation of AlCl involves the use of aluminum and $AlCl_3$ under reduced pressure at high temperatures (Eq. (1)).

$$2Al_{(s)} + AlCl_{3(g)} \underset{600\,°C}{\overset{1200\,°C}{\rightleftarrows}} 3AlCl(g) \quad (1)$$

Inorganic Chemistry in Focus II. Edited by G. Meyer, D. Naumann, and L. Wesemann
Copyright © 2005 WILEY-VCH Verlag GmbH & Co. KGaA, Weinheim
ISBN: 3-527-30811-3

This method is preferentially applied for the preparation of high purity aluminum, due to the reverse reaction at 600 °C. This route can also be used for the preparation of AlF and AlBr.

In 1989, Tacke and Schnöckel [3] reported on a high yield synthesis of AlCl starting from aluminum and hydrogen chloride at low pressure (<0.2 mbar) (Eq. (2)).

$$Al_{(s)} + HCl_{(g)} \xrightarrow{1200 K} AlCl_{(g)} + \tfrac{1}{2} H_2 \qquad (2)$$

The yield of AlCl exceeds 90%. The IR spectrum of AlCl shows a broad absorption with a maximum at 320 cm^{-1} and a half-width of about 100 cm^{-1}. At about 180 K, the vitreous AlCl on the cold finger changes to black under formation of aluminum and aluminum trichloride (Eq. (3)).

$$3AlCl_{(s)} \rightarrow AlCl_{3(s)} + 2Al_{(s)} \qquad (3)$$

However, when AlCl is co-condensed at 77 K with Et$_2$O in toluene, an adduct of the composition AlCl·xEt$_2$O is formed. In solution, this adduct is stable at up to room temperature for a few hours. It is quite obvious that the stability of the adduct AlCl·xEt$_2$O is increased in comparison to solutions of AlCl in weakly coordinating solvents such as toluene.

Recently, AlF was prepared from Al and CHF$_3$ and trapped in solid argon matrices [4]. In summary, the monomeric aluminum halides AlX (X = F, Cl, Br, I) are known. Adducts with donor molecules increase the stability of these monovalent aluminum halides.

6.3
Reactions of Aluminum(I) Halides

In solid argon matrices, HCl and HBr form adducts with AlF and AlCl. The products, AlF·HCl, AlF·HBr, and AlCl·HBr, are intermediates in photoactivation processes yielding hydrides of the compositions HAlFCl, HAlFBr, and HAlClBr, respectively [4]. Photolytic reactions of AlX (X = F, Cl, Br) with O$_2$ in solid argon matrices give the peroxo species XAlO$_2$ and XAl(O$_2$)$_2$ as the main products. The presence of (AlX)$_2$ as a by-product in this system yields O$_2$XAl(μ-O)$_2$AlX [5]. The photoactivation has also been used in the reaction of AlCl with H$_2$ in a solid argon matrix to give ClAlH$_2$ [6].

The adduct of AlCl with a donor turned out to be a very useful system for the preparation of aluminum(I) compounds. In Scheme 6.1, some representative examples of aluminum(I) halides are shown. A summary of these reactions is given in a recent review article [7].

It is worth mentioning that reactions of AlI with LiN(SiMe$_3$)$_2$ result in the large cluster dianion of the composition [{(Me$_3$Si)$_2$N}$_{20}$Al$_{77}$]$^{2-}$ (1) [8], while the reaction of t-BuLi with AlCl followed by reduction gives the stable radical [t-BuAl]$_6^{\bullet}$ (7) [9]. Moreover, it has been reported that the use of AlCl instead of AlI in the reaction

6.3 Reactions of Aluminum(I) Halides

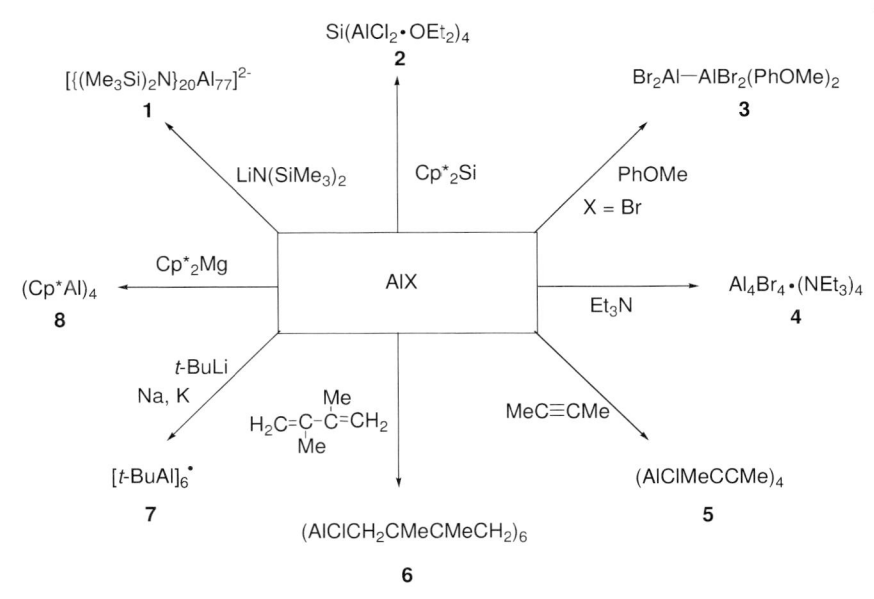

Scheme 6.1 Products from the reaction of AlX with various substrates.

[{(Me$_3$Si)$_2$N}$_{20}$Al$_{77}$]$^{2-}$ **1**

Si(AlCl$_2$·OEt$_2$)$_4$ **2**

Br$_2$Al—AlBr$_2$(PhOMe)$_2$ **3**

Al$_4$Br$_4$·(NEt$_3$)$_4$ **4**

(AlClMeCCMe)$_4$ **5**

(AlClCH$_2$CMeCMeCH$_2$)$_6$ **6**

[t-BuAl]$_6^{\bullet}$ **7**

(Cp*Al)$_4$ **8**

Reagents: LiN(SiMe$_3$)$_2$; Cp*$_2$Si; PhOMe (X = Br); Et$_3$N; MeC≡CMe; H$_2$C=C-C=CH$_2$ (Me, Me); t-BuLi, Na, K; Cp*$_2$Mg

with LiN(SiMe$_3$)$_2$ results in an [Al$_{69}${N(SiMe$_3$)$_2$}$_{18}$]$^{3-}$ cluster [10]. Evidently, very small changes in the reaction conditions yield different products.

A rather surprising result is the product of the reaction of Cp*$_2$Si with AlCl in the presence of Et$_2$O. Compound **2** contains four Si–Al bonds and each of the four AlCl$_2$ groups is further coordinated by a diethyl ether molecule [11]. Compound **4** forms a four-membered Al$_4$ ring with terminally arranged bromine atoms and NEt$_3$ molecules (Fig. 6.1) [11, 12].

Particularly interesting is the reaction of AlCl with MeC ≡ CMe yielding two isomers of **5** with an Al$_4$C$_8$ framework [13]. The reaction may proceed according to the route given in Scheme 6.2.

The most spectacular compound of Scheme 6.1 is (Cp*Al)$_4$ (**8**), prepared from Cp*$_2$Mg and AlCl [14–16]. The molecular structure of **8** is shown in Fig. 6.2.

An alternative method for the preparation of (Cp*Al)$_4$ (**8**) is the reduction of [Cp*AlX(μ-X)]$_2$ with K metal or Na/K alloy. The starting material [Cp*AlX(μ-X)]$_2$ (X = Cl, Br, I) is prepared from Cp*SiMe$_3$ and AlX$_3$. The synthetic strategy is shown in Scheme 6.3.

Previously, the author's group has reported on the reduction of **9** with potassium to give (Cp*Al)$_4$ **8** in 20–28% yield [17]. However, when compounds **9–11** were reduced with Na/K alloy, the formation of **8** in respective yields of 38%, 48%, and 53% was observed. The decreasing aluminum-halide bond energy in the series AlCl (331 kJ mol^{-1}) > AlBr (256 kJ mol^{-1}) > AlI (172 kJ mol^{-1}) results in higher yields of **8** using the iodide in the reduction process [18]. Less bulky substituents than Cp* on aluminum(I) compounds give rather unstable derivatives.

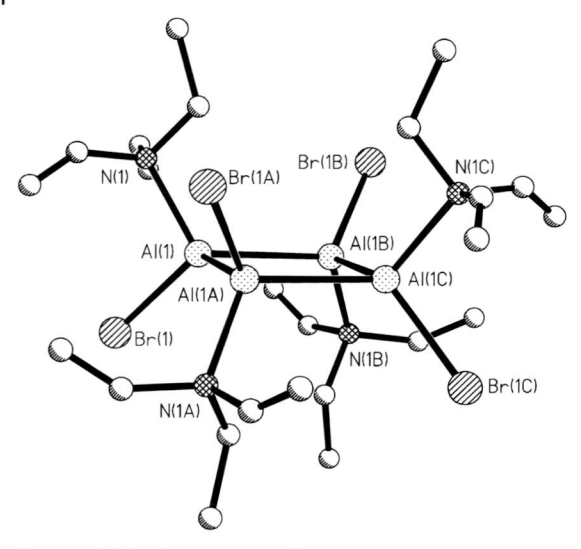

Fig. 6.1 Molecular structure of **4**.

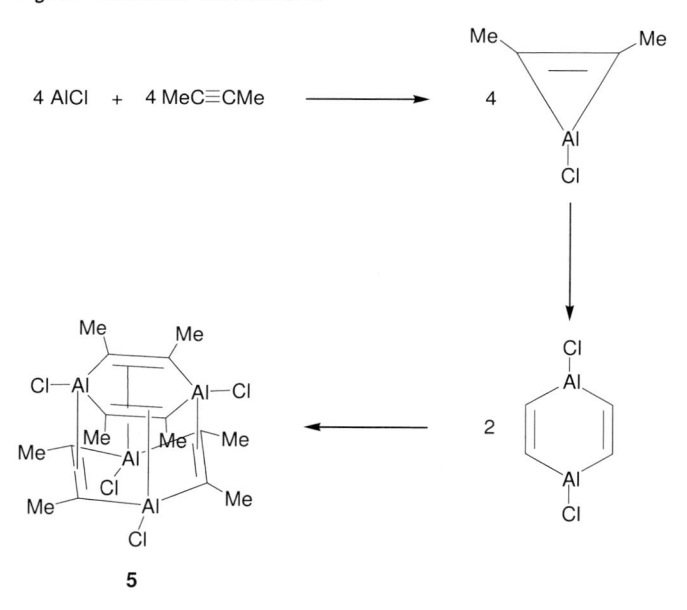

Scheme 6.2 Synthesis of **5**.

Schwarz et al. [19] have provided spectroscopic evidence for MeAl, which has only transient existence in the gas phase. Phenylaluminum, PhAl, has not been isolated but has been invoked as a feasible intermediate in the photolysis and other organic transformations of triphenylaluminum [20]. Schnöckel et al. have tried various routes to synthesize 'CpAl', but with only limited success [21]. This

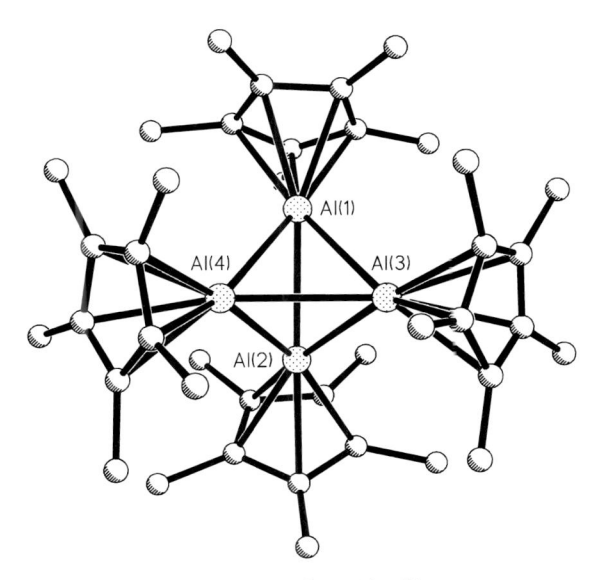

Fig. 6.2 Molecular structure of (Cp*Al)$_4$ (**8**).

$$2\,Cp^*SiMe_3 + 2\,AlX_3 \xrightarrow{-2\,Me_3SiX} \text{[Cp*Al(}\mu\text{-X)}_2\text{AlCp*]}$$

X = Cl **9**; Br **10**; I **11**

Scheme 6.3 Preparation of the starting materials for the synthesis of (Cp*Al)$_4$.

species, for which ^{27}Al NMR spectral evidence could be obtained [22], proved to be extremely thermally sensitive and decomposed within seconds above $-60\,°C$. In contrast to CpAl, the monomeric Cp*Al does exist and has been characterized by ^{27}Al NMR spectroscopy [22]. Cp*Al is known to equilibrate in solution with its tetramer [16]. A few reactions also provide supportive evidence for the existence of monomeric Cp*Al.

In addition to the stable yellow colored (Cp*Al)$_4$ (**8**), (Me$_3$Si)$_3$C-substituted Al$_4$ (**12**) was obtained by the reduction of (Me$_3$Si)$_3$CAlI$_2$. In contrast to **8**, which has η^5-coordination, [(Me$_3$Si)$_3$CAl]$_4$ (**12**) has Al-C σ-bonds. An analogous compound was reported to be formed upon reduction of (Me$_3$CCH$_2$)$_2$AlCl with potassium, which led to the tetramer [(Me$_3$CCH$_2$)Al]$_4$ (**13**) and (Me$_3$CCH$_2$)$_3$Al [23] (Eq. (4)).

$$2(Me_3CCH_2)_2AlCl + 2K \rightarrow \tfrac{1}{4}[(Me_3CCH_2)Al]_4 + (Me_3CCH_2)_3Al \quad (4)$$

Al$_4$ clusters with Al-Si bonds have been obtained from AlI · NEt$_3$ and t-Bu$_3$SiNa (Eq. (5)) [24].

$$4\text{AlI} \cdot \text{NEt}_3 + 4t\text{-Bu}_3\text{SiNa} \rightarrow [t\text{-Bu}_3\text{SiAl}]_4 + 4\text{NaI} + 4\text{Et}_3\text{N} \tag{5}$$
$$\mathbf{14}$$

The corresponding adduct AlBr · NEt$_3$ reacts with (Me$_3$Si)$_3$SiLi · 3THF to give [(Me$_3$Si)$_3$SiAl]$_4$ (**15**) (Eq. (6)) [25].

$$4\text{AlBr} \cdot \text{NEt}_3 + 4(\text{Me}_3\text{Si})_3\text{SiLi} \rightarrow [(\text{Me}_3\text{Si})_3\text{SiAl}]_4 + 4\text{LiBr} \tag{6}$$
$$\mathbf{15}$$

It has been reported that one of the Cp* groups in (Cp*Al)$_4$ can be substituted by an N-based ligand to give [(Cp*Al)$_3$AlN(SiMe$_3$)$_2$] [22]. A completely N-substituted Al$_4$ cluster **16** was obtained by reacting [2,6-i-Pr$_2$C$_6$H$_3$NSiMe$_3$AlI$_2$]$_2$ with Na/K alloy (Scheme 6.4). Compound **16** is a pale-yellow solid, while **13–15** are intensely colored (brown to violet) and decompose at 270 °C. Single-crystal X-ray structure analysis shows a nearly perfect tetrahedral Al$_4$ framework for **16**. No Al–Al–Al angle deviates by more than 1° from the ideal angle of 60°. The average Al–Al bond length (2.619 Å) is relatively short compared to those in the previously reported compounds (2.592 to 2.773 Å).

The EI mass spectrum of **16** reveals the molecular ion peak with a relative intensity of 16%. This result is quite surprising since the mass spectrum of (Cp*Al)$_4$ exhibits only the peak due to the monomeric species Cp*Al under comparable conditions [26].

The first example of a room temperature stable Al(I) compound of composition [HC(CMeCNAr)$_2$]Al (**17**) was obtained from the reaction of the diiodide [HC(CMeCNAr)$_2$]AlI$_2$ with potassium (Scheme 6.5) (Ar = 2,6-i-Pr$_2$C$_6$H$_3$).

Single-crystal X-ray structural analysis of **17** has been performed and the molecular structure is shown in Fig. 6.3 [27]. Red crystalline **17** is soluble in organic solvents such as toluene, benzene, and hexane, and is stable both thermally and under an inert atmosphere. The decomposition point is above 150 °C.

Compound **17** is monomeric and no remarkable close contacts are found in the unit cell. This is the first dicoordinate aluminum(I) compound to be prepared and

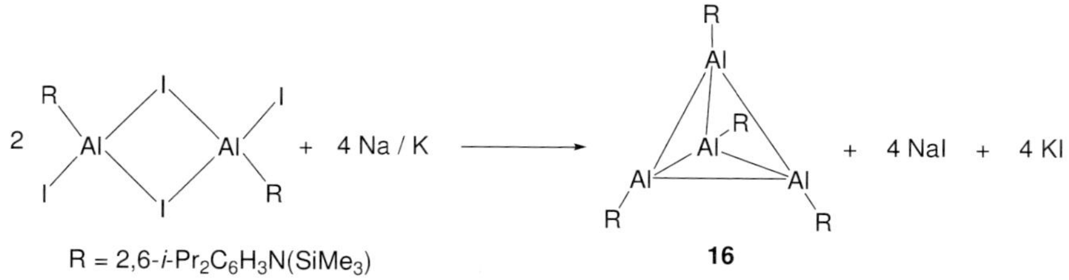

Scheme 6.4 Synthesis of compound [2,6-i-Pr$_2$C$_6$H$_3$N(SiMe$_3$)Al]$_4$ (**16**).

Scheme 6.5 Synthesis of a room temperature stable monomeric Al(I) compound (**17**).

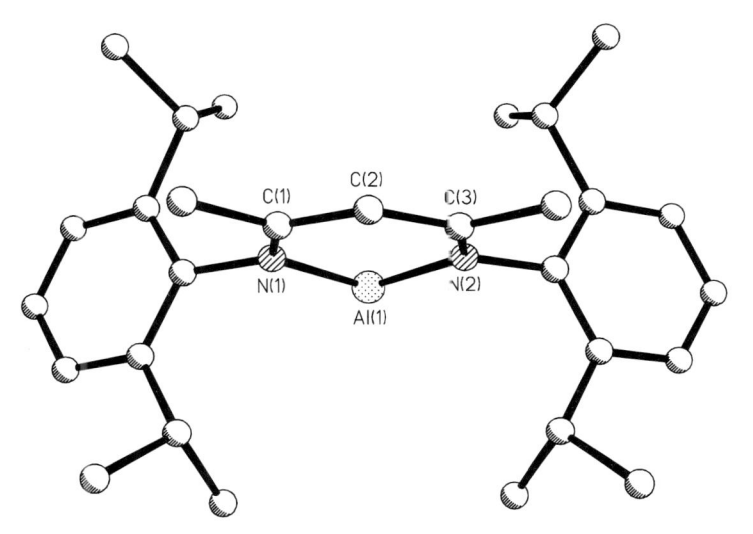

Fig. 6.3 Molecular structure of the monomer **17**.

structurally characterized. The aluminum atom in **17** is part of a planar six-membered heterocycle along with the skeletal atoms of the NCCCN ligand. The Al-N bonds (1.957 Å) are slightly longer than those in comparable Al(III) compounds due to the larger ionic radius of Al(I) compared to Al(III).

On the basis of ab initio calculations [27], it was shown that the lone pair on Al(I) in **17** is stereochemically active and possibly has a quasi-trigonal-planar orientation. Charge depletion of the aluminum atom into the semiplane of the ring is also noticed. These features provide scope for observing both Lewis acid and Lewis base behavior for **17**.

6.4
Chemistry of (Cp*Al)₄ (8)

Extensive use of ^{27}Al NMR spectroscopy indicated the dissociation of **8** in solution [16, 22] and supported the observation of the formation of monomeric Cp*Al in the gas phase [17]. Cp*Al is a two-electron donor and can be regarded as being isolobal to a carbene. The molecular structure of monomeric Cp*Al was determined in the gas phase [28]. Cp*Al is stable at temperatures up to 100 °C. Solutions of Cp*Al in toluene slowly decompose at 100 °C with the formation of finely divided aluminum [28].

Cowley et al. reacted (Cp*Al)₄ with B(C₆F₅)₃ to yield the first Lewis acid–base adduct of composition Cp*AlB(C₆F₅)₃ (**18**) [29].

$$\tfrac{1}{4}(Cp^*Al)_4 + B(C_6F_5)_3 \rightarrow \underset{\mathbf{18}}{Cp^*AlB(C_6F_5)_3} \qquad (7)$$

In compound **18**, the Cp* group is attached in an η^5 fashion and the ring centroid–Al–B moiety is essentially linear [172.9(1)°]. The Al–B bond distance in **18** is 2.169(3) Å.

An organometallic approach for stabilizing the monomeric Cp*Al species has been demonstrated [30–33]. Thus, Cp*Al-Fe(CO)₄ was obtained from the reaction of Cp*AlCl₂ with K₂Fe(CO)₄ [32]. A comparable adduct with Cr(CO)₅ of the composition Cp*AlCr(CO)₅ was structurally characterized by single-crystal X-ray analysis. A bis(dicyclohexylphosphino)ethane-platinum adduct containing two Cp*Al groups was prepared according to Eq. (8) [33] under elimination of CMe₄.

(8)

Compound **19** is a yellowish solid, stable at room temperature, with relatively short Pt–Al bonds (2.33 Å). In binary alloys of platinum and aluminum, the Pt–Al distance is 2.52 Å.

The respective reactions of (Cp*Al)₄ with BBr₃ and AlCl₃ yield very different products. Thus, the reaction with BBr₃ leads to an exchange of the Cp* from aluminum to boron to yield [Cp*BBr][AlBr₄] [34], whereas that with AlCl₃ gives the [Cp*₂Al]⁺ cation. The latter species is isoelectronic with the magnesium analogue and has a sandwich-like structure [35]. In the complex [(CpNi)₂(Cp*Al)₂] (**20**), the Cp*Al does not function as a terminal ligand but rather as a bridging one between two nickel atoms [36] (Scheme 6.6).

6.4 Chemistry of (Cp*Al)₄ (8)

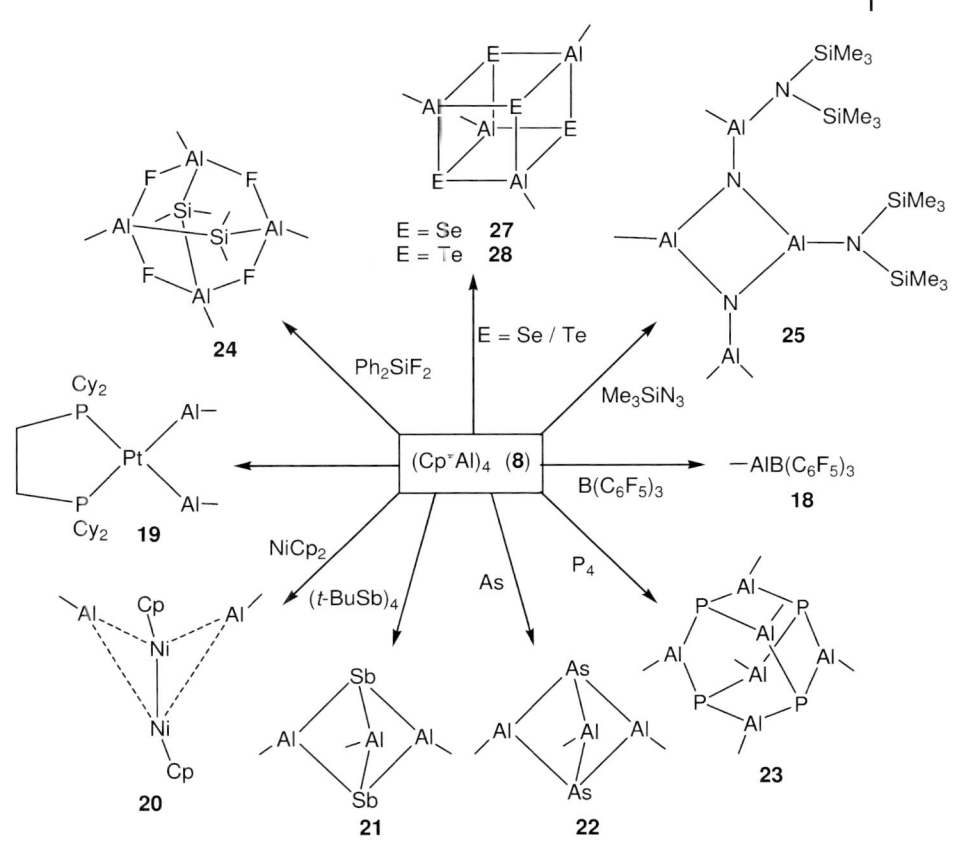

Scheme 6.6 Products of reactions using (Cp*Al)₄ as a starting material. Cp* groups have been omitted for clarity.

The polyhedral [(Cp*Al)₃Sb₂] (**21**) was prepared from (*t*-BuSb)₄ and (Cp*Al)₄. Compound **21** is an orange-brown crystalline solid. In this reaction, (*t*-BuSb)₄ is decomposed with the formation of elemental antimony. However, (Cp*Al)₄ and antimony do not react under the same conditions [37]. The analogous arsenic compound **22** [38] was obtained from (*t*-BuAs)₄ and (Cp*Al)₄.

Phosphorus, the congener of arsenic and antimony, reacts with (Cp*Al)₄ to give **23**, a polyhedral species with a P_4Al_6 core structure [39]. The tendency of the tetramer **8** to convert to a variety of rings, cages, and clusters is noteworthy. The reaction of Ph₂SiF₂ and **8** [37] afforded an unexpected cluster composition, **24**, having novel structural features. Particularly interesting are the Al–Si–Al bonds in **24**.

The expected formation of Cp*AlF₂ was not observed. An unusual product **25** resulted from the reaction of **8** with Me₃SiN₃. The central core is a four-membered, almost ideally planar Al_2N_2 ring. All the aluminum and nitrogen atoms are three-coordinate. As regards its formation, one can envisage that **25** originates from an Al_4N_4 cube of which the top layer is broken open and the remaining ring represents the base. A silyl group migration and rearrangement of a Cp* group take place simultaneously [40]. A comparable four-membered Al_2N_2 ring was obtained (**26**) when t-Bu_3SiN_3 was reacted with $(Cp*Al)_4$ (Eq. (9)) [41].

$$(Cp^*Al)_4 + 4(t\text{-}Bu)_3SiN_3 \rightarrow 2[Cp^*AlNSi(t\text{-}Bu)_3]_2 + 4N_2 \tag{9}$$
$$\mathbf{26}$$

A neutral $SiAl_{14}$ cluster was formed when Cp*Al was treated with $SiCl_4$ [42]. When a solution of **8** in toluene is treated with an excess of either selenium or tellurium, yellow-green solutions are obtained. From these solutions, colorless crystals of $[Cp^*AlSe]_4$ (**27**) and pale-green shiny octahedral crystals of $[Cp^*AlTe]_4$ (**28**) can be obtained. Both compounds are tetrameric, with the parent skeleton comprising an Al–E heterocubane (E = Se, Te). The respective bond angles at Al, Se, and Te deviate very little from the ideal 90° angle (av. Se–Al–Se 94.6°, Te–Al–Te 95.0°, Al–Se–Al 85.2°, Al–Te–Al 84.7°). The facile formation of **27** and **28** under mild conditions illustrates the desire of aluminum to attain the more stable oxidation state +III [17].

6.5
Reactions of [HC(CMeNAr)$_2$]Al (17); Ar = 2,6-i-Pr$_2$C$_6$H$_3$

The reaction of **17** and N_3-2,6-$Trip_2C_6H_3$ (Trip = 2,4,6-i-$Pr_3C_6H_2$) occurs near room temperature in toluene and affords the monomeric imide of aluminum (Eq. (10)).

$$[HC(CMeNAr)_2]Al + N_3\text{-}2,6\text{-}Trip_2C_6H_3 \rightarrow [HC(CMeNAr)_2]AlN\text{-}2,6\text{-}Trip_2C_6H_3 + N_2 \tag{10}$$
$$\mathbf{17} \qquad\qquad\qquad\qquad\qquad\qquad\qquad\qquad \mathbf{29}$$

Compound **29** is the first stable monomeric imide of aluminum. In general, imides of aluminum display a strong tendency to oligomerize. The large size of the ligands and the mild reaction conditions are notable features of this synthetic success [44].

In contrast, the smaller Me_3SiN_3 reacts via the intermediate $[HC(CMeNAr)_2Al=NSiMe_3$ in a [2+3] cycloaddition to give the tetrazole **30** (Scheme 6.7) [45].

Compound **30** is rather stable and decomposes at 130 °C. Interestingly, the AlN_4 five-membered ring in **30** is essentially planar.

Ph_3SiN_3 reacts with **17** to yield the same structure as the one described for **30**. However, the methyl groups at the silicon atom are replaced by phenyl substituents. Quite surprising is the result of the reaction of **17** with t-$BuSi(N_3)_3$ (Scheme 6.8) [46].

Compound **31** contains two six-membered and one four-membered ring. Obviously, one of the azide groups on each silicon is transferred to aluminum and

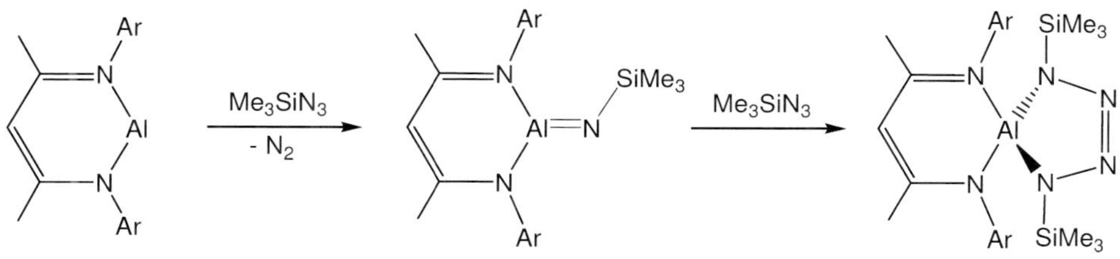

Scheme 6.7 Synthesis of **30**. Ar = 2,6-i-Pr$_2$C$_6$H$_3$

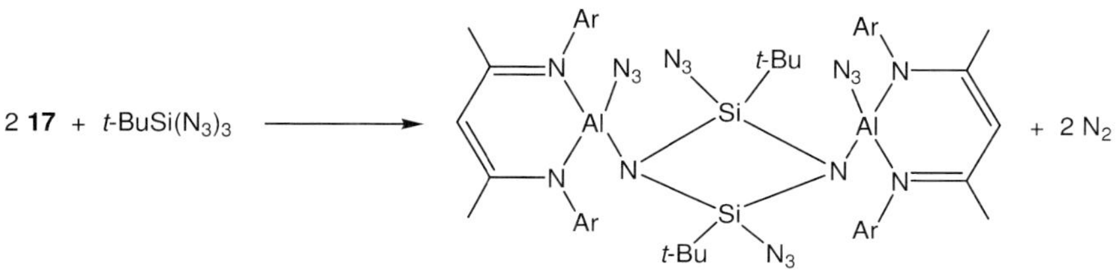

Scheme 6.8

two others eliminate nitrogen to yield the N$_2$Si$_2$ ring. The total number of azide groups in compound **31** is four.

In Eq. (11), the reaction of **17** with an N-heterocyclic carbene to yield **32** is shown. Compound **32** is formed under the migration of a hydrogen atom from a methyl group to aluminum.

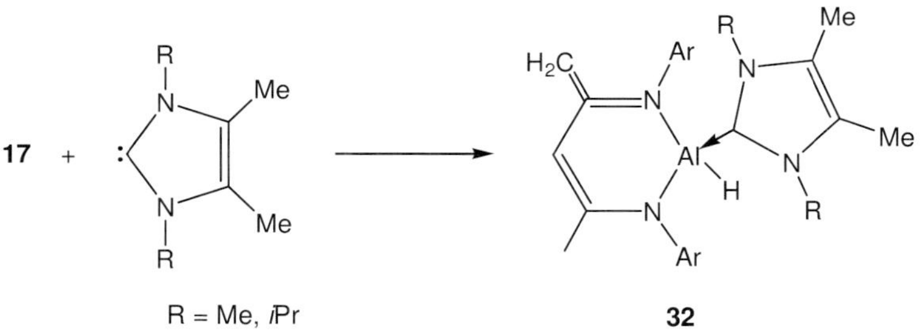

R = Me, iPr   **32**   (11)

The desired formation of a derivative with an Al=C double bond was not observed. Likewise, the reaction of **17** with white phosphorus gave, unexpectedly, the four-membered P$_4$ ring attached to two aluminum-containing units (Scheme 6.9).

The P$_4$ ring in **33** is highly puckered and can be considered as bearing the negative charges in an ionic model due to the electron transfer from Al(I) to Al(III). Subsequently, during the electron transfer two of the six P–P bonds in the P$_4$ tetrahedron are cleaved [47].

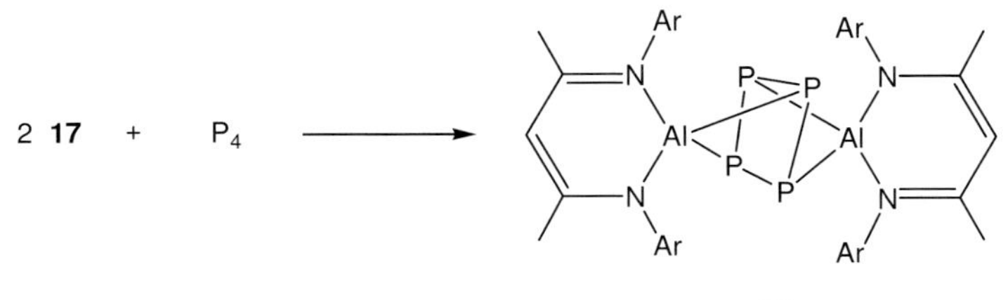

6 Aluminum(I) Chemistry

Scheme 6.9 Preparation of a phosphorus–aluminum compound with a P_4Al_2 core.

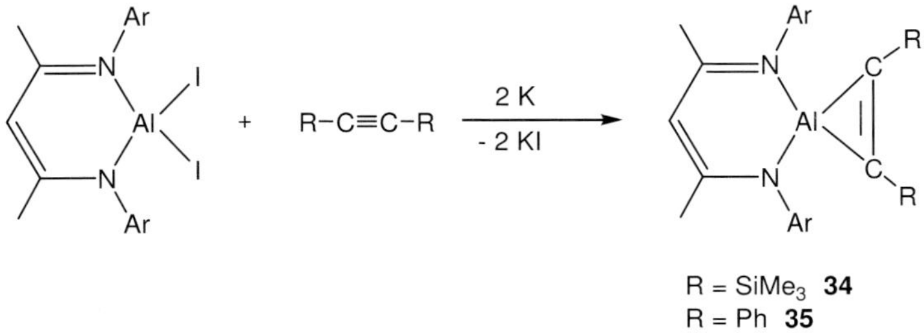

Scheme 6.10 Synthesis of a cyclopropene derivative containing one aluminum atom.

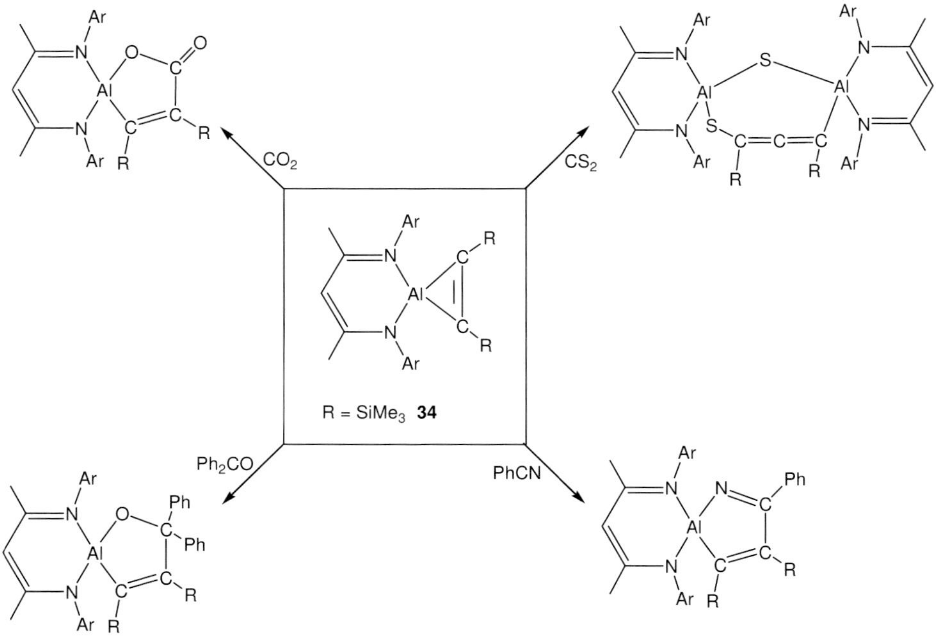

Scheme 6.11 Insertion reactions into the reactive Al–C bond of the three-membered AlC_2 ring.

The addition of an alkyne during the reduction of [HC(CMeNAr)$_2$]AlI$_2$ with potassium resulted in a compound containing an AlC$_2$ three-membered ring **34**, which has a very acute C–Al–C angle of 42.56° (Scheme 6.10) [48].

Compounds **34** and **35**, which are extremely sensitive to air and moisture, exhibit facile insertion reactions with CO$_2$, CS$_2$, ketones, and nitriles. A summary of the reactions of the three-membered ring compound **34** is given in Scheme 6.11 [49].

6.6
Conclusions

This account reveals convincingly that the chemistry of organoaluminum compounds with aluminum of formal oxidation state +I has evolved significantly in the last decade. These are very fascinating results which promise many more exciting compounds for the future. The formation of stable RAl(I) compounds indicates that the use of suitably shielding R groups clearly offers tremendous potential and scope for further notable contributions to preparative chemistry. This topic of research will undoubtedly continue to develop into an important domain of aluminum chemistry.

Acknowledgements

I am very thankful to my students who have contributed to this field of research. Their names are listed in the references. The *Deutsche Forschungsgemeinschaft* and the *Göttinger Akademie der Wissenschaften* are gratefully acknowledged for their financial support.

References

1   a) A. F. HOLLEMAN, E. WIBERG, N. WIBERG, *Lehrbuch der Anorganischen Chemie*, Walter de Gruyter, Berlin, 1995; b) F. A. COTTON, G. WILKINSON, C. A. MURILLO, M. BOCHMANN, *Advanced Inorganic Chemistry*, John Wiley, New York, 1999.
2   W. KLEMM, E. VOSS, K. GEIERSBERGER, *Z. Anorg. Allg. Chem.* **1948**, *256*, 15.
3   M. TACKE, H. SCHNÖCKEL, *Inorg. Chem.* **1989**, *28*, 2895.
4   H.-J. HIMMEL, J. BAHLS, M. HAUSSMANN, F. KURTH, G. STÖSSER, H. SCHNÖCKEL, *Inorg. Chem.* **2002**, *41*, 4952.
5   J. BAHLO, H.-J. HIMMEL, H. SCHNÖCKEL, *Inorg. Chem.* **2002**, *41*, 2678; *Inorg. Chem.* **2002**, *41*, 4488.
6   H.-J. HIMMEL, C. KLAUS, *Z. Anorg. Allg. Chem.* **2003**, *629*, 1477.
7   M. N. S. RAO, H. W. ROESKY, G. ANATHARAMAN, *J. Organomet. Chem.* **2002**, *646*, 4.
8   A. ECKER, E. WECKERT, H. SCHNÖCKEL, *Nature* **1997**, *387*, 379.
9   C. DOHMEIER, M. MOCKER, H. SCHNÖCKEL, A. LÖTZ, U. SCHNEIDER, R. AHLRICHS, *Angew. Chem.* **1993**, *105*,

1491; *Angew. Chem. Int. Ed. Engl.* **1993**, *32*, 1428.

10 H. Köhnlein, A. Purath, C. Klemp, E. Baum, J. Krossing, H. Schnöckel, *Inorg. Chem.* **2001**, *40*, 4830.

11 A. Schnepf, H. Schnöckel, *Adv. Organomet. Chem.* **2001**, *47*, 235.

12 Ch. Klemp, H. Schnöckel, *Inorg. Chem. Highlights* (Eds.: G. Meyer, D. Naumann, L. Wesemann), Wiley-VCH Weinheim **2002**, 245.

13 Ch. Üffing, A. Ecker, R. Köppe, K. Merzweiler, H. Schnöckel, *Chem. Eur. J.* **1998**, *11*, 2142.

14 C. Dohmeier, C. Robl, M. Tacke, H. Schnöckel, *Angew. Chem.* **1991**, *103*, 594; *Angew. Chem. Int. Ed. Engl.* **1991**, *30*, 564.

15 C. Dohmeier, D. Loos, H. Schnöckel, *Angew. Chem.* **1996**, *108*, 141; *Angew. Chem. Int. Ed. Engl.* **1996**, *35*, 129.

16 J. Gauss, U. Schneider, R. Ahlrichs, C. Dohmeier, H. Schnöckel, *J. Am. Chem. Soc.* **1993**, *115*, 2402.

17 S. Schulz, H.W. Roesky, H.-J. Koch, G. M. Sheldrick, D. Stalke, A. Kuhn, *Angew. Chem.* **1993**, *105*, 1828; *Angew. Chem. Int. Ed. Engl.* **1993**, *32*, 1729.

18 M. Schormann, K.S. Klimek, H. Hatop, S.P. Varkey, H.W. Roesky, C. Lehmann, C. Röpken, R. Herbst-Irmer, M. Noltemeyer, *J. Solid State Chem.* **2001**, *162*, 225.

19 R. Srinivas, D. Süzle, H. Schwarz, *J. Am. Chem. Soc.* **1990**, *112*, 8334.

20 J.J. Eisch, J.L. Considine, *J. Am. Chem. Soc.* **1968**, *90*, 6257.

21 A. Ecker, H. Schnöckel, *Z. Anorg. Allg. Chem.* **1998**, *624*, 813 and references therein.

22 H. Sitzmann, M.F. Lappert, C. Dohmeier, C. Üffing, H. Schnöckel, *J. Organomet. Chem.* **1998**, *561*, 203.

23 E.P. Schram, N. Sudha, *Inorg. Chim. Acta* **1991**, *183*, 213.

24 A. Purath, C. Dohmeier, A. Ecker, H. Schnöckel, K. Amelunxen, T. Passler, N. Wiberg, *Organometallics* **1998**, *17*, 1894.

25 A. Purath, H. Schnöckel, *J. Organomet. Chem.* **1999**, *579*, 373.

26 M. Schiefer, N.D. Reddy, H.W. Roesky, D. Vidovic, *Organometallics* **2003**, *22*, 3637.

27 C. Cui, H.W. Roesky, H.-G. Schmidt, M. Noltemeyer, H. Hao, F. Cimpoesu, *Angew. Chem.* **2000**, *112*, 4444; *Angew. Chem. Int. Ed.* **2000**, *39*, 4274.

28 A. Haaland, K.-G. Martinsen, S.A. Shlykov, H.V. Volden, C. Dohmeier, H. Schnöckel, *Organometallics* **1995**, *14*, 3116.

29 J.D. Gorden, A. Voigt, C.L.B. Macdonald, J.S. Silverman, A.H. Cowley, *J. Am. Chem. Soc.* **2000**, *122*, 950.

30 Q. Yu, A. Purath, A. Donchev, H. Schnöckel, *J. Organomet. Chem.* **1999**, *584*, 94.

31 M.M. Schulte, E. Herdtweck, G. Raudaschl-Sieber, R.A. Fischer, *Angew. Chem.* **1996**, *108*, 489; *Angew. Chem. Int. Ed. Engl.* **1996**, *35*, 424.

32 J. Weiss, D. Stetzkamp, B. Nuber, R.A. Fischer, C. Boehme, G. Frenking, *Angew. Chem.* **1997**, *109*, 95; *Angew. Chem. Int. Ed. Engl.* **1997**, *36*, 70.

33 D. Weiss, T. Steinke, M. Winter, R.A. Fischer, N. Fröhlich, J. Uddin, G. Frenking, *Organometallics* **2000**, *19*, 4583.

34 C. Dohmeier, R. Köppe, C. Robl, H. Schnöckel, *J. Organomet. Chem.* **1995**, *487*, 127.

35 C. Dohmeier, H. Schnöckel, C. Robl, U. Schneider, R. Ahlrichs, *Angew. Chem.* **1993**, *105*, 1124; *Angew. Chem. Int. Ed. Engl.* **1993**, *32*, 1655.

36 C. Dohmeier, H. Krautscheid, H. Schnöckel, *Angew. Chem.* **1994**, *106*, 2570; *Angew. Chem. Int. Ed. Engl.* **1994**, *33*, 2482.

37 S. Schulz, T. Schoop, H.W. Roesky, L. Häming, A. Steiner, R. Herbst-Irmer, *Angew. Chem.* **1995**, *107*, 1015; *Angew. Chem. Int. Ed. Engl.* **1995**, *34*, 919.

38 C.K.F. von Hänisch, C. Üffing, M.A. Junker, A. Ecker, B.O. Kneisel, H. Schnöckel, *Angew. Chem.* **1996**, *108*, 3003; *Angew. Chem. Int. Ed. Engl.* **1996**, *35*, 2876.

39 C. Dohmeier, H. Schnöckel, C. Robl, U. Schneider, R. Ahlrichs, *Angew. Chem.* **1994**, *106*, 225; *Angew. Chem. Int. Ed. Engl.* **1994**, *33*, 199.

40 S. Schulz, L. Hämig, R. Herbst-Irmer, H. W. Roesky, G. M. Sheldrick, *Angew. Chem.* **1994**, *106*, 1052; *Angew. Chem. Int. Ed. Engl.* **1994**, *33*, 969.

41 S. Schulz, A. Voigt, H. W. Roesky, L. Hämig, R. Herbst-Irmer, *Organometallics* **1996**, *15*, 5252.

42 A. Purath, C. Dohmeier, A. Ecker, R. Köppe, H. Krautscheid, H. Schnöckel, R. Ahlrichs, C. Stoermer, J. Friedrich, P. Jutzi, *J. Am. Chem. Soc.* **2000**, *122*, 6955.

43 C. Cui, H. W. Roesky, H.-G. Schmidt, M. Noltemeyer, *Angew. Chem.* **2000**, *112*, 4705; *Angew. Chem. Int. Ed.* **2000**, *39*, 4531.

44 N. J. Hardman, C. Cui, H. W. Roesky, W. H. Fink, P. P. Power, *Angew. Chem.* **2001**, *113*, 2230; *Angew. Chem. Int. Ed.* **2001**, *40*, 2172.

45 C. Cui, H. W. Roesky, H.-G. Schmidt, M. Noltemeyer, *Angew. Chem.* **2000**, *112*, 4705; *Angew. Chem. Int. Ed.* **2000**, *39*, 4531.

46 H. Zhu, H. Fan, H. Zhu, H. W. Roesky, J. Magall, C. E. Hughes, *Angew. Chem.* **2004**, *116*, 3525–3527; *Angew. Chem. Int. Ed.* **2004**, *43*, 3443–3445.

47 Y. Peng, H. Fan, H. Zhu, H. W. Roesky, J. Magull, C. E. Hughes, *Angew. Chem.* **2004**, *116*, 3525–3527; *Angew. Chem. Int. Ed.* **2004**, *43*, 3443-3445.

48 C. Cui, S. Köpke, R. Herbst-Irmer, H. W. Roesky, M. Noltemeyer, H.-G. Schmidt, B. Wrackmeyer, *J. Am. Chem. Soc.* **2001**, *123*, 9091.

49 H. Zhu, J. Chai, Q. Ma, V. Jancik, H. W. Roesky, H. Fan, R. Herbst-Irmer, *J. Am. Chem. Soc.* **2004**, *126*, 10194–10195

7
Divalent Scandium

Gerd Meyer, Liesbet Jongen, Anja-Verena Mudring, and Angela Möller

7.1
Introduction

Scandium is the first element in order of ascending atomic numbers which incorporates a "3d electron". Thus, the electronic configuration of the scandium atom is $[Ar]4s^2 3d^1$. With ionization potentials of 631, 1235, and 2389 kJ mol^{-1} [1], compounds of mono-, di-, and trivalent scandium should, in principle, be available. However, the chemistry of scandium appears to be primarily that of the trivalent cation [2]. In solid-state chemistry, there are only a few reports of scandium(II) compounds in the literature. Note that we exclude from our discussion so-called scandium-rich halides such as "ScCl", which contain exclusively interstitial atoms (e.g., H, B, C, N; "ScCl" is in fact ScClH$_x$) that "stabilize" octahedral clusters of the M_6X_{12} type [3].

Based on thermochemical measurements, the phase $ScI_{2.15}$ was characterized in 1968 [4]. Thereafter, from a single-crystal X-ray structure determination, it was concluded that this iodide is in fact scandium-deficient with an intact iodide sublattice exhibiting essentially a CdI_2-type structure, and should therefore be better formulated as $Sc_{0.93}I_2$ [5].

After a first account of "CsScCl$_3$" [6], a number of ternary halides, RbScX$_3$ (X=Cl, Br) and CsScX$_3$ (X=Cl, Br, I) [7] as well as LiScI$_3$ [8] and Na$_{0.5}$ScI$_3$ [8] have since been reported. As in the case of $Sc_{0.93}I_2$, the RbScX$_3$ and CsScX$_3$ halides have scandium-deficient compositions, in sharp contrast to the neighboring titanium iodides, TiI$_2$ and CsTiI$_3$, for example, which have "line" compositions [9].

That all of the above-mentioned halides of divalent scandium are scandium-deficient may have two reasons. First, there could be a thermodynamic reason and, secondly, the reason could lie in the electronic structure of these compounds.

Inorganic Chemistry in Focus II. Edited by G. Meyer, D. Naumann, and L. Wesemann
Copyright © 2005 WILEY-VCH Verlag GmbH & Co. KGaA, Weinheim
ISBN: 3-527-30811-3

7.2
Synthesis

The scandium trihalides, ScX_3 (X=Cl, Br, I), are starting materials and may be prepared either by the ammonium halide route, preferably for the chloride and bromide, from scandium sesquioxide, Sc_2O_3, and ammonium halide, $(NH_4)X$ (X=Cl, Br) [10], or from the elements as in the case of ScI_3 [11]. The latter is subsequently reduced with (surplus) scandium metal in sealed tantalum ampoules at 750 °C to yield $Sc_{0.9}I_2$.

The ternary halides $A_wSc_xX_3$ (A=Li, Na, K, Rb, Cs; X=Cl, Br, I) may be prepared by a variety of routes. The ternary scandium(III) halides, for example $Cs_3Sc_2Cl_9$, may be reduced by scandium metal. Alternatively, a mixture of $3CsCl + 2ScCl_3$ may be likewise reduced, also in sealed tantalum or niobium containers at elevated temperatures, at around 700 °C. Another feasible way is the reduction of the trihalides ScX_3 with alkali metals, for example of $ScCl_3$ with caesium [12]. For $CsSc_xI_3$, one may also oxidize scandium metal with CsI_3.

7.3
Scandium Diiodide, Sc_xI_2

Scandium diiodide, Sc_xI_2, crystallizes essentially with the CdI_2-type structure, as shown in Fig. 7.1. Closest-packed layers of iodide anions are stacked in the fashion of a hexagonal closest-packing of spheres; the stacking sequence in the direction of the c axis is ...AB... Octahedral holes between every second of these iodide layers are, in the present case not fully, occupied by scandium atoms. Were these layers fully occupied, they would also follow the feature of a two-dimensional closest-packing of scandium atoms, named γ. One may also describe this arrangement as a 3^6 net. However, the crystal structure of Sc_xI_2 may be de-

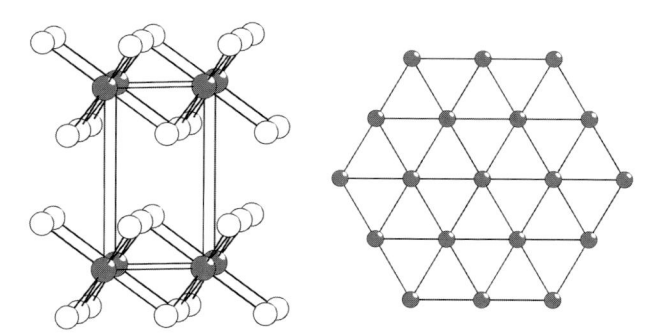

Fig. 7.1 Crystal structure of scandium diiodide, Sc_xI_2. Left: A perspective view of the unit cell exhibiting the ...BγA☐BγA... packing. Right: View of the scandium γ layer, a 3^6 net of equal atoms.

picted in the nomenclature of closest-packed structures as ...γA☐Bγ... The identity period, the crystallographic c axis, extends between identical scandium atoms in the γ layers. Values reported in the literature range from $c = 696.2(2)$ to $c = 698.24(7)$ pm [5,13]. The c axis is, of course, identical to the interlayer Sc–Sc distance. It is rather long and interactions between these will not play an important role. More important are interactions within the γ layers (Fig. 7.1). The intralayer Sc–Sc distances correspond to the crystallographic a axis, which was found within the limits of $a = 408.51(9)$ and $a = 410.2(1)$ pm [5, 13].

The most important feature of the crystal structure of Sc_xI_2 is the under-occupancy of the scandium 3^6 net. Values of x between 0.87 and 0.93 were determined from single-crystal X-ray structure analyses. Superstructures have not been observed (note that the crystal quality is usually relatively poor for cadmium halide type structures as there are only van der Waals contacts between the BγA packages and stacking disorder frequently occurs). In the case of $x = 0.89$, one-ninth of the scandium positions would be empty. In other words, the correct formula would be $9 \times Sc_{0.89}I_2 = Sc_8\square_1I_{18}$. A representation of such a hypothetical superstructure is shown in Fig. 7.2.

Electronically, this would mean, according to $(Sc^{3+})_8(e^-)_6(I^-)_{18}$, that there would be six excess electrons per formula unit of $Sc_8\square_1I_{18}$. These might be delocalized so that scandium-deficient scandium diiodide would be a two-dimensional 3d metal. Alternatively, the "excess" electrons could be localized in 3d states, and thus $Sc_8\square_1I_{18}$ would be a mixed-valence compound according to $(Sc^{2+})_6(Sc^{3+})_2\square_1(I^-)_{18}$.

The first conductivity measurements on $Sc_{0.89}I_2$ had shown that it is indeed metallic above 100 K and an insulator below 100 K [5]. Temperature-dependent electron spin resonance (ESR) spectra confirm this result (Fig. 7.3). Therefore,

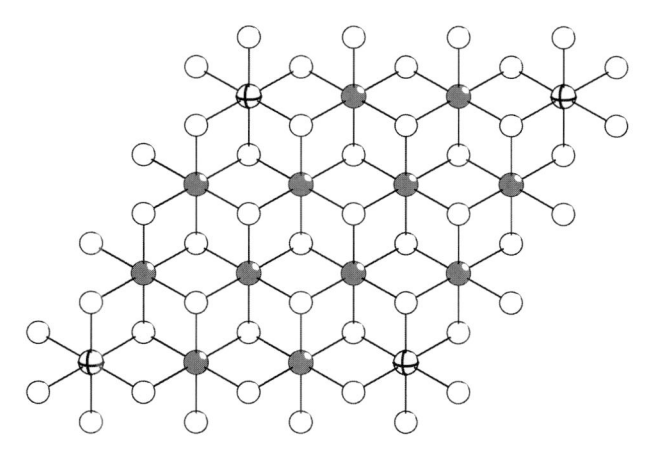

Fig. 7.2 Nine-fold superstructure of $Sc_{0.89}I_2$, i.e. view from top of one BγA package with one (crossed circles) out of nine scandium positions (the other eight are black circles) unoccupied.

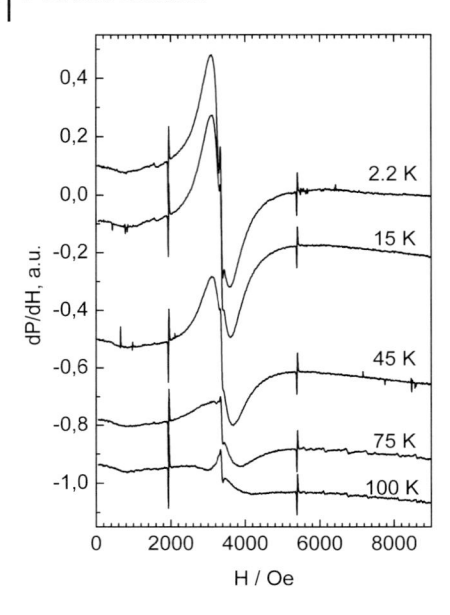

Fig. 7.3 ESR spectra of $(Sc^{3+})_8(e^-)_6(I^-)_{18}$ at various temperatures. The two sharp resonances at low and high fields are due to the internal standard (ruby).

the excess electrons must be localized below 100 K, because an almost isotropic ESR spectrum of Lorentzian shape with a constant bandwidth of 600 Oe is observed below 50 K, with g values around 1.99. The intensities of the signals approximately follow a Curie–Weiss type law, indicating magnetic interactions between the Sc^{2+} ions. At higher temperatures, no signal is detected so that the excess electrons must be delocalized.

Therefore, the formulations $(Sc^{3+})_8(e^-)_6(I^-)_{18}$ and $(Sc^{2+})_6(Sc^{3+})_2(I^-)_{18}$ rationalize very well the physical properties of $Sc_{0.89}I_2$ above and below, say, 100 K. Such a transition is also observed by X-ray powder diffraction at temperatures between 25 and 293 K. Fig. 7.4 shows the temperature-dependence of the molar volume. At high temperatures, in the metallic regime, one observes the usual volume decrease with decreasing temperature. At low temperatures, a negative volume expansion is seen with decreasing temperature. This can be interpreted in terms of increasing localization of the excess electron in the 3d state. In other words, "true" divalent scandium is now produced. Sc^{2+} has, of course, a larger volume than Sc^{3+}. The volume increase between 100 and 25 K, although at 0.61 $cm^3\ mol^{-1}$ only small, is well in accord with the assumption that only six out of nine octahedral holes now bear Sc^{2+}, two still retain Sc^{3+}, and one is empty.

Although we can now explain some of the properties of so-called scandium diiodide, we still do not understand the scandium deficiency. To put it in other words, we may ask the question why does "ScI_2" or Sc_9I_{18} only tolerate six excess electrons and not nine per formula unit, and hence is $(Sc^{3+})_8(e^-)_6(I^-)_{18}$ and not $(Sc^{3+})_9(e^-)_9(I^-)_{18}$? Fig. 7.5 shows the density of states (DOS, Fig. 7.5, left) and crystal overlap Hamiltonian population (COHP, Fig. 7.5, right) curves, which are used to describe the interaction of the atom type pairs chosen for non-deficient hypo-

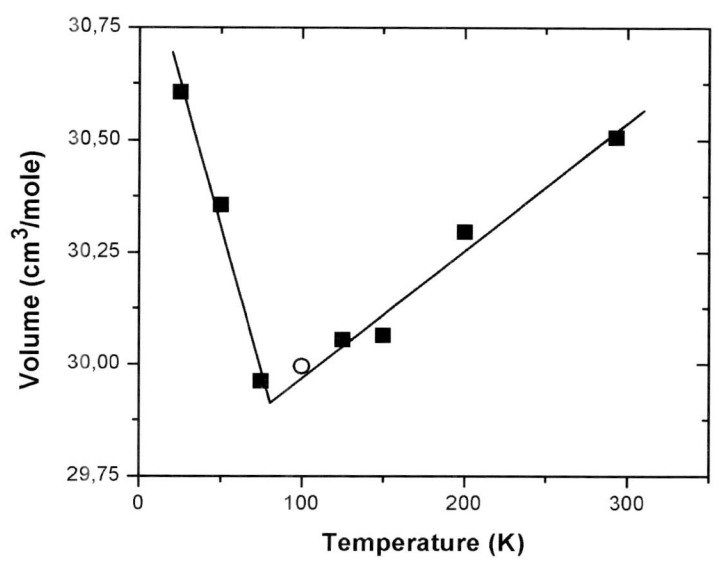

Fig. 7.4 Temperature dependence of the molar volume of $Sc_{0.89}I_2$. The molar volume is defined as $V_m = a^2 \cdot c \cdot \sqrt{3/2} \cdot N_A \cdot 1/Z$.

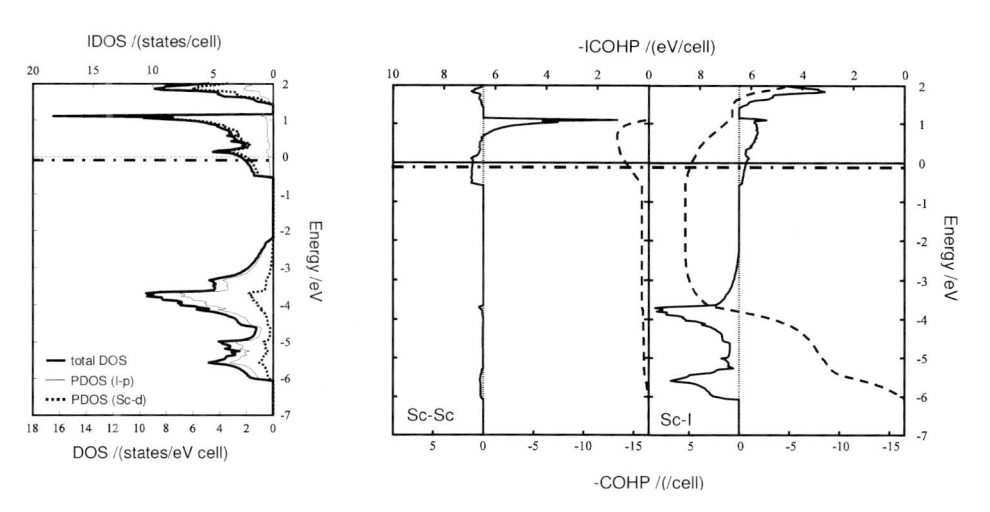

Fig. 7.5 Left: Density of states (DOS) as well as integrated DOS (IDOS). Right: Crystal overlap Hamiltonian population (COHP) curves and integrated COHP (ICOHP) as the result of DFT-LMTO-ASA calculations (density functional theory – linear muffin tin orbitals – atomic sphere approximation) for ScI_2. The dashed horizontal lines indicate the hypothetical Fermi level for $Sc_{0.9}I_2$; the dashed vertical curves give the integrated COHP as a measure of bonding strength.

thetical ScI_2. The density of states at the lower energy levels is mainly of iodine p-character and around the Fermi-level becomes dominated by Sc d orbitals. The COHP curve for Sc–Sc shows bonding interactions around and above the Fermi level. On the other side, the Sc–I interactions at this energy are strongly antibonding and thus destabilize the whole compound. Qualitatively, a net stabilization is achieved when these states are not occupied, with a lower electron count in $Sc_{0.89}I_2$ (as shown by the dashed line in Fig. 7.5, right), as the Sc–I interactions are far more antibonding than the Sc–Sc interactions are bonding. These are weak when compared to metal–metal interactions among other 3d metals as a result of shell-structure effects (first electron in a 3d orbital!). This also becomes obvious when the ionic radii and ionization potentials are compared.

The excess electrons may still be localized (which occurs at low temperatures, below 100 K) or delocalized (at higher temperatures). At low temperatures, $Sc_{0.89}I_2 = (Sc^{2+})_6(Sc^{3+})_2(I^-)_{18}$ then truly contains *divalent scandium* in the sense that six individual 3d electrons are localized on six out of eight scandium atoms.

7.4
Other Scandium Dihalides?

There are no reports in the literature on other solid scandium dihalides, $ScCl_2$ or $ScBr_2$, neither line phases nor scandium- or halide-deficient. This observation would be in accordance with our interpretation of the bonding situation in ScI_2 as the Sc–X interactions around the hypothetical Fermi level become even more antibonding for the lighter halides (*cf.* ICOHP (M-X)/(eV/cell), $ScCl_2$: 7.27; ScI_2: 8.29).

There have been, however, reports on scandium monohalides, ScCl and ScBr [14]. These are in fact interstitially stabilized "monohalides", $ScClH_x$ [15] or $ScClC_{0.5}$ [16], for example. They have essentially ZrCl- or ZrBr-type structures, where ...A–B... packages are stacked in a trigonal-rhombohedric fashion with the threefold axis in the direction of the c axis being either left- or right-handed. Between the double layer of metal atoms, hydrogen or carbon are incorporated either in tetrahedral or octahedral interstices, and, therefore, contribute to the electronic structures of ScClH and Sc_2Cl_2C. The former would then have one excess electron according to $(Sc^{3+})(e^-)(Cl^-)(H^-)$, while the latter would have no excess electrons if the carbon were to be considered as "carbidic", $(Sc^{3+})_2(Cl^-)_2(C^{4-})$. Theoretical calculations confirm this description of the chemical bonding.

7.5
The Ternary Scandium Halides ASc_xX_3 (A = Alkali Metal, X = Cl, Br, I)

In considering the binary scandium iodide $Sc_{0.89}I_2$ we have dealt with a two-dimensional problem where scandium deficiency occurs to avoid Sc–I antibonding interactions. If, for example, CsI were to be reacted with ScI_2, which would yield $CsScI_3$, we would proceed from a two-dimensional deficient 3^6 net to a one-dimensional chain of scandium atoms. Indeed, $CsScI_3$ would have the so-called hexagonal perovskite structure ($CsNiCl_3$-type), with cesium and iodine forming a two-dimensional closest-packed layer. These $[CsI_3]$ layers are stacked in the fashion of a hexagonal closest packing of spheres. With the incorporation of cesium atoms in the iodide layers, two kinds of octahedral holes are produced in the stacking direction (equivalent to the hexagonal c axis), three-quarters made of $[Cs_2I_4]$ vertices and one-quarter being $[I_6]$ octahedra, which might be fully occupied by scandium atoms. Thus, one out of four octahedra might be occupied, which would lead to the formula $CsScI_3$. As we have hexagonal stacking (...A B...) of the CsI_3 layers, all octahedra $[ScI_6]$ share common faces according to the Niggli formulation $[ScI_{6/2}]$. In the crystal structure of $CsScI_3$ (Fig. 7.6), the Sc–Sc distance is equal to $c/2$, i.e. 341.6(1) pm.

As is the case for $Sc_{0.89}I_2$, the scandium positions in "$CsScI_3$" are not fully occupied. From single-crystal X-ray diffraction analysis, we have obtained the formula $CsSc_{0.80}I_3$ (Tab. 7.1). This means that starting from $CsSc_{0.67}I_3$ (which is $Cs_3Sc_2I_9$ with only trivalent scandium and every third octahedral hole empty), only 0.13 scandium atoms per formula unit $CsSc_{0.67}I_3$ are added. Again, the question is, why is so little scandium taken in and where are the additional electrons, or, in other words, is $CsSc_{0.80}I_3$ a one-dimensional metal according to $(Cs^+)(Sc^{3+})_{0.8}(e^-)_{0.4}(I^-)_3$ with the excess electrons delocalized, or is it a mixed-valence compound with Sc^{3+} and Sc^{2+}?

Besides $CsSc_{0.80}I_3$, a number of ternary scandium halides ASc_xX_3 exist, where X denotes Cl, Br, or I, and A might be an alkali metal. Tab. 7.1 gives an overview. Three structure types are observed. The first one is the so-called hexagonal

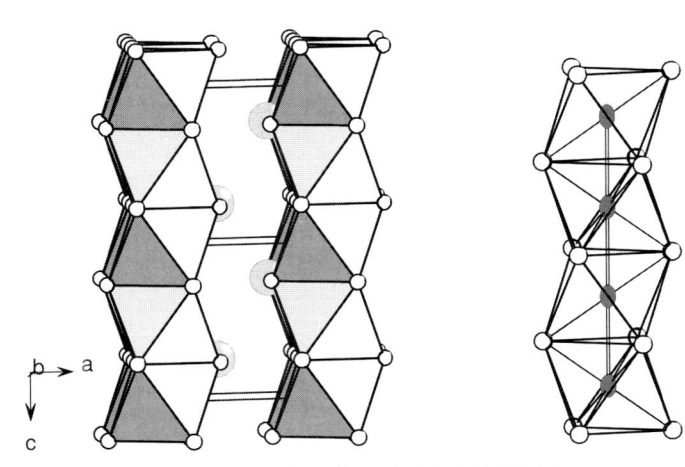

Fig. 7.6 Crystal structure of (hypothetical) $CsScI_3$ highlighting chains of face-sharing $[ScI_6]$ octahedra (left) and the one-dimensional chains of scandium atoms with rather high temperature factor coefficients U_{33} (right).

Tab. 7.1 Ternary alkali metal scandium halides ASc_xX_3.

Compound	Scandium occupation	$d(Sc-Sc)/pm$	Remarks	CSD/Ref.
$CsSc_xCl_3$	$x=0.70(1)$	302.44(5)	$CsNiCl_3$ type	CSD-413914
$CsSc_xBr_3$	$x=0.70(1)$	318.37(6)	$CsNiCl_3$ type	CSD-413913
$CsSc_xI_3$	$x=0.80(2)$	341.6(1)	$CsNiCl_3$ type	CSD-413912
$RbSc_xCl_3$	$x=0.70(1)$	297.54(8)	$CsNiCl_3$ type	CSD-413915
$RbSc_xBr_3$	$x=0.79(1)$	316.56(9)	$CsNiCl_3$ type	CSD-413916
$RbSc_xI_3$	$x=0.8$?	337–343	$KNiCl_3$ type	Unpublished
$K_{0.75}Sc_xI_3$	$x=0.91(3)$	336–344	$KNiCl_3$ type	CSD-413917
$Na_{0.5}ScI_3$	$x=1$	327.8(5), 357.2(5)	$FeTiO_3$ type	Ref. [8]
$LiScI_3$	$x=1$	338.4(1)	$FeTiO_3$ type	Ref. [8]

perovskite type as just described. This is always observed when the ratio of the ionic radii of A^+ and X^- is such that the A^+ ion is incorporated in a closest packing of X^- spheres (one-quarter of which it substitutes) so that the coordination number (C.N.) of 12 can be achieved for A^+. When the $r(A^+)/r(X^-)$ ratio becomes unfavorable or, in other words, the A^+ cation is too small for C.N. 12, a structural variant is observed, sometimes called the $KNiCl_3$-type structure, in which the $[ScI_{6/2}]$ chains are shifted relative to each other (note that each chain is surrounded by six equal chains) by some part of the hexagonal c axis. In this way, the C.N. of the A^+ cation is reduced to 8 in $K_{0.75}Sc_{0.91}X_3$. When even this C.N. is no longer possible, i.e. with the small alkali metal ions Na^+ and Li^+, the ilmenite-type structure ($FeTiO_3$) is observed. Here, hexagonal closest packing of I^- is observed (as in the CdI_2 type, as above with $Sc_{0.89}I_2$) and both the alkali metal and scandium atoms occupy one-third each of the octahedral interstices (Fig. 7.7).

When members of the series of alkali metal scandium iodides are compared, it is interesting to note that in $CsSc_{0.80}I_3$ only the scandium position is under-occupied; in $K_{0.75}Sc_{0.91}I_3$, however, it is both the potassium and scandium positions, while in $Na_{0.5}ScI_3$ it is only the sodium position while the scandium position is fully occupied, all according to single-crystal X-ray structure determinations. The oxidation states of scandium are approximately the same in all three compounds, +2.5, +2.46, and +2.5, respectively. This means that once again there must be electronic reasons for such behavior.

Before we deal with the electronic structures of these compounds, let us first pay some attention to their physical properties. The halides $CsSc_{0.7}Cl_3$, $CsSc_{0.7}Br_3$, and $CsSc_{0.8}I_3$ all have intense colors, being blue, green, and purple, respectively. They exhibit broad bands in the visible to near-infrared (VIS/NIR) region, Fig. 7.8, with maxima at about 12500, 11200, and 7800 cm^{-1} for the chloride, bromide, and iodide, respectively. The colors of the bromide and the iodide stem from a mixture of two absorptions, the first of which originates from a d-d transition and the second from an intense charge-transfer band that

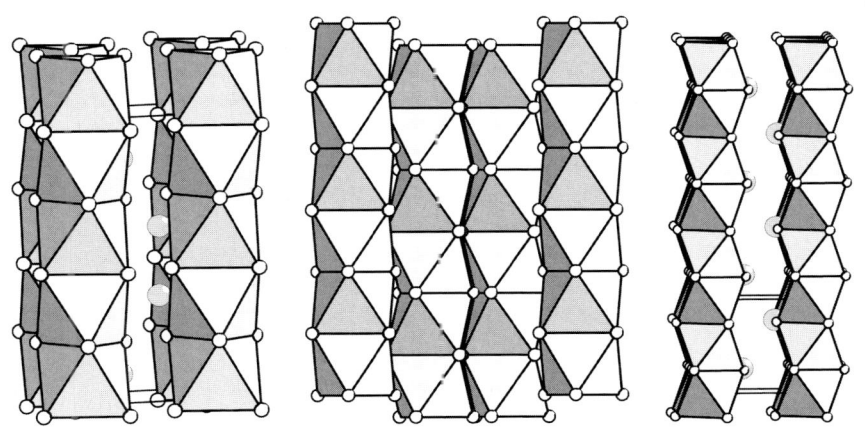

Fig. 7.7 A comparison of the crystal structures of $Na_{0.5}ScI_3$, $K_{0.75}Sc_{0.91}I_3$, and $CsSc_{0.80}I_3$ (from left to right). The [ScI_6] polyhedra are shaded whereas the alkali metal and halide ions are illustrated as grey and white circles, respectively.

extends from the ultraviolet into the visible region. The absorption maxima were calculated with the aid of the angular overlap model (AOM) for a single d^1 ion (Sc^{2+}). At 12 600, 11 100, and 7 800 cm^{-1}, the values obtained are in excellent agreement with the measurements.

Absorption spectra can, in general, only be observed when electronic transitions (accompanied with vibrational and rotational modes) occur. In the present case (Fig. 7.8), spectra typical for a 3d transition metal cation in an octahedral ligand field are observed. Furthermore, the expected red shift of the absorption maxima in the sequence Cl^-, Br^-, I^- is also observed, in accord with the spectrochemical series. Therefore, the electrons must be localized, or, in other words, Sc^{2+} (d^1) is present.

This assumption can be further verified by electron spin resonance spectroscopy. Compounds with localized and an odd number of unpaired electrons are expected to give an ESR signal. Thus, in the case of $CsSc_{0.7}Cl_3$ one would expect an ESR spectrum if Sc^{2+} (d^1) is present. Indeed, an almost isotropic ESR spectrum (Gaussian linewidth, $\Delta H = 194$ Oe) is observed in the temperature range 4.6–293 K (Fig. 7.9) with a typical average g value of 1.98 for a d^1 ion in an axially distorted environment. The analysis of the observed hyperfine splitting with eight lines is in accord with an electronic spin of $S = \frac{1}{2}$; coupled to the nuclear spin of scandium of $I = 7/2$ ($A_{hf} = 47$ Oe, $g = 1.978$, $\Delta H_{hf} = 30$ Oe). Only ~20% of the total intensity of the signal can be described by the isotropic hyperfine structure that develops at higher temperatures.

A closer inspection of the ESR spectrum at somewhat lower temperatures reveals a second splitting, which might result from either axial and/or equatorial hyperfine interactions. Unfortunately, this anisotropy is not sufficiently well resolved so that no further attempts have been made to fit the data in this respect.

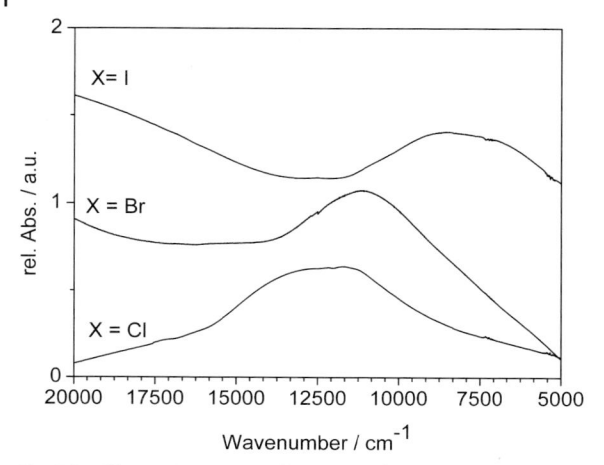

Fig. 7.8 Absorption spectra for $CsSc_{0.7}Cl_3$, $CsSc_{0.7}Br_3$, and $CsSc_{0.8}I_3$.

As the temperature is lowered further, the structure of the hyperfine interaction vanishes. This may be understood in terms of the just mentioned increasing anisotropy resulting from local distortions of the [$ScCl_6$] octahedra, possibly caused by a transition from a dynamical to a static Jahn–Teller effect. Such a phenomenon is well known for many octahedral copper(II) compounds [17]. The linewidth of the ESR signal remains almost constant over the whole tem-

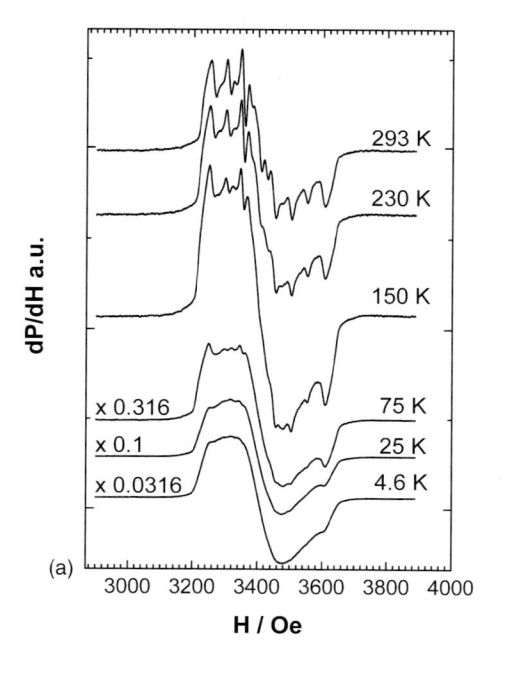

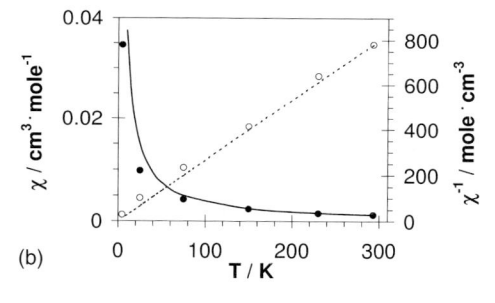

Fig. 7.9 (a) Temperature dependence of the ESR spectrum of $CsSc_{0.7}Cl_3$ (enlargement factors are indicated for low temperatures). (b) The Curie behavior (full and dashed line for an $S=½$ system) of the normalized intensities of the ESR signal represented by χ_{ESR} (●) and χ_{ESR}^{-1} (○) per mole Sc^{2+}.

perature range. If one were to assume motional narrowing at high temperatures, this could well be responsible for the development of the more pronounced isotropic hyperfine structure with decreasing linewidth for ΔH_{hf} but constant A_{hf} above 150 K. Furthermore, the temperature dependence of the ESR signal intensities closely follows a Curie law.

In summary, on the basis of absorption as well as ESR spectroscopy, we state that the excess electrons in the ternary reduced scandium halides must be localized. In order to create a structural model, we assume that three-quarters of the scandium positions are occupied, and hence consider the composition as $CsSc_{0.75}Cl_3$ or $Cs_4Sc_3Cl_{12}$. As we do not see a crystallographic superstructure, we assume that the scandium atoms are ordered in one strand but that there is no long-range order between the strands. Therefore, we consider triple units of octahedra occupied with scandium separated from each other by empty octahedra (Fig. 7.10).

According to $(Cs^+)_4(Sc^{3+})_3(e^-)(Cl^-)_{12}$, we have one excess electron per octahedral triple unit, hence we could assume that the three scandium atoms have $3d^0$, $3d^1$, and $3d^0$ configurations, respectively. This is a plausible model, and indeed there is an analogy in titanium chemistry. $K_4Ti_3Br_{12}$ was recently obtained and its crystal structure determined; it contains $[Ti_3Br_{12}]^{4-}$ trimers. The distances within this trimer clearly show that the electronic configurations of the three titanium atoms are $3d^1$, $3d^2$, and $3d^1$, respectively [18].

Quantum chemical calculations for fully reduced $CsScI_3$ and for "real" $CsSc_{0.8}I_3$ indicate that in both cases there are bonding Sc–Sc interactions below the Fermi level, and they would be more favorable in $CsScI_3$. It appears quite plausible that Sc–Sc interactions become stronger if more electrons participate. On the other hand, Sc–I antibonding becomes more pronounced around the Fermi level when more electrons are filled in. Thus, it is again the avoidance of Sc–I antibonding that contributes more to the minimization of the energy of the whole system than metal–metal bonding interactions. Fig. 7.11 illustrates these considerations: Fig. 7.11a shows the bonding situation for a hypothetical $CsScI_3$ and Fig 7.11b that for a model structure as shown in Fig. 7.10 with the overall composition $CsSc_{0.75}I_3$.

Fortunately, the case of $Na_{0.5}ScI_3 = NaSc_2I_6$ is even easier to handle. It exhibits the ilmenite type of structure, see Fig. 7.7 (left). The octahedral interstices in the [001] direction of the hexagonal unit cell are fully occupied and alternating

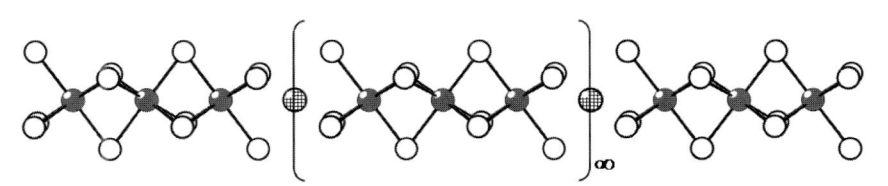

Fig. 7.10 Assumed triple-octahedral units in $CsSc_{0.75}Cl_3$, for example (● scandium, ⊕ empty sites, ○ chlorine).

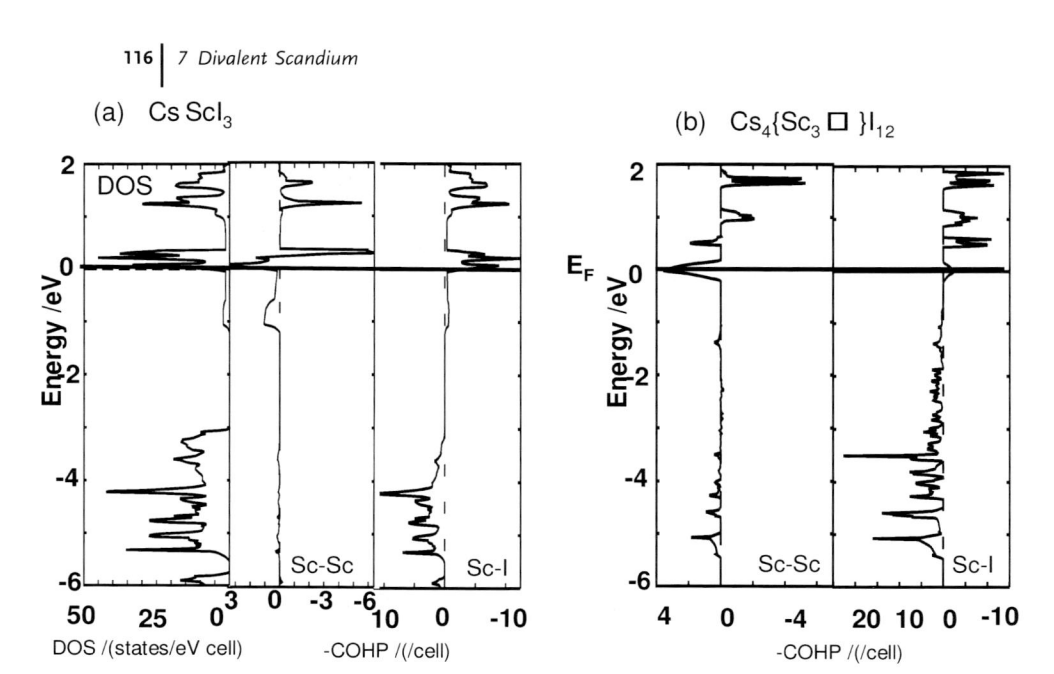

Fig. 7.11 Density of states (DOS) and crystal overlap Hamiltonian population (COHP) curves for $CsScI_3$ and $CsSc_{0.75}I_3 = Cs_4Sc_3I_{12}$.

longer and shorter Sc–Sc distances of 355 and 331 pm, respectively, in the sense of a Peierls distortion are observed by single-crystal X-ray structure determination.

According to $NaSc_2I_6 = (Na^+)(Sc^{3+})_2(e^-)(I^-)_6$ or $(Na^+)(Sc^{3+})(Sc^{2+})(I^-)_6$, it should be metallic or paramagnetic. However, preliminary magnetic measurements revealed diamagnetism. This result, together with the crystal structure, can be understood in terms of a two-center–two-electron bond. If one assumes that a metal–metal bond between two Sc^{2+} ions is present, a rather different view of the chain structure (Fig. 7.12) has to be considered. The sequence within one chain should then be $\cdots Sc^{2+}-Sc^{2+}\cdots Sc^{3+}\cdots Sc^{3+}\cdots Sc^{2+}-Sc^{2+}\cdots$ with short $Sc^{2+}-Sc^{2+}$ and longer $Sc^{3+}\cdots Sc^{3+}$ and $Sc^{3+}\cdots Sc^{2+}$ distances. It follows that only every fourth interatomic distance should be short. The structure determination, however, reveals alternating distances, i.e., every second Sc–Sc distance is short. This observation can now only be understood if neighboring chains are shifted by 0,0,1/2 relative to each other. The averaged picture obtained from the X-ray structure solution is then in reasonable agreement with the model for typical interatomic distances as $d(Sc^{2+}-Sc^{2+}) \approx 320$ pm, $d(Sc^{3+}-Sc^{3+}) \approx 340$ pm (averaged: $d_1 = 330$ pm), and $d(Sc^{3+}-Sc^{2+}) = 355$ pm $= d_2$.

In conclusion, we can assume for $NaSc_2I_6$ that the one excess electron per formula unit resides in a two-center–two-electron bond. Scandium–scandium interactions ought to be "observed" by quantum-chemical calculations.

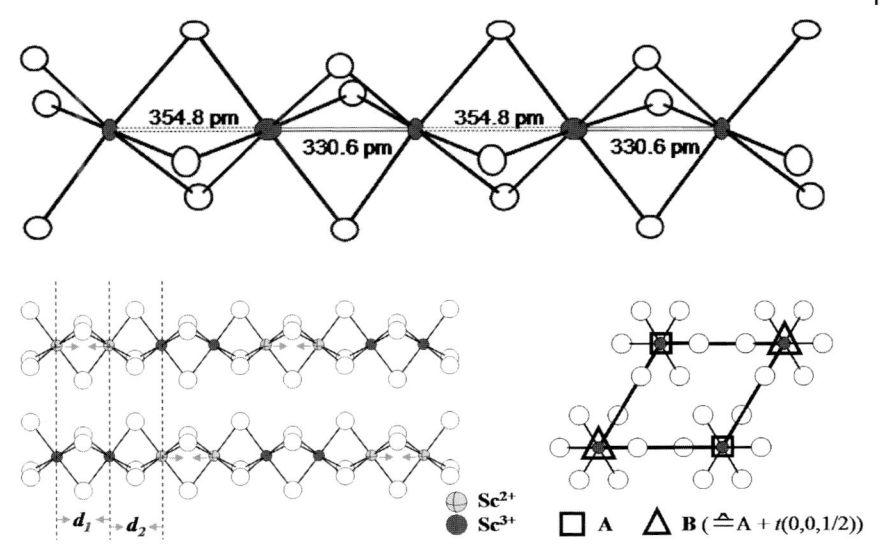

Fig. 7.12 Part of the infinite chain of face-sharing [ScI$_6$] octahedra in NaSc$_2$I$_6$ with alternating short and long Sc–Sc distances (above) as observed from single-crystal X-ray structure refinement. The model used for the description of the physical properties and interatomic distances is depicted below.

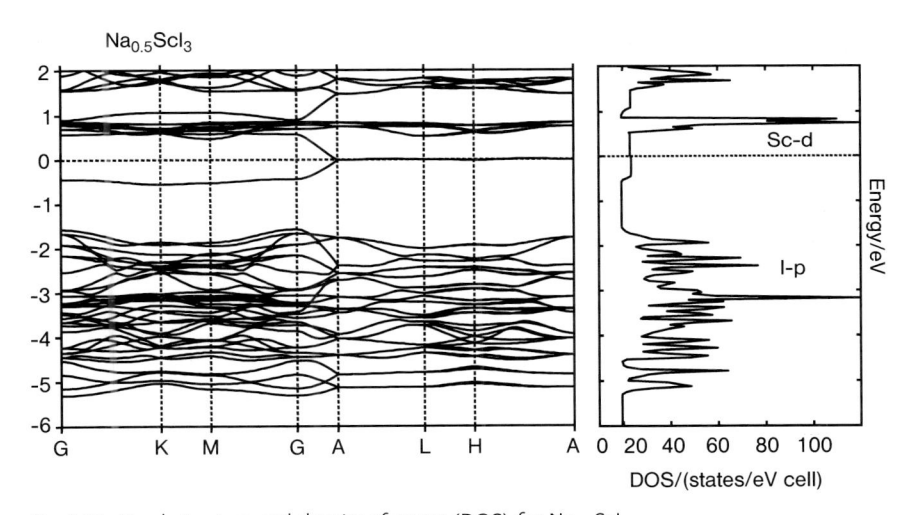

Fig. 7.13 Band structure and density of states (DOS) for Na$_{0.5}$ScI$_3$.

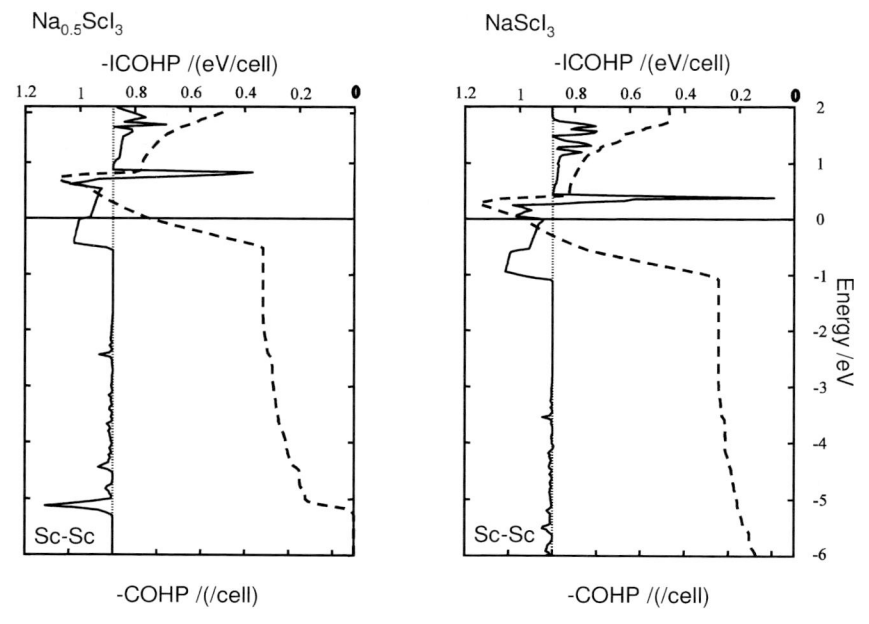

Fig. 7.14 Crystal overlap Hamiltonian population (COHP) curves and integrated crystal overlap Hamiltonian population (ICOHP, dashed vertical curves) for $Na_{0.5}ScI_3$ and $NaScI_3$.

For these calculations, no further assumptions for a structural model have been made, other that the results of the crystal structure determination for $NaSc_2I_6$ were used. Thereby, the electronic structure can be reliably calculated (*cf.* Fig. 7.13). Again, bonding analysis by crystal Hamiltonian overlap populations shows that the difference between a hypothetical $NaScI_3$ and the truly existing $Na_{0.5}ScI_3$ is such that through depopulation of the Sc levels overall bonding gets strengthened. Although more electrons would lead to more Sc–Sc bonding interactions (Fig. 7.14), at the same time these interactions are more than outweighed by antibonding Sc–I interactions.

7.6
Conclusions

Binary and ternary trihalides of scandium, ScI_3 and $Cs_3Sc_2X_9$ (X = Cl, Br, I), may be reduced by scandium metal yielding, for example, scandium-deficient $Sc_{0.9}I_2$ and $CsSc_{0.7}Cl_3$. From X-ray structure determinations and according to other physical properties such as magnetic measurements, ESR, and absorption spectroscopy, the excess electrons according to $9 \times Sc_{0.89}I_2 = Sc_8I_{18} = (Sc^{3+})_8(e^-)_6(I^-)_{18}$ are delocalized above about 100 K, making $Sc_{0.9}I_2$ a two-dimensional metal. Below 100 K, $Sc_{0.9}I_2$ is an insulator. Then, out of nine octahedral positions within the closest-packed iodide layers, eight are occupied, six by divalent Sc^{2+} and two by still trivalent Sc^{3+}. The reason for this rather strange behavior has its origins in the electronic structure of $Sc_{0.9}I_2$. Bonding Sc–Sc interactions compete with (stronger) antibonding Sc–I interactions around the Fermi level. The occupation of both bonding and antibonding states is optimized by under-occupation of scandium positions. Similar reasons are responsible for the scandium deficiency of, for example, $CsSc_{0.7}Cl_3$. In a reasonable structural model, the excess electron with respect to $4 \times CsSc_{0.75}Cl_3 = (Cs^+)_4(Sc^{3+})_3(e^-)(Cl^-)_{12}$ is localized at the middle atom of an $[Sc_3Cl_{12}]$ trimer composed of three octahedra sharing faces. This model is in agreement with the observed ESR spectrum and the red shifts of the absorption bands for the halides $CsSc_xX_3$ (X = Cl, Br, I). In $Na_{0.5}ScI_3 = NaSc_2I_6$, Sc–Sc σ-bonds occur in a rather complicated way.

Thus, the optimization of the bonding situation in these "divalent" scandium halides leads to delocalization of the excess $3d^1$ electron(s), leading to metallic behavior ("$Sc^{3+}e^-$"), or localization at sufficiently low temperatures leading to insulating and paramagnetic properties (Sc^{2+}), or even dimerization through covalent σ-bonding (Sc–Sc) whenever the matrices in which the scandium ions (Sc^{3+}) are reduced are chosen properly.

Acknowledgements

This work has been supported through stipends to LJ by the Alexander von Humboldt foundation and to AVM by the Fonds der Chemischen Industrie (Liebig-Stipendium). Generous support by the State of Nordrhein-Westfalen and the Universität zu Köln and by the Deutsche Forschungsgemeinschaft (Sonderforschungsbereich 608: Komplexe Übergangsmetallverbindungen mit Spin- und Ladungsfreiheitsgraden und Unordnung) is also gratefully acknowledged. We are also indebted to Dr. V. Kataev, now of IWF Dresden, for the ESR measurements.

References

1 J. EMSLEY, *The Elements*, 2nd edition, Clarendon Press, Oxford, 1991.
2 S. COTTON, *Lanthanides and Actinides*, McMillan Education, London, 1991.
3 S.-J. HWU, J. D. CORBETT, *J. Solid State Chem.* **1986**, *64*, 331.
4 B.C. MCCOLLUM, J. D. CORBETT, *J. Chem. Soc., Chem. Commun.* **1968**, 1666.
5 B.C. MCCOLLUM, D. J. DUDIS, A. LACHGAR, J. D. CORBETT, *Inorg. Chem.* **1990**, *29*, 2030.
6 K. R. POEPPELMEIER, J. D. CORBETT, T. P. MCMULLEN, D. R. TORGESON, R. G. BARNES, *Inorg. Chem.* **1980**, *19*, 129.
7 G. MEYER, J. D. CORBETT, *Inorg. Chem.* **1981**, *20*, 2627.
8 A. LACHGAR, D. S. DUDIS, P. K. DORHOUT, J. D. CORBETT, *Inorg. Chem.* **1991**, *30*, 3321.
9 T. GLOGER, *Dissertation*, Universität zu Köln, **1998**.

10 G. Meyer, Th. Staffel, S. Dötsch, Th. Schleid, *Inorg. Chem.* **1985**, *24*, 3504; G. Meyer, in: *Advances in the Synthesis and Reactivity of Solids* (Ed.: T. E. Mallouk), JAI Press, Vol. 2, **1994**.

11 J. D. Corbett, *Inorg. Synth.* **1983**, *22*, 31; G. Meyer, in: *Synthesis of Lanthanide and Actinide Compounds* (Eds.: G. Meyer, L. R. Morss), Kluwer Acad. Publ., 1991, p. 135.

12 J. D. Corbett, in: *Synthesis of Lanthanide and Actinide Compounds* (Eds.: G. Meyer, L. R. Morss), Kluwer Acad. Publ., 1991, p. 159; G. Meyer, Th. Schleid, in: *Synthesis of Lanthanide and Actinide Compounds* (Eds.: G. Meyer, L. R. Morss), Kluwer Acad. Publ., 1991, p. 175.

13 N. Gerlitzki, *Dissertation*, Universität zu Köln, **2002**.

14 K. R. Poeppelmeier, J. D. Corbett, *Inorg. Chem.* **1977**, *17*, 294.

15 G. Meyer, S.-J. Hwu, S. Wijeyesekera, J. D. Corbett, *Inorg. Chem.* **1986**, *25*, 4811.

16 S.-J. Hwu, P. P. Ziebarth, S. von Winbush, J. E. Ford, J. D. Corbett, *Inorg. Chem.* **1986**, *25*, 283.

17 H. Bill, in: *The Dynamical Jahn-Teller Effect in Localized Systems* (Eds.: E. Perlin Yu, M. Wagner), Elsevier, Amsterdam, 1984.

18 L. Jongen, G. Meyer, *Z. Anorg. Allg. Chem.* **2004**, *630*, 1732.

8
Rare-Earth Metal-Rich Tellurides – A Spectrum of Solid-State Chemistry, Metal–Metal Bonding, and Principles

John D. Corbett

8.1
Introduction

Motivation and justification for doing research can take many forms. One general and powerful driving force, the desire to discover new chemistry, has led to much of what we know. An example of this is the research described here, not because of its overarching value or applicability, but from the sheer newness and novelty of the compounds and structures that seem to be waiting to be discovered. Chemistry in solid-state systems is a relatively young field, and these seem to be prolific and productive places to search. The general thrust of the research described herein arose from a curiosity about the metal–salt interface, that is, the region of metal-rich compounds in which properties of the metal and salt extremes could be expected to meet and blend. Herein, the consideration is restricted to the orbitally-richer transition metal systems. Halides (X) among simple salts were prominent in the earliest investigations, well typified by the pioneering and extensive studies of niobium and tantalum cluster halides by Schäfer, Schnering, and others [1]. The fact that glass vessels served as adequate containers for these studies was important to their development. On the other hand, similar studies of well-reduced systems of the earlier and more active transition (or alkaline earth) metals proceed much more cleanly with the use of niobium or the more refractory tantalum as nonreducible containers. For example, these allowed the discovery of a great many new metal cluster halides, particularly of zirconium, and later of the rare-earth metals R (for reviews, see refs. [2–4]). The compositions of many of the products containing isolated octahedral metal clusters were close to MX_2, and the electronic states of the clusters could be described semi-quantitatively by fairly simple MO schemes. However, these systems highlight an important feature owing to the relative paucity of valence electrons for the earlier elements relative to those of Nb, Ta, etc.; thus, essentially all cluster compounds of the earlier metals contain and require a cluster-centering atom, in other words an interstitial Z that contributes electrons and central bonding. A good number of the rare-earth metal systems (meaning lanthanides plus Y, Sc) also form a variety of ternary phases with lower halide proportions ($X:R \geq 1:1$) in which they exhibit condensation to form mainly oligomers, but also infinite 1-D chains, some 2-D (layered)

Inorganic Chemistry in Focus II. Edited by G. Meyer, D. Naumann, and L. Wesemann
Copyright © 2005 WILEY-VCH Verlag GmbH & Co. KGaA, Weinheim
ISBN: 3-527-30811-3

constructions, and a few 3-D examples. Iodides seem to be the most numerous. (Of course, the breadth of and trends in the results may be biased by where the most efforts have been applied.) Most constructions can be envisaged as originating from cluster octahedra that are further condensed via shared *trans* edges. In general, halide anions always appear to be tightly bound to edges or vertices of the rare-earth metal clusters (as opposed to more unbound states), and as such may be so numerous as to greatly limit the degree of metal cluster condensation and metal–metal bonding possible. In parallel with this, halide phases containing discrete clusters are essentially all poor semiconductors, whereas metallic characteristics (open valence bands) are generally found for the condensed chain and sheet compounds.

A change to chalcogenide anions with twice as large a charge (oxidation state) seemed an obvious way to attain more extensive metal–metal bonding. In fact, many of the new rare-earth metal tellurides described herein contain metal arrays that, although structurally very different in detail, still exhibit about the same range of average metal oxidation states as do the cluster halides (+1.0±0.4). The choice of Sc and Te as the starting elements was based on the facts that (a) the little-studied element Sc is the smallest R, and (b) the largest possible anion Te^{2-} would allow manifestation of the greatest degree of lower dimensional metal–metal bonding inasmuch as these anions customarily function as spacers between, or sheaths around, R–R aggregates. The neighboring group 4 (Ti–Zr) chalcogenides have been better studied, particularly for the lighter S and Se, and some similarities with these will be noted later.

A significant hindrance to earlier investigations of the highly reduced rare-earth metal chalcogenide systems was evidently again the lack of a suitable container among the traditional choices. Thus, no pure rare-earth metal chalcogenide (Ch) with R:Ch proportions >1.0, that is, more reduced than the fairly common RCh (NaCl type), has evidently ever been obtained without the use of an Nb or Ta container. (Mo or W would serve this purpose, but these are harder to fabricate.) Of course, metal-rich compounds of the same metals are also found as pnictides, with Sb and Bi in particular [5] (and with earlier main group element anions as well), but in the author's opinion these are rather different and will not be considered here. It is also true that intermetallic characteristics are increasingly approached in the pnictide and tetrelide (Si–Pb) systems, whereas Te^{2-} anions essentially act purely as inert spacers with filled and low-lying valence levels that only contribute to some polar covalent bonding with adjoining metals. Our rather productive extensions to ternaries R–Tn–Te have generally been successful with the later transition metals Tn from the 3d, 4d, and 5d series. These take advantage of a known trend, already recognized in the cluster halide systems, that early–late transition metal combinations in intermetallic systems are especially stable. This follows on from concepts first advanced by Brewer and co-workers about 30 years ago [6]. Nevertheless, the challenging yet exciting and rewarding aspects of these studies is that many of the discoveries have been unforeseen and without precedent. It is very difficult to design the synthesis of something you cannot imagine!

Four groups of compounds will be presented to illustrate some of the remarkable features found in telluride systems, mainly with Sc and Lu, but also with a few examples with Y, Gd, and Dy:

1. Binary metal-rich tellurides. These afford some remarkable combinations for Sc and completely unprecedented compounds with Lu. In some instances, analogous electron-richer phases with Ti or Zr were either known or have since been found. Some lessons in matrix (size) effects also come out of these, for example, that in some circumstances distances alone may be poor measures of relative metal–metal bond strengths as inferred from overlap populations.
2. The polymerization of the metal cluster ribbons in Sc_2Te into layers by a late transition metal through a structurally simple displacement/coupling reaction.
3. The versatility of the Fe_2P structure type among ternary rare-earth-transition metal tellurides, etc., in organizing rather diverse chemistries into a common lattice type.
4. Two new and contrasting ways in which hexagonal-shaped columns of R atoms that enclose zigzag chains of Tn may be condensed into either pillar or sheet structures, both with $R_5Tn_2Te_2$ compositions. Furthermore, a conceptual relationship can be found not just between these two, but also with a seemingly common parent, an unusual structure of a condensed ternary gadolinium iodide.

All of these tellurides appear to be metallic, meaning that no classical electronic rules arising from closed-shell (octet or other) bonding are operative, rather the highest-lying electrons are bound in open bands. Notwithstanding, many parts of the metal–metal bonding appear to be effectively optimized in quite narrow ranges of composition and electron counts, such that nearly all compounds appear to be close to if not truly stoichiometric "line" phases. Bonding in these is appreciably delocalized, as appropriate for infinite solids, but covalent R–Tn as well as R–Te bonding is still important. Furthermore, the Madelung contributions to the total energies, the coulombic part of the bonding between (and among) the metal components and telluride anions, must be highly significant with regard to structure selection and stability, although this is often overlooked and seldom quantified. Given this circumstance, one should expect that efficient packing and the relative sizes of ions, clusters, and other charged units will also be important in relation to the stability of a given structure (see, for example, ref. [7]). The synthetic reactions all require elevated temperatures, and so the products pertain to thermodynamic stability.

8.2
Novel Binary Tellurides

Relatively electron-poor and orbitally-rich compounds of the nature of the known Sc_2Te, Sc_8Te_3, and Sc_9Te_2 would be expected to exhibit delocalized bonding and some incompletely filled bonding states, and this is indeed the case. However, the ways in which a stable composition and structure may be achieved are impossible to predict a priori, in the face of a presumably large number of possible and often unknown alternatives and the aforementioned complexities of what determines stability.

Fig. 8.1 depicts a [010] projection of the structure of Sc_2Te down a short (3.92 Å) axis, with the Sc atoms represented by dark circles and the Te atoms by unfilled circles [8]. The marked two-center Sc–Sc bonds have been selected according to their overlap population values (OP), specifically with a break at 3.5 Å corresponding to a minimum (and arbitrary) OP cut-off at ≥ 0.05 (except for the Sc4 chain). For clarity, no Sc–Te bonds (pairwise contacts) are marked, although polar covalent bonding between the Sc polyatomic cations and Te anions certainly exists (in addition to coulombic interactions). Two metal frameworks are evident in this structure: some larger complex "blades" and fairly isolated zigzag [Te_2] chains, both quasi-infinite along the projection direction. Their perception is aided if it is noted that the Sc atoms all lie on mirror planes and that adjoining members around the outside all differ in depth by $b/2 = 3.92/2$ Å. The separate units are shown in Fig. 8.2 with the distances marked (in Å). (Centers of symmetry lie at the center of the short Sc6–Sc6 contacts, 3.05 Å.) Fragments that aid understanding of the organization can be seen in the middle of the larger unit, namely distorted octahedra that share *trans* edges and also some

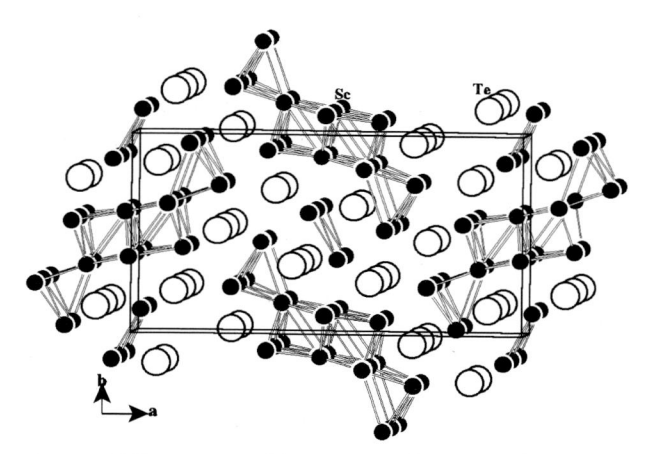

Fig. 8.1 [010] Projection of the structure of Sc_2Te along its short 3.92 Å axis. All atoms lie at $y=0$, 1, or 1/2;. Black atoms are Sc, white are Te. Only Sc–Sc bonds of less than 3.50 Å (OP>0.04) are marked.

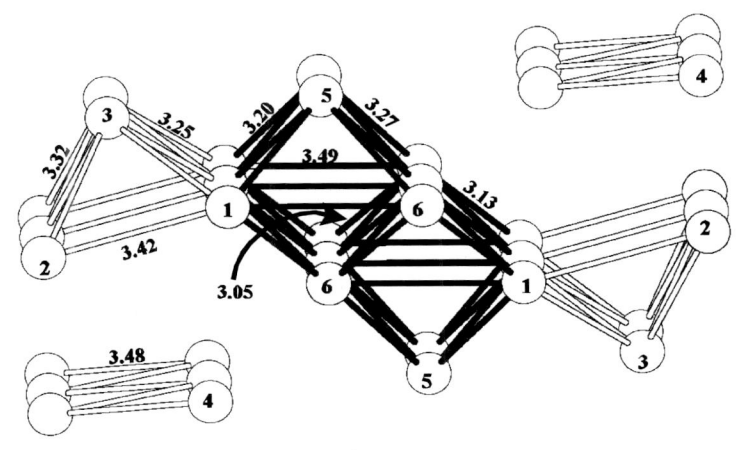

Fig. 8.2 The metal units in Sc$_2$Te with Sc–Sc separations marked (Å). The pairwise overlap populations for each of these are detailed in the text. The larger OP values (≥0.12) occur within the central unit outlined in black.

side edges in the projection. This block unit is augmented at the extremes by triangles or trigonal prisms of metal atoms.

Before these features are detailed, it is worthwhile to look ahead at some descriptions of the bonding. Extended Hückel band calculations yield the DOS (densities of states) for an infinite solid as a function of energy, as shown on the left in Fig. 8.3 for the later-discovered isotypic Dy$_2$Te [9]. The number of valence electrons available is sufficient to fill only the lower part of the large valence (conduction) band (to ε_F) that is generated principally by Sc 3d or Dy 5d orbitals. Not surprisingly, the phases are predicted to be metallic and this has been confirmed for Dy$_2$Te. (The Te 5p states are off-scale at the bottom of Fig. 8.3.) Calculations of the integrated overlap populations (OP) for pairwise interactions, and the related COOP curves (overlap-weighted populations) as a function of energy (Fig. 8.3, right) show the same general effect for both R$_2$Te compounds – that R–R states actually remain bonding (positive) up to ca. 0.9 eV above ε_F. In other words, the metal–metal bonding is not optimized, a fact that seems appropriate in such an electron-poor system.

If we match the distances in Sc$_2$Te (Fig. 8.2) with the respective overlap populations, we can get a better idea of where the stronger bonding interactions are in the central unit. Thus, the shorter Sc–Sc distances in Sc$_2$Te and the corresponding OP values are 3.05 Å and 0.37 in the center (Sc6–Sc6), 3.13–3.27 Å and 0.23–0.15 around the outer parts of the central block, 3.25–3.42 Å and 0.12–0.07 in the outer triangles, 3.48 Å and 0.03 in the zigzag Sc4 chain, and 3.53 and 3.68 Å with 0.04 and 0.01 for the shortest separations between the large and small metal units (not marked). For emphasis, note that the 3.48 Å Sc–Sc separation in the zigzag chain and the very similar 3.49 Å "across the waist" of

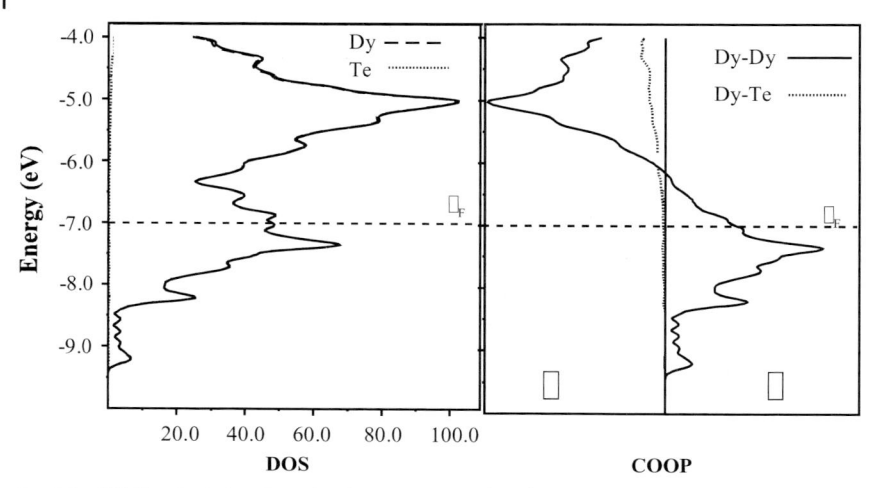

Fig. 8.3 EHTB calculational results for Dy$_2$Te, isostructural with Sc$_2$Te. Left: The densities of states (DOS) continuum for Dy$_2$Te as a function of energy. Electrons occupy the states up to the Fermi energy (ε_F). Right: The corresponding COOP curves, the overlap-weighted bond populations for Dy–Dy (solid line) and Dy–Te (dotted) as a function of energy. Data to the right (+) side are bonding, to the left (–) antibonding.

the shared octahedra (Sc1–Sc6) have very contrasting overlap populations of 0.03 and 0.23, respectively. Delocalized electrons are clearly concentrated in the interior of the metal array. Note also that the Sc–Sc distances in the pure metal (3.24 Å, CN12) or Pauling's single-bond metal–metal distance (2.88 Å [10]) have little relevance to the "bonds" in this low-dimensional and electron-poor example. (Note that the average oxidation state of Sc and Dy in these R$_2$Te systems is +1, not zero.)

Differences of this sort are repeatedly found for compounds of the types discussed here, which really bring into focus the contrasts that may exist between distances and intrinsic bond strengths as a result of matrix effects. Electrons are logically bound among the metal cores and avoid tellurium anions. This in fact means that metal–metal overlap populations (qualitatively considered to be related to *bond energies*) between metal atoms diminish more or less in parallel with increases in the number of Te atoms about a pair of neighboring metal atoms. Thus, the isolated Sc4–Sc4 chain in Sc$_2$Te can be envisaged as existing principally only because of "matrix effects" generated by coulombic Te and Sc atom interactions and packing, irrespective or in spite of any electron population between them. (The absence of Li–Li bonding in LiF(s) is a well-known parallel example, in this case there effectively being no electrons for such bonds considering how much more tightly bound are the electrons on fluoride.) The effect of tellurium on neighboring Sc–Sc bonding is easy to explain: covalent bonding interactions between filled low-lying 5p states on Te and empty higher-lying 3d orbitals on Sc will cause the latter to be raised ("pushed up") in energy, thus causing the outer R atoms on the surfaces of metal aggregates to be generally less well bonded to one another. The reader can imagine how such matrix effects might appear on a plot of OP data *vs.* less meaningful R–R separations. The OP data for interior bonds within the metal framework would fall around a distinctly higher curve relative to OP at similar distances on the outside of the metal array where there are Te neighbors. The latter may in fact even fall on a regular and distinctly lower curve, trends of the sort clearly seen with Sc_9Te_2 [11] and for Zr_2Te [12] *vs.* Sc_2Te. These characteristics also mean that the most positive charges (the lowest calculated atom populations) will be found among more isolated metal atoms, e.g. Sc4 in Fig. 8.1, whereas the lowest, perhaps even negative, charges so approximated will be for metal atoms in the interior of a metal array. (See the examples in Section 8.4 on Fe_2P below.)

Overall changes seen on going from Sc_2Te to Dy_2Te reflect the larger core and the larger orbitals on the latter, which broaden the valence band and generally raise OP values (which, by the way, should not be intercompared quantitatively for different elements). The Te^{2-} anion in the presence of the larger metal in Dy_2Te is a relatively poorer spacer than it is in Sc_2Te, and this affords more R–R bonding (lesser matrix effects) both between and within the chains. Thus, it is not surprising that increases in comparable R–R distances on going from Sc_2Te to Dy_2Te are less than the differences in the usual estimates of core radii for Sc^{3+} *vs.* Dy^{3+} or for the metals. It is worth noting that Dy bonding in Dy_2Te principally involves 5d orbitals, as it does in the metal, and this occurs before we reach the traditional 5d transition metals. The 4f core states are only truly important with regard to magnetic properties and the like.

The isostructural β-Ti_2Se [13] and Zr_2Te [12], with 50% more metal-based electrons, were subsequently also shown to exist. The radii of Sc and Zr are very similar, and the distances in Sc_2Te and Zr_2Te are roughly similar throughout. Notwithstanding, the broader valence (conduction) band in the latter is filled to a higher level, these electrons occupying more of the bonding states and thereby increasing the overlap populations, especially for contacts corresponding to the more marginal bonds in Sc_2Te. The Zr–Zr bonding is still not optimized, however, judging from the COOP curves. Detailed consideration was also given to the effects of the greater valence electron concentration in Zr_2Te on bond strengths, the lessened influences of Te on the bonding, and relationships with other zirconium and hafnium chalcogenides [12].

As an aside, the syntheses of Dy_2Te and Gd_2Te are considerably more difficult than that of Sc_2Te. The latter is obtained in good yield after arc-melting of a pressed pellet of an appropriate mixture of Sc and Sc_2Te_3 or ScTe followed by annealing for a few days at around 1125 °C and slow cooling. In this system, the neighboring ScTe melts at a relatively low temperature, near 1000 °C. The Dy/Dy_2Te_3 system contains the customary DyTe (NaCl-type), but in contrast this now melts at ~1850 °C, whereas the relatively less stable Dy_2Te decomposes incongruently into Dy and DyTe near 1060 °C. Arc-melting thus gives a Dy+DyTe product that becomes much too segregated on cooling to achieve sufficient reaction during subsequent annealing below 1060 °C. Rather, successive reactions of powdered Dy and Dy_2Te_3 in pressed pellets interspersed with grinding are much more effective in achieving single-phase Dy_2Te because of the reduced diffusion pathlengths between mixed phases.

Two more reduced binary Sc–Te phases have also been found, Sc_8Te_3 and Sc_9Te_2, both of which have analogues among the group 4 chalcogenides. The view of Sc_8Te_3 [14] in Fig. 8.4a, again projected along a short 3.85 Å axis, illustrates its complexity, although the general character of the condensation into sheets is similar to that seen for Sc_2Te. Note, however, that the two puckered sheets therein are different. The DOS and COOP data and the break between marked bonding and unmarked "nonbonding" separations of 3.50 Å in Sc_8Te_3 are very similar to those already presented for Sc_2Te. The environments around and lack of any bonding between Te^{2-} ions are also comparable. An essentially similar situation is found in Y_8Te_3 [15].

The considerable contrasts between the largely 2-D bonding in Sc_8Te_3 and the much more 3-D bonded metal arrays in the isotypic Ti_8S_3 [16] and Ti_8Se_3 [17] are quite striking. These arise from both the 44% greater number of metal-based valence electrons available and the presence of smaller anion spacers. This

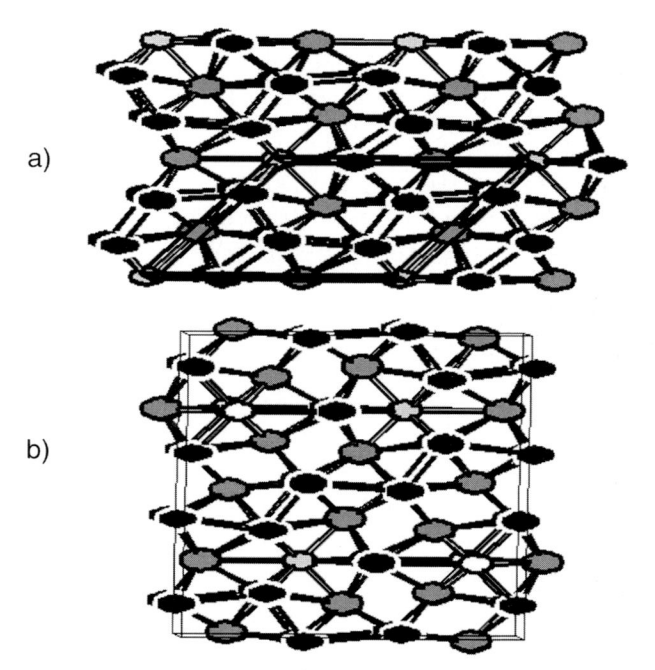

Fig. 8.4 (a) [010] Projection of the structure of Sc_8Te_3 with Sc–Sc bonds drawn for overlap populations greater than 0.03 (≤ 3.50 Å). The two chains (sheets in 3-D) are different. The general character of the cluster condensation resembles that in Sc_2Te, Fig. 8.1.
(b) Projection of the two lower chains in the isotypic Ti_8S_3 with bonds drawn for overlap populations above 0.04 (≤ 3.20 Å). The open connections represent the additional bonds formed in Ti_8S_3 by this criterion, presumably because of additional metal electrons and the smaller S spacer.

is emphasized in Fig. 8.4b, which shows portions of the equivalent pair of sheets in Ti_8S_3. The "bonds" therein are now marked for Ti–Ti separations of 3.20 Å and less, the corresponding cut-off according to calculated overlap populations. The open bonds represent the additional ones gained over Sc_2Te according to the OP criterion because of the smaller S atoms and the electron-richer Ti.

The rather complex Sc_9Te_2 structure [11] is not illustrated, but it shows the result of an apparently distortive transition relative to the related but higher symmetry Ti_9Se_2 [18]. There seems to be no precedent that would have enabled one to anticipate the remarkable discoveries made in the equivalent Lu–Te system. Fig. 8.5 illustrates these surprises in projections down the short ∼ 3.7 Å axes of the structures of (a) hexagonal Lu_8Te and (b) orthorhombic Lu_7Te [19]. Both exhibit simple hexagonal close-packing (AB...) of atom layers along the projection axis so that all atoms have fractional coordinates of either 1/2 or 0, 1 along this axis. Atoms that lie in the middle layer are colored gray; viz., black and dark gray for Lu, white and light gray for the smaller Te, respectively. The Lu layers in these views are far from close-packed, but regular sequences can be recognized along

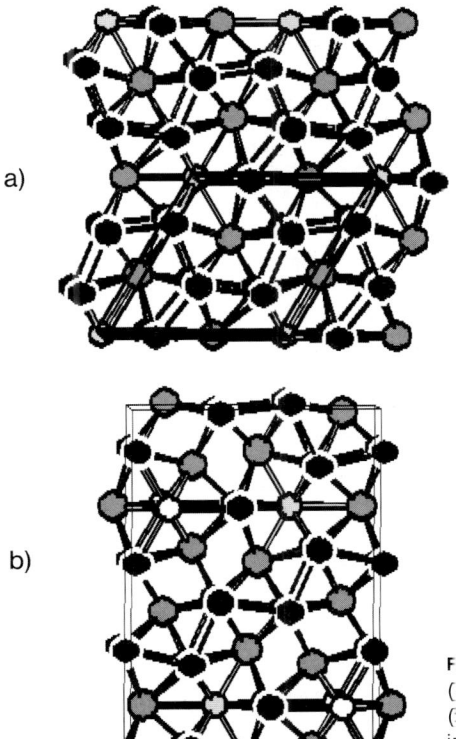

Fig. 8.5 Projections down the short (∼ 3.7 Å) axes of (a) hexagonal Lu_8Te and (b) orthorhombic Lu_7Te. Simple close packing applies along the projection axes. Black and dark gray atoms are Lu, white and light gray are Te, the gray one in each pair being midway along the projection.

the vertical axis. The novel results can best be viewed in terms of regular substitutions of smaller Te atoms in a rather distorted h.c.p. Lu metal, specifically for every fourth atom in every third horizontal row of Lu in Lu_8Te and for every other atom in every fourth row of Lu in Lu_7Te. The Te atoms center regular and distorted tricapped trigonal prisms of Lu in Lu_8Te and Lu_7Te, respectively. The volumes per atom and the coordination numbers of all Lu decrease in this order, the latter from 12 in the pure metal to 9–11 in these remarkable compounds. It is very difficult to imagine what might be found next in these very unusual systems. Analogous R_8Te or R_7Te products have not been found for R = La, Pr, Gd, Ho, Tm or Yb, that is, even for the closest neighbors of Lu! With the protocol established, these syntheses do not appear especially difficult: Lu_7Te is formed in high yield after arc-melting of a suitable composition followed by annealing of the sample at around 1300 °C. Lu_8Te was first formed through disproportionation of Lu_7Te at 1000–1200 °C, but it can also be synthesized directly by heating a suitable pelletized Lu/Lu_2Te_3 mixture at 1000 °C for 2 weeks.

8.3
The Conceptual Polymerization of Sc_2Te by Transition Metals

$$4Sc_2Te + Tn \rightarrow Sc_6TnTe_2 + 2ScTe; Tn = Pd, Ag, Cu, Cd$$

The discoveries of essentially all of the compounds reported herein came as the unexpected outcomes of numerous exploratory reactions, most of which produced nothing new or novel. Occasionally, however, running the right reaction for the wrong reason can give an exciting result, as in this instance. The reaction as written has actually never been carried out starting with Sc_2Te, rather a product was discovered that shows this relationship following an attempt to make something else, in this case to replace Ni with Pd in the orthorhombic $Sc_5Ni_2Te_2$ (see Section 8.5). The structural evolution in the formal conversion of Sc_2Te to Sc_6PdTe_2 is illustrated in Fig. 8.6 [20]. In this case, the particular Te atoms circled in the left view are formally displaced by d^{10} Pd during a reductive coupling reaction, some of the starting phase in turn being oxidized to ScTe. This reaction converts what have been called "blades" in Sc_2Te into rumpled heterometal sheets, the attack taking place at bonds for which the former bonding (OP) was relatively weak (low). Thus, the weakly bonded zigzag chains of Sc4–Sc4 (Fig. 8.2) are strengthened by this substitution, such that the OPs for Sc–Sc around the added Tn become comparable to those elsewhere in the structure. Some of the driving force for this reaction must be the particular stability of early–late transition metal bonding (Section 8.4), the Pd becoming bound in a tricapped trigonal prism of Sc.

Isostructural products have been obtained in subsequent studies in which the Pd in Sc_6PdTe_2 has been replaced first by Tn = Ag, Cd or $Cu_{0.8}Te_{0.2}$ [21], and later in the isostructural Y_6TnTe_2 phases by Tn = Ru, Os, Rh, Ir, Pt, Ag or Cu [22]. The

8.3 The Conceptual Polymerization of Sc_2Te by Transition Metals | 131

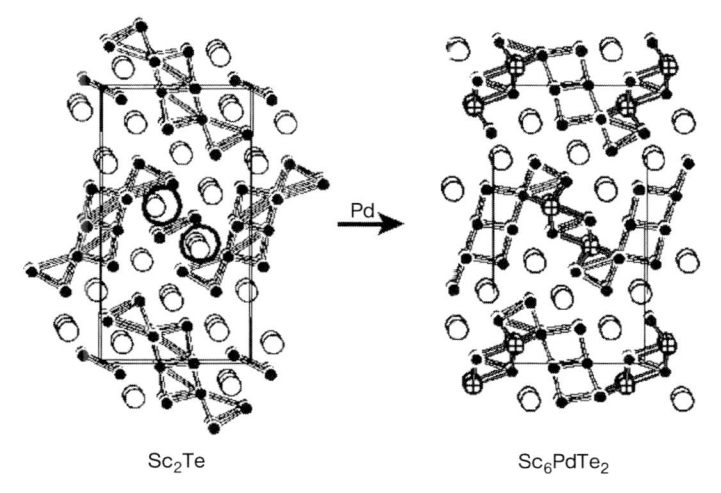

Fig. 8.6 The conceptual polymerization of the Sc_2Te ribbons into orthorhombic Sc_6PdTe_2 sheets by Pd metal with displacement of ScTe. Sc is black, Te white, Pd hatched. The Te displaced by Pd are circled on the left.

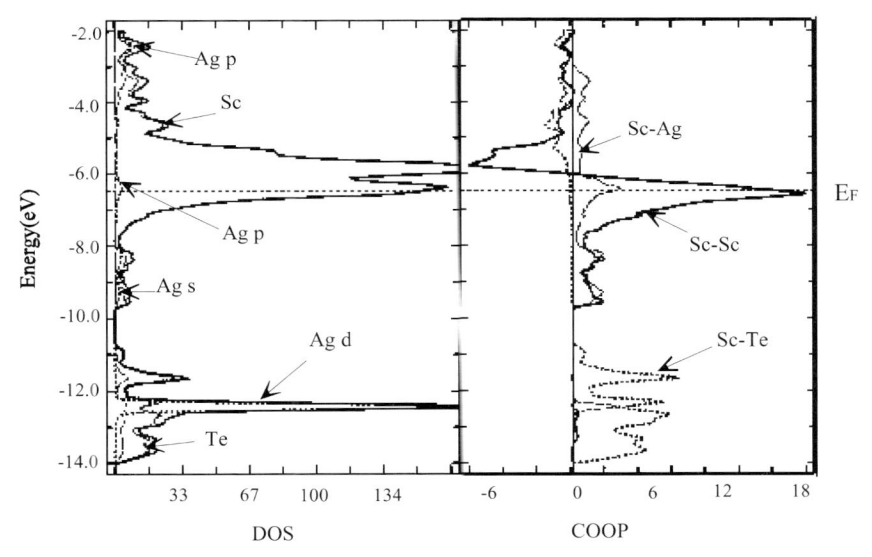

Fig. 8.7 Left: The DOS for orthorhombic Sc_6AgTe_2, with the Sc, Ag, and Te projected out in dotted, dashed, and dot-dash lines, respectively. Right: The corresponding COOP data for all Sc–Sc distances < 3.6 Å and all Sc–Ag and Sc–Te < 3.0 Å.

latter is different in that the conceptual starting phase Y_2Te does not exist. Theoretically, the 5s and 5p orbitals on Ag, for example, contribute in the higher occupied levels in the broadened valence band in Sc_6AgTe_2 (Fig. 8.7). It is worth noting that the formal electron counts of these Tn atoms, which of course increase on going from Ru to Cd, do not appear to be of great importance. The open valence band in the metallic substrate has continuous filled and empty bonding states that meet at E_F and so the latter can instead act as an electron reservoir. The stabilities of *alternate* phases are usually much more important in such questions of phase stability. Finally, atom charges according to the Mulliken approximation are informative. The four surface Sc lose charge to the interior and become somewhat positive (+0.6 to 0.9 in the case of Ag), the two interior Sc are negative (–0.8 and –1.0), the Ag is intermediate ($\sim$–0.1), and the Te is of course negative (–0.6). Remember that bonding electrons are partitioned equally between the atom pairs in the Mulliken approximation, even though the bonds may be quite polar.

The foregoing orthorhombic phases should not be confused with the higher symmetry hexagonal Sc_6TnTe_2 examples in the following section. Greater charge transfers are found among the latter.

8.4
Versatility of the Fe$_2$P-Type Structure and Some Amazing Substitutions

This versatile hexagonal structure features two types of tricapped trigonal prisms defined by the more active metal, and these are usually each centered by a differ-

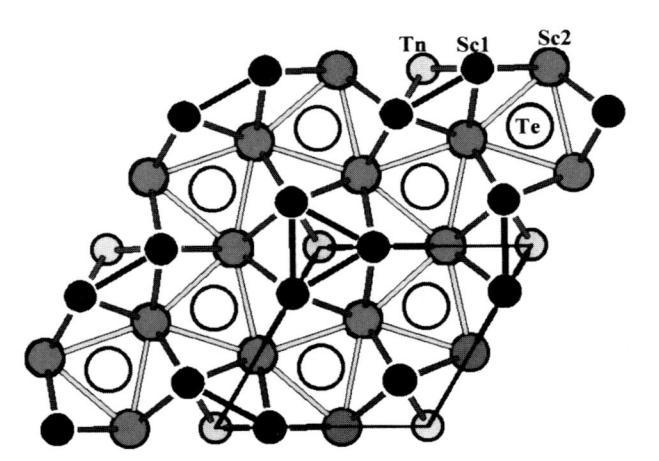

Fig. 8.8 The structural characteristics of the hexagonal Fe$_2$P-type structure for Sc$_3$TnTe$_2$ [001]. Smaller trigonal prisms of Sc(1) (black) around the origin are centered by the smaller transition metal Tn (light gray) and capped on the three rectangular faces by Sc(2) (dark gray). The latter, in turn, generate larger trigonal prisms that share vertical edges and are displaced by $b/2$ from those of Sc(1). These are centered by Te (white) and, in turn, are face-capped by a distant Sc(1).

ent element, commonly a main group member. In the parent structure there are equal numbers of three-fold Fe sites (Fe(1), Fe(2)) and two P sites, a one-fold and a two-fold site, which according to atom types make the structural formula $Fe_3 \cdot Fe_3PP_2$. In Fig. 8.8, the disposition of these is shown along [001] for Sc_6TnTe_2, in which smaller trigonal prisms (TP) of Sc(1) (black) are centered by Tn (light gray) at the cell origin and face-capped by Sc(2) atoms (dark gray). (All atoms again lie on mirror planes at $z=0$ or $1/2$). The latter Sc(2) atoms displaced by $c/2$ lie on cell edges and within the cell, and these generate an array of larger trigonal prisms that share parallel edges (along view), are centered by Te (white), and are in turn face-capped by Sc(1) (black). The arrangement thus results in two types of trigonal prisms of the active metal that are capped on all rectangular faces by the other type. Of course, both TP types share their basal triangular faces above and below in this view to generate infinite arrays along [001]. According to theory as well as interatomic distances, stronger bonding of Sc occurs along [100]. Differentiation between the two types of three-fold metal sites that define the trigonal prisms occurs in some intermetallic phases, but this need not concern us here. Rather, differentiation of the main group elements (G in place of P) instead gives the principal motif we find here, R_6TnG_2 phases in which Tn is customarily a transition metal. (Among ternary intermetallics these are sometimes known as Zr_6CoAl_2 types.)

Tab. 8.1 summarizes the diversity possible in the phases of present interest, starting with the more conventional R_6TnTe_2. It may be presumed that the absence of the next end members of the Tn series, for example, Mn, Re, and Cu, Pt in the first two rows in the table, means that these reactions yielded other products. The R_6TnTe_2 category is not new; there are also a variety of group 4 metal examples of this type, Zr_6TnTe_2, Tn=Mn–Ni, Ru, Pt [27] as well as $Hf_6Tn_{1-x}Sb_{2+x}$, Tn=Fe, Co, Ni [28], in which a partial (mixed) Sb occupancy occurs at the Tn site as well. The last three rows record some novel discoveries in which nonmetals may occupy the nominal Tn site, and the site commonly occu-

Tab. 8.1 Fe$_2$P-type R$_6$TnG$_2$ phases and their variations

R$_6$	Tn [a]	G [b]	Ref.
Sc	Fe–Ni	Te	23
Sc	Ru, Os, Rh, Ir	Te	24
Y, La	Co	Te	25
Dy	Fe–Ni	Te	25
Gd	Co, Ni	Te	26
Er	Ru	Te	26
Sc	Te$_{0.80}$	Bi$_{1.68}$	24
Lu	Mo	Sb	24
Lu	Te	Lu	19

[a] Normally a transition metal within the smaller [R(1)]$_6$ trigonal prismatic site (3f).
[b] Usually a main-group element that centers the larger [R(2)]$_6$ trigonal prismatic site (3g).

pied by Te (G) is occupied by the pnictogen Bi or Sb or even by a rare-earth metal. Any simple assignment of periodic characteristics to a particular centering site thus becomes difficult, although relative sizes appears to be an important consideration.

The more common R_6TnTe_2 phases consistently place the large Te in the larger trigonal prism defined by R(2). The R(2) elements in this TP are also relatively close to the R(1) atoms in the smaller TP, where they have a face-capping role, but the converse is not true. The R(1)–Tn interactions within the smaller TP appear to be strong by two measures: the R–Tn distances are quite close to Pauling's metallic single-bond radii sums, especially for Sc–Os, Sc–Rh, and Lu–Mo, and the overlap populations are also substantial (see below). In the new regime illustrated by $Sc_6Te_{0.80}Bi_{1.68}$, Lu_6MoSb_2, and Lu_6TeLu_2, the smaller interstitial atom is still found in the smaller TP, but now this is Te or Mo. This follows from the fact that the other nominal main group component, Sb, Bi, or Lu, is now the larger and is bound in the larger two-fold TP. The low stoichiometries for Te and Bi in the bismuth example are rather inexplicable, and its synthesis was irregular too in that the compounds appeared to have been formed by an autogenous chemical transport reaction, naturally in lower yields. Otherwise, all the examples considered here appear to be equilibrium products that are obtained in high yields at a reasonable temperature and in the correct stoichiometry.

A notable electronic flexibility for compounds of this structure type is already implied by the distribution of examples in Tab. 8.1. Computational results in the form of DOS and COOP plots illustrate this in a somewhat more quantitative manner as these take into account the intrinsic differences in the orbital valence energies and sizes of the component R, Tn, and G (usually Te) atoms. However, such calculations require detailed examinations to verify the inferences. A more useful guide to some of the contrasts between polar interactions and intermetallic bonding that are both present in these compounds can be found in the approximate charges determined from atom populations. These have been estimated from extended Hückel calculations with the aid of the Mulliken approximation, by which the electrons in formal two-center bonds are split equally. These numbers should be construed as only semi-quantitative charge or polarity estimates as the extended Hückel bonding model in itself is a rather simple one. Nevertheless, the chemical trends indicated seem to be quite useful.

The atom charges so approximated for representatives of the Fe_2P-type compounds noted here, starting with the isoelectronic Tn elements Fe, Ru, and Os in the Sc_6TnTe_2 family, are collected in Tab. 8.2. (Orbital energies used were generally those iterated to charge consistency, and the listed Wyckoff symbols 3f, 3g, 1b, and 2c are the formal crystallographic designations for the respective R1, R2, Tn, and G (or Te) positions employed heretofore.)

Two opposing bonding characteristics are operative in the more regular Sc_6TnTe_2. (a) The large differences in orbital energies (and electronegativities) between Sc(2) and Te in the larger TP mean that these also afford the larger coulombic contributions, whereas the orbital energies of the Sc-dominated states are raised through covalent interactions, as before. The Sc(2)–Sc(2) bond-

Tab. 8.2 Effective atom charges in some R_6TnG_2 phases as calculated by extended Hückel means [24]

Wyckoff site	Sc_6FeTe_2		Sc_6RuTe_2		Sc_6OsTe_2	
	Atom	Charge	Atom	Charge	Atom	Charge
3f	Sc1	−0.71	Sc1	−0.7	Sc1	−0.58
3g	Sc2	0.47	Sc2	0.48	Sc2	0.69
1b	Fe	1.98	Ru	1.92	Os	0.86
2c	Te	−0.62	Te	−0.61	Te	−0.6

Wyckoff site	Sc_6TeBi_2 [a]		Lu_6TeLu_2		Lu_6MoSb_2	
	Atom	Charge	Atom	Charge	Atom	Charge
3f	Sc1	−0.08	Lu1	0.29	Lu1	−0.41
3g	Sc2	0.05	Lu2	−0.01	Lu2	0.29
1b	Te	−0.36	Te	−0.2	Mo	2.14
2c	Bi	0.21	Lu3	−0.32	Sb	−0.89

a) Calculated with full occupancies.

ing is less than that with Sc(1). (b) More similar orbital energies and more mixing between Sc(1) and Tn = Fe, Ru, Os pertain to the smaller TP. Here, the electron-rich Tn elements transfer some charge to the surrounding more numerous and electron-poor Sc(1) atoms (−0.7 to −0.6 e⁻) and thereby become somewhat positive, less so in the case of Tn = Os as the valence orbitals contract appreciably. Interestingly, this is the direction of electron transfer that Brewer postulated in mixed electron-poor–electron-rich intermetallic systems [6]. For Lu(1) trigonal prisms and Mo and Sb interstitials, the Mo → Lu(1) charge transfer is in the same direction, whereas that for Lu(1) → Sb is comparable to that for Sc(2) → Te, the Sb p orbitals lying about as low as Te 5p. The Lu–Lu, Lu–Mo, and Lu–Sb bonds appear stronger and the bands wider than with Sc, paralleling the 5d orbital size effects noted earlier for Sc_2Te vs. Dy_2Te.

The inverse populations in Sc_6TeBi_2 (necessarily calculated for full occupancies), with the larger and less electronegative Bi in the Sc(2) TP, reflect a diminished polarity and mixing at that site, and the charge transfer in a broader band well below E_F is small and in the other direction. The more polar Sc(2)–Te interactions now result in charge transfer to Te, but less than before with tighter bonding of Te in the 2c site. The largest charge difference is seen between the interstitials Bi and Te.

Finally, Lu_6TeLu_2 (Lu_8Te), with little parallel to the bismuth case, shows more interesting effects, for now the customarily larger interstitial site is occupied by Lu, and Te is again in the smaller TP of Lu(1). (Compare Figs. 8.5a and 8.8.) The DOS and Lu–Lu COOP curves are broader in energy and higher owing to stronger interactions. Charge transfer Lu(1) → Te takes place, of course, but also from Lu(1) toward the other Lu with fewer Te neighbors, first to Lu2 (−0.01 e⁻), which is face-capping to the $[Lu(1)]_6$ TP, and then to Lu3 (−0.32 e⁻) in the larger

TP, in which it has no Te neighbors. These are precisely the bonding effects that were deduced earlier in relation to the binary scandium ribbon and sheet tellurides. One would expect the same general effects in Lu_7Te, but this is of lower symmetry and more complex to analyze. In particular, one waist-capping Lu is 0.36 Å further away from Te than the other two, evidently because of crowding (Fig. 8.5 b).

8.5
Diverse, Contrasting, and Yet Related Cluster Chain Motifs

Predicting or understanding the choices of structures among condensed cluster compounds is exceedingly difficult, especially because one usually cannot foresee or imagine what the alternatives may be. Since most phases are metallic, there are no closed-shell or other simple and transferable bonding rules to aid this selection, rather an aggregate of complex factors [7, 29]. These can be sorted into, firstly, delocalized metal–metal bonding that is often optimized, that is, the bonding states are essentially all filled in an open band. This is complemented by covalent metal–nonmetal bonding, e.g. for R–Te, but the Madelung energies of such an assembly of polar components (cations, cluster anions, monoanions, etc.) are always likely to be of major importance as well. Although one can usually expect fairly regular geometries for halide around metal aggregates, viz., three- or four-bonded to edges and apices on the metal structure, telluride is apparently always higher coordinate and less clear-cut in its functionality, and usually resides in an augmented (face-capped) trigonal-prismatic environment. Finally, there must be quite good packing and space filling in solid materials with good metal–metal bonding and containing polar components.

Out of necessity, one tends to look for and then quickly grasp any discernible relationships among these structures, that is, in the form of building blocks, such as for the apparent polymerization of the Sc_2Te structure through telluride displacement by certain metals (see Section 8.3, Fig. 8.6). Such seeking out of similarities also pertains, albeit in a more distant manner, to two other rather contrasting structures, both with $R_5Tn_2Te_2$ compositions, for R = Sc or Y and Tn = Ni (among others). A conceptual relationship of each of these to the unusual architecture of the halide Gd_3MnI_3 [30] is a common thread.

a) $Y_5Tn_2Te_2$; Tn = Fe, Co, Ni [31]
This novel orthorhombic structure (*Cmcm*) was once again the product of an exploratory synthetic investigation, with nothing out of the ordinary beyond the need to run the reactions at ~1050 °C in welded Ta tubing. A projection of the structure for Tn = Fe is given in Fig. 8.9, again along the short *a* axis of ~3.95 Å. This view gives the impression of columns of yttrium hexagons (black) condensed into sheets at *trans* vertices, but these and the internal chains of Fe atoms are in fact both puckered, such that neighboring atoms in both chains are alternately displaced by *a*/2. The result is columns of puckered Y

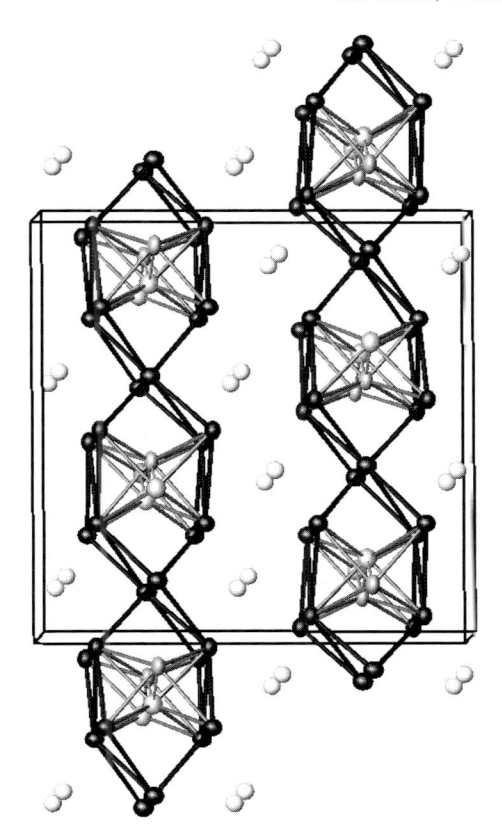

Fig. 8.9 [100] Projection of the structure of orthorhombic $Y_5Fe_2Te_2$ along the short axis (horizontal). Zigzag chains of Fe (small gray) are surrounded by puckered hexagonal columns of Y (black) and condensed at *trans* vertices to yield vertical metal sheets. Te is white.

atoms sharing all vertices that surround a zigzag Fe atom chain, as shown more clearly in the side view of two adjoining columns in Fig. 8.10. Considerations of distances and overlap populations indicate that all of the marked two-center interactions are strong, including that for Fe–Fe, $d = 2.30$ Å, OP = 0.29. In a simple

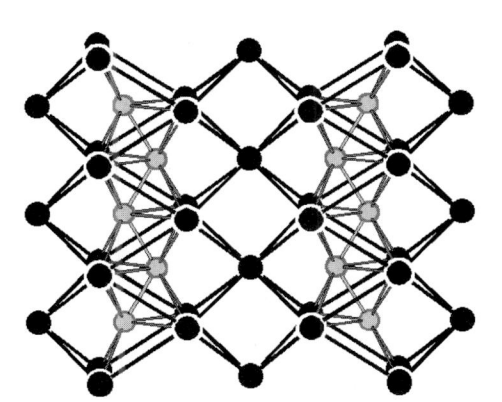

Fig. 8.10 Side view [001] of two Y columns (vertical) and the enclosed Fe chains in $Y_5Fe_2Te_2$. Black – Y, gray – Fe.

view, there may be some matrix effects [8, 9, 11] from the Y–Fe and Y–Y bonded network that hold the Fe atoms together, irrespective of their bonding, as their separations are actually 0.03 Å less than Pauling's single-bond metallic distance [10]. Theoretically, there appears to be good mixing of Y and Fe orbitals in a broad conduction band, and the Y–Fe bonding (COOP) is nearly optimized in this complex arrangement. The bonding is notably less for Ni–Ni, however, as this core is contracted appreciably, but the lattice along $\vec{a}$ or $\vec{b}$ is not. Moreover, a monohydride exists only in the case of Ni, in which the added atom is bound in a tetrahedral site formed by bending of the sheet at each shared vertex (Fig. 8.9), so as to create a smaller cavity.

b) $Sc_5Ni_2Te_2$ [32], $Y_5Cu_2Te_2$ [27]

This orthorhombic phase, obtained by a similar procedure, shows a distinctly different architecture (Fig. 8.11 – *Pnma*), although a qualitative relationship with that of $Y_5Tn_2Te_2$ can still be seen. Here, quite similar columns surround a zigzag chain of Ni, but a pair of these are instead condensed in a side-by-side manner. This column has lower symmetry, there being only centers of symmetry on the central Sc–Sc vertical bonds together with mirror planes perpendicular to the view at $y = 0$, 1/2 that contain all atoms. An alternative description can also be found. The top and bottom halves of the figure can be viewed as being constructed from pairs of rectangular pyramids of scandium that share side edges and are infinitely condensed along the view direction by sharing opposite edges. Two independent Ni atoms are strongly bonded to, and lie 0.54 Å below, the bases of these. Opposed pairs of these double-pyramidal chains displaced by $b/2$ are further interbonded base-to-base along vertical Sc–Sc bonds to generate the Ni–Ni chains. In this case, the Ni–Ni separations appear to be dictated largely by the strong Sc–Sc and Sc–Ni bonding; the inter-nickel distance is 2.66 Å as compared with a single-bond value of 2.30 Å and has a corresponding overlap

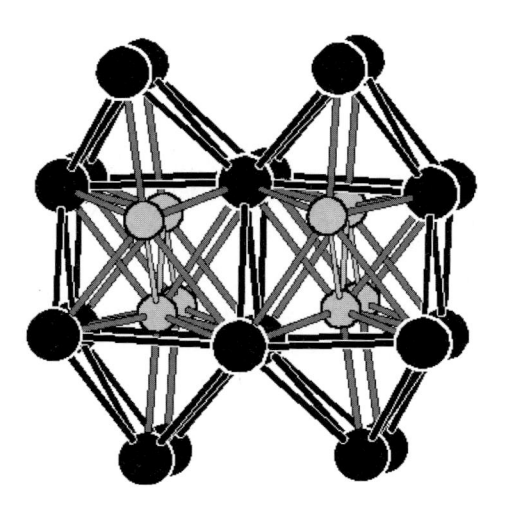

Fig. 8.11 [010] Projection along the chain structure of orthorhombic $Sc_5Ni_2Te_2$ in which Sc columns surround zigzag Ni chains. Such columns are now condensed in pairs (left and right) through sharing of side edges. The top and bottom halves can also be viewed in terms of rectangular pyramids condensed in pairs and also along the view, the basal faces of all pyramids being capped by Ni.

population of only 0.02. Now the conduction band is nearly all Sc in origin, and a somewhat broadened Ni (+Sc) valence band is centered at ~ 1.4 eV below ε_F. The upper levels in the latter are antibonding, as expected for a nominal d^{10} shell on Ni. This structure does share one common feature with the foregoing $Y_5Tn_2Te_2$, i.e. zigzag 3d metal chains within a sheath of rare-earth metal. This suggests a possible common parentage.

c) Gd_3MnI_3

The Gd_3MnI_3 structure was a quite singular example when it was discovered in 1994 [30], but it now affords a common structural pattern and model for the construction of the two aforementioned $R_5Tn_2Te_2$ structures. However, the relationships are largely geometric in character; the "before" and "after" chains are by no means isoelectronic, and some chemical prestidigitation is necessary in the transformation as well.

The conceptual conversion is shown in Fig. 8.12, in which the three metal structures and their neighboring anions are all viewed in simple projections along the cluster chains. The constructions of the individual columns are all very similar, with a common zigzag chain of 3d interstitials bonded within columns of a rare-earth metal. The starting metal column in the iodide has also been described in terms of rectangular pyramids, condensed along the view, analogous to those in the top and bottom halves of Fig. 8.11. The derivation of the two tellurides follows if iodides in the relatively electron-poorer Gd_3MnI_3 are first replaced by half as many tellurides, and the chains are then condensed as shown (along with suitable transmutations of the framework atoms Gd → Y, Sc; Tn interstitials Mn → Ni) to form the two products with elimination of one mole of RTe, i.e.:

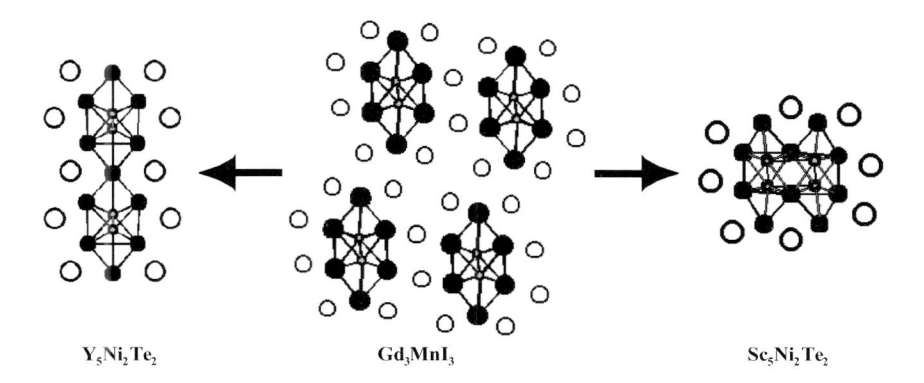

Fig. 8.12 The geometric conversions of the isolated R chains in Gd_3MnI_3 (center) into the two different modes of condensation found in (left) $Y_5Ni_2Te_2$ via shared R vertices to generate sheets and (right) $Sc_5Ni_2Te_2$ with shared side edges to generate columns. All structures are depicted along the infinite metal chains together with the nearest nonmetal positions.

$$2R_3TnI_3 \rightarrow R_5Tn_2Te_2 + RTe$$

As indicated on the left in Fig. 8.12, Y_5NiTe_2 is formed when chains of metal columns similar to those in Gd_3MnI_3 are condensed into sheets at opposite vertices, whereas Sc_5NiTe_2 arises if these columns are instead condensed in a side-by-side manner. The former mode gives sheets, while the latter ends with double chains, even though the two products are isoelectronic.

Although a good theoretical explanation or more quantitative models for these transformations are not evident, some differentiation between the two routes is possible on a thermodynamic basis. The formation of the layers on the left from the monochain parent yields four additional Y–Y bonds per formula unit, whereas the double-chain product on the right results in two more Sc–Sc and two additional Sc–Ni bonds. Bonds between early 4d elements (Y) are known to be stronger than for the 3d Sc [33], whereas stronger R–Tn bonds are favored for Sc–Ni as a result of better orbital mixing [34], bond strengths in both cases being inferred from overlap populations.

The iodide parent also has a distant relationship to a series of monoclinic R_3TnI_3 compounds with Tn = Ru or Os, in which pairs of chains of *trans*-edge-sharing R octahedra appear to have been condensed through sharing of side edges, as in the center of Fig. 8.1 [35]. According to an insightful analysis by Köckerling and Martin [34], these chains distort toward the Gd_3MnI_3 extreme, in which the R–R and R–Tn bonds become appreciably shorter in those cases with improved orbital mixing for the R and Tn elements. This is favored for those rare-earth metal components that have relatively high first ionization energies, Sc, Y, Gd, Tm, and Yb when coupled with 3d Tn. The proposal was for halides, but it would appear that it also pertains to Sc and Y tellurides as well. It is not clear as to whether all of the possible combinations have been tested.

Although these pictorial relationships may appear to somehow explain the simple condensation alternatives described herein, we are still no nearer to understanding why these particular structures are the most stable for these particular compositions. An inability to imagine let alone analyze alternative, and often unknown, structures and compositions is a major limitation. Experimentation still leads the way in providing answers, but not necessarily the underlying reasons.

Acknowledgements

The several skillful and insightful coworkers who are listed in the references made major contributions to the discoveries and interpretations presented here. This research has been supported by the U.S. National Science Foundation, Division of Material Science, most recently through grants DMR-9809850 and -0129785, and it has been carried out in the facilities of the Ames Laboratory, U.S. Department of Energy.

References

1. H. Schäfer, H.-G. Schnering, *Angew. Chem.* **1964**, *76*, 833.
2. J. D. Corbett, *Acc. Chem. Res.* **1981**, *14*, 239.
3. A. Simon, Hj. Mattausch, G. J. Miller, W. Bauhofer, R. K. Kremer, in *Handbook on the Physics and Chemistry of Rare Earths* (Eds.: K. A. Gschneidner, L. Eyring), Vol. 15, Kluwer Academic Publishers, Dordrecht, **1992**, p. 191.
4. J. D. Corbett, *J. Alloys Compd.* **1995**, *229*, 10.
5. A. M. Mills, R. Lam, M. J. Ferguson, L. Deakin, A. Mar, *Coord. Chem. Rev.* **2002**, *233*, 207.
6. L. Brewer, P. R. Wengert, *Metall. Trans.* **1973**, *4*, 83.
7. D.-K. Seo, J. D. Corbett, *J. Am. Chem. Soc.* **2000**, *122*, 9621.
8. P. A. Maggard, J. D. Corbett, *Angew. Chem. Int. Ed. Engl.* **1997**, *36*, 1974.
9. P. S. Herle, J. D. Corbett, *Inorg. Chem.* **2001**, *40*, 1858.
10. L. Pauling, *The Nature of the Chemical Bond*, 3rd ed., Cornell University Press, Ithaca, NY, **1960**, p. 402.
11. P. A. Maggard, J. D. Corbett, *J. Am. Chem. Soc.* **2000**, *122*, 838.
12. G. Örlygsson, B. Harbrecht, *Inorg. Chem.* **1999**, *38*, 3377.
13. T. E. Weirich, X. Zhou, R. Ramlau, A. Simon, G. L. Cascarano, C. Giacovazzo, S. Hovmoller, *Acta Crystallogr.* **2000**, *A56*, 29.
14. P. A. Maggard, J. D. Corbett, *Inorg. Chem.* **1998**, *37*, 814.
15. P. A. Maggard, J. D. Corbett, unpublished research.
16. J. P. Owens, H. F. Franzen, *Acta Crystallogr.* **1974**, *B30*, 427.
17. T. E. Weirich, R. Pöttgen, A. Simon, *Z. Kristallogr.* **2000**, *211*, 929.
18. T. E. Weirich, R. Pöttgen, A. Simon, *Z. Anorg. Allg. Chem.* **1996**, *622*, 630.
19. L. Chen, J. D. Corbett, *J. Am. Chem. Soc.* **2003**, *125*, 7794.
20. P. A. Maggard, J. D. Corbett, *J. Am. Chem. Soc.* **2000**, *122*, 10740.
21. L. Chen, J. D. Corbett, *Inorg. Chem.* **2002**, *41*, 2146.
22. L. Castro, L. Chen, J. D. Corbett, unpublished research.
23. P. A. Maggard, J. D. Corbett, *Inorg. Chem.* **2000**, *39*, 4143.
24. L. Chen, J. D. Corbett, *Inorg. Chem.* **2004**, *43*, 436.
25. N. Bestaoui, S. Herle, J. D. Corbett, *J. Solid State Chem.* **2000**, *155*, 9.
26. F. Meng, C. Magliocchi, T. Hughbanks, *J. Alloy Compd.* **2003**, *358*, 103.
27. C. Wang, T. Hughbanks, *Inorg. Chem.* **1996**, *35*, 6987.
28. H. Kleinke, *J. Alloys Compd.* **1998**, *270*, 136.
29. U. Häussermann, S. Amerioun, L. Eriksson, C.-S. Lee, G. J. Miller, *J. Am. Chem. Soc.* **2002**, *124*, 4371.
30. M. Ebihara, J. D. Martin, J. D. Corbett, *Inorg. Chem.* **1994**, *33*, 2078.
31. P. A. Maggard, J. D. Corbett, *Inorg. Chem.* **2004**, *43*, 2556.
32. P. A. Maggard, J. D. Corbett, *Inorg. Chem.* **1999**, *38*, 1945.
33. H. F. Franzen, M. Köckerling, *Prog. Solid State Chem.* **1995**, *23*, 265.
34. M. Köckerling, J. D. Martin, *Inorg. Chem.* **2001**, *40*, 389.
35. M. W. Payne, P. K. Dorhout, S.-J. Kim, T. R. Hughbanks, J. D. Corbett, *Inorg. Chem.* **1992**, *31*, 1389.

9
Zintl Phases of Tetrelides – Old Problems and Their Solutions

Reinhard Nesper

9.1
Introduction and Concept

In the overwhelming majority of cases of metal semiconductor compounds, the Zintl-Klemm concept works extremely well [1–7]. However, there are long-known problem cases of valence electron mismatch, and over a period of more than 20 years, no satisfactory proposals for their solution have been advanced. This has been mainly due to too strict an interpretation of the Zintl-Klemm concept with respect to novel bonding situations. In particular, the conjecture that Zintl phases should exhibit semiconducting properties has led to searches for semiconducting behavior in cases where metallicity was found, and these have frequently been abandoned due to "obvious experimental mistakes or mysteries". For many of these cases, we know now that metallic conductivity is wholly compatible with a slightly extended and still powerful Zintl-Klemm concept.

The Zintl-Klemm concept utilizes simple counting rules assuming full charge transfer from the metal to the semi-metal components. This approach was often criticized as being much too crude to be realistic. However, it is not the effective charge transfer that really matters, and not even the question as to where the electrons are, but only: which entities "own the electrons"? Or more precisely put: to which bonding and nonbonding pattern do the valence electrons contribute? In Zintl phases, these are always the structural patterns of the semi-metal components and metal–metal bonding does not occur [8]. This is not to say that the metal components and their spatial arrangements merely play the role of space fillers. They do have most important stabilization effects [9], in the absence of which just a few lowly charged Zintl anions can survive, i.e. X_4^{4-} and $X_9^{2-/3-}$. Up to now, practically all attempts to transfer Zintl anions with charges higher than –1 per anion center into solution or into other environments have met with failure. The only "solvents" that allow for huge cation stabilization effects comparable to those in the pristine Zintl phases are salt melts of the corresponding cation salts, preferably of metal halides [10].

The power of the Zintl-Klemm concept lies in the sound and easy-to-understand relationships between atomic and electronic structure of Zintl phases. These have been extended in different ways into valuable expressions through

Inorganic Chemistry in Focus II. Edited by G. Meyer, D. Naumann, and L. Wesemann
Copyright © 2005 WILEY-VCH Verlag GmbH & Co. KGaA, Weinheim
ISBN: 3-527-30811-3

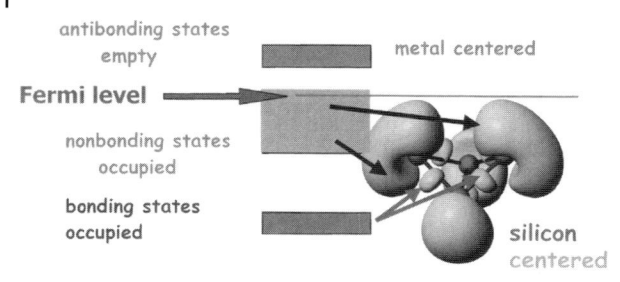

Fig. 9.1 Representative scheme of blocks of electronic states in Zintl phases; the two blocks at lower energy denote bonding as well as nonbonding and antibonding states of the Zintl anions. The Fermi level may be found above or within the second block. The upper rectangle represents the unoccupied metal-centered states.

which more predictive conclusions have been drawn from the concept [11, 12], as exemplified by Eq. (1):

$$E = (8 - N)(a + x) + N^*d + E'_c + E''_c \tag{1}$$

(E = valence electron number per formula unit; N = average bond number per semi-metal atom in a chosen defect-free reference structure; N^* = additional electrons that have to be provided upon formation of one defect in the reference structure, for example a defect in a phosphorus-type base structure localizes three electrons to form three lone electron pairs in the neighborhood of the defect; E_c = electrons located in electron-poor clusters, which may be determined according to Wades rules, for example [13], to the 18e rule, or to Teos' rules [14], respectively; E'_c = electron number in electron-poor cluster composed of semi-metal atoms; E''_c = electron number in electron-poor clusters composed of metal atoms.) By definition, Zintl phases do not contain metal-centered occupied electronic states; for these cases, Eq. (1) may be rewritten as [11]:

$$E = (8 - N)x + N^*d + E'_c \tag{2}$$

(E'_c = electron number in electron-poor clusters composed of semi-metal atoms.) Eq. (2) states that bonds only occur between semi-metal atoms and that lone electron pairs are also only localized at these centers. In ref. [12], much broader correlations between counting rules and structure are given.

9.2
Li$_{13}$Si$_4$ – One-Dimensional Metallicity vs. Cage Orbitals

Li$_{13}$Si$_4$, though seemingly constructed like a Zintl phase, does not fit the counting rules because two different Si$_2$ dumbbells have to be postulated (i.e. Si_2^{4-} and Si_2^{6-}; cf. Tab. 9.1), for which there is no experimental evidence. One of the theoretical approaches to solve the obvious mismatch between structure, valence electron

Tab. 9.1 Overview over tetrelide Zintl phases for which Li-centered cage orbitals have been postulated [15]

Compound	Cluster arrangement	Formal charges on anionic parts of the cluster	
Li_7Ge_{12}	$Li_7^2{}_\infty[Ge_{12}]$	Ge^0, Ge^{1-}, Ge^{2-}	
LiGe	$Li_\infty^3[Ge]$	Ge^{1-}	
$Li_{12}Si_7$	$(Li_6Si_5)_2(Li_{12}Si_4)$	$2Si_5^{6-}$, Si_4^{10-}	$+2e$
$Li_{12}Ge_7$	$(Li_6Ge_5)_2(Li_{12}Ge_4)$	$4Ge_5^{6-}$, Ge_4^{10-}	$+2e$
Li_9Ge_4	$2 \times Li_9(Ge_2)_2$	$4Ge_2^{4-}$	$+2e$
$Li_{14}Si_6$	$Li_{14}(Si_2)_3$	$3Si_2^{4-}$	$+2e$
$Li_{14}Ge_6$	$Li_{14}(Ge_2)_3$	$3Ge_2^{4-}$	$+2e$
$Li_{13}Si_4$	$2 \times Li_{13}(Si_2)(Si)_2$	$2Si_2^{4-}$, $4Si^{4-}$	$+2e$
$Li_{14}Ge_4$	Li_1		$+2e$
$Li_{15}Ge_4$	Li_1		
$Li_{21}Si_5$	Li		$+2e$
$Li_{21}Ge_5$	$(Li_{20}Ge_6)(Li_{22}Ge_4)$	$6Ge^{4-}$, $4Ge^{4-}$	$+2e$

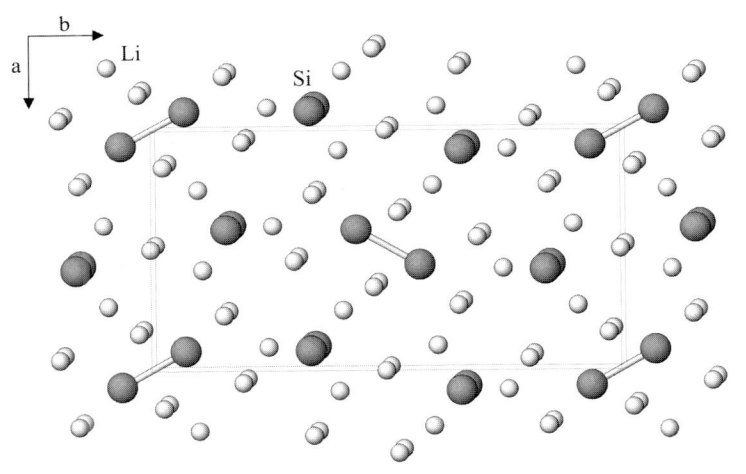

Fig. 9.2 Crystal structure of $Li_{13}Si_4$ in skew [001] projection. The stacking separation (perpendicular to projection plane) is $c = 443$ pm.

number, bonding models, and conductivity was to get rid of excess electrons by postulating selected cation-centered occupied orbitals that were stabilized by a cluster-like cation arrangement. These were termed cage orbitals and were derived mainly from INDO calculations with a local view of chemical bonding [15]. Tab. 9.1 shows such cases and the corresponding valence electron distributions. One of the examples given is that of $Li_{13}Si_4$, which contains two isolated Si^{4-} anions and one Si_2 pair per formula unit. The crystal structure is shown in Fig. 9.2. By application of the standard Zintl-Klemm concept, a VEC for the Si_2 dumbbell of 13, a formal charge of (5–), and thus a (1.5b)Si situation is postulated, which is

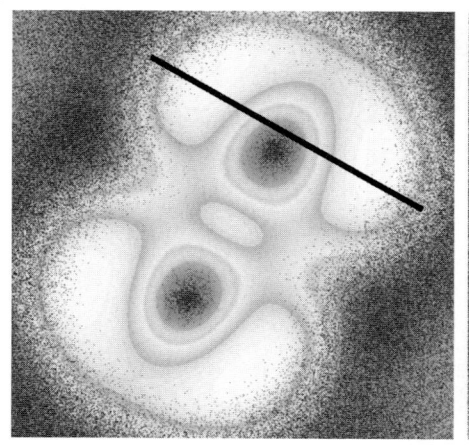

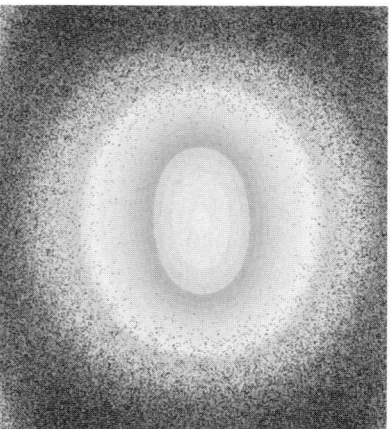

Fig. 9.3 ELF sections through the Si$_2$ dumbbell unit in Li$_{13}$Si$_4$: Left: (001) plane showing large horseshoe-shaped nonbonding regions (black line ~440 pm). Right: elliptical profile perpendicular to the bond vector with the large axis directed in the stacking direction of the Si$_2$ dumbbells.

exactly in between a single and a double bond. In general, no real help is to be expected from bond-length–bond-strength arguments for tetrel Zintl phases. Due to pronounced matrix effects of the cation packing, which is governed by coulombic interactions, only qualitative distance comparisons can be applied based on vague trends.

The Si–Si separation of 239 pm is very close to the normal single-bond distance, but much larger values have also been observed in the case of single bonds. The formulation given in ref. [15] implies the existence of a double bond and of an Li-centered cage orbital according to (Li$_{13}$)$^{11+}$(Si^{4-})$_2$(Si$_2$)$^{4-}$. A recent thorough investigation of the band structure by Extended Hückel (EH) [16, 17] and linear muffin tin orbital (LMTO) [18] methods clearly favors an inter-dumbbell interaction generating a conduction band due to π^*–π^* overlap in the stacking direction [19].

Fig. 9.3 displays two sections of the electron localization function (ELF) [20, 21] perpendicular to the bond vector and in the a-b plane. Due to the nature of the EH method, no core localization is visible. Quite clearly, there is an enormous expansion of the ELF regions of the nonbonding electron pairs, which extend by about 220 pm from the silicon sites into space (white-grey ELF border in Fig. 9.3). Geometrically, this vast extension allows for an overlap of orbitals in the c stacking direction of the dumbbells in the immediate vicinity. The states in question derive from p orbitals of silicon, which constitute the underoccupied π^* set extending along [001]. The Extended Hückel band structure without charge-iterated H$_{ii}$ values (Si: H$_{ss}$=–17.3 eV; H$_{pp}$=–9.2 eV; Li: H$_{ss}$=–5.4 eV; H$_{pp}$=–3.5 eV) exhibits a clear band gap at a valence electron count of 28 e.

In general, one would expect such a filling level to be achieved for a stable compound. Obviously, the electronegativity and electron affinity of silicon only allow for the aforementioned filling level and corresponding negative charge, but not for a completely σ-bonded single bound state. However, if additional stabilization is introduced by replacing a few percent of the Li$^+$ by doubly-charged Mg^{2+} cations, then the high band filling is reached just as for the larger and somewhat more electronegative germanium and the Li$_{13}$Ge$_4$ structure is assumed long before the full electron number according to Li$_{12}$MgSi$_4$ is reached [22] (see below).

If charge iteration is applied, an energetic approach of the Si- and Li-centered states becomes visible, which closes the band gap completely (Si: H$_{ss}$=−12.7 eV; H$_{pp}$=−5.4 eV; Li: H$_{ss}$=−5.3 eV; H$_{pp}$=−3.9 eV [19]). In both cases, the Fermi level lies at a low density of states (DOS), which largely arises from π^* states combined from Si orbitals. These results are supported by the LMTO band structure at the bottom of Fig. 9.4, which is based on a self-consistent DFT formalism. In all three cases, a high dispersion of the π^* bands is found for k vectors along the 0–Z, T–Y, and S–R reciprocal directions, which represent interactions along the c axis, this being the stacking direction. Although for both of the refined band structures (the lower two in Fig. 9.4) a close approach of silicon- and lithium-centered bands occurs, the latter may be considered as unoccupied. This is just a necessary prerequisite of an advanced Zintl-Klemm concept, namely that valence electrons only reside in anion-centered states.

Without destroying the applicability of the Zintl-Klemm concept, the presented analysis of the electronic situation in Li$_{13}$Si$_4$ not only solves an old problem of bonding in tetrel Zintl phases, but also describes a very frequent case of imprecise electronic filling in these compounds.

A general situation of the kind outlined above arises under the following conditions:

1. The structure contains planar or partially planar Zintl anions, which are stacked ecliptically at distances between 400 and 500 pm.
2. The cations surrounding the stack are also aligned in an untypical ecliptical manner.
3. The sigmoid overlap of π^* states along the stacking direction generates a conduction band that intersects the Fermi level.
4. The partial filling of this conduction band is due to both structure-specific stabilization conditions and the limited charge stabilization capability of the Zintl anions.
5. In other words, planarity, ecliptic stacking, and metallic conductivity with a one-dimensional characteristic are the result of the optimal charge stabilization capacity of the Zintl anions having such geometries.
6. Any attempt to isolate the corresponding Zintl clusters in molecular form must fail if the charge stabilization is lost at any point during the process.
7. Salt melts are suitable solvents for reactions with Zintl anions, especially those in which the metal in question exhibits a certain solubility, such that the medium provides metal and electron buffer capabilities.
8. A typical overlap situation is represented in Fig. 9.5 for the special case of Li$_{13}$Si$_4$.

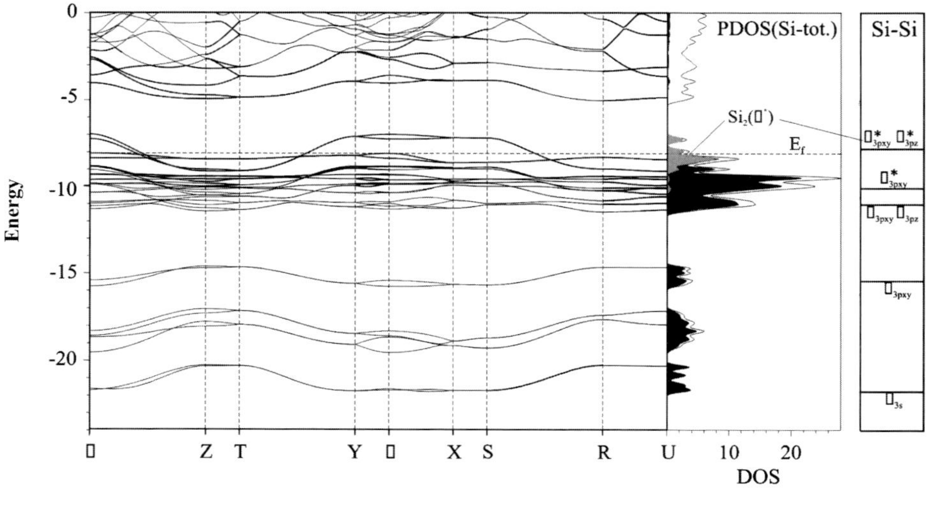
Energy
0
-5
-10
-15
-20
Z T Y X S R U
10 20
DOS
PDOS(Si-tot.)
Si-Si
E_f
3pxy
3pz
3s

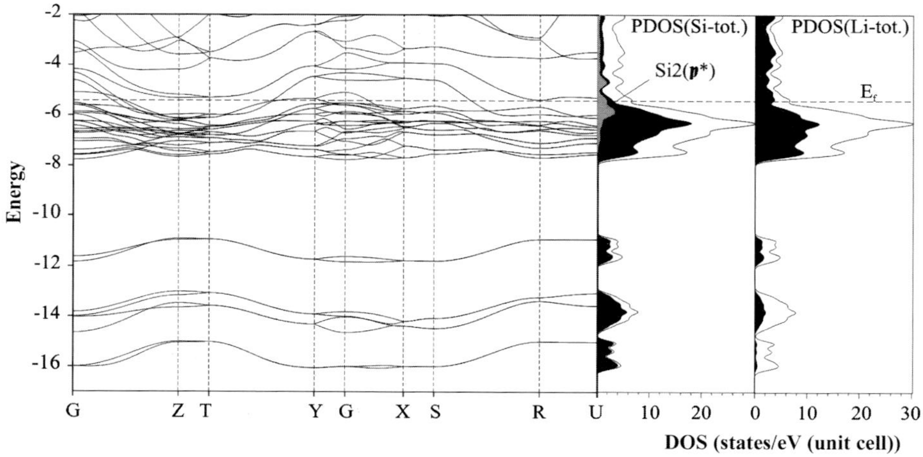
Energy
-2
-4
-6
-8
-10
-12
-14
-16
G Z T Y G X S R U
10 20 0 10 20 30
DOS (states/eV (unit cell))
PDOS(Si-tot.)
PDOS(Li-tot.)
E_f

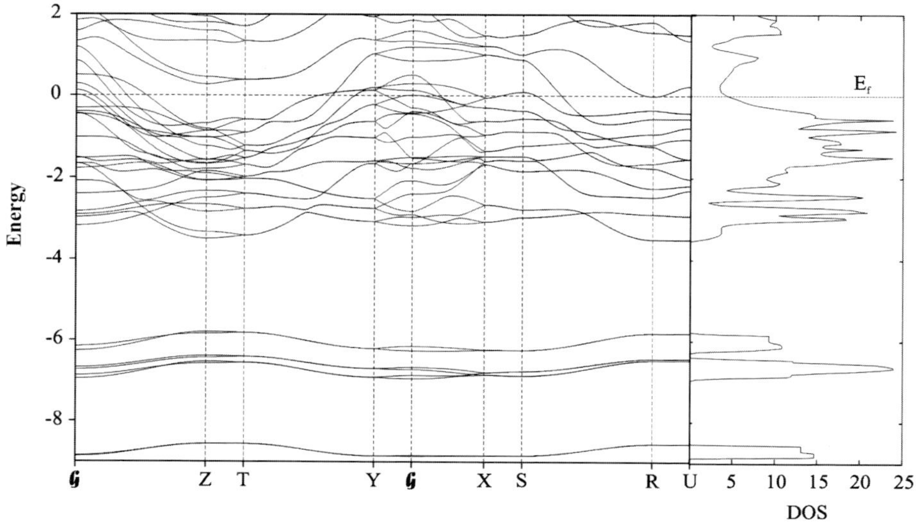
Energy
2
0
-2
-4
-6
-8
Z T Y X S R U
5 10 15 20 25
DOS
E_f

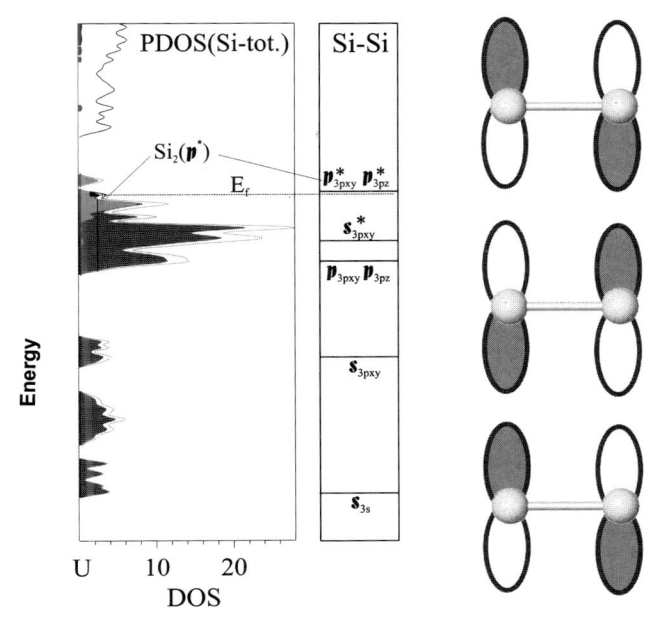

Fig. 9.5 Extended Hückel density of states plot, approximate orbital energies of silicon, and orbital overlap scheme along the stack direction of Si_2 dumbbells (black arrow) for $Li_{13}Si_4$.

The special coupling of electronic and geometric structures in $Li_{13}Si_4$, which leads to a metallic state, is being found in many tetrelide Zintl phases [8]. However, cases like $Li_{21}Si_4$, which is a γ-brass phase, may still be considered as having Li-centered cage orbitals [19, 20].

9.3
$Li_{13}Si_4$ vs. $Li_{13}Ge_4$ Structure Type

The structure of so-called $Li_{14}Ge_4$ [23] was reinvestigated by single-crystal X-ray diffraction and it was actually found to have the composition $Li_{13}Ge_4$ [22]. As shown in Fig. 9.6, the two structure types have some similarities, but also exhibit significant differences. The $Li_{13}Ge_4$-type also occurs for the silicide when some lithium is exchanged for magnesium, through which the coulombic stabilization is enhanced [22]. Indeed, there is a series of compounds $Li_{13-x}Mg_xSi_4$ with the $Li_{13}Ge_4$ structure, which are regarded as Zintl phases $(Li_{13-x}Mg_x)^{14+}(X^{4-})_2(X_2)^{6-}$ (X = Si, Ge) without π-bonding contributions and with singly-bonded dumbbells for the ideal composition $x = 1$. This interpretation is based on two features:

Fig. 9.4 Band structure and densities of states plots for $Li_{13}Si_4$: top – Extended Hückel calculation without charge iteration; middle – with charge-iterated H_{ii} orbital values [16, 17]; bottom: LMTO band structure [18].

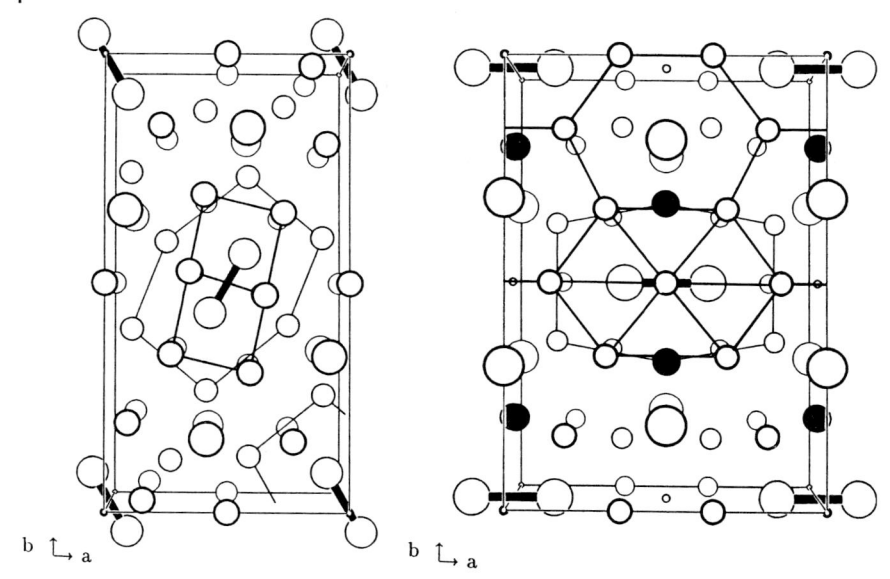

Fig. 9.6 Crystal structures of $Li_{13}Si_4$ (left) and of $Li_{13}Ge_4$ (right) in [001] projection (Si, Ge – large circles; Li – small circles; Mg sites – filled black circles). Note the intermediate Li atom in the stacking direction of the X–X dumbbells for $Li_{13}Ge_4$ (cf. arrow).

1. Comparison of X–X bond distances, which – for $Li_{13}Si_4$ is 239 pm and some double-bond contribution is assumed – but was determined to be around 248 pm in the series $Li_{13-x}Mg_xSi_4$ ($x=0.05, 0.2, 0.3$) with the $Li_{13}Ge_4$ structure [22].
2. There is an ecliptic stacking of the dumbbells along the c direction in both structures. However, for $Li_{13}Ge_4$ an intermediate cation is found in the stacks according to [X_2–Li–X_2–Li], destroying the direct π^*–π^* overlap. This is not the case for $Li_{13}Si_4$ and all other tetrelide Zintl phases, for which the π^*–π^* overlap situation and resulting metallic conductivity is found.

Considering the local environment of the X–X dumbbell in Fig. 9.6 (marked by thin black lines), it seems that the two Mg sites (right-hand side; black circles in the $Li_{13}Ge_4$ type) polarize the electron states of the pairs in the ab plane such that the overlap along [001] is not possible anymore and consequently an intermediate Li^+ cation is introduced. The electronic situation is thus interpreted as an only partial filling of the third lone pairs at each of the Ge sites in the dumbbells. This is also an example of the extremely high polarizability of tetrel Zintl anions. The non-existence of the ideal phase $Li_{13}MgSi_4$ could be due to thermodynamic reasons, i.e. the competition of other Mg-based silicides in the ternary phase system.

9.4
$Li_{12}Si_7$ – Revisited

Another old problem relates to the phase $Li_{12}Si_7$, which exhibits a beautiful crystal structure containing two different multinuclear Zintl clusters – a planar Si_5 ring and a planar Si_4 star [24]. Formal application of the Zintl-Klemm concept offers several interpretations, the most probable of which are $(Li_{24})^{24+}(Si_5^{8-})_2(Si_4^{8-})$ and $(Li_{24})^{24+}(Si_5^{6-})_2(Si_4^{12-})$. The assumption of a quasi-aromatic pentasilapentadienyl group was supported by the existence of the compound Li_8MgSi_6 [25], which was consistently understood as $(Li_8)^{8+}Mg^{2+}(Si_5^{6-})Si^{4-}$ on the basis of INDO calculations [26]. The problem of a 28-electron three-bonded Si_4 cluster could conceivably have found analogy in interhalogen compounds such as ClF_3, having one two-center and one three-center bond, but the geometry of the Si_4 star does not allow for a corresponding orbital mixing.

Already some time ago the solution to that was worked out from quantum mechanics postulating thirteen silicon centered occupied states and one so-called cage orbital that extends over the lithium cage surrounding the silicon cluster according to $Li_{12}^{10-} Si_4^{10-} (Si_5^{6-})_2$ [27, 28]. At the same time, this result was supported by proof of the existence of lithium-centered cage orbitals in the γ-brass type $Li_{21}Si_5$ [29, 30] and in the hypothetical Li_6C [31] on the basis of quantum mechanical calculations. However, at that time, the internal charge transfer between an $(Li_6Si_5)_2$ stack and an $Li_{12}Si_4$ sheet could not be assessed in a proper way because of the complexity of the structure.

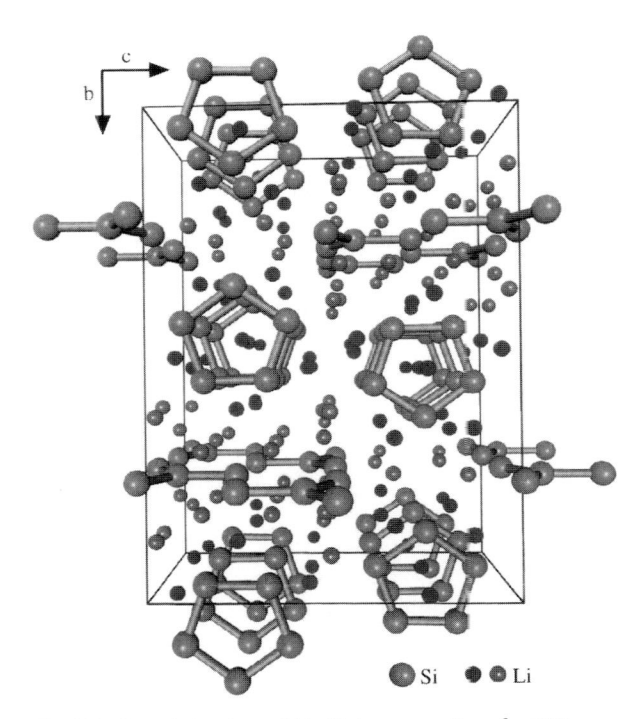

Fig. 9.7 Crystal structure of $Li_{12}Si_7$ in perspective of a [100] projection after ref. [19]. The stacking separation of the five-membered rings is $a = 428$ pm.

About a decade ago, van Leuken et al. performed ab initio local spherical wave calculations [32] on the complete unit cell of $Li_{12}Si_7$ and on the fragments $Li_{12}Si_4$ and Li_6Si_5 [33]. They postulated a charge transfer between the fragments leading to $(Li_{12}Si_4)^{4+}(Li_6Si_5^{2-})_2$. The former cluster would then contain an Si_4^{8-} anion isoelectronic and isosteric with CO_3^{2-}. However, in this case, an Li-centered doubly-populated s-state at the sandwiched Li atoms between the Si_5 rings has to be postulated for the $(Li_6Si_5^{2-})$ stack. Furthermore, a band gap in the DOS of the compound was confirmed.

A recent full ab initio LMTO (linear muffin tin orbital) band structure calculation on $Li_{12}Si_7$ provides new insights into this compound [19]. The LMTO band structures for the fragments $Li_{12}Si_4$ and Li_6Si_5 are very similar to those published by van Leuken et al. [32]. This is not surprising because LSW and LMTO-ASA are very similar methods.

Comparing the electronic structures of the two fragments, a charge transfer becomes immediately obvious. However, contrary to ref. [32], we see a charge transfer of only two electrons, giving rise to the partial structures $(Li_{12}Si_4)^{2+}$ and $Li_6Si_5^{2-}$ and to the Zintl anions Si_4^{8-} and Si_5^{8-} because then the individual Fermi levels fall into a region of low DOS in each case (Fig. 9.8, left and middle). It should be noted that this cannot be achieved by raising the electronic filling of the $Li_{12}Si_4$ fragment, which has another pseudo-gap for 30 valence electrons. Furthermore, we do not find real band gaps to support semiconducting properties. This is neither true for the fragments nor for the full band structure of

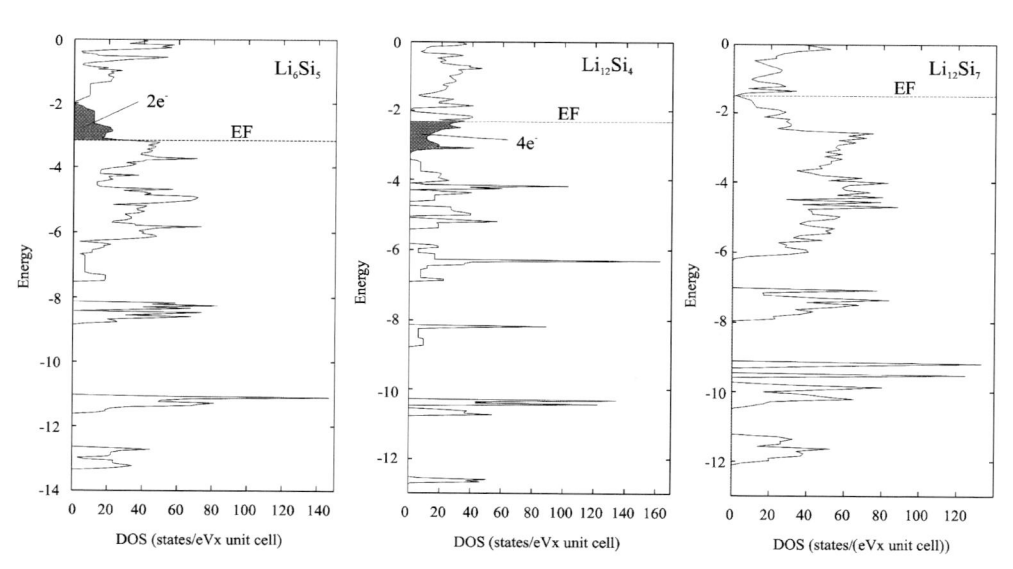

Fig. 9.8 Density of states plots of the full structure of $Li_{12}Si_7$ (right), and of the fragments $Li_{12}Si_4$ (middle) and Li_6Si_5 (left). Two additional electrons on Li_6Si_5 and four less on $Li_{12}Si_4$ gain filling rates which lead to low DOS as found for the complete structure [19].

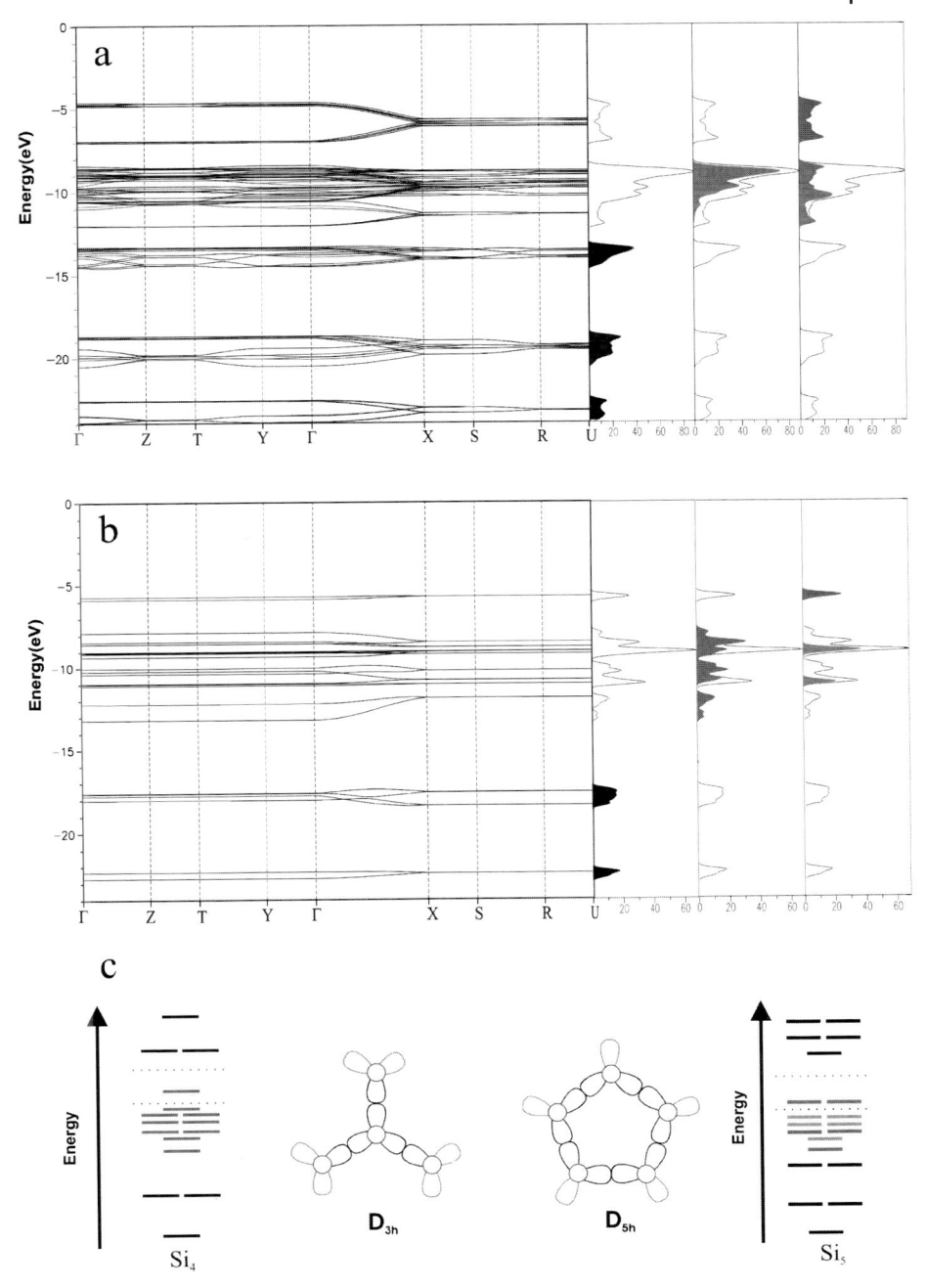

Fig. 9.9 Extended Hückel band structure DOS and partial DOS of the fragments Li$_6$Si$_5$ (a) and Li$_{12}$Si$_4$ (b). Shaded partial DOS in (a) and (b) arise from states denoted in the same way in the MO schemes in (c).

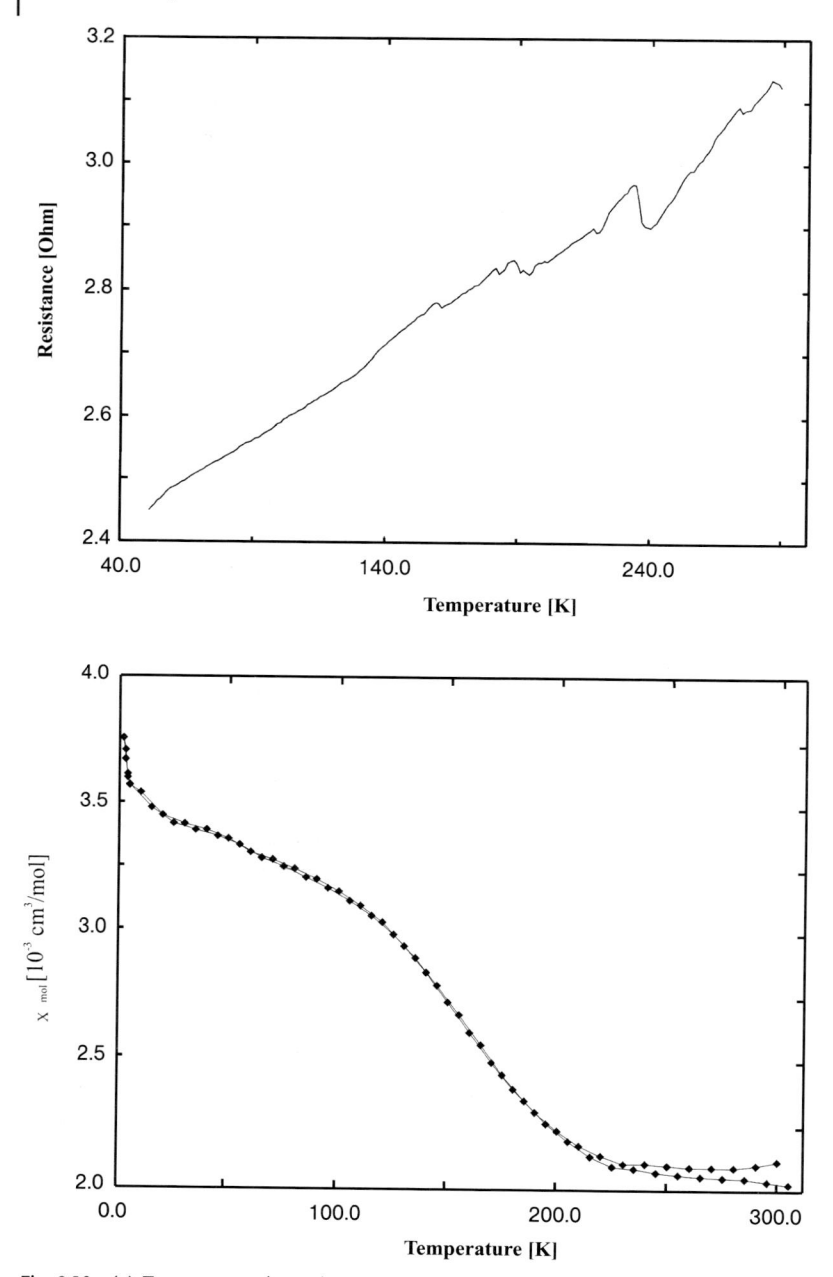

Fig. 9.10 (a) Temperature-dependent resistance and (b) magnetic behavior of $Li_{12}Si_7$ at an external field of 100 G.

Li$_{12}$Si$_7$, and not even for the EH band structure (cf. Fig. 9.4), although EH notoriously overestimates the gaps.

Our recent physical measurements fully confirm these findings: weakly metallic character and paramagnetic behavior with temperature-dependent changes are found (Fig. 9.10). Bearing in mind the strong diamagnetic contributions of the Zintl anions due to their very polarizable electronic structures, the strength of the paramagnetic behavior appears to be quite pronounced. Even though the origins of the contributions to the paramagnetic response are not yet clear, there are pronounced temperature-dependent changes indicating that the electronic structure changes to a certain extent under cooling. A clear-cut Mott transition, i.e., a localization of conduction electrons at low temperatures, which might be expected, is not visible in the investigated temperature range from 4 to 300 K, but some structural features are apparent. Concerning metallic conductivity, it should be noted that this may be blurred by semiconducting contacts in a multicrystalline sample, as found in a number of comparable cases. There is a direct band gap Γ at in the band structure, which may explain why optical measurements have shown a considerable band gap of ~ 0.5 eV since these methods excite only via the direct gap.

Before entering into a more detailed discussion, we can already state that the band structure is almost completely dispersionless, except for the Γ to X direction, which corresponds to the [100] lattice zone and thus describes the stacking of the planar Si$_5$ rings. It is exactly this dispersion which is responsible for the DOS at the Fermi level.

In Fig. 9.11, ELF sections are displayed, which have been calculated on the basis of the LMTO band structure of the complete unit cell of Li$_{12}$Si$_7$. The principal cross-sections are shown in Fig. 9.11a. One matches the plane of the Si$_5$ rings (Fig. 9.11b) and the other the densely packed plane of the Si$_4$ stars (Fig. 9.11c). Atomic cores and bonding and nonbonding electron pairs are nicely visible as white to light-grey ELF regions. The latter two are exclusively localized at silicon atoms. All lithium centers appear to be core-like. This supports the interpretation according to an expanded Zintl-Klemm concept invoking complete formal charge transfer to rationalize the electronic structure of the Zintl phase according to $(Li^+)_{24}(Si_{14})^{24-}$ [6,.8]. Based on our experience with the ELF [21], we can clearly state that there are no indications of Li-centered states, neither locally at a single lithium atom, not even at the Li atom sandwiched between the Si$_5$ rings, nor in the form of cage orbitals, contrary to our early results [15] and also to the interpretation of van Leuken et al. [32]. The bond cross-sections of the Si$_4$ star (Fig. 9.11d) show a slightly elliptical character and thus hint at a contribution to the bonds. Furthermore, the cross-section perpendicular to the mean plane of the star shows two Si atoms, of which only one terminal has pronounced lone-pair regions (ellipse in Fig. 9.11b). If an Si$_4^{10-}$ anion were present, lone-pair regions would have to appear at all atoms, not only at the terminal but also at the central one. This can readily be seen in Fig. 9.12, in which a similar cross-section of two planar Si$_4^{10-}$ stars in the compound Ba$_5$Mg$_{18}$Si$_{13}$ is displayed. Although these units are slightly pyramidal, we calculated the planar case for a better comparison. Consequently, the bond sections

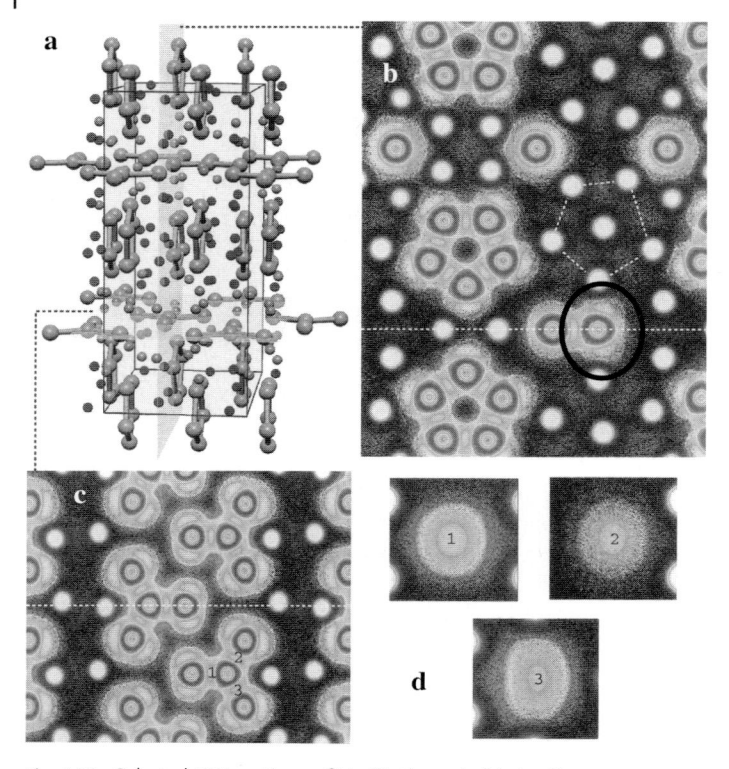

Fig. 9.11 Selected ELF sections of $Li_{12}Si_7$ through (b) the Si_5 ring planes, (c) the Si_4 stars, and (d) the three bonds of the star unit which exhibit weak elliptical distributions indicating π bond contributions.

in this case are much more circular, hinting at the σ character of the bonds, and there are lone pairs at each of the two intersected atoms.

The dispersion in the band structure of $Li_{12}Si_7$ arises from σ-type interactions in the double-chain sheets, while the dispersion of the pure Li_6Si_5 part is due to the eclipticly stacked Si_5 rings. The stacking direction contains two five-membered ring units per unit cell, but the ideal packing ($a'=a/2$) is slightly broken by a Peierls distortion. Therefore, the degeneracies vanish which form by folding of the Brillouin zone. This picture is preserved in the DOS of the full band structure, where the Fermi level just falls into the low DOS region. The latter can be traced back to the loss of these degeneracies, which, in the molecular picture, correspond to the occupation of the first degenerate π level of the Si_5^{8-} ring by two unpaired electrons. As these states interact to form a band along the stacking direction, both Curie and Pauli paramagnetism may occur at different ratios at different temperatures.

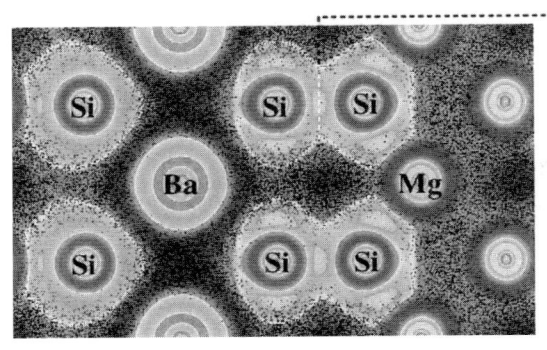

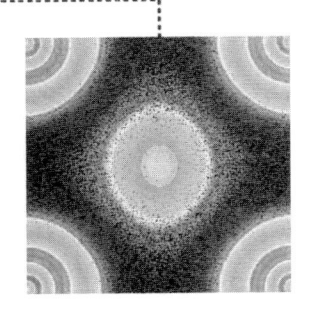

Fig. 9.12 Selected ELF sections of Ba$_5$Mg$_{18}$Si$_{13}$ containing the star-shaped 26 e Zintl anion Si$_4^{10-}$. There are obviously lone-pair regions at both the central (left) and terminal Si atoms, as indicated by lighter spots.

9.5
The Lanthanide Disilicide Problem

For many decades, lanthanide disilicides have been the subject of studies for all the rare-earth metals, and three different structure types have been repeatedly reported: the tetragonal ThSi$_2$, the orthorhombic GdSi$_{2-x}$, and the hexagonal AlB$_2$ structures [34–36]. Occasionally, defect formation has been found, and compositions LnSi$_{2-x}$ with $0.25 < x < 0.5$ have been given for GdSi$_{2-x}$ types and with $x \sim 0.6$ for structures related to AlB$_2$ [37, 38]. However, there are discrepancies concerning the assignment of compounds and structure types, especially for the undistorted and stoichiometric a-ThSi$_2$ and the distorted a-GdSi$_{2-x}$ variants. In the case of compounds of the smaller Er and Yb, the hexagonal phase is found. For the trivalent lanthanides Nd, Sm, Gd, Ho, and Tb, both a-GdSi$_{2-x}$ and AlB$_2$ forms have been reported as being preferentially adopted. A thorough study of the synthesis conditions revealed that the frequently mentioned tetragonal "defect-free" phase (a-ThSi$_2$) only occurs when the synthesis is performed in an arc furnace with a copper sample holder [39]. Later chemical syntheses with and without the addition of Cu, Ag or Au clearly proved that in the case of Nd, Sm, Gd, Ho, and Tb, the a-ThSi$_2$ type only forms if noble metals are incorporated into the structure, and that these are exclusively located at the silicon sites [39]. This may result in occupancies in excess of one Si atom per silicon site. There have been numerous investigations on this compound class. However, in many instances, the possibility of impurities arising from the copper holders used in arc furnace experiments was not taken into consideration. These have been shown to be a source of considerable Cu contamination and consequently of phase and compositional changes [39].

In our investigations, samples were prepared from the pure elements under dry argon atmosphere in sealed tantalum ampoules at 1820 K (5 h). In Tab. 9.2, the changes in the lattice constants of copper-doped GdSi$_{2-x}$Cu$_y$ phases are displayed.

Tab. 9.2 Compositions and lattice constants of selected GdSi$_{2-x}$M$_y$ phases (M=Cu except where indicated, Ag[a)] and Au[b)]) according to ref. [39]

y (Cu)	a (pm)	b (pm)	c (pm)	V [pm$^3 \times 10^6$]	c/a
0.0	406.8(1)	398.7(1)	1338.6(3)	217.11	3.290
0.03	406.5(1)	399.1(2)	1340.7(5)	217.51	3.298
0.05	405.9(3)	403.7(3)	1350.3(6)	221.26	3.327
0.07	404.5(1)	404.3(1)	1357.1(5)	221.23	3.355
0.43	404.5(4)	404.5(2)	1383.5(7)	226.37	3.420
0.43[a)]	409.2(2)	409.0(2)	1402.8(8)	234.78	3.428
0.43[b)]	410.1(2)	409.9(2)	1407.2(8)	236.55	3.432

At higher copper concentrations, a convergence of the dimensions of the a and b axes occurs and the tetragonal a-ThSi$_2$ phase is found. At roughly a 20% replacement of Si by the noble metal, the a and b axes become equal in size.

Further exploratory doping experiments revealed that beside the noble metals, Ni, As, and Ga may also occupy the silicon sites, and probably many other components or component combinations as well. Only for Eu did we find full occupancy of the silicon site and a a-ThSi$_2$ structure without any significant impurity at this site [39].

All structural data were taken from single-crystal diffraction experiments (NdSi$_{2-x}$, $x=0.5$; GdSi$_{2-x}$, $x=0.4$). GdSi$_{2-x}$, $x=0.4$ was heated from 295 to 970 K in a Guinier camera and did not show any indication of a phase transition. However, depending on the synthesis conditions, compositional changes in x were found. Only at Gd:Si ratios larger than 1.8 were no further changes in the lattice constants observed, indicating that an upper silicon content is reached at this ratio. Tab. 9.2 lists crystal lattice data on Cu-doped LnSi$_{2-x}$Cu$_y$ compounds at $y \sim 0.4$ with the a-ThSi$_2$ structure, and undoped LnSi$_{2-x}$ at $x=0.3$ with the a-GdSi$_{2-x}$ structure, as well as LnSi$_{2-z}$ at $z=0.6$ with the AlB$_2$ type arrangement (cf. Tab. 9.3).

Europium behaves differently from the other lanthanides, with practically the same large volumes for a-ThSi$_2$ and a-GdSi$_{2-x}$ structure types, revealing its divalent nature and full silicon occupancy in both cases. For each of the three different cases, the typical volume shrinking due to the lanthanide contraction is seen. Comparison of those metals in which two different cases are found reveals that the general volume development is a shrinkage on going from a-ThSi$_2$ to the a-GdSi$_{2-x}$ and AlB$_2$ types. However, this is not the case for neodymium. Only for gadolinium are all three phases found, and they do follow the general trend.

Although there have been numerous investigations on the so-called rare-earth disilicides, even some recent ones, the results are difficult to analyze because many authors did not consider the above-mentioned preparative aspects and thus in many cases it cannot be clarified whether real binary or contaminated ternary compounds were produced. From our investigations, we can draw the following conclusions:

9.5 The Lanthanide Disilicide Problem

Tab. 9.3 Lattice constants and unit cell volumes of lanthanide "disilicides" of the α-ThSi$_2$ (Cu-doped), the $\bar{\alpha}$-GdSi$_2$ (Si defects) and the AlB$_2$ type (Si defects) structures (volumes in bold letters refer to the formula RE$_3$Si$_{(6-x)}$)

Structure type		Pr	Nd	Sm	Eu	Gd	Tb	Dy	Ho	Er	Yb
α-ThSi$_2$	a=	417.8 (1)	411.3 (1)	406.3 (1)	429.7 (1)	404.38 (6)	399.02 (8)	397.39 (9)	396.1 (1)	393.7 (1)	
	c=	1365.5 (6)	1376.0 (1)	1373.1 (5)	1370.4 (3)	1380.2 (5)	1369.2 (4)	1367.6 (5)	1364.5 (10)	1361.6 (5)	
	V=	**143.6 (1)**	**140.20 (1)**	**136.54 (8)**	**153.0 (1)**	**135.94 (7)**	**131.30 (7)**	**130.07 (8)**	**129.0 (1)**	**127.13 (9)**	
$\bar{\alpha}$-GdSi$_2$	a=		416.5 (3)	411.5 (1)		408.31 (5)	403.09 (7)	401.8 (1)	399.1 (3)		
	b=		414.5 (2)	406.0 (2)		400.41 (7)	393.86 (6)	390.61 (8)	389.3 (3)		
	c=		1364.2 (6)	1342.2 (4)		1342.9 (2)	1327.8 (2)	1324.6 (2)	1319.4 (8)		
	V=		**141.86 (13)**	**135.06 (8)**		**132.23 (3)**	**126.97 (3)**	**125.20 (5)**	**123.5 (2)**		
AlB$_2$	a=				405.2 (2)	385.25 (7)				376.8 (1)	376.19 (9)
	c=				448.2 (3)	414.70 (9)				407.1 (2)	409.2 (1)
	V=				38.40 (3)	32.10 (1)				30.15 (2)	30.21 (2)
	4V=				**153.6 (1)**	**128.80 (4)**				**120.6 (8)**	**120.82 (8)**

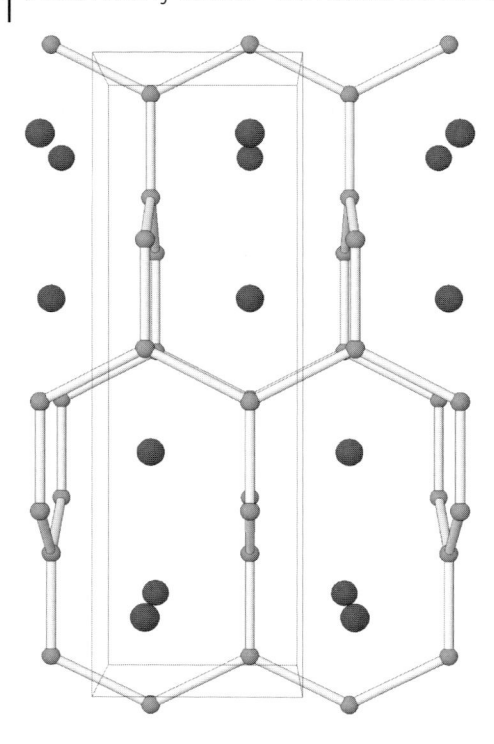

Fig. 9.13 ThSi$_2$ structure in skew [100] projection (small circles – Si; large circles – cations). Copper-doped phases are observed for Pr to Sm and Gd to Er; pure binary phases occur for Eu and for Nd to Sm as well as for Gd to Ho in the form of slightly orthorhombically distorted structures.

1. A pure binary disilicide only exists for europium.
2. a-GdSi$_{2-x}$ compounds are silicon-deficient with x around 0.3(1).
3. AlB$_2$-type compounds are silicon-deficient with z around 0.6(1).

Some major questions still remain unanswered, for example:
1. Why do the a-GdSi$_{2-x}$ type phases possess orthorhombic symmetry despite showing no geometrical evidence for an orthorhombic structure beyond a divergence of the lattice constants a and b? From doping experiments with copper, and more especially with silver and gold, it is known that superstructure reflections occur, which arise from multiple lattice extensions. For a high Au-doping, we found a 5×5×2 superstructure, which, of course, constitutes a severe super-symmetry problem in the extended cell. From such observations, one may deduce that the undoped phases also show a vacancy ordering. However, the observed pattern, which in this case does not show super structure reflections, cannot be refined due to directional differences.
2. Why does defect formation occur? Within the framework of the Zintl–Klemm concept, we would argue that the excess charge of the trivalent cations has to be accommodated by the Zintl anions. If this is indeed the case – and the overwhelming majority of Ln tetrelides seem to follow an extended Zintl–Klemm concept – void formations must be present. Each singular void in such a three-bonded (3b) framework will stimulate the formation of three lone electron pairs on the neighboring framework atoms for

reasons of valence electron shell completion. A general scheme for deriving reasonable void formulae and distributions was given in ref. [11] some time ago. Applying the concept to the Ln(III) silicides, we have to write $LnSi_{2-x}$, where $E = (8 - N)(2) + N^* x = (8 - N)(2) + 3x = 11 = (5 \cdot 2) + 3x$ and $x = 0.3333$, if we assume an undistorted three-connected base network into which isolated voids x are introduced ($N^* = 3$ for isolated Si defects, where each x localizes three additional electrons at the three broken bonds). Thus, according to the Zintl-Klemm concept, a composition of $LnSi_{1.67}$ is expected by isolated void formation in any three-bonded Si framework. If exclusively Ln(II) were to occur, then $E = (8 - N)2 + 3x = 10 = 10 + 3x$ with $x = 0$, and consequently defect-free AlB_2 or $ThSi_2$ types would form. However, if defect clusters are predicted, then the value of N^* reduces with increasing clustering of voids.

3. If no Ln(III) species occur, then what is the reason for the formation of defective (3b) frameworks? This seems to be the case, at least for $YbSi_{1.4}$.
4. Why do rare-earth metals prefer planar Zintl anions of silicon, which are normally indicative of unsaturated bond systems in the Zintl anions, while alkaline-earth metals give rise to (at least partially) typically saturated and puckered geometries, i.e. $CaSi_2$?

9.6
YbSi$_{1.4}$ – A Beautiful Solution to a "Disorder Problem"

The phase diagram of the binary system Yb/Si has recently been reinvestigated [40, 41]. Six phases have been found and structurally characterized, namely Yb_5Si_3, Yb_5Si_4, YbSi, Yb_3Si_4, Yb_3Si_5, and $YbSi_{2-x}$.

Yb_5Si_3, the Si-poorest phase, has the hexagonal Mn_5Si_3 structure with isolated Si^{4-} anions assuming mixed-valent $Yb^{2+/3+}$ cations [42]. Recently, Yb_5Si_4 was synthesized and suggested to be isotypic to Sm_5Ge_4 [41]. YbSi is isotypic to CrB, with zigzag chains of Si atoms stacked in an ecliptic fashion [43]. It should be noted that this is typical for unsaturated Zintl anions and thus constitutes another unsolved problem as already for Yb^{2+} formally saturated σ-bonded chains should result. In the Si-rich part of the Yb/Si system, three phases exist. Yb_3Si_5 crystallizes in the Th_3Pd_5 type [44–46]. The Si-richest phase, $YbSi_{2-x}$ ($0.14 > x > 0.3$), contains graphite-like sheets of Si as in the AlB_2 type [41, 44]. As with other rare-earth metals of this kind, randomly distributed defects in the Si sublattice are supposed to occur here.

As the structures of YbSi and of the Si-richer phases are closely related, it is conjectured that many structural intermediates may exist in between these compositions. At the composition $YbSi_{1.41}$ two phases are found, a hexagonal defect form of AlB_2 and a new orthorhombic structure. Above 1240 K, the hexagonal form is stable and can be quenched to room temperature. The new form can be prepared in high amounts after tempering for one to three days at 1220 K. In the course of systematic thermochemical investigations of this part of the Yb/Si system, we recently reinvestigated the phase with the approximate composition

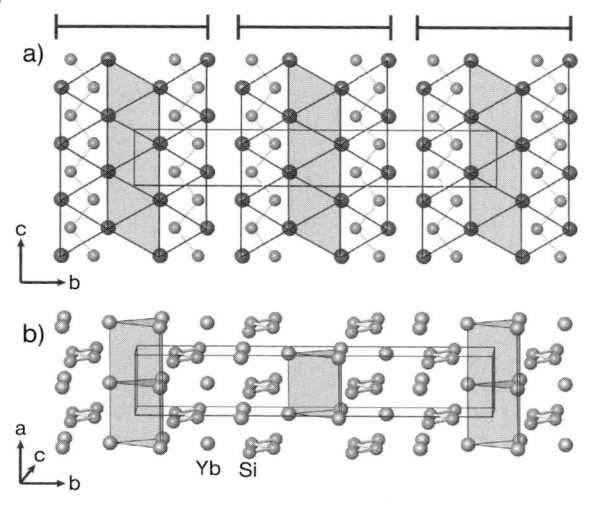

Fig. 9.14 Model of the small orthorhombic structure of YbSi$_{1.4}$ (Yb atoms – large spheres; Si atoms – small spheres). Projections in (a) [100] and (b) skew [001] directions. Yb$_6$ prisms only partially occupied by Si are shaded in grey. Chains of Si occur in the fully occupied Yb$_6$ prisms. AlB$_2$-type blocks are marked by bars at the top. Adjacent blocks are mutually shifted by $a/2$.

YbSi$_{1.4}$ [47]. Similarity of composition and lattice parameters suggest that this is basically the same phase as Yb$_3$Si$_4$ reported just recently by Palenzona et al. [41]. According to X-ray powder investigations, the structure of Yb$_3$Si$_4$ was found to be isotypic to the Ho$_3$Si$_4$-type [41]. The results of a comprehensive structural characterization by single-crystal X-ray and transmission electron microscopy studies show that the reality is much more complicated and that indeed fascinating local structures may hide behind seemingly partially occupied positions of an average structure.

Our transmission electron microscopy (TEM) investigations showed the presence of two distinct ordering variants: YbSi$_{1.4}$-I can be described by a unit cell with doubled volume (space group: $Imm2$; $a=4.16$, $b=7.56$, $c=23.51$ Å), while YbSi$_{1.4}$-II shows an incommensurate modulation that can be described in the 3+1 dimensional superspace group $Cmcm(10\gamma)$ (no. 63.3). From X-ray diffraction measurements, only a small orthorhombic unit cell can be derived ($Cmcm$; $a=4.159(1)$, $b=23.510(5)$, $c=3.775(1)$ Å). In Fig. 9.14, the main structure of YbSi$_{1.4}$ is shown with under-occupied "Si chains" in the grey blocks. The principal difference to AlB$_2$ is a shifting of adjacent AlB$_2$-like blocks of three chains width, each by $a/2$.

A model for YbSi$_{1.4}$-I was derived from Fourier-filtered TEM images and model simulations. The most reasonable trial structure is shown in Fig. 9.15, and corresponding measured and simulated TEM images are displayed in Fig. 9.16. The structure in Fig. 9.14 has a composition of Yb$_8$Si$_{11}$ with a formal charge of $x=2.25+$ on Yb. This was derived from application of the Zintl-Klemm concept, giving rise to the formulation $(Yb^{x+})_8[\{(3b)Si^-\}_4\{(2b)Si^{2-}\}_7]$. If purely doubly-charged Yb^{2+} cations are assumed, then there should be additional filled Si sites

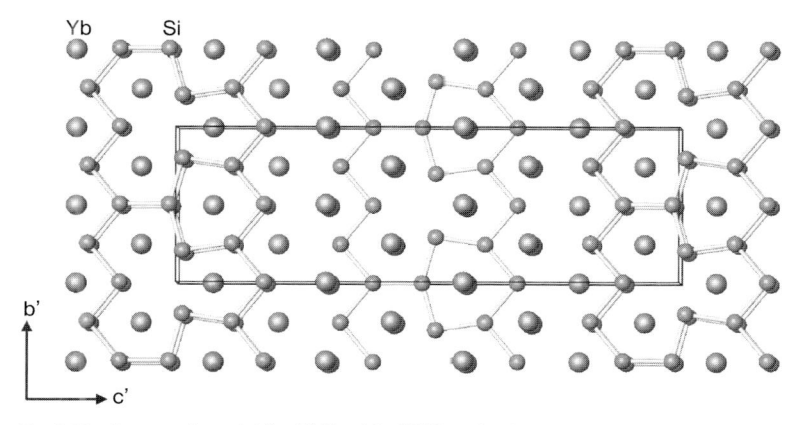

Fig. 9.15 Structural model for YbSi$_{1.4}$-I in [100] projection. The marked sections contain the locally ordered chain blocks of the under-occupied silicon sites.

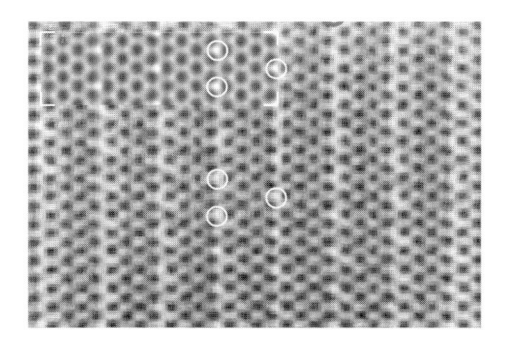

Fig. 9.16 Fourier-filtered HRTEM images of YbSi$_{1.4}$-I along [100], recorded close to Scherzer defocus. The simulated image is shown in the inset. Bright contrasts (three of them encircled in both images) indicate unoccupied Yb$_6$ prisms.

such that a period of $(Yb^{x+})_{3\times 8}[\{(3b)Si^-\}_{16}\{(2b)Si^{2-}\}_{18}]$ according to a composition Yb$_{24}$Si$_{34}$ or YbSi$_{1.417}$ occurs. This is very close to the analytical composition of YbSi$_{1.41}$.

However, a corresponding charge balance may also be achieved by introducing complete slabs of the defect-free AlB$_2$ form. If both ways of charge cancellation compete, irregular or non-commensurate sequences of the different kinds may occur (see Fig. 9.17). The evaluation of HRTEM images of YbSi$_{1.4}$-II indicates parallel domains with the structure of YbSi$_{1.4}$-I. The modulation is due to a shift of the domains against each other. There is an older Mössbauer investigation on YbSi$_{2-x}$, which is completely in agreement with divalent ytterbium [44].

This raises even more the question why a defect-free AlB$_2$ does not form for the divalent cation. In this case, the answer may be geometrical rather than

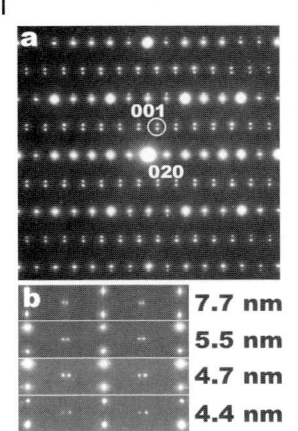

Fig. 9.17 (a) SAED pattern of YbSi$_{1.4}$-II along [100]. Indices of two reflections of the basic structure are given below the corresponding spot. Additional split reflections (encircled) appear, indicating the incommensurately modulated structure. (b) Sections of SAED patterns. The distances between the additional reflections vary and correspond to the real space modulation values given on the right.

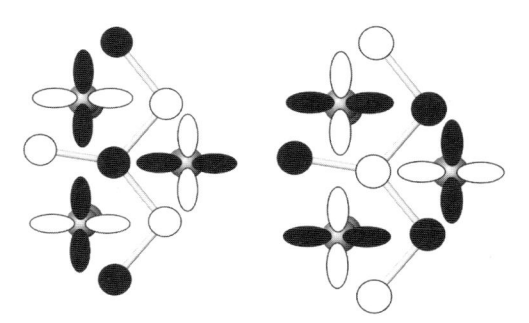

7.7 nm
5.5 nm
4.7 nm
4.4 nm

Fig. 9.18 Schematic model of the stabilization of completely filled π^* orbitals of silicon (bonded sites) by interaction via the d-states of the rare-earth metals (isolated sites). Because of the odd symmetry of the silicon p_z orbitals, the orbital signs (black – white) are switched above and below the plane of the Zintl anion.

electronic in origin: such a model structure would have reasonable distances ($d_{Si-Si}=236$ pm, $d_{Yb-Si}=300$ pm, and $d_{Yb-Yb} \sim 370$ pm) at a volume of roughly $67 \cdot 10^6$ pm^3 per formula unit, while the observed average structure has a volume of $51 \cdot 10^6$ pm^3 and the aforementioned superstructure of YbSi$_{1.41}$ one of only $46.2 \cdot 10^6$ pm^3 per formula unit.

The CaSi$_2$ structure types (with puckered Si$_6$ ring layers) would easily accommodate the hypothetical YbSi$_2$ for geometric reasons but have not hitherto been observed. From a more general point of view, an electronic factor would seem to be responsible the planar Si–Si coordinations in all LnSi$_{2-x}$ phases. It may be conjectured that there is a considerable Si–Si interaction between different Zintl anion layers through fully occupied π^* systems enhanced via the d states of the rare-earth metals, as shown schematically in Fig. 9.18.

Fig. 9.19 shows a comparison of the structures of a hypothetical YSi$_2$ (AlB$_2$-type), Yb$_3$Si$_5$, YbSi$_{2-x}$, and of YbSi (CrB-type). It is obvious that all observed structures derive from AlB$_2$ by defect formation and subsequent shift of the blocks against each other.

A corresponding experimental structure for the GdSi$_{2-x}$ structures is still lacking, but there is good evidence to suggest that it will also be a valid structural pattern of the Zintl-Klemm concept. In particular, because $x=0.33$ is observed, normal iso-

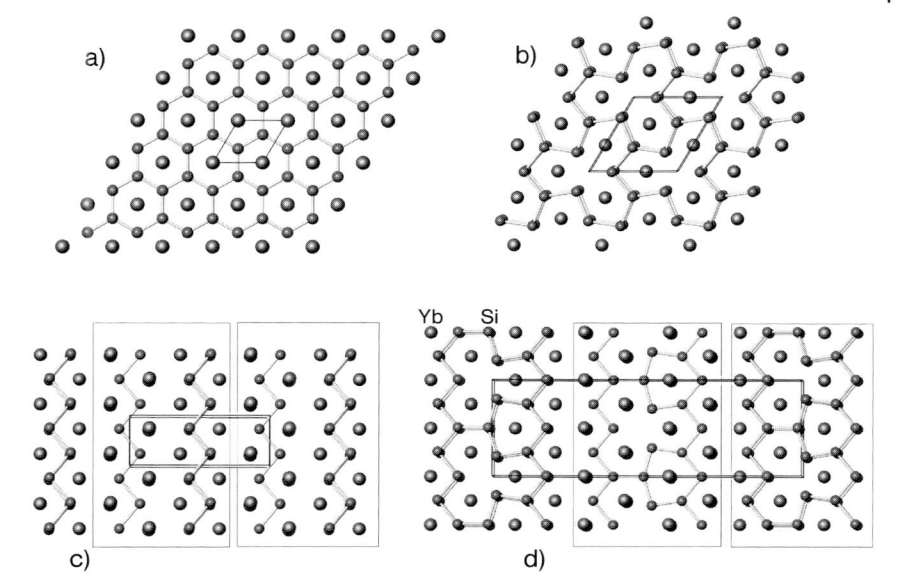

Fig. 9.19 Comparison of the hypothetical (a) YSi$_2$ (AlB$_2$-type) with the observed structures of (b) Yb$_3$Si$_5$, (d) YbSi$_{2-x}$, and of (c) YbSi (CrB-type) in orientations perpendicular to the planes of the Zintl anions. The geometrical relationships and the defect formations are clearly visible. Rectangular boxes indicate blocks which are shifted against their neigbors by ½ of the lattice constant in the projection direction.

lated void patterns for the larger trivalent rare-earth cations can be assumed. However, these are already considerably smaller than Eu^{2+}, which forms the defect-free ThSi$_2$ type [11, 48]. The tetragonal a axes can be compared to the hexagonal c axes as they constitute a pseudo-layer distance. It seems that there is a volume-to-distance problem which reduces the symmetry to orthorhombic allowing for a volume shrinkage without affecting all interatomic bonds in the same manner.

References

1 W. Klemm, *Proc. Chem. Soc. London* **1958**, 329.
2 E. Mooser, W. B. Pearson, *Progress in Semiconductors*, Vol. 5, Wiley & Sons, Inc., New York, **1960**.
3 W. Klemm, *Festkörperprobleme*, Vieweg, Braunschweig, **1963**.
4 E. Busmann, *Z. Anorg. Allg. Chem.* **1961**, *313*, 90.
5 W. Klemm, E. Busmann, *Z. Anorg. Allg. Chem.* **1963**, *319*, 297.
6 R. Nesper, *Prog. Solid State Chem.* **1990**, *20*, 1.
7 S. M. Kautzlarich (Ed.) *Chemistry, Structure, and Bonding of Zintl Phases and Ions: Selected Topics and Recent Advances*, Wiley-VCH, Weinheim, 1996.
8 R. Nesper, "Zintl Phases Revisited", in P. Jutzi, U. Schubert (Eds.), *Silicon Chemistry*, Wiley-VCH, Weinheim, 2004.
9 A. Currao, J. Curda, R. Nesper, *Z. Anorg. Allg. Chem.* **1996**, *622*, 85.
10 S. Wengert, R. Nesper, *J. Solid State Chem.* **2000**, *152*, 460.
11 R. Nesper, H. G. von Schnering, *Tschermaks Min. Petr. Mitt.* **1983**, *32*, 195.

12 E. Parthé, "Elements of Inorganic Structural Chemistry", K. Sutter Parthé Publisher, Petit Lancy, Switzerland, 1996.
13 K. Wade, Adv. Inorg. Chem. Radiochem. 1976, 18, 1.
14 B.K. Teo, H. Zhang, Y. Kean, H. Dang, X. Shi, J. Chem. Phys. 1993, 99, 2929.
15 R. Ramirez, R. Nesper, H.G. von Schnering, M. Böhm, J. Phys. Chem. Solids 1987, 48, 51.
16 M.-H. Wangbo, M. Evan, T. Hughbanks, M. Kertesz, S. Wijeyesekera, C. Wilker, C. Zheng, R. Hofmann, Programme EHMACC: Extended Hückel Molecular and Crystal Calculations.
17 M.-H. Wangbo, M. Evan, T. Hughbanks, M. Kertesz, S. Wijeyesekera, C. Wilker, C. Zheng, R. Hofmann, Programme EHPC: Extended Hückel Property Calculations.
18 G. Krier, O. Jepsen, A. Burkhardt, O.K. Andersson, TB-LMTO-ASA Program, Stuttgart, Germany, 1994.
19 S. Wengert, Dissertation, ETH-Nr. 12070, ETH Zürich, 1997.
20 A.D. Becke, K.E. Edgecombe, J. Chem. Phys. 1990, 92, 5397.
21 A. Savin, R. Nesper, S. Wengert, T.F. Fässler, Angew. Chem. 1997, 109, 1892; Angew. Chem. Int. Ed. Engl. 1997, 36.
22 R. Nesper, Habilitationsschrift, Universität Stuttgart, 1989.
23 V. Hopf, W. Müller, H. Schäfer, Z. Naturforsch. 1972, 30b, 1157.
24 H.G. von Schnering, R. Nesper, J. Curda, K. Tebbe, Angew. Chem. 1980, 92, 1070; Angew. Chem. Int. Ed. Engl. 1980, 19, 1033.
25 R. Nesper, J. Curda, H.G. von Schnering, J. Solid State Chem. 1986, 62, 199.
26 R. Ramirez, R. Nesper, H.G. von Schnering, M. Böhm, Chem. Phys. 1985, 95, 17.
27 M. Böhm, R. Ramirez, R. Nesper, H.G. von Schnering, Phys. Rev. 1984, B30, 4870.
28 M. Böhm, R. Ramirez, R. Nesper, H.G. von Schnering, Ber. Bunsenges. Phys. Chem. 1985, 89, 465.
29 R. Nesper, H.G. von Schnering, J. Solid State Chem. 1987, 70, 48.
30 R. Ramirez, R. Nesper, H.G. von Schnering, M. Böhm, Z. Naturforsch. 1986, 41a, 1267.
31 P. von Ragué Schleyer, E.-U. Würthwein, E. Kaufmann, T. Clark, J.A. Pople, J. Am. Chem. Soc. 1983, 105, 5930.
32 G.A. de Wijs, Ph.D. Thesis, Rijksuniversiteit Groningen, 1995.
33 H. van Leuken, G.A. de Wijs, W. van der Lugt, R.A. de Groot, Phys. Rev. 1996, B53, 10599.
34 G.V. Samsonov, Tugoplavkie soedineniya redkozemel'nikh metallov s nemetallami Isdatel'stvo Mettallurgiya, Moskva, 1964.
35 B. Aronson, T. Lundström, S. Rundquist, Borides, Silicides and Phosphides, J. Wiley & Sons, Inc., New York, 1965.
36 H. Nowotny, Silicides and Borides, in Inorg. Chem. Ser. 1, 1972, 10, 151.
36 J.A. Perri, E. Banks, B. Post, J. Phys. Chem. 1959, 63, 2073.
37 J.A. Perri, I. Binder, B. Post, J. Phys. Chem. 1959, 63, 616.
38 I. Gladyshevskii, Dop. Akad. Nauk Ukrain. 1963, 7, 886.
39 R. Nesper, H.G. von Schnering, J. Curda, Solid Compounds of Transition Elements IV, Internat. Conf., Stuttgart, 1979, Coll. Abstracts pp. 150.
40 A. Grytsiv, D. Kaczorowski, A. Leithe-Jasper, V.H. Tran, A. Pikul, P. Rogl, M. Potel, H. Noël, M. Bohn, T. Velikanova, J. Solid State Chem. 2002, 163, 178.
41 A. Palenzona, P. Manfrinetti, S. Brutti, G. Balducci, J. Alloys Compds. 2003, 348, 100.
42 E.I. Gladyshevskii, Neorg. Mater. 1965, 1, 797.
43 F. Merlo, M.L. Fornasini, J. Less-Common Met. 1967, 13, 603.
44 A. Iandelli, A. Palenzona, G.L. Olcese, J. Less-Common Met. 1979, 64, 213.
45 R. Pöttgen, R.D. Hoffmann, D. Kussmann, Z. Anorg. Allg. Chem. 1998, 624, 945.
46 I. Mayer, I. Shidlovsky, Inorg. Chem. 1969, 8, 1240.
47 C. Kubata, F. Krumeich, M. Wörle, R. Nesper, Z. Anorg. Allg. Chem. 2002, 628, 2201; F. Krumeich, C. Kubata, M. Wörle, R. Nesper, Microsc. Microanal. 2003, 9, Suppl. 3, 340.
48 J. Evers, G. Oehlinger, A. Weiss, F. Hulliger, J. Less-Common Met. 1983, 90, L19–L23.

10
Nonclassical Sb–Sb Bonding in Transition Metal Antimonides

Holger Kleinke

10.1
Introduction: Violations of the Lewis Octet Rule?

Molecules comprising only main group elements that seemingly violate Lewis' octet rule [1] are often described as being hypervalent [2–4]. Standard examples are PCl_5, SF_6, ClF_3, and XeF_2, all of which seem to have more than four electron pairs around the central atom, if drawn as conventional Lewis diagrams (Fig. 10.1).

Initially, explanations of these "violations" invoked valence shell expansion by allowing d orbitals to participate in sp^3d^n hybrid orbitals of the central atom, based on the classical valence bond (VB) theory [5]. This allows for predicting/rationalizing the shapes of the molecules by applying the valence shell electron repulsion theory (VSEPR) [6]: in PCl_5, for example, one assumes sp^2 hybrid orbitals in the equatorial plane and pd hybrids (formed by the p_z and d_z^2 orbitals) in the apical positions of the trigonal bipyramid. Although ab initio calculations revealed that the d orbitals can only play a minor role because of the high promotion energies [7–10], as was discussed as early as the 1960s [11], these sp^3d^n hybrid orbitals continue to be invoked in many basic inorganic chemistry textbooks [12, 13]. In most textbooks [14–16], however, these weaknesses are discussed, and this model is compared with simple molecular orbital descriptions involving only the s and p orbitals of the central atom. For example, in Housecraft-Sharpe's Inorganic Chemistry textbook, both a VB model with d orbitals (i.e., an sp^3d^2 hybridization) as well as an MO model invoking delocalized multi-center bonding are introduced for octahedral SF_6.

Such an MO diagram can be qualitatively developed based on symmetry considerations, applying the O_h character table, and utilizing only the 3s and 3p orbitals of the S atom and those 2p orbitals of F that point directly towards the S atom (Fig. 10.2). This leaves three lone pairs at each F atom.

The MO diagram is comprised of four fully occupied σ bonding molecular orbitals (the a_{1g} and the three degenerate t_{1u} orbitals) and two occupied nonbonding e_g orbitals. Thus, there are four bonds per S atom. The two nonbonding e_g orbitals do not violate the octet rule as they are located exclusively at the F atoms [3]. As the outer 3d orbitals of the S atom split in octahedral symmetry

Inorganic Chemistry in Focus II. Edited by G. Meyer, D. Naumann, and L. Wesemann
Copyright © 2005 WILEY-VCH Verlag GmbH & Co. KGaA, Weinheim
ISBN: 3-527-30811-3

168 | *10 Nonclassical Sb–Sb Bonding in Transition Metal Antimonides*

Fig. 10.1 Lewis diagrams of selected hypervalent molecules (from left to right: PCl_5, SF_6, ClF_3, XeF_2).

into t_{2g} and e_g orbitals, the latter can mix with the nonbonding e_g orbitals of Fig. 10.2, rendering them slightly bonding [17]. Because of the larger energy differences, however, this effect is rather small and cannot be compared to an sp^3d^2 hybridization. Reed and Weinhold quantified the overall occupancy of the 3d orbitals to be 0.16 by calculating ab initio SCF wavefunctions [9].

XeF_2 [18] is more relevant for this chapter than SF_6, for it is isoelectronic with Sb_3^{7-} and I_3^-. All three of these molecules/ions are linear and comprise 22 valence electrons. Sb_3^{7-} occurs in the $Ca_{14}AlSb_{11}$ structure type [19], which has many representatives, some of which exhibit colossal magnetoresistance [20–24]. I_3^- anions are found in a variety of different compounds, including inorganic binaries such CsI_3 [25], RbI_3, and TlI_3 [26]. In an early review of interhalides, three different bonding models were compared, one based on electrostatic interactions, one based on sp^3d^n hybridization, and the other on delocalized p orbitals [27].

As the Lewis diagram reveals, five electron pairs apparently surround the Xe atom in XeF_2. Since this would violate the octet rule, XeF_2 is often classed as a

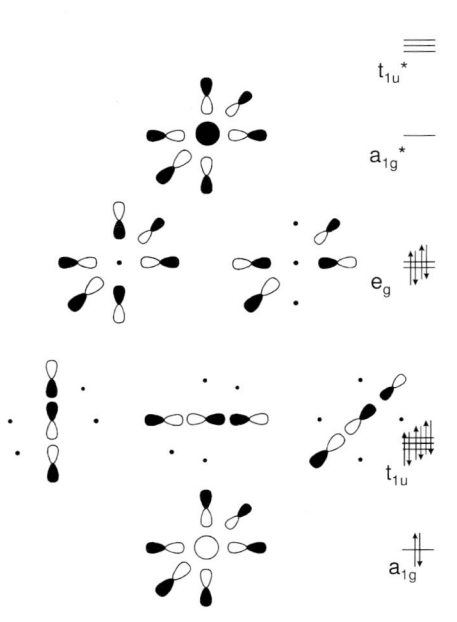

Fig. 10.2 Schematic MO diagram for SF_6. The participating orbitals are s and p for S, and p for F.

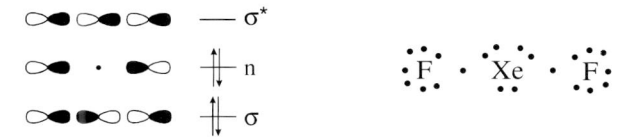

Fig. 10.3 Left: Schematic MO diagram for XeF$_2$. Right: The corresponding Lewis diagram.

hypervalent molecule. The rationalization of the structure of XeF$_2$ with the VSEPR theory is straightforward, if one accepts the presence of "low-lying" d orbitals, and subsequently of sp^{3}d hybridization. As mentioned above for the case of PCl$_5$, this would explain the trigonal-bipyramidal arrangement of the five orbitals. The three lone pairs would be situated in the equatorial plane, leaving the two apical pd hybrid orbitals for Xe–F bonds. This is in agreement with the 180° F–Xe–F angle. This approach, however, is particularly problematic for XeF$_2$ because of the especially high promotion energies of the Xe d orbitals.

A simplifying MO approach comparable to the case of SF$_6$ utilizes only three p orbitals of XeF$_2$, treating the s and the other p orbitals as lone pairs. Putting the z axis along the F–Xe–F bonds, the orbitals under consideration are the three p$_z$ orbitals of Xe and of the two F atoms. The resulting MO diagram then comprises three molecular orbitals. Two of them are filled, namely a σ bonding and a nonbonding molecular orbital. The resulting bonding situation may be described as a three-center–four-electron (3c–4e) bond, as was already discussed back in 1963 [11]. Two of these four electrons are used for Xe–F bonding, and two are delocalized over nonbonding F orbitals located exclusively on the F atoms. Altogether, there are eight electrons in Xe orbitals. Therefore, the octet rule is not violated. In Fig. 10.3, the MO diagram is compared with the corresponding Lewis formula.

A (non-relativistic) ab initio calculation of the electronic structure of XeF$_2$, including the 5d orbitals of Xe, resulted in a Mulliken population [28] of the Xe d orbitals of only 0.26 [29], which again is much less than predicted for pd hybrid orbitals, but more than postulated in the somewhat simplifying model depicted in Fig. 10.3. Considering the high polarity of the S–F and Xe–F bonds in SF$_6$ and XeF$_2$, respectively, different resonance structures involving ionic interactions have been discussed as well. Fig. 10.4 gives two examples for XeF$_2$ [30].

The polarity of the bonds in the ions Sb$_3^{7-}$ and I$_3^-$ is much smaller than in XeF$_2$, rendering the ionic models obsolete for these cases. Assuming the validity of the MO model shown in Fig. 10.3 for these two anions, one would predict linearity as well as equal bond lengths within each ion, and thus the presence of an inversion center, unless an asymmetric surrounding in the crystal structure destroys the symmetry of the anion. Furthermore, as only two electrons are available for two bonds, one presumes (half) bonds significantly longer than single bonds. Sb$_3^{7-}$ in Ca$_{14}$AlSb$_{11}$ is centrosymmetric, with two Sb–Sb bonds of 320 pm, compared to Sb–Sb single bonds of 280–285 pm.

On the other hand, many triiodides are asymmetric, with the two I–I bonds in each I$_3^-$ unit differing in length by up to 20 pm. For example, the I–I distances of

$:\ddot{F}\!-\!\ddot{X}\overset{\cdot\cdot\oplus}{e}\!:\;\;:\overset{\cdot\cdot\ominus}{F}\!:\qquad\qquad:\overset{\cdot\cdot\ominus}{F}\!:\;\;:\overset{\cdot\cdot\oplus}{X}\overset{\cdot\cdot}{e}\!-\!\ddot{F}\!:$

Fig. 10.4 Selected ionic resonance structures of XeF$_2$.

the I$_3^-$ ion in CsI$_3$ are 284 and 304 pm, with a bond angle of 178.0°. This asymmetry was ascribed to the irregular Cs coordination of the I atoms [27]. Generally, triiodides will be more symmetric with larger cations, a trend that is supported by the structures of [Ph$_4$As]I$_3$ [31] (centrosymmetric I$_3^-$, I–I bonds of 290 pm) and [Me$_4$As]I$_3$ [32]. Three symmetry-independent I$_3^-$ ions, all of different symmetry (point groups $D_{\infty h}$, C_{2v}, and C_1), occur in the latter structure, and they all exhibit very similar I–I distances (291–293 pm) as well as bond angles close to 180°, regardless of whether they are fixed by symmetry or not. All these bonds are significantly longer than the typical I–I single bond of 276 pm in I$_2$, which is qualitatively in agreement with the half-bond proposal shown in Fig. 10.4. Applying Pauling's bond order formula (Eq. (1)), we obtain 294 pm for a half I–I bond (ignoring the one negative charge per polyiodide anion):

$$d(n) = d(1) - 60\,\text{pm}\log n \tag{1}$$

where $d(n)$ = observed distance, $d(1)$ = single-bond distance, n = bond order [33], with $d(1) = 276$ pm and $n = 0.5$: → $d(0.5) = 294$ pm.

As covered in the Shriver-Atkins textbook [34], many higher polyiodides I$_n^-$ exist, which may be viewed as adducts of I$_2$ to I$_3^-$. Among others, Tebbe et al. have reported on a vast number of different polyhalides, see, e.g. [35–38], and references cited therein.

10.2
Sb–Sb Interactions in Zintl Phases

10.2.1
General Remarks about Zintl Phases

For a more detailed introduction to Zintl phases, the reader is referred to the first two chapters of Volume 1 of *Inorganic Chemistry Highlights* [39]. Numerous earlier reviews dealing with different Zintl phases are also available [40–42]. Zintl phases AQ$_x$ are usually comprised of an electropositive element A (e.g., an alkali metal) and a more electronegative element Q of the later (post transition) main groups. A principal concept for understanding the resulting structures is based on a formal charge transfer of all valence electrons of A to Q. Then, cation A possesses a full octet, and the (formal) anion Q achieves the same by forming homonuclear Q–Q bonds in addition to the reduction. The vast majority of Zintl phases consist in part of classical single (two-center–two-electron, 2c–2e) bonds, so that exactly one bond is formed for each electron missing to complete the octet of Q. The standard example is NaTl [43], which is written in this context as Na$^+$Tl$^-$. Each Tl$^-$ "ion" then possesses four valence electrons and, therefore, needs four additional electrons to fulfil the octet rule, which is achieved by forming four Tl–Tl single bonds per Tl. This results in the Tl atoms forming a diamond-analogous network, which is in perfect agreement with the VSEPR theory (tetrahedral coordination for four bonds, sp^3 hybridization, no lone pairs).

10.2.2
Molecular Antimony Units in Zintl Phases

Many different antimony atom substructures in Zintl phases ASb$_x$ have been found in the last three decades. Most of these consist of classical single bonds (2c–2e bonds). To rationalize these Sb atom substructures, one can determine the formal charges by transferring all valence electrons of the electropositive elements A to the Sb atoms. Then, the number of Sb–Sb bonds per Sb atom can be calculated by applying the (8–N) rule, with N=main group number (i.e. 5 for Sb) + formal charge. A neutral Sb atom can form 8–N=3 Sb–Sb bonds (as found, e.g., in elemental antimony), an Sb$^-$ as in K$^+$Sb$^-$ [44] and Co^{3+}(Sb$^-$)$_3$ [45] forms 8–N=2 bonds, an Sb^{2-} as in (Na$^+$)$_2$(Sr^{2+})$_3$(Sb^{2-})$_4$ [46] forms 8–N=1 bond, and an Sb^{3-} as in (K$^+$)$_3$Sb^{3-} [47] cannot participate in any Sb–Sb bonds (8–N=0). Selected finite Sb atom units are shown in Fig. 10.5.

When the formal charge per Sb atom is fractional, as in (Cs$^+$)$_3$(Sb$^{3/7-}$)$_7$ [48] and (Sr^{2+})$_2$(Sb$^{4/3-}$)$_3$ [49], at least two differently charged and therefore differently connected Sb atoms will be present. In (Cs$^+$)$_3$(Sb$^{\pm 0}$)$_4$(Sb$^-$)$_3$, one observes four three-bonded (3-b) and three 2-b Sb atoms, and in (Sr^{2+})$_2$(Sb$^-$)$_2$Sb^{2-} two 2-b and one 1-b Sb. A more complex example is Ca$_{11}$Sb$_{10}$ [50], which contains three differently charged Sb atoms, as can be seen in the formulation (Ca^{2+})$_{11}$(Sb$^-$)$_{2-}$

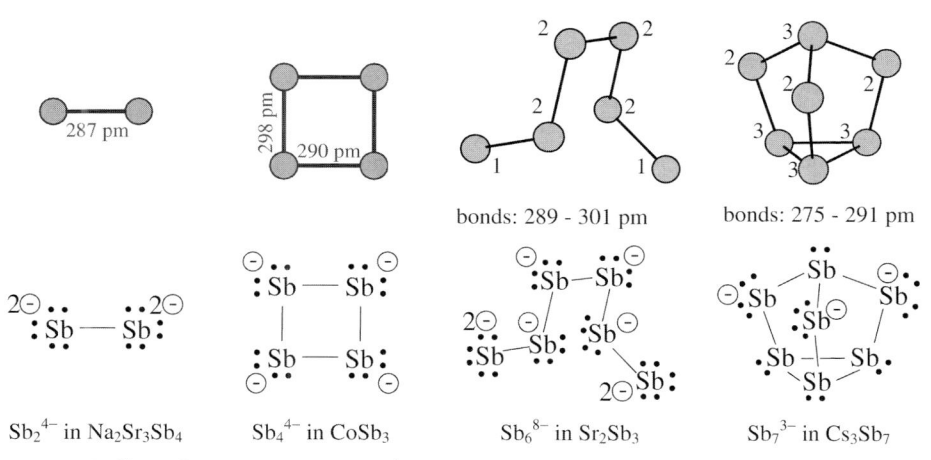

Fig. 10.5 Different finite antimony atom substructures in Zintl phases.

$(Sb^{2-})_4(Sb^{3-})_4$. Therein, as expected based on the (8–N) rule, the Sb^- atoms each form two Sb–Sb bonds, resulting in Sb_4^{4-} squares, the Sb^{2-} form pairs, and the Sb^{3-} atoms do not have any Sb–Sb contacts.

10.2.3
Extended Antimony Units in Zintl Phases

One weakness of the electron-counting procedures is the lack of predictability regarding the long-range order of the Sb atom substructures. The Sb^- atoms, for example, may form rings of different sizes (like elemental sulfur) or zigzag/helical chains (like elemental selenium and tellurium). Furthermore, there are often different possibilities in assigning charges; e.g., instead of two 2-b Sb^- one could find one 3-b $Sb^{\pm 0}$ and one 1-b Sb^{2-}, which would increase the complexity of the Sb atom substructure. A four-membered ring is shown in Fig. 10.5, a folded five-membered Sb_5^{5-} ring was found coordinated by two Li cations in $[Li_2(NH_3)_2Sb_5]^{3-}$ [51], and Fig. 10.6 shows sections of the infinite helical Sb^- chain in the structure of KSb [44] (left) and of the infinite planar zigzag Sb^- chain in $CaSb_2$ [52] (right).

Assuming sp^3 hybridization, one would expect either a helical chain, as found in KSb and in isoelectronic elemental tellurium, or nonplanar folded rings as in $[Li_2(NH_3)_2Sb_5]^{3-}$ and in elemental sulfur. The experimentally observed deviations may be attributed to packing effects, or, more generally, to the influence of the cations. Since the cation:Sb ratio is 1:2 in $CaSb_2$, compared to 1:1 in KSb, one might accept the necessity for the Sb chains to come closer to each other in $CaSb_2$, which, in turn, will have an impact on their geometry. That this may indeed be the case is supported by the shorter interchain Sb–Sb distances of 346 pm in $CaSb_2$, compared to 492 pm in KSb. Similarly, each Sb atom of the Sb_4 rings in $CoSb_3$ exhibits four interchain Sb–Sb contacts of 341 pm. Furthermore, the intrachain Sb–Sb distances in $CaSb_2$ and $CoSb_3$ (290–298 pm) are all elongated, compared to typical Sb–Sb single bonds of 280–285 pm. The Sb–Sb distances of 341–346 pm definitely have an impact on the bonding situation [53, 54], considering the Sb–Sb van der Waals contact of 440 pm. Significant force constants ranging from 0.125 to 0.18 N cm^{-1} have been calculated for the so-called secondary Sb–Sb interactions of 363 to 399 pm in thermochromic distibanes, compared to 1.1 N cm^{-1} for the short intermolecular Sb–Sb distance of 284 pm [55]. These bonding interactions were calculated to stem from correlation effects [56].

The predictability of compounds with two or more differently charged Sb atoms is even lower, as discussed above in the case of $Ca_{11}Sb_{10}$. One can readily

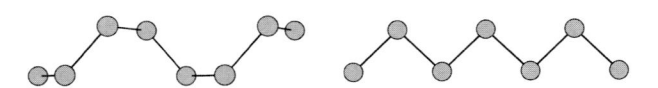

Fig. 10.6 Left: Helical Sb^- chain of KSb. Right: Planar Sb^- zigzag chain of $CaSb_2$.

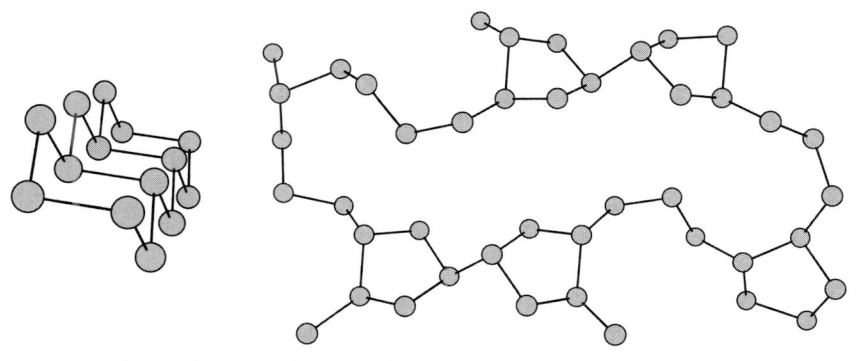

Fig. 10.7 Left: One-dimensional Sb unit of KSb$_2$. Right: Two-dimensional Sb unit of Cs$_5$Sb$_8$.

rationalize the presence of one 3-b (Sb$^{\pm 0}$) and one 2-b Sb atom (Sb$^-$) per formula unit in KSb$_2$ [57], but which Sb atom substructures may result? The formation of six-membered Sb rings (in the chair conformation) condensed to form infinite chains was observed experimentally (left part of Fig. 10.7), but many other valid structural motifs can be envisaged. The same is true for the Sb atom substructure of Cs$_5$Sb$_8$ [58], which can be rationalized based on three 3-b Sb$^{\pm 0}$ and five 2-b Sb$^-$ per formula unit. In this case, puckered layers comprising Sb$_5$ and Sb$_{28}$ rings are present (right part of Fig. 10.7), most likely not a predicted variant.

10.2.4
Beyond Classical Sb–Sb Single Bonds in Zintl Phases

The Zintl concept fails to account for the Sb atom substructure of K$_5$Sb$_4$ [59] (and isostructural Rb$_5$Sb$_4$ [60]). Assigning valence electrons results in a formal 5– charge for the Sb$_4$ units. A flat zigzag tetramer is found, resembling a section of the chains found in CaSb$_2$. Such a chain-like finite Sb$_4$ unit would contain two 1-b and two 2-b Sb atoms, corresponding to a charge of 6– (two 1-b Sb^{2-} and two 2-b Sb$^-$, cf. the Sb$_6^{8-}$ unit of Sr$_2$Sb$_3$ shown in Fig. 10.5), if all bonds were classical single bonds. The geometry of the tetramer is notably different to that expected for classical bonds, as reflected in the planarity as well as in the lengths of the Sb–Sb bonds, which at 279 and 281 pm are somewhat shorter than those in the chains of KSb and CaSb$_2$. This points towards a partial π bonding character, which is supported by the results of ab initio calculations [61]. However, the experimentally determined Pauli paramagnetism and metallic conductivity indicate that the Sb$_4$ unit may not be a radical anion, as apparently some electrons are delocalized throughout the whole crystal structure [60].

That Sb atoms are capable of forming double bonds was proven in 1998 with the discovery of a distibene R–Sb=Sb–R containing an unbridged Sb–Sb double bond with a length of 264 pm [62]. A prior example contained a bridged Sb–Sb

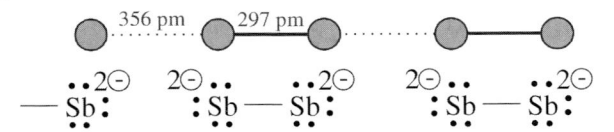

Fig. 10.8 Linear (distorted) Sb atom chain of Li_2Sb with alternating distances.

double bond of 277 pm [63]. A bridged formal triple bond of 266 pm exists in the carbonyl [Sb≡Sb{W(CO)$_5$}$_3$] [64].

(2,2,2-crypt-K)$_2$Sb$_4$ contains the square-planar anion Sb_4^{2-} with interatomic distances of 275 pm [65], which are significantly smaller than in the Sb_4^{4-} ring of CoSb$_3$ as a consequence of the additional π bonding. The anion Sb_4^{2-} is isovalent with the aromatic $C_4H_4^{2-}$ as found in $Li_2C_4H_4$, and the more commonly observed cation Te_4^{2+} [66, 67]. Both the Sb_4^{2-} and the Te_4^{2+} ions are regular 6 π Hückel aromatic ions, as both electronic structure calculations and photoelectron spectroscopy experiments have revealed [68, 69]. Furthermore, a triple-decker sandwich complex has been characterized that contains an almost planar Sb$_5$ ring in the middle, named a pentastibacyclopentadienyl ligand [70]. This Sb$_5$ ring contains different Sb–Sb bond lengths, ranging from 276 to 285 pm, the latter corresponding to a typical single bond.

Besides K$_5$Sb$_4$, a second alkali metal antimonide that does not conform to the Zintl concept is Li$_2$Sb [71]. The ionic model (Li$^+$)$_2$Sb^{2-} points towards 1-b Sb atoms, which should form Sb_2^{4-} pairs as found in (Na$^+$)$_2$(Sr^{2+})$_3$(Sb^{2-})$_4$ and CdSb [72]. In contrast to this prediction, the structure of Li$_2$Sb contains two different Sb atom chains that are both linear. The first one (Fig. 10.8) exhibits alternating Sb–Sb distances of 297 and 356 pm, while all the Sb–Sb distances of the second chain are 326 pm. Thus, one can regard the first chain as a distorted variant of the second.

In a crude approximation, one may consider the distorted chain as consisting of linearly arranged Sb_2^{4-} pairs, neglecting the facts that: i) the distances of 297 pm are longer than expected for a single bond, and ii) the distances of 356 pm between the pairs are much shorter than the van der Waals distance of 440 pm. Sb_2^{4-} pairs in related compounds exhibit distances between 281 pm in CdSb [72] and 287 pm in Na$_2$Sr$_3$Sb$_4$. Since the second chain carries the same charge, one may postulate the presence of one single bond per Sb atom as well [41]. Since there are two equivalent distances of 326 pm per Sb atom, both can be treated as having a bond order of ½, i.e. a half bond (Fig. 10.9). Drawing the second chain in a conventional Lewis diagram demonstrates that it may be regarded as a nonclassical (hypervalent) chain, with three lone pairs and two bonds per Sb atom [73].

To gain more insight, we calculated the electronic structures for both chains of the Li$_2$Sb structure utilizing the Extended Hückel approximation [74, 75] with the Sb parameters that were published earlier [76]. Both chains run parallel to the crystallographic c axis, and contain two Sb atoms per repeat unit. Therefore,

10.2 Sb–Sb Interactions in Zintl Phases

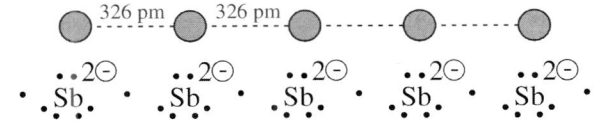

Fig. 10.9 Linear (undistorted) chain of equidistant Sb atoms in Li$_2$Sb.

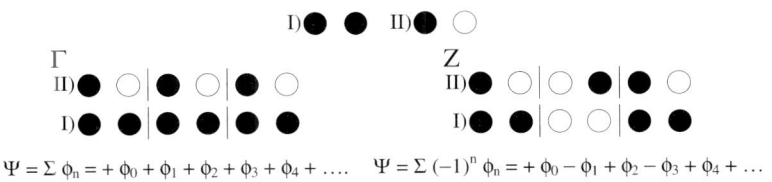

Fig. 10.10 The s orbitals of the undistorted chain at the zone center Γ and the zone border Z.

$$\Psi = \Sigma \phi_n = + \phi_0 + \phi_1 + \phi_2 + \phi_3 + \phi_4 + \ldots \qquad \Psi = \Sigma (-1)^n \phi_n = + \phi_0 - \phi_1 + \phi_2 - \phi_3 + \phi_4 + \ldots$$

eight valence orbitals, namely two s and six p orbitals, are to be considered. The s and p$_z$ orbitals (with z being parallel to the c axis) can participate in σ interactions, and the (degenerate) p$_x$ and p$_y$ orbitals in π interactions, as the latter are perpendicular to the chain direction.

For a general construction of bands from crystal orbitals the reader is referred to a review [77], or to two books dealing with the electronic structure of solids [78, 79]. In LCAO (Linear Combination of Atomic Orbitals) theory, the wavefunction Ψ can be obtained by integrating over all basis orbitals ϕ_n utilizing Bloch's theorem. For a one-dimensional case study, Eq. (2) applies:

$$\Psi = \Sigma e^{iknc}\phi_n = \Sigma\{\cos(knc) + i\sin(knc)\}\phi_n \tag{2}$$

where k=wavenumber, ϕ_n: nth basis orbital, c=lattice parameter

To construct the band structure $E(k)$, we first investigate the two border cases, the origin (zone center Γ) and the border (zone border Z) of the Brillouin zone, i.e. $k=0$ and $k=\pi/c$. At $k=0$, one finds $\cos(knc)=1$ and $\sin(knc)=0$, thus $\Psi=\Sigma\phi_n$. At $k=\pi/c$, we obtain $\cos(knc)=\cos(\pi n)=(-1)^n$ and $\sin(knc)=\sin(\pi n)=0$, thus $\Psi=\Sigma(-1)^n\phi_n$.

We start by considering the s orbitals of the Sb chain. With two atoms per repeat unit, the two basis orbitals result from the in-phase (I in Fig. 10.10) and the out-of-phase (II in Fig. 10.10) combination of the two s orbitals. To construct the wavefunction at the zone border Γ, one needs to add up all the basis orbitals in the same phase. At the zone edge Z, the signs of the basis orbitals alternate. From this simple construction alone, one can gather some important facts. At Γ, I) is all bonding (σ), and II) is all antibonding (σ^*) and thus much higher in energy; at Z, I) and II) are degenerate.

With this knowledge, one can deduce the whole band structure of the undistorted Sb atom chain (Fig. 10.11) by connecting the points between Γ and Z with a smooth S-shaped curve (cf. Eq. (2)). The lowest-lying bands are the two s

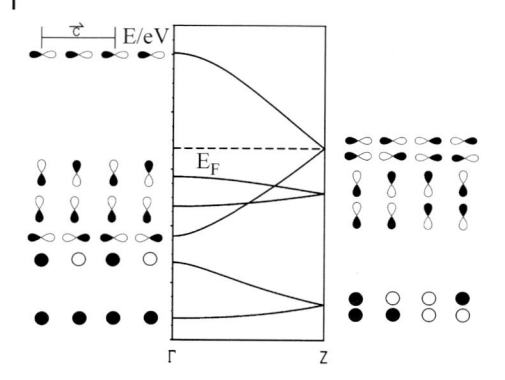

Fig. 10.11 Calculated band structure of the undistorted Sb chain of Li$_2$Sb.

bands, which are completely filled. As their bonding and antibonding contributions roughly cancel each other out, the s orbitals may be treated as lone pairs. At Γ, one p$_z$-based band is σ bonding, and the other σ antibonding. Like all the other bands, they are degenerate at the zone border Z. It is important to note that the p$_x$ and p$_y$ bands, which exhibit π character, are lower in energy than the upper p$_z$ band (over the whole Brillouin zone), which is a consequence of significant s,p mixing.

As Sb^{2-} comprises seven valence electrons, and we have two thereof per unit cell, 14 valence electrons are available to fill seven bands, namely both s, both p$_x$ and both p$_y$, and the lower p$_z$ band. In this way, the Fermi level (dashed line in Fig. 10.11) cuts right through the point at which the two p$_z$-based bands touch. Being comparable to the s bands, the filled p$_x$ and p$_y$ bands may be considered as lone pairs as well, for they exhibit equally (π) bonding and antibonding contributions. This is only an approximation, however, because the antibonding contributions are slightly stronger than the bonding ones.

To summarize the bonding situation in the undistorted Sb atom chain of Li$_2$Sb, the net interaction stems solely from the completely filled lower p$_z$ band, which is σ bonding throughout the whole Brillouin zone (all bonding at the center, and nonbonding at the border), while the empty upper p$_z$ band is antibonding. Therefore, we describe the bonds as delocalized two-center–one-electron bonds, i.e. as half bonds, in agreement with the assessment above (Fig. 10.9). The strong resemblance between Fig. 10.11 and the band structure obtained for a hypothetical Sb chain with equidistant interactions of 320 pm is obvious [73].

Since two degenerate states occur directly at the Fermi level, this Sb chain is susceptible to a Peierls distortion [80], which may be regarded as the solid-state analogue to the Jahn-Teller distortion. (Strictly speaking, to qualify for a Peierls distortion, this condition must occur with a decrease of translational symmetry. That would be the case if one considers a linear chain with one Sb atom per repeat unit as the starting point.) A distortion of the chain, as realized in the actual Li$_2$Sb structure with alternating Sb–Sb distances of 297 and 356 pm, can lower its energy by lowering the filled bonding states and raising the empty

Fig. 10.12 The forms of the s orbitals of the distorted chain.

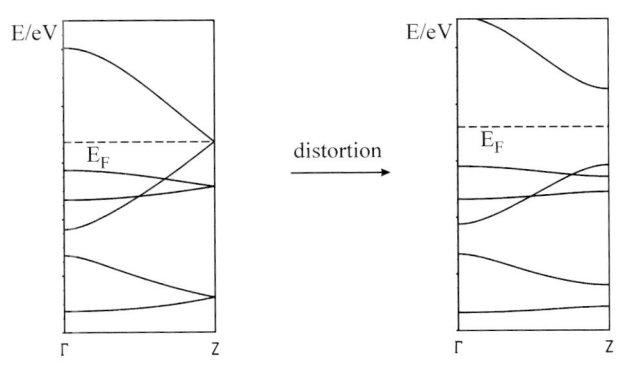

Fig. 10.13 Calculated band structures of the two Sb chains of Li_2Sb.

antibonding states. The consequence of such a distortion for the s orbitals is shown in Fig. 10.12.

The impact is higher at the zone border, where the bonding interactions become stronger than the antibonding ones for case I), and weaker for II). Therefore, the two bands are not degenerate at the zone border Z, in contrast to the situation in the undistorted chain. The same formalism can be applied to the other bands, leading to the creation of a band gap at the Fermi level (Fig. 10.13). Overall, this results in a lower total energy as the highest occupied states are lowered.

10.3
Antimony Atom Substructures in Transition Metal Antimonides

10.3.1
Valence-Electron-Rich Transition Metal Antimonides

The assignment of charges in transition metal compounds is principally not as unambiguous as in those of the alkali metals, for the transition metal elements exhibit more than one possible oxidation state. In addition to skutterudite $CoSb_3$ [45], the iron group metals form diantimonides of the marcasite type, $FeSb_2$ and $NiSb_2$ [81], and $CoSb_2$ [82]. The aristotype is FeS_2, for which the assignment of oxidation states leads to an S_2^{2-} dumbbell and thus Fe^{2+}. Assuming that Fe^{2+} is also present in the isostructural $FeSb_2$, we find two Sb^-, and therefore expect two Sb–Sb bonds per Sb atom. This is realized in $FeSb_2$ by the formation

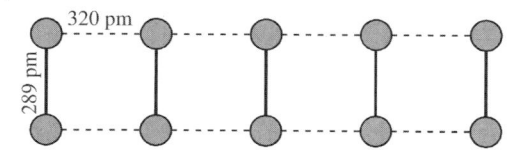

Fig. 10.14 Antimony ladder in the structure of FeSb$_2$.

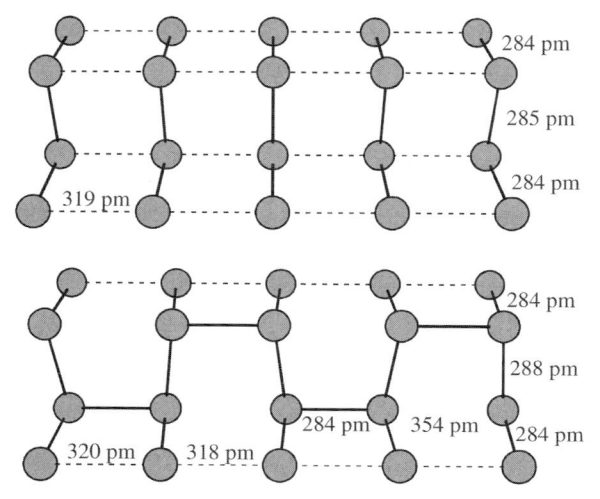

Fig. 10.15 Intercondensed antimony ladders (top) and their distorted variant (bottom) in the structure of MoSb$_2$S.

of one Sb–Sb single bond (289 pm) and two Sb–Sb half bonds (320 pm) per Sb atom, creating a ladder-like arrangement, with the chain direction being reminiscent of the undistorted chain of the Li$_2$Sb structure (Fig. 10.14). Therefore, this chain involves nonclassical (hypervalent) bonding as well. The more metal-rich antimonides, such as MnSb, FeSb, CoSb, and NiSb, all occur in the NiAs type without any noticeable Sb–Sb bonding [83].

A topologically equivalent Sb atom ladder, together with a Peierls distorted variant thereof, was found in MoSb$_2$S [84]. Each Sb atom of the ladder and its distorted variant is additionally connected via a typical single bond to another Sb atom (Fig. 10.15). This further reduces the formal charge, for now each Sb atom forms two half and two single bonds, corresponding to three full bonds, thus to an Sb$^{\pm 0}$. The outside Sb atoms carry a (formal) negative charge of one, as they form one bond to the ladder plus two half bonds parallel to the ladder. Altogether, this chain can be viewed as three interconnected ladders, or a zigzag Sb$_4$ strip forming an equidistant chain. Its distorted variant, found in the same structure, exhibits a large distortion in the central ladder (284/354 pm), and a rather negligible one in the outside chains (318/320 pm). It is interesting to note that only the latter variant occurs in the related selenide MoSb$_2$Se [85].

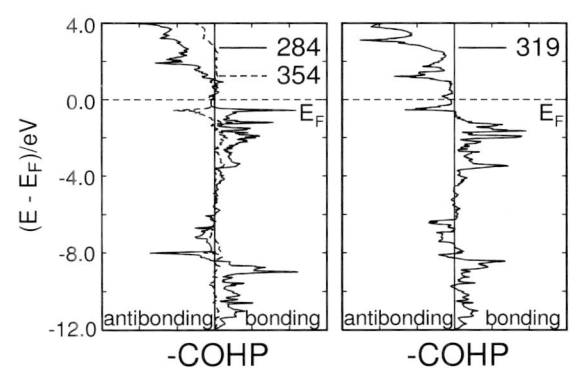

Fig. 10.16 Selected Sb–Sb COHP curves of MoSb$_2$S. Left: distorted chain. Right: undistorted.

Calculations of the electronic structure of MoSb$_2$S were performed by applying the self-consistent TB-LMTO-ASA method, which utilizes density functional theory with the local density approximation (LDA) [86–88]. One can extract the densities of states (DOS) from the band structure calculation, which, in turn, can be weighed into its pairwise bonding and antibonding contributions, a procedure which yields the crystal orbital Hamilton population (COHP) [89]. The COHP curves resemble the well-established crystal orbital overlap population (COOP) curves [90], which are usually extracted from Extended Hückel calculations; bonding interactions are depicted on the right-hand side of the diagram and antibonding ones on the left. Fig. 10.16 compares the trace for the distorted central chain with alternating Sb–Sb distances of 284 pm and 354 pm with that for the undistorted one (2 × 319 pm). In the latter case, the two bonds per cell were summed to give a single plot.

The 284 pm bond exhibits the strongest overlaps and the least filled antibonding states, as only a small part of the s contribution around –8 eV is antibonding. All curves have roughly comparable shapes below –1 eV, and the interactions of 354 and 319 pm show an antibonding peak directly below the Fermi level formed by the Sb p states. It is evident that the 354 pm distance corresponds more or less to a nonbonding interaction, while the other two are clearly net bonding interactions, for more bonding than antibonding states are filled. This can be quantified with the integrated COHP values (ICOHPs), which may be used to compare relative bond strengths in analogy to the longer established Mulliken overlap populations (MOPs) [28] obtained from Extended Hückel calculations. Three major differences between ICOHPs and MOPs are noteworthy. First, bonding interactions are reflected in positive MOPs, but negative ICOHPs. Second, ICOHPs and MOPs have different units (eV vs. electrons per bond). Third, ICOHPs exhibit higher absolute values than the MOPs in the cases studied thus far. The first published ICOHPs are –1.53 and –1.00 eV/bond for the shortest metal–metal bonds in elemental (bcc) iron and nickel, respectively [91]. We calculated –1.63 for the short bond (291 pm) and –0.23 eV/bond for the long bond (335 pm) of elemental antimony. The averaged ICOHP for the Sb–Sb single bonds of $MoSb_2S$ is -1.67 eV/bond, compared to -0.61 for the half bond of the central ladder (right part of the ladder). A comparison of two of the latter interactions ($2 \times -0.61 = -1.22$) with the sum of the two different bonds in the chain of alternating 284 and 354 pm interactions ($-1.56 + -0.12 = -1.68$) confirms that the distorted variant exhibits stronger bonding.

A second ternary exists in the Mo/Sb/S system, namely Mo_2SbS_2 [92]. This compound is a ternary variant of Mo_2S_3, with one Sb^{2-} replacing one S^{2-}. As in the $FeSb_2/FeS_2$ comparison, this formal replacement enables the formation of Sb–Sb bonds, realized here in the formation of a linear Sb atom chain with interatomic distances in the half-bond range (318 pm). In this case, the ICOHP value was calculated to be -0.53 eV per bond. Aside from the nonclassical Sb–Sb bonds, Mo_2SbS_2 undergoes a transition to the superconducting state upon cooling, while Mo_2S_3 was investigated for its phase transformations driven by the clustering of the d^3 Mo atoms [93, 94].

10.3.2
Valence-Electron-Poor Transition Metal Antimonides

The antimonides of the valence-electron-poor ("early") transition metal elements differ from those of the late ones in the fact that metal–metal (M–M) bonds are common in the former. TiSb, VSb, and NbSb all crystallize in the NiAs type, wherein M–M bonds occur in the M atom chains parallel to the c axis. The late transition metal atoms are too small to form M–M bonds in antimonides of the NiAs type. Like the late transition elements, the early ones form diantimonides as well, albeit in different, more complex structure types. As these materials, e.g. $TiSb_2$, $ZrSb_2$, and VSb_2 [95–97], are metallic, they are not covered in the original Zintl concept, but in the following it is demonstrated how useful the assignment of valence electrons in partly nonclassical Sb networks can be.

$TiSb_2$ and VSb_2 adopt the $CuAl_2$ structure type, with Sb–Sb pair bond lengths of 284 and 289 pm, respectively [83]. These short bonds occur between the Sb atoms of neighboring columns of face-sharing MSb_8 square antiprisms. Additional longer Sb–Sb contacts of 331 pm occurring within the columns may not be insignificant. However, by neglecting the latter, and assigning a charge of 4– for the Sb pairs, one obtains Ti^{4+} and V^{4+}, respectively. For VSb_2, this leaves one d electron available per V atom for V–V bonding, and the V atoms form linear chains with two interatomic V–V distances of 282 pm per V atom. On the other hand, Ti^{4+} cannot participate in Ti–Ti bonding, but the Ti atom chains in $TiSb_2$ exhibit Ti–Ti distances of 290 pm, which is within the normal bonding range: Ti metal (hexagonal close-packed) comprises Ti–Ti bonds of 289 and 295 pm. Therefore, an assignment of 4+ for Ti is unlikely and, consequently, the Sb pairs are less reduced than 4–. This explains the additional occurrence of the longer Sb–Sb contacts as needed to fulfil the octet.

The diantimonides of the heavier group 4 metals, $ZrSb_2$ [97] and $HfSb_2$ [98], form unique (isostructural) crystal structures. The Sb atom substructure of

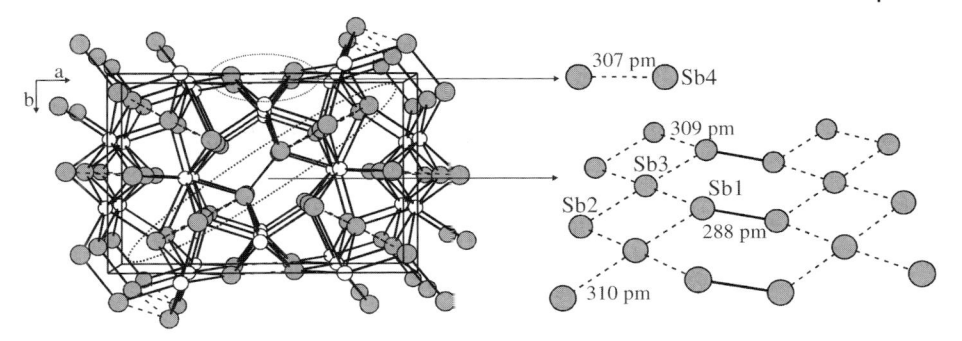

Fig. 10.17 Structure of ZrSb$_2$ in a projection along the c axis (right) and its Sb substructures (right).

ZrSb$_2$ was analyzed based on Sb$_2$ pairs and infinite "Sb$_6$ strips" [99]. The values for the interatomic distances are 307 pm in the Sb$_2$ pair and 288, 309, and 310 pm in the Sb$_6$ strip (Fig. 10.17), all of which fit very well into the scheme of half (307–310 pm) and full (288 pm) bonds applied herein. Assigning charges based on the (8–N) rule results in 5– for the Sb4$_2$ pair, 2– for the terminal Sb atoms (Sb2) of the Sb$_6$ strip, for they form two half bonds, 1– for Sb3 (four half bonds), and 1– for Sb1 (one full and two half bonds). This yields the overall formula (Zr$^{3.25+}$)$_2$Sb1$^-$Sb2$^-$Sb3^{2-}Sb4$^{2.5-}$. The 0.75 4d electrons per Zr atom are available for the experimentally determined metallic conductivity.

Earlier reports on the existence of a second modification, "β-ZrSb$_2$", were recently corrected. This compound is actually a ternary silicide-antimonide [100], and comprises a mixed Si/Sb substructure which will not be discussed here.

The group 5 diantimonides NbSb$_2$ and TaSb$_2$ crystallize in the OsGe$_2$ structure type with an interesting nonclassical (hypervalent) layered Sb atom substructure (Fig. 10.18) [101]. Therein, the Nb (or Ta) atoms are located in trigonal prisms formed by Sb atoms, the three rectangular faces of which are capped by one Nb and two Sb atoms.

Since the Nb atoms are arranged in a pairwise fashion, a 4d^1 configuration is likely, making them Nb^{4+}. Two different puckered Sb atom layers are found parallel to the b,c plane. The first one (formed by Sb1) cuts through the Nb–Nb bonds, and has three Sb–Sb contacts of 330 pm per Sb atom. The second one (Sb2) has shorter Sb–Sb contacts, specifically one of 277 pm and two of 304 pm. Following the above-mentioned counting scheme, we ignore in a first approximation the 330 pm distances, assign a bond order of one to the 277 pm bond, and treat the 304 pm interactions as a half bond. Applying the (8–N) rule, the resulting formal charges are Sb1^{3-} (no Sb–Sb bonds) and Sb2^{1-} (one full and two half bonds, which corresponds to two full bonds), thus resulting in the ionic formula Nb^{4+}Sb1^{3-}Sb2^{1-}.

According to this formula, semiconducting properties appear possible, but experiments revealed metallic conductivity, in agreement with our electronic structure calculations using the DFT-based program package WIEN2K [102]. On the

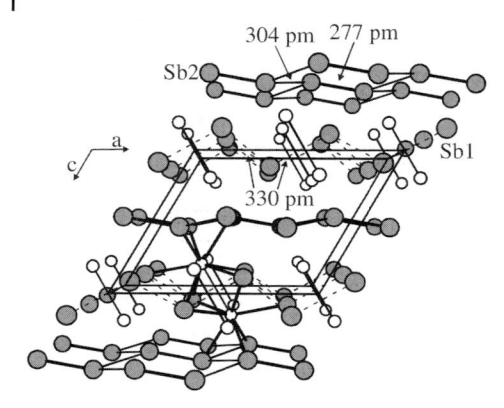

Fig. 10.18 Structure of NbSb$_2$ in a projection along the *b* axis. Small, white circles: Nb; large, gray: Sb.

other hand, a calculation of the ternary variant HfMoSb$_4$ predicts semiconductivity, if the Hf and Mo atoms order as to avoid Hf–Hf and Mo–Mo bonds [103], implying that this structure may indeed be treated as a transition metal Zintl phase with nonclassical Sb–Sb bonding (e.g., like Li$_2$Sb). Fig. 10.19 shows the densities of states (DOS) of NbSb$_2$ and HfMoSb$_4$ with exclusively Hf–Mo pairs. Only the latter case exhibits a (small) band gap, as required in the original Zintl phase definitions.

A similar case is that of Mo$_3$Sb$_7$ [104], the structure of which (Ir$_3$Ge$_7$ type) also comprises three kinds of Sb–Sb contacts, with distances of 291, 310, and 338 pm. Here, the first distance is of the order of a single bond, the second of a half bond, and the third corresponds to a rather small bond order that might be negligible in a crude approximation. Sb1 participates exclusively in the longest interactions (338 pm), and each Sb2 forms three half bonds and one single bond in a peculiar fashion: the Sb2 atoms form cubes through the half bonds of 310 pm, and Sb2 pairs connect the cubes to a three-dimensionally extended Sb atom network (Fig. 10.20).

Adding up these bond order numbers to obtain the formal charges according to the (8–N) rule results in (Mo$^{3.67+}$)$_3$(Sb1^{3-})$_3$(Sb2$^{0.5-}$)$_4$, which is in agreement with

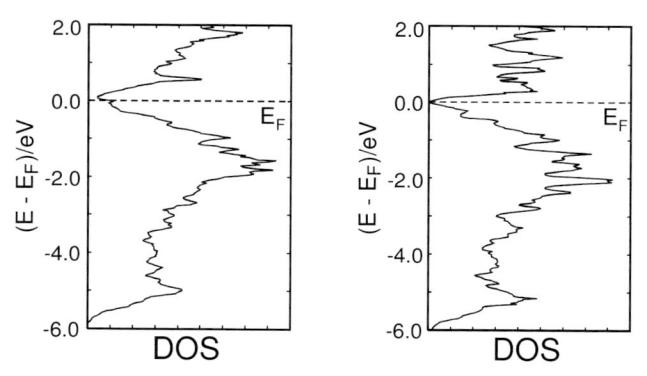

Fig. 10.19 Density of states of NbSb$_2$ (left) and HfMoSb$_4$ (right).

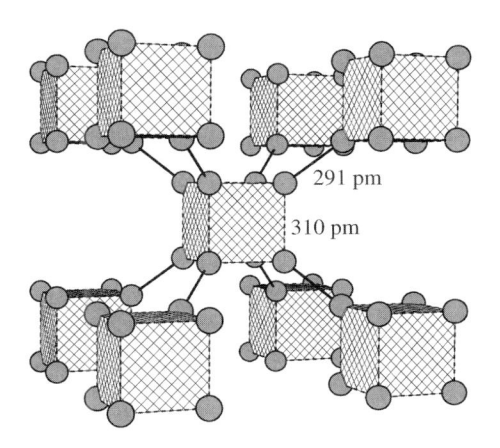

Fig. 10.20 Substructure of Sb2 in Mo_3Sb_7.

the metallic properties. This material was predicted to become semiconducting when two Sb1 atoms are replaced with two Te atoms [105], which formally reduces the charge of the Mo atoms to Mo^{3+} (($Mo^{3+})_3(Sb1^{3-})(Te^{2-})_2$ $(Sb2^{0.5-})_4$). Since there is one Mo–Mo bond per Mo atom, all electrons can be localized, a postulate supported by the experimentally observed semiconducting properties (Fig. 10.21) [106].

$V_{8-\delta}Sb_9$ [107] and its ternary variant $Zr_2V_6Sb_9$ [108] comprise Sb atom substructures that may be described as two-dimensional segments of the Sb2 motifs of Mo_3Sb_7. In $V_{8-\delta}Sb_9$, four-membered square rings of Sb atoms with Sb–Sb distances of 319 pm are interconnected via shorter Sb–Sb bonds of 304 pm to form puckered layers (Fig. 10.22). In addition, there is another symmetry-inequivalent square Sb_4 ring, with interatomic distances of 324 pm, plus Sb atoms without any Sb–Sb contacts shorter than 330 pm. In this case, we refrain from counting electrons because of the number of fractional V–V interactions in addition to metallic conductivity.

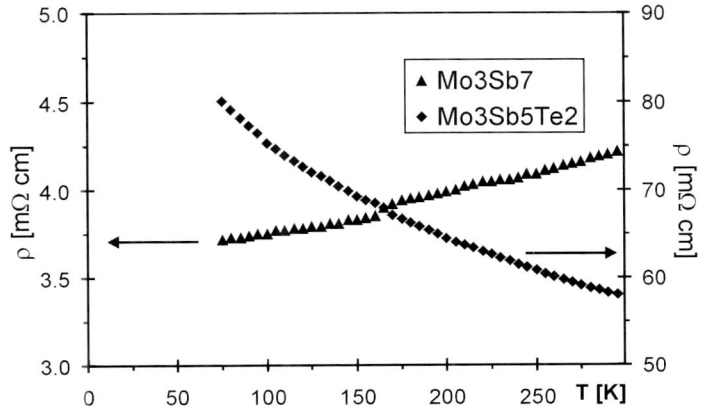

Fig. 10.21 Electrical resistivities of Mo_3Sb_7 and $Mo_3Sb_5Te_2$.

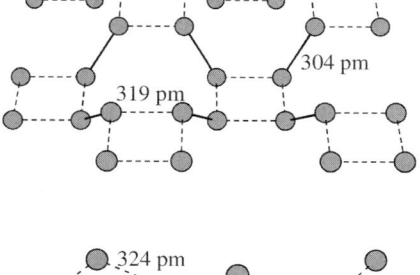

Fig. 10.22 Puckered antimony layer of $V_{8-\delta}Sb_9$.

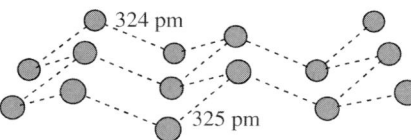

Fig. 10.23 Sb2 layers in ZrSb.

The interactions within the Sb atom substructures of the recently discovered Ti_5Sb_8 [109], $Zr_{11}Sb_{18}$ [110], and their ternary variants [111] cannot be unambiguously classified as in the above examples, for too many different bond lengths between 310 and 340 pm are found. Their main motifs are Sb_6 anchors with distances between 310 and 320 pm, which are interconnected through the longer contacts to form a three-dimensional network of apparently fractional Sb–Sb bonds.

A borderline case is ZrSb [97], the crystal structure of which comprises Sb layers (Fig. 10.23) intermediate between the two different layers in $NbSb_2$. The Sb–Sb bonds of 324 and 325 pm are longer than most Sb–Sb half bonds, but still within their range, the upper limit for which may be realized in Li_2Sb at 326 pm. Comparable bonds can even be found in compounds that are as metal-rich as Hf_6TiSb_4 (328 pm) [112]. The structure of V_3Sb_2 [113] consists in part of planar Sb layers comprising Sb_6 hexagons with Sb–Sb distances of 320 pm, which may be described as flattened Sb layers of ZrSb.

Considering the unique nature of the ZrSb structure in comparison with the NiAs structure of TiSb, it is not unexpected that a new structure type is observed on the quasi-binary section TiSb–ZrSb. $Zr_{1-\delta}Ti_\delta Sb$ forms a new type for $0.38(3) \leq \delta \leq 0.549(6)$ [114]. Fragments of this structure are reminiscent of the TiSb structure, in which (distorted) $TiSb_6$ octahedra are interconnected via common faces to form linear chains, which, in turn, are intercondensed by sharing edges (Fig. 10.24).

The transition metal atom sites are occupied by either Zr or Ti, except for the part related to the NiAs structure type, which contains exclusively Ti atoms. There is no part reminiscent of the ZrSb structure; instead we find 4^4 nets formed by Sb atoms with weak Sb–Sb bonding (>330 pm) and Zr/Ti channels filled with two parallel Sb atom chains forming a ladder. The ladder shows a short Sb–Sb bond of 284 pm along the chain direction, and a longer contact of 347 pm perpendicular to this, i.e. within the tread of the ladder. This is more or less the reverse of the ladders in $FeSb_2$ and $MoSb_2S$ (Figs. 10.14 and 10.15), in

10.3 Antimony Atom Substructures in Transition Metal Antimonides

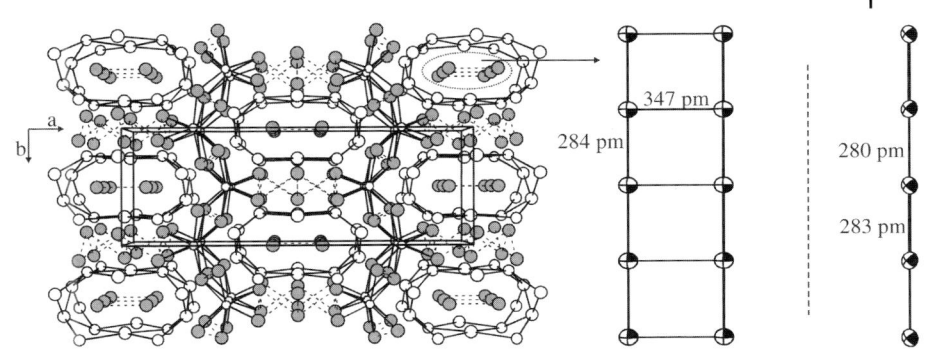

Fig. 10.24 Left: Structure of (Zr,Ti)Sb in a projection along the c axis. White circles: Zr/Ti; gray: Sb. The smaller the white circle, the higher the Ti content. Middle: Sb ladder of (Zr,Ti)Sb. Right: Sb chain of $(Zr,V)_{11}Sb_8$.

which the Sb–Sb distance along the chain direction is much longer than that along the tread (320 vs. 289 pm, and 318 vs. 284 pm, respectively). Consequently, a different bonding situation is likely. The experimental results seem to contradict each other, as the length of the interaction points towards regular Sb–Sb single bonds, while the geometry (linearity) suggests hypervalency with half bonds, as in the linear Sb chains of Li_2Sb.

This contradiction will be tackled further below. It is important to note that (Zr,Ti)Sb is not the only compound with this anomaly; closely related Sb atom substructures have been uncovered in ternary Zr–V–Sb systems. $(Zr,V)_{13}Sb_{10}$ [115] (new type) and $(Zr,V)_{11}Sb_8$ [116] ($Cr_{11}Ge_8$ type) contain similarly distorted VSb_6 octahedra chains and Zr/V channels filled with a linear Sb atom chain, the latter being reminiscent of the square-antiprism chains of W atoms filled with a linear Si atom chain in the W_5Si_3 structure type. In addition to the different compositions of the Zr/V compounds, the Sb chains are different, for they do not form ladders, and they exhibit alternating Sb–Sb distances of 280 and 288 pm in $Zr_{7.5}V_{5.5}Sb_{10}$, and 283 and 288 pm in $Zr_{7.5}V_{3.5}Sb_8$. Since the distance along the tread of the Sb ladder of (Zr,Ti)Sb (347 pm) is even longer than all of the half bonds discussed above, the major difference between the Sb chains in the Zr/V antimonides and those in (Zr,Ti)Sb is the occurrence of the alternating distances, i.e. a distortion reminiscent of the distorted Li_2Sb chain, albeit of smaller size (280/288 pm and 283/288 pm, compared to 297/356 pm). It is noted that the ellipsoid parameters of the chains are inconspicuous, as indicated in Fig. 10.24.

The linear Sb chains of (Zr,Ti)Sb are situated in columns of face-sharing Zr/Ti octahedra, and those of $(Zr,V)_{13}Sb_{10}$ and $(Zr,V)_{11}Sb_8$ in columns of face-sharing defect $(Zr/V)_7\square_1$ square antiprisms, with $\square$ = vacancy. The chain of octahedra is topologically equivalent to the linear chains of U_3TiSb_5 [117] and La_3TiSb_5 [118], but the latter antimonides exhibit much longer Sb–Sb bonds of 306 and 314 pm, respectively. Both of the latter distances were discussed as half bonds in the original articles, and are in accord with the range for half bonds used herein.

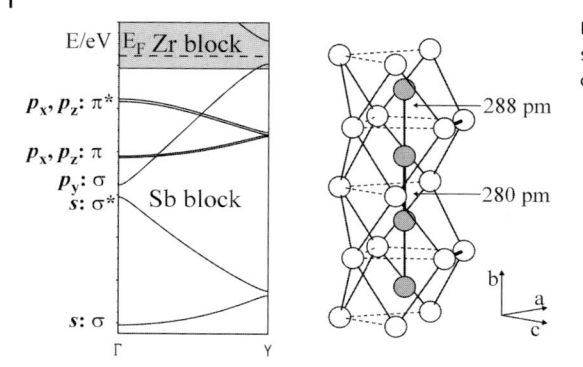

Fig. 10.25 Left: Schematic band structure of an infinite Zr_7Sb_2 chain. Right: Structure.

As the bonds of the linear chains of $(Zr,Ti)Sb$, $(Zr,V)_{13}Sb_{10}$, and $(Zr,V)_{11}Sb_8$ are about 20–40 pm longer than those of the other Sb chains of the same geometry, an investigation of their electronic structure will be instructive. Are they only half bonds as well, despite their short bond lengths, or rather full bonds, despite their linearity? Since the Zr–Sb bonds are somewhat more covalent than the Li–Sb bonds (of Li_2Sb), we included the Zr channel atoms of the $(Zr,V)_{13}Sb_{10}$ structure in the Extended Hückel calculation. This structure model has the formula Zr_7Sb_2 per repeating unit, with Sb–Sb distances of 280 and 288 pm (Fig. 10.25). Comparing this model with the Li_2Sb models (Figs. 10.12 and 10.13), we note that the chain direction is parallel to the b axis, instead of the c axis as in Li_2Sb. Therefore, the p_y orbitals are engaged in σ interactions, and the p_x and p_z orbitals in π interactions. Both band structures are quite comparable, despite the large differences in interatomic separations. All s states are filled, as well as all the π and π^* ones. As in Li_2Sb, one may regard them in a first approximation as lone pairs.

One difference between the chains of Li_2Sb and $(Zr,V)_{13}Sb_{10}$ is that the two π bands are not degenerate in the case of the Zr_7Sb_2 model, for the a and c directions are inequivalent with respect to their Zr coordination. Second, the energy differences of the respective bonding and antibonding bands at the zone border are much smaller than in Li_2Sb, because the difference in interatomic distances of the Zr_7Sb_2 model is much smaller than in the distorted Sb chain of Li_2Sb (8 vs. 59 pm). The small distortion of the Sb chains in $(Zr,V)_{13}Sb_{10}$ and $(Zr,V)_{11}Sb_8$ is not necessarily electronically driven, for the longer bond cuts through a plane of four Zr atoms and the shorter through a plane of only three Zr atoms. Another argument is that the closely related Sb chain of $(Zr,Ti)Sb$ is not distorted, each Sb–Sb bond being symmetry-equivalent with the Sb atoms situated on mirror planes perpendicular to the chain direction.

Like in Li_2Sb, the p band of σ-bonding character (here p_y) is filled and the p_{σ^*} band is empty. Therefore, the net interaction again stems from one σ-bonding band for two Sb atoms, i.e., we again observe delocalized two-center–one-electron bonds. The bond orders of ½ are associated with a mismatch between bond orders and bond lengths, for the latter are in the range of normal Sb–Sb single bonds (i.e., a bond order of 1). This mismatch is most likely caused by steric pressure, stemming from the parallel running chains of face-condensed VSb_6 octahedra with strong V–V bonding in a linear V chain, as well as from the Zr bonds within the $Zr_7\square_1$ square antiprism.

To ascertain whether this Zr_7Sb_2 model can accurately describe the Sb–Sb bonding, we have calculated the electronic structure of the whole unit cell. Because of its large size, we utilized the Extended Hückel method instead of LMTO or WIEN2K. Consequently, the measure for relative bond strengths is the Mulliken overlap population, not the ICOHP value. For comparison, the MOPs for the classical Sb–Sb single bonds in KSb are 0.6–0.7 electrons per bond. The values for the half Sb–Sb bonds of the linear chains under consideration are, on average, 0.36 in $(Zr,V)_{13}Sb_{10}$, 0.34 in $(Zr,V)_{11}Sb_8$, and 0.38 in $(Zr,Ti)Sb$. As these values are roughly half as big as those for KSb, they confirm the conceptual assignment of a bond order of ½, despite their short lengths.

We experimentally attempted to increase the mismatch between bond orders and lengths by further shortening the Sb–Sb distances of the linear chains. This was achieved through a decrease in the unit cell dimensions by replacing all Zr and V atoms with Ti. $Ti_{11}Sb_8$ [119] is isostructural with $(Zr,V)_{11}Sb_8$ and exhibits a significantly shorter b axis (560 pm vs. 571 pm), which runs parallel to the chain direction. Since there are two Sb atoms per repeating unit, the average interatomic distance of these chains must be 560/2 pm = 280 pm. Again, we observe a distorted chain, with alternating distances of 276 and 284 pm. The 276 pm contacts are even shorter than most single bonds. Since this is too short for Sb–Sb half bonds, part (about 13%) of the Sb atoms are replaced with Ti atoms to reduce the steric pressure as 276–284 pm is a typical Ti–Sb bond length.

While it is quite unusual to find mixed occupancies of the cationic and anionic components in a binary pnictide, we found mixed V/Sb occupancies in $Hf_{10}VSb_5$ [120]. This compound forms an ordered variant of W_5Si_3, with the linear chain consisting of 50% V and 50% Sb with equidistant interatomic distances of <280 pm. A number of isostructural antimonides of the general formula $M_{10}M'Sb_5$, with M=Ti, Zr, Hf; M'=V, Mn, Fe, Co, Ni, Cu, are known to exist [121–124]. Within the linear chain, the M':Sb ratio may vary between 3:1 and 2:3, as shown for M'=Fe in the Hf antimonide [120]. Consequently, Sb–Sb bonds shorter than 280 pm cannot be avoided in the latter case with more than 50% Sb atoms in the chain, but low-range order is possible to prevent the formation of two short Sb–Sb bonds per Sb atom. Therefore, the concept of half bonds is not needed in this last case.

10.4
Concluding Remarks

This chapter gives an overview of the partly unusual antimony atom substructures that occur in a variety of different materials. Several antimonides are introduced that seem to violate the octet rule. As in well-known molecular main group examples such as XeF_2 or SF_6, including the d orbitals is evidently not necessary to rationalize the bonding situation. The concept of two-center–one-electron (half) bonds is demonstrated to be particularly helpful for an understanding of the occurrence of linear extended or finite chains. This explains the geometry, and moreover it enables valence-electron counting that can be used to predict physical properties, in particular to identify possible semiconductivity.

While the antimonides obviously provide many examples, the neighboring elements (Ge, Sn, As, Bi, Te) may also participate in such nonclassical geometries, albeit arguably to a lesser extent. This is especially true for linear chains, as in $(K^+)_4(Ca^{2+})_{0.5}(Te^{2-})_2(Te^-)_2$ with Te–Te half bonds of 314 pm in linear Te^- chains, and for square nets, as are found in a variety of compounds, e.g., the $SmSb_2$, $ZrSiS$, and $HfCuSi_2$ structure types. Overviews of the latter two can be found in a number of articles dealing with these nets and distortions thereof caused by Peierls instabilities [73, 125, 126].

Acknowledgements

This work was partially supported by the Natural Sciences and Engineering Research Council of Canada, the Canada Foundation for Innovation, the Ontario Innovation Trust (Ontario Distinguished Researcher Award), the Province of Ontario (Premier's Research Excellence Award), and the Government of Canada (Canada Research Chair).

References

1 G. N. Lewis, *J. Am. Chem. Soc.* **1916**, *38*, 762–785.
2 J. I. Musher, *Angew. Chem. Int. Ed. Engl.* **1969**, *8*, 54–68.
3 O. J. Curnow, *J. Chem. Educ.* **1998**, *75*, 910–915.
4 R. J. Gillespie, B. Silvi, *Coord. Chem. Rev.* **2002**, *233/234*, 53–62.
5 L. Pauling, *J. Am. Chem. Soc.* **1931**, *53*, 1367–1400.
6 R. J. Gillespie, *Angew. Chem. Int. Ed. Engl.* **1969**, *6*, 819–896.
7 T. Kiang, R. N. Zare, *J. Am. Chem. Soc.* **1980**, 4025–4029.
8 W. Kutzelnigg, *Angew. Chem. Int. Ed. Engl.* **1984**, *23*, 272–295.
9 A. E. Reed, F. Weinhold, *J. Am. Chem. Soc.* **1986**, *108*, 3586–3593.
10 A. E. Reed, P. von Ragué Schleyer, *J. Am. Chem. Soc.* **1990**, *112*, 1434–1445.
11 R. E. Rundle, *J. Am. Chem. Soc.* **1963**, *85*, 112–113.
12 F. A. Cotton, G. Wilkinson, P. L. Gaus, *Basic Inorganic Chemistry*, 3rd ed., John Wiley & Sons, Inc., New York, NY, USA, **1995**.
13 A. F. Hollemann, E. Wiberg, *Lehrbuch der Anorganischen Chemie*, 91.–100. ed., Walter de Gruyter & Co., Berlin, Germany; New York, NY, USA, **1985**.
14 J. E. Huheey, E. A. Keiter, R. L. Keiter, *Inorganic Chemistry: Principles of Structure and Reactivity*, 4th ed., Harper Collins College Publishers, New York, NY, USA, **1993**.

15 C. E. Housecraft, A. G. Sharpe, *Inorganic Chemistry*, 1st ed., Pearson Education Limited, Essex, UK, **2001**.
16 B. Douglas, D. McDaniel, J. Alexander, *Concepts and Models of Inorganic Chemistry*, 3rd ed., John Wiley & Sons, Inc., New York, NY, USA, **1994**.
17 Y.-S. Cheung, C.-Y. Ng, S.-W. Chiu, W.-K. Li, *J. Mol. Struct. (Theochem)* **2003**, *623*, 1–10.
18 S. Siegel, E. Gebert, *J. Am. Chem. Soc.* **1963**, *85*, 240.
19 G. Cordier, H. Schäfer, M. Stelter, *Z. Anorg. Allg. Chem.* **1984**, *519*, 183–188.
20 J. Y. Chan, S. M. Kauzlarich, P. Klavins, R. N. Shelton, D. J. Webb, *Chem. Mater.* **1997**, *9*, 3132–3135.
21 J. Y. Chan, S. M. Kauzlarich, P. Klavins, J. Z. Liu, R. N. Shelton, D. J. Webb, *Phys. Rev.* **2000**, *61B*, 459–463.
22 H. Kim, J. Y. Chan, M. M. Olmstead, P. Klavins, D. J. Webb, S. M. Kauzlarich, *Chem. Mater.* **2002**, *14*, 206–216.
23 H. Kim, P. Klavins, S. M. Kauzlarich, *Chem. Mater.* **2002**, *14*, 2308–2316.
24 H. Kim, M. M. Olmstead, P. Klavins, D. J. Webb, S. M. Kauzlarich, *Chem. Mater.* **2002**, *14*, 3382–3390.
25 H. A. Tasman, K. H. Boswijk, *Acta Crystallogr.* **1955**, *8*, 59–60.
26 K. F. Tebbe, U. Georgy, *Acta Crystallogr.* **1986**, *42C*, 1675–1678.
27 E. H. Wiebenga, E. E. Havinga, K. H. Boswijk, *Adv. Inorg. Chem. Radiochem.* **1961**, *3*, 133–169.
28 R. S. Mulliken, *J. Chem. Phys.* **1955**, *23*, 2343–2346.
29 P. S. Bagus, B. Liu, D. H. Liskow, H. F. Schaefer III, *J. Am. Chem. Soc.* **1975**, *97*, 7216–7219.
30 R. D. Harcourt, T. M. Klapötke, *J. Fluor. Chem.* **2003**, *123*, 5–20.
31 R. C. Mooney, *Acta Crystallogr.* **1959**, *12*, 187.
32 U. Behrens, H. J. Breunig, M. Denker, K. H. Ebert, *Angew. Chem. Int. Ed. Engl.* **1994**, *33*, 987–989.
33 L. Pauling, *The Nature of the Chemical Bond*, 3rd ed., Cornell University Press, Ithaca, NY, USA, **1948**.
34 D. Shriver, P. Atkins, *Inorganic Chemistry*, 3rd ed., W. H. Freeman and Co., New York, NY, USA, **1999**.
35 K. F. Tebbe, R. Buchem, *Z. Kristallogr.* **1995**, *210*, 438–441.
36 K. F. Tebbe, R. Buchem, *Z. Kristallogr.* **1996**, *211*, 689–694.
37 K. F. Tebbe, R. Buchem, *Z. Anorg. Allg. Chem.* **1998**, *624*, 679–684.
38 I. Pantenburg, K. F. Tebbe, *Z. Naturforsch.* **2001**, *56B*, 271–280.
39 G. Meyer, D. Naumann, L. Wesemann (Eds). *Inorganic Chemistry Highlights*, Vol. 1, Wiley-VCH, Weinheim, Germany, **2002**.
40 H. Schäfer, B. Eisenmann, W. Müller, *Angew. Chem. Int. Ed. Engl.* **1973**, *12*, 694–712.
41 R. Nesper, *Prog. Solid State Chem.* **1990**, *20*, 1–45.
42 M. Kauzlarich, *Chemistry, Structure, and Bonding of Zintl Phases and Ions*, VCH, New York, NY, USA, **1996**.
43 J. E. Inglesfield, *J. Phys.* **1971**, *4C*, 1003–1012.
44 W. Hönle, H.-G. von Schnering, *Z. Kristallogr.* **1981**, *155*, 307–314.
45 A. Kjekshus, T. Rakke, *Acta Chem. Scand.* **1974**, *28*, 99–103.
46 L. Chi, J. D. Corbett, *J. Solid State Chem.* **2001**, *162*, 327–332.
47 G. Brauer, E. Zintl, *Z. Phys. Chem.* **1937**, *37B*, 323–352.
48 C. Hirschle, C. Röhr, *Z. Anorg. Allg. Chem.* **2000**, *626*, 1992–1998.
49 B. Eisenmann, *Z. Naturforsch.* **1979**, *34B*, 1162–1164.
50 K. Deller, B. Eisenmann, *Z. Naturforsch.* **1976**, *31B*, 29–34.
51 N. Korber, F. Richter, *Angew. Chem. Int. Ed. Engl.* **1997**, *36*, 1512–1514.
52 K. Deller, B. Eisenmann, *Z. Anorg. Allg. Chem.* **1976**, *425*, 104–108.
53 P. Pyykkö, *Chem. Rev.* **1997**, *97*, 597–636.
54 H. Kleinke, *Chem. Soc. Rev.* **2000**, *29*, 411–418.
55 H. Burger, R. Eujen, G. Becker, O. Mundt, M. Westerhausen, C. Witthauer, *J. Mol. Struct.* **1983**, *98*, 265–276.
56 K. W. Klinkhammer, P. Pyykkö, *Inorg. Chem.* **1995**, *34*, 4134–4138.
57 A. Rehr, F. Guerra, S. Parkin, H. Hope, S. M. Kauzlarich, *Inorg. Chem.* **1995**, *34*, 6218–6220.

58 F. Emmerling, C. Hirschle, C. Röhr, Z. Anorg. Allg. Chem. **2002**, *628*, 559–563.
59 M. Somer, M. Hartweg, K. Peters, H. G. von Schnering, Z. Kristallogr. **1991**, *195*, 103–104.
60 F. Gascoin, S. C. Sevov, Inorg. Chem. **2001**, *40*, 5177–5181.
61 K. Seifert-Lorenz, J. J. Hafner, Phys. Rev. **1999**, *59B*, 829–842.
62 N. Tokitoh, Y. Arai, T. Sasamori, R. Okazaki, S. Nagase, H. Uekusa, Y. Ohashi, J. Am. Chem. Soc. **1998**, *120*, 433–434.
63 A. H. Cowley, N. C. Norman, M. Pakulski, D. L. Bricker, D. H. Russell, J. Am. Chem. Soc. **1985**, *107*, 8211–8218.
64 G. Huttner, U. Weber, B. Sigwarth, O. Scheidsteger, Angew. Chem. Int. Ed. Engl. **1982**, *21*, 215–216.
65 S. C. Critchlow, J. D. Corbett, Inorg. Chem. **1984**, *23*, 770–774.
66 J. Barr, R. J. Gillespie, D. P. Santry, J. Am. Chem. Soc. **1968**, *90*, 6855–6856.
67 J. Beck, F. Steden, A. Reich, H. Fölsing, Z. Anorg. Allg. Chem. **2003**, *629*, 1073–1079.
68 A. E. Kuznetsov, H.-J. Zhai, A. I. Boldyrev, Inorg. Chem. **2002**, *41*, 6062–6070.
69 K. Tanaka, T. Yamabe, H. Teramae, K. Fukui, Inorg. Chem. **1979**, *18*, 3591–3595.
70 H. J. Breunig, N. Burford, R. Rösler, Angew. Chem. Int. Ed. **2000**, *39*, 4148–4150.
71 W. Müller, Z. Naturforsch. **1977**, *32B*, 357–359.
72 K. E. Almin, Acta Chem. Scand. **1948**, *2*, 400–407.
73 G. A. Papoian, R. Hoffmann, Angew. Chem. Int. Ed. **2000**, *39*, 2408–2448.
74 R. Hoffmann, J. Chem. Phys. **1963**, *39*, 1397–1412.
75 M.-H. Whangbo, R. Hoffmann, J. Am. Chem. Soc. **1978**, *100*, 6093–6098.
76 H. Kleinke, Z. Anorg. Allg. Chem. **1998**, *624*, 1272–1278.
77 R. Hoffmann, Angew. Chem. Int. Ed. Engl. **1987**, *26*, 846–878.
78 R. Hoffmann, Solids and Surfaces, VCH, Weinheim, Germany, **1988**.
79 J. K. Burdett, Chemical Bonding in Solids, Oxford University Press, New York, NY, USA, **1995**.
80 R. E. Peierls, Quantum Theory of Solids, Clarendon Press, Oxford, UK, **1955**.
81 H. Holseth, A. Kjekshus, Acta Chem. Scand. **1968**, *22*, 3284–3292.
82 A. K. Abrikosov, L. I. Petrova, Inorg. Mater. **1978**, *14*, 346–351.
83 P. Villars, Pearson's Handbook, Desk Edition, American Society for Metals, Materials Park, OH, USA, **1997**.
84 C.-S. Lee, H. Kleinke, Eur. J. Inorg. Chem. **2002**, 591–596.
85 H. Kleinke, Chem. Commun. (Cambridge) **2000**, 1941–1942.
86 O. K. Andersen, Phys. Rev. **1975**, *B12*, 3060–3083.
87 L. Hedin, B. I. Lundqvist, J. Phys. **1971**, *4C*, 2064–2083.
88 H. L. Skriver, The LMTO Method, Springer, Berlin, Germany, **1984**.
89 R. Dronskowski, P. E. Blöchl, J. Phys. Chem. **1993**, *97*, 8617–8624.
90 T. Hughbanks, R. Hoffmann, J. Am. Chem. Soc. **1983**, *105*, 3528–3537.
91 G. A. Landrum, R. Dronskowski, Angew. Chem. Int. Ed. **2000**, *39*, 1560–1585.
92 C.-S. Lee, A. Safa-Sefat, J. E. Greedan, H. Kleinke, Chem. Mater. **2003**, *15*, 780–786.
93 A. K. Rastogi, Philos. Mag. **1985**, *52B*, 909–919.
94 E. Canadell, A. LeBeuze, M. A. El Khalifa, R. Chevrel, M. H. Whangbo, J. Am. Chem. Soc. **1989**, *111*, 3778–3782.
95 E. E. Havinga, H. Damsma, J. M. Kanis, J. Less-Common Met. **1972**, *27*, 281–291.
96 J. D. Donaldson, A. Kjekshus, D. G. Nicholson, T. Rakke, J. Less-Common Met. **1975**, *41*, 255–263.
97 E. Garcia, J. D. Corbett, J. Solid State Chem. **1988**, *73*, 452–467.
98 A. Kjekshus, Acta Chem. Scand. **1972**, *26*, 1633–1639.
99 G. Papoian, R. Hoffmann, J. Am. Chem. Soc. **2001**, *123*, 6600–6608.
100 N. Soheilnia, A. Assoud, H. Kleinke, Inorg. Chem. **2003**, *42*, 7319–7325.
101 A. Rehr, S. M. Kauzlarich, Acta Crystallogr. **1994**, *50C*, 1177–1178.
102 P. Blaha, K. Schwarz, G. K. H. Madsen, D. Kvasnicka, J. Luitz, An augmented plane wave plus local orbitals program for calculating crystal properties, Vienna University of Technology, Vienna, Austria, **2001**.

103 E. Dashjav, Y. Zhu, H. Kleinke, in *Chemistry, Physics, and Materials Science of Thermoelectric Materials: Beyond Bismuth Telluride*, Traverse City, MI, USA, **2003**, pp. 89–106.
104 A. Brown, *Nature (London)* **1965**, *206*, 502–503.
105 E. Dashjav, A. Szczepenowska, H. Kleinke, *J. Mater. Chem.* **2002**, *12*, 345–349.
106 E. Dashjav, H. Kleinke, *Mater. Res. Soc. Symp.* **2002**, *730*, 131–136.
107 S. Furuseth, H. Fjellvag, *Acta Chem. Scand.* **1995**, *49*, 417–422.
108 H. Kleinke, *Eur. J. Inorg. Chem.* **1998**, 1369–1375.
109 Y. Zhu, H. Kleinke, *Z. Anorg. Allg. Chem.* **2002**, *628*, 2233.
110 I. Elder, C.-S. Lee, H. Kleinke, *Inorg. Chem.* **2002**, *41*, 538–545.
111 H. Kleinke, *Inorg. Chem.* **2001**, *40*, 95–100.
112 H. Kleinke, *Inorg. Chem.* **1999**, *38*, 2931–2935.
113 J. Steinmetz, B. Malaman, B. Roques, *Compt. Rend.* **1977**, *284C*, 499–502.
114 H. Kleinke, *J. Am. Chem. Soc.* **2000**, *122*, 853–860.
115 H. Kleinke, *Chem. Commun. (Cambridge)* **1998**, 2219–2220.
116 H. Kleinke, *J. Mater. Chem.* **1999**, *9*, 2703–2708.
117 M. Brylak, W. Jeitschko, *Z. Naturforsch.* **1994**, *49B*, 747–752.
118 G. Bolloré, M.J. Ferguson, R.W. Hushagen, A. Mar, *Chem. Mater.* **1995**, *7*, 2229–2231.
119 S. Bobev, H. Kleinke, *Chem. Mater.* **2003**, *15*, 3523–3529.
120 H. Kleinke, C. Ruckert, C. Felser, *Eur. J. Inorg. Chem.* **2000**, 315–322.
121 Y.U. Kwon, S.C. Sevov, J.D. Corbett, *Chem. Mater.* **1990**, *2*, 550–556.
122 N.O. Koblyuk, L.P. Romaka, O.I. Bodak, *J. Alloys Compd.* **2000**, *309*, 176–178.
123 A. Tkachuk, Y. Gorelenko, Y. Stadnyk, B. Padlyak, A. Jankowska-Frydel, O. Bodak, V. Sechovsky, *J. Alloys Compd.* **2001**, *319*, 74–79.
124 H. Kleinke, *Can. J. Chem.* **2001**, *79*, 1338–1343.
125 W. Tremel, R. Hoffmann, *J. Am. Chem. Soc.* **1987**, *109*, 124–140.
126 R. Patschke, M.G. Kanatzidis, *Phys. Chem. Chem. Phys.* **2002**, *4*, 3266–3281.

11
Transition Metal Organosulfur Coordination Polymers

Maochun Hong, Rong Cao, Weiping Su, and Xintiao Wu

11.1
Introduction

Transition metal coordination polymers are aggregates of metal atoms and organic ligands joined by covalent or noncovalent bonds. The increasing interest in coordination polymers of this type has resulted in a great deal of effort being directed towards the development of new functional materials [1–3], because they possess fascinating properties such as electrical conductivity [4], magnetism [5], and ion-exchange capability [6], and have potential applications in separation technology and catalysis [7]. Studies in this field have provided numerous examples of rationally designed one-dimensional (1-D), two-dimensional (2-D), and three-dimensional (3-D) polymeric structures having specific pore sizes and types through the assembly of molecular building units.

The coordination polymeric materials can be characterized by the tunability of their structures and properties [1–8]. For example, the size and shape of cavities and channels in zeolite-like porous coordination polymers can be controlled through the selection of organic ligands; the array of inorganic layers in inorganic–organic hybrid multilayered perovskites can be modified by organic components such that the physical properties of these compounds are tunable [9, 10]. Strategies for the construction of these solid materials have utilized metal-ligand and metal-cluster copolymerization reactions to link the appropriate molecular components. Much work in this field relies upon the use of N- or O-donor organic molecules to act as *exo*-bidentate ligands to linearly bridge metal ions, generating polymeric metal complexes with infinite chain structures and two- or three-dimensional networks [11]. However, relatively few efforts have been made to investigate transition metal organosulfur coordination polymers because it is difficult to control the reactions of thiolates with metal ions and to obtain single crystals of polymeric metal thiolates suitable for X-ray diffraction analysis. Recently, we have explored the design and syntheses of polymeric metal complexes with chain structures by using coinage metal ions and bridging ligands. The strategy that we have employed to build self-assembling aggregates has been to use multi-bridge ligands and coordinatively unsaturated metal ions as the key elements for soluble species. Our studies have revealed that the reac-

Inorganic Chemistry in Focus II. Edited by G. Meyer, D. Naumann, and L. Wesemann
Copyright © 2005 WILEY-VCH Verlag GmbH & Co. KGaA, Weinheim
ISBN: 3-527-30811-3

tions of N-donor-containing thiols with silver ions lead to soluble species, which may be converted to silver-thiolate or silver-thione polymers under appropriate conditions.

Coordination polymeric complexes containing N- and O-donor ligands have been described in many reports and reviews [1, 2]. This chapter is therefore mainly focused on our recent investigations of transition metal coordination polymers containing organosulfur ligands.

11.2
Chain-Like Polymers of Silver with Metal–Metal Interactions

Simple monothiolate ligands acting as μ_2- or μ_3-bridges are very common in transition metal chemistry. However, those acting as μ_4-bridges, thereby linking four metal atoms via sulfur atoms to form polymeric complexes, are limited. As far as we are aware, only five examples have been reported in the literature [12–15], of which four are polymeric complexes with 1-D chain structures, specifically [{Ag$_2$(μ-SPyH)$_2$(μ_4-SPyH)}(NO$_3$)$_2$]$_n$ (**1**) [12], [Ag$_6$(μ_3-SPy)$_4$(μ_4-SPy)$_2$]$_n$ (**2**) [13], [{Ag$_8$(μ_4-SC$_2$H$_4$NH$_3$)$_6$Cl$_6$}Cl$_2$]$_n$ (**3**) [14], and [Cu$_8$Cl$_6$(μ_3-SC$_2$H$_4$NH$_3$)$_2$(μ_4-SC$_2$H$_4$NH$_3$)$_4$Cl$_2$]$_n$ (**4**) [15 b] (SPy = 2-mercaptopyridine, SPyH = N-hydro-2-mercaptopyridine, SC$_2$H$_4$NH$_3$ = N-hydro-2-mercaptoethylamine). All four compounds possess metal–metal interactions that are rarely reported in studies of transition metal polymeric complexes.

The reaction of AgNO$_3$ and 2-mercaptopyridine gave compound **1** [12]. As shown in Fig. 11.1, the cation of **1** can be viewed as a 1-D chain comprising two individual metal atom chains linked by a μ_4-bridging ligand HPyS. Each silver atom is coordinated by four sulfur atoms from two μ_2-bridging ligands and two μ_4-bridging ligands. The Ag–S bond distance involving the μ_2-ligand (2.503 Å) is much shorter than that involving the μ_4-ligand (2.826 Å), and S–Ag–S bond an-

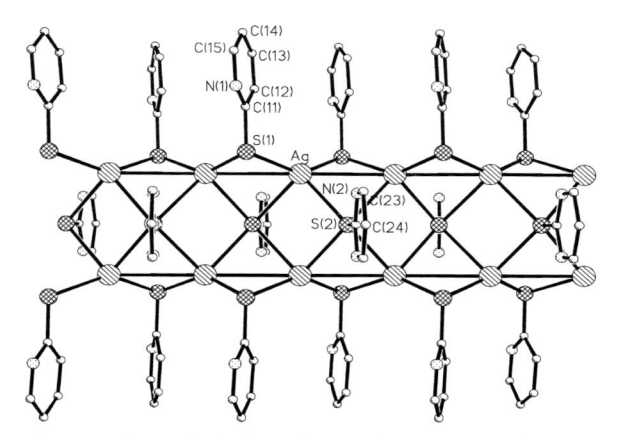

Fig. 11.1 Two individual metal atom chain structure of the cation in **1**.

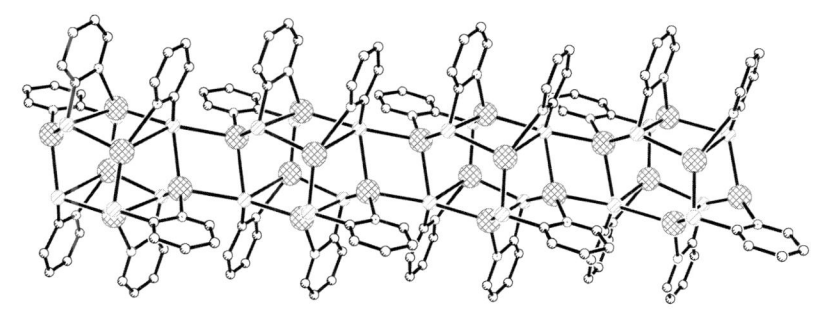

Fig. 11.2 Chain structure in **2**.

gles range from 88.23° to 142.63°. It is noteworthy that HPyS acts as a μ_4-bridge to link four metal atoms via sulfur atoms. As far as we are aware, this is the first observation of this coordination mode.

The assembly reaction of AgNO$_3$ with pyridine-2-thiol in DMSO as a solvent in the presence of Et$_3$N gave **2** at room temperature [13]. The hexanuclear building units of [Ag$_6$(SPy)$_6$] in **2** are linked by Ag–S bonds and Ag–Ag interactions to form a 1-D chain structure (Fig. 11.2). Two kinds of coordination modes of the mercaptopyridine ligands are present in **2**: the sulfur atoms in four of the six mercaptopyridine ligands exhibit μ_3-bridging linking three metal centers (Fig. 11.3a), while those in the other two exhibit μ_4-bridging connecting four metal centers (Fig. 11.3b).

By changing the monothiolate bridging ligands, a new type of polymeric silver(I) complex **3** was obtained [14]. The octanuclear cation [Ag$_8$-(SC$_2$H$_4$NH$_3$)$_6$Cl$_6$]$^{2+}$ acts as a building block unit and these are linked by Ag–S bonds and Ag–Ag interactions to form a 1-D chain structure (Fig. 11.4). All six zwitterions $^-$SC$_2$H$_4$NH$_3^+$ exhibit μ_4-S bridges linking four Ag(I) centers (Fig. 11.3c). By using a copper(I) salt, Parish obtained a polymeric copper(I) complex **4** that has a similar structure as **3** [15b].

To obtain polymers with mixed thiolate ligands, the reaction systems AgNO$_3$/K$_2$-i-mnt/pyridine-2-thiol and AgNO$_3$/NaS$_2$CNEt$_2$/pyridine-2-thiol were investigated (i-mnt=2,2-dicyanoethene-1,1-dithiolate). Two polymers containing Ag$_4$ square cores, [Ag$_4$(μ_4-i-mnt)$_2$(μ-SPyH)$_4$]$_n$ (**5**) and [Ag$_4$(μ_3-Et$_2$NCS$_2$)$_2$(SPy)$_{4/2}$]$_n$ (**6**),

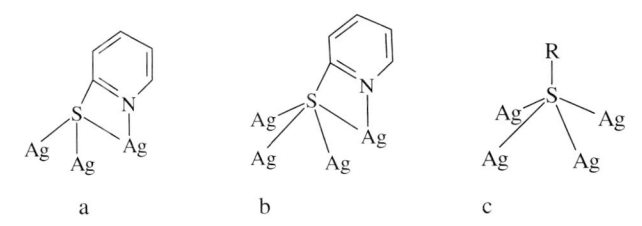

Fig. 11.3 Coordination modes of thiolato ligands.

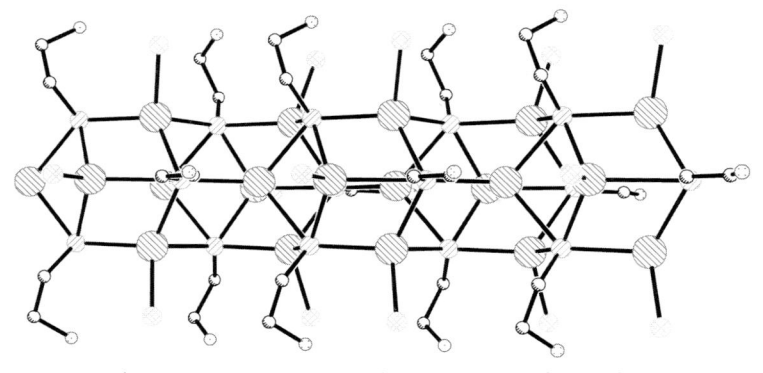

Fig. 11.4 Chain structure in **3**. Reprinted with permission from Ref. [14].

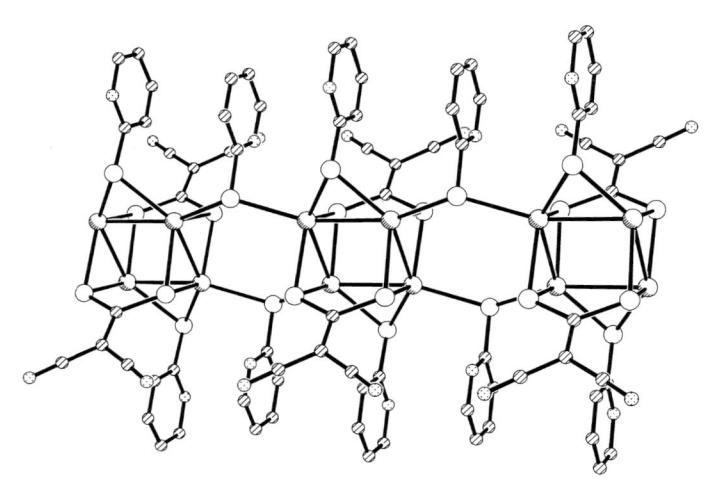

Fig. 11.5 Chain structure in **5**.

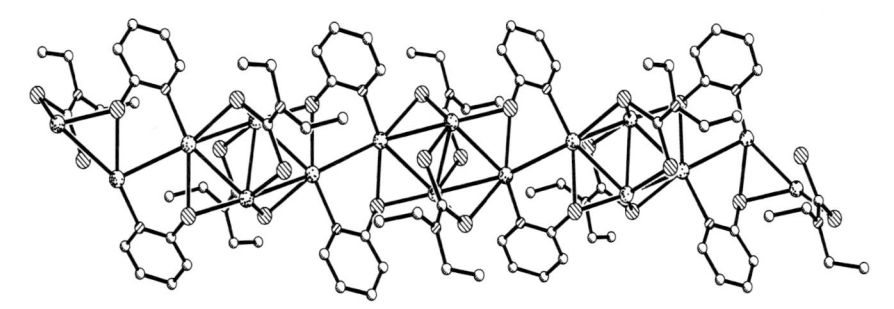

Fig. 11.6 Chain structure in **6**.

were isolated [16]. During the formation of **6**, pyridine-2-thiol undergoes deprotonation, while pyridine-2-thiol in **5** acts as a neutral ligand (thione form), which can be viewed as a consequence of the requirement of electrical neutralization of the products. The structure of **5** is a 1-D chain consisting of [Ag$_4$(i-mnt)$_2$] cluster units linked by SPyH ligands (Fig. 11.5). Each square-planar Ag$_4$ cluster in complex **6** is linked to adjacent clusters through SPy$^-$ bridging ligands to form a 1-D chain (Fig. 11.6).

11.3
Lamellar Polymers with Thiolato Bridges

The array of inorganic layers in inorganic–organic hybrid multi-layered polymers can be modified by organic components [17]. Based on nitronyl-nitroxide ligands, Luneau reported a 2-D manganese(II) complex in which the metal coordination sphere is free from electron-withdrawing groups and is occupied exclusively by paramagnetic bridging ligands [18]. Monfort reported a metamagnetic 2-D molecular material prepared from nickel(II) and an azide bridging ligand [19]. When the reaction of AgBF$_4$ with HSPy was carried out at room temperature, a polymer [{Ag$_5$(-SPy)$_4$(SPyH)}(BF$_4$)]$_n$ (**7**) with a layered structure was isolated. However, at 0 °C, the same reaction resulted in a polymer [Ag(SPy)]$_n$ (**8**) with a lamellar structure [20]. **8** is metastable and can be transformed into its isomer **2** at room temperature [15].

The cationic structure in **7** is an extended 2-D motif (Fig. 11.7). BF$_4^-$ anions are embedded in the interlayer region. In the solid-state structure of **7**, the array of silver ions is irregular and Ag–Ag distances range from 2.995 to 3.010 Å. As shown in Fig. 11.8a, **8** has a 2-D lamellar structure wherein the silver atoms are linked by PyS$^-$ ligands to form inorganic layers and the pyridyl groups of the PyS$^-$ ligands protrude into the interlayer region. The interlayer distance is

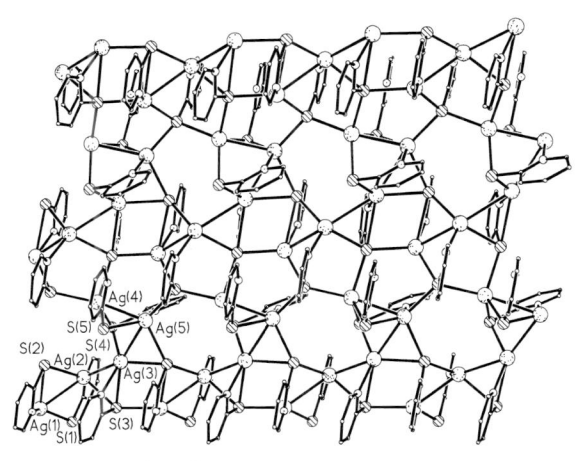

Fig. 11.7 2-D lamellar structure in **7**.

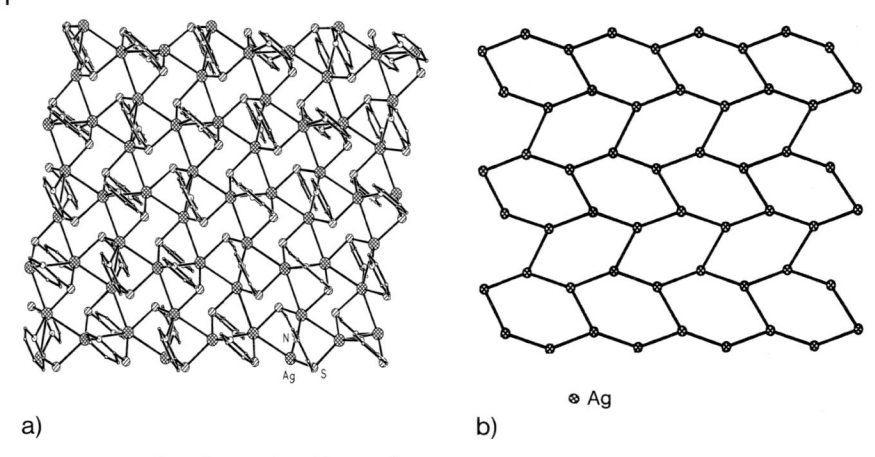

Fig. 11.8 2-D lamellar graphite-like motif in **8**.

17.1 Å. All silver atoms of each layer are nearly coplanar, with deviations of 0.13 Å, and are arranged to form a graphite-like hexagonal motif (Fig. 11.8b). Conductivity measurements show that **8** is a semiconductor (Fig. 11.9), which provides evidence for interactions between the formally closed-shell d^{10} cations in the extended solid structure [21, 22].

Considering the influence of the anion on self-assembly, AgClO$_4$, a silver salt with the poorly coordinating perchlorate anion, was used. The reaction of AgClO$_4$ with the bpsb ligand (bpsb = 1,2-bis[(2-pyrimidinyl)sulfanylmethyl]benzene) in a metal-to-ligand ratio of 2:1 gave rise to a 2-D lamellar polymer [{Ag$_2$(bpsb)$_3$}(ClO$_4$)$_2$]$_n$ (**9**) [23]. In the structure of **9**, three Ag atoms related by a

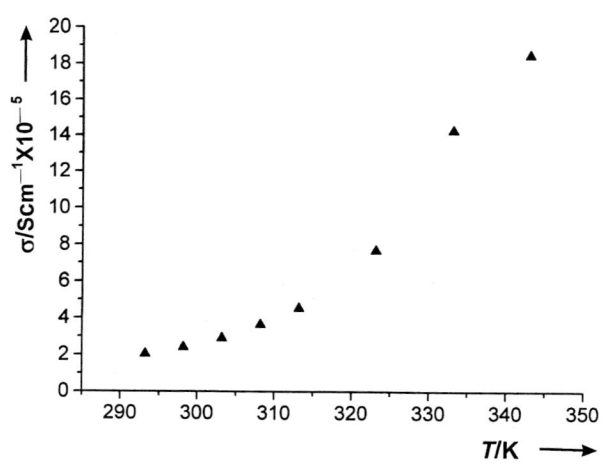

Fig. 11.9 Temperature-dependent electrical conductivity of **8**.

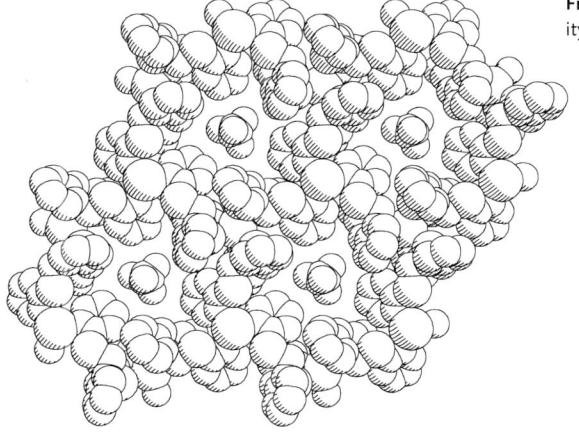

Fig. 11.10 View of a nano-cavity with [ClO$_4$]$^-$ in **9**.

three-fold axis of symmetry are located in one plane perpendicular to the *c* axis, and the other three Ag atoms lie in another plane. Therefore, the repeating unit forms a crown-like nano-cavity of [Ag$_6$(bpsb)$_6$]. The silver atoms in two adjacent layers are arranged in a staggered way, as shown in Fig. 11.10.

This network comprising isolated cavities of [Ag$_6$(bpsb)$_6$] is quite different from the usual honeycomb structure. Each of the [ClO$_4$]$^-$ ions is partly encapsulated inside the cavity of [Ag$_6$(bpsb)$_6$] and weakly bound to Ag(I) centers of another layer (Fig. 11.11).

Recently, we have prepared the multifunctional organic ligand Hspcp (spcp = 4-sulfanylmethyl-4′-phenylcarboxylate pyridine). The intriguing features of the ligand spcp$^-$ arise not only from its flexibility but also from the presence of hybrid pyridyl and benzoic carboxylate moieties that are connected by the –CH$_2$S– spacer. As shown in Fig. 11.12, the ligand possesses four possible bonding sites, i.e., the

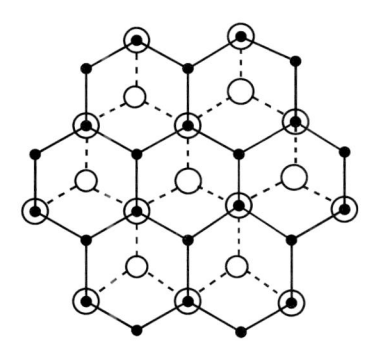

○ Ag atom in below layer
● Ag atom in upper layer

Fig. 11.11 Staggered arrangement of the silver atoms in **9**.

Fig. 11.12 The spcp⁻ multidentate ligand.

nitrogen atom, the sulfur atom, and the two oxygen atoms of the carboxylate group. Rich coordination modes may be expected because of the presence of the carboxylate group, which can adopt monodentate, chelating-bidentate, bridging-bidentate, and bridging multidentate modes. If the carboxylate end adopts the chelating-bidentate mode, spcp⁻ may be used as a mono-anionic μ_2-bridging ligand, and grid-like structures may be expected by introducing metal ions favoring hexacoordination with the advantage that no co-ligand is needed for saturating the coordination sphere of the metal ion owing to the chelating function of carboxylate. Thus, Naspcp was reacted with $Ni(OAc)_2$ in a metal-to-ligand ratio of 1:2 under solvothermal conditions. As expected, we succeeded in isolating the coordination polymer $[Ni(spcp)_2]_n$ (**10**) [24]. Crystallographic analysis revealed that complex **10** is a 2-D grid-like polymer. As shown in Fig. 11.13, each ligand acts as a μ_2-bridge linking two nickel ions through its nitrogen and carboxylate ends, and each nickel ion connects four ligands to form the infinite 2-D structure. The grid-like $Ni_4(spcp)_4$ species can be viewed as the basic building blocks of the structure, in which the apices are occupied by nickel ions and the sides are formed by spcp⁻ ligands. Four $Ni_4(spcp)_4$ grids are joined together by sharing the nickel apices to give the final 2-D layer structure consisting of puckered rhombus-like grids with dimensions of 11.72 × 11.09 Å based on the metal–metal distances. The most intriguing feature of the structure is that the basic grid is puckered,

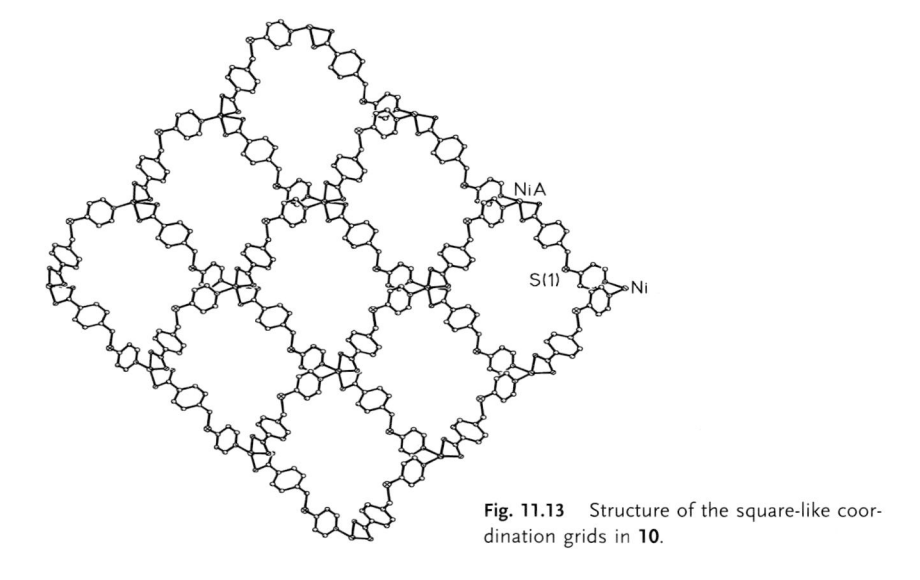

Fig. 11.13 Structure of the square-like coordination grids in **10**.

in contrast to the reported grid-like structures, in which the basic grids are usually coplanar or quasi-coplanar. The puckered shape of the basic grid in **10** is understandable, because the sp^3 configurations of C and S of the $-CH_2S-$ spacer force the spcp$^-$ ligand to become nonlinear, generating the nonlinear grid sides and thereby the puckered grids. It should be noted that the actual structure of the two-dimensional layer is wave-like, and the convex regions of one layer are immersed in the concave regions of an adjacent layer to produce a tightly packed structure without any guest molecules. A variable-temperature magnetic study indicated that the interactions between the nickel atoms are very weak, suggesting that there is no efficient orbital pathway for super-exchange between the Ni(II) ions through the spcp$^-$ bridge, which may be due to its length and flexibility.

Through hydrothermal reactions of bivalent transition metal ions with pyrimidine-2-thiol, we have succeeded in the isolation of coordination polymers with lamellar structure, e.g., $[Ni_2(C_4H_3N_2S)_4]_n$ (**11**) ($C_4H_3N_2S$ = pyrimidine-2-thiolate) [25]. The basic unit develops into a honeycomb-like structure (Fig. 11.14). Therefore, the solid-state polymer can be viewed as a 2-D lamellar structure composed of $[Ni_2(C_4H_3N_2S)_4]$ dimeric units, wherein the nickel(II) centers are linked by sulfur atoms of thiolate groups and by the pyrimidine rings of the pyrimidine-2-thiolate ligands so as to form inorganic and organic layers. The interlayer distance is 8.119 Å.

The most interesting features of **11** are its paramagnetic and semiconducting properties. The cryomagnetic behavior demonstrates the existence of ferromagnetic interactions between the two nickel(II) centers of the dimeric unit and of antiferromagnetic interactions, both between adjacent dimeric units and be-

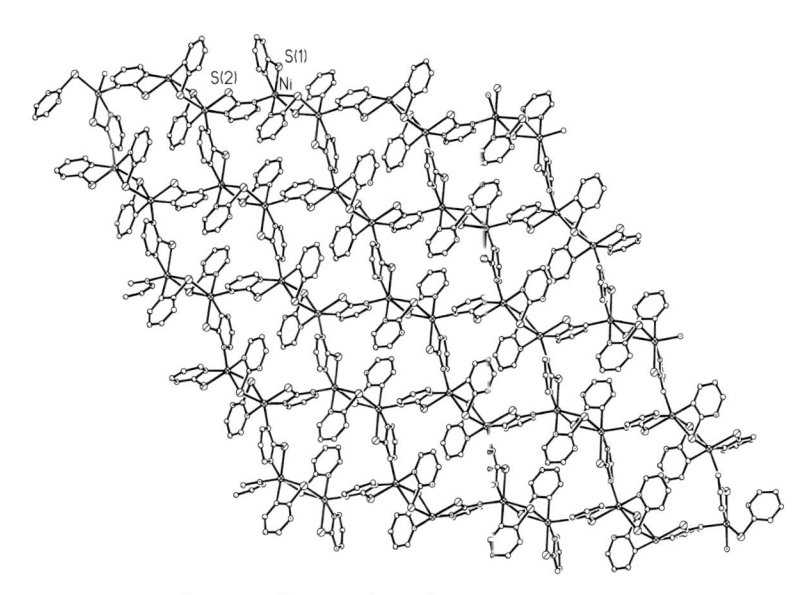

Fig. 11.14 Two-dimensional structural motif in **11**.

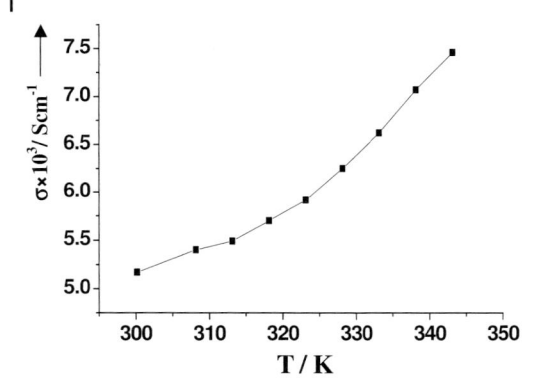

Fig. 11.15 Temperature-dependent electrical conductivity of **11**.

tween adjacent layers. The magnetic behavior of **11** is quite different from that of the nickel(II)-azide 2-D polymer, in which each nickel(II) center is in a high-spin configuration and normally shows ferromagnetic interactions [18]. The semiconducting properties of **11** (Fig. 11.15) can be attributed to the characteristic structural features of the array of nickel(II)-pyrimidine rings and are indicative of interactions between nickel and the pyrimidine rings.

11.4
Organosulfur Coordination Polymers with Tubular Structures

We have designed a multidentate ligand, 2,4,6-tri[(4-pyridyl)sulfanylmethyl]-1,3,5-triazine (tpst), which possesses *exo*-tridentate bonding sites. The three pyridyl rings and the central spacer in the ligand are connected by sulfanylmethyl groups (Fig. 11.16) [26].

Considering the symmetry and stereochemistry of the tpst ligand, coordination polymers with nanometer-sized tubes of different lengths can be expected if the assembly reaction of metal ions possessing linear coordination geometry, such as gold(I), silver(I) or copper(I), is carried out with this ligand. Thus, the reaction of $AgNO_3$ with tpst in DMF/MeCN, followed by the addition of solid $AgClO_4$, yielded a single-stranded 1-D coordination polymer, $[Ag_7(tpst)_4(ClO_4)_2$-

Fig. 11.16 The tpst multidentate ligand.

(NO$_3$)$_5$]$_n$ (**12**), containing nanometer-sized tubes [27]. Crystallographic analysis of the complex revealed that its crystal structure consists of a 1-D chain-like polymer containing a nanometer-sized tube [Ag$_7$(tpst)$_4$] as the basic unit. Each tpst ligand binds to three silver(I) centers through the N atoms of its three pyridyl groups, and each silver(I) is in turn coordinated by the pyridyl group of another tpst ligand to form an [Ag$_3$(tpst)$_2$] nanometer-sized ring. Two such rings are linked by Ag–N and Ag–S bonds to form the basic nanometer-sized tubular unit with dimensions of 13.4 × 9.6 × 8.9 Å, which accommodates two DMF molecules and two perchlorate anions. The tubular units share silver atoms to form a 1-D chain, as shown in Fig. 11.17.

Coordination polymers based on asymmetric bridging ligands have attracted much attention owing to their potential applications as second-order nonlinear optical (NLO) materials [28]. The introduction of chemically interacting sites in asymmetric bridging ligands is a challenging subject in developing new polar and chiral systems. However, helical and other recurring coordination networks used in NLO materials still remain relatively unexplored [29]. With the goal of designing NLO materials with new, non-centrosymmetric MOFs containing chemically interacting sites, we are currently investigating self-assembly reactions between metal ions and sulfur-containing asymmetrical linking ligands [30], the sulfide moieties being well known as redox-active functional groups that could enhance electronic asymmetry. By reaction of Hspcp and Zn(NO$_3$)$_2$ · 6H$_2$O in H$_2$O/MeOH in the presence of triethylamine, the novel 2-D tubular coordination polymer [Zn(spcp)(OH)]$_n$ (**13**) was obtained [30]. One striking structural feature of this polymer is the alternating assembly of two distinct homochiral helices (Fig. 11.18). One helix is formed by hydroxo-bridging of Zn(II) atoms. The other type of helix is constructed by spcp$^-$ bridges between the Zn(II) centers, displaying an opposite helical orientation to the former helix. The Zn···Zn distance bridged by spcp$^-$ is about 11.527 Å. The dimensions of the [Zn-spcp] helical tube are about 1.0 × 0.7 nm. These two distinct homochiral helices are in an ordered arrangement, with the zinc atoms functioning as

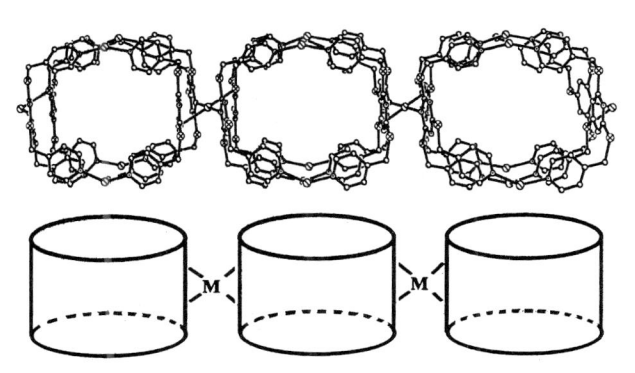

Fig. 11.17 View of the 1-D polymer chain structure in **12**.

a)

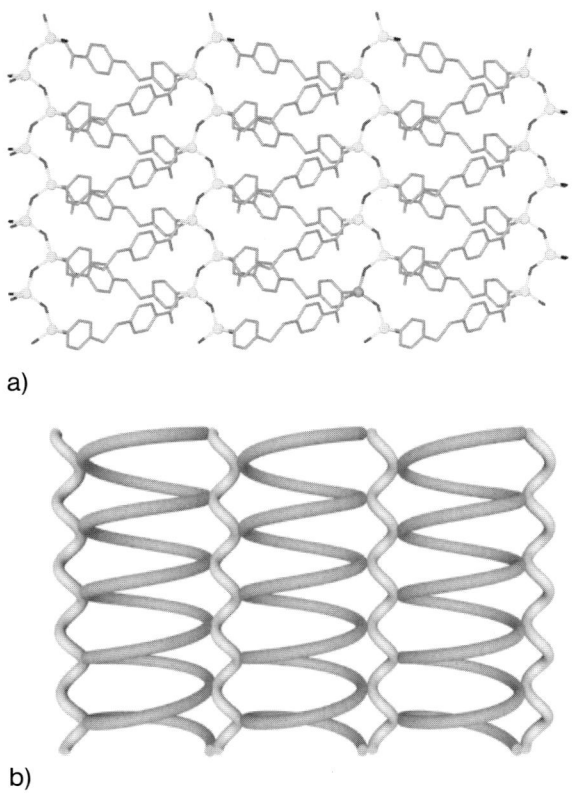

b)

Fig. 11.18 A view of the two types of homochiral helices in **13**: (a) a view of the two types of homochiral helices, and (b) a schematic view of the regular structure.

hinges. For each tubular helix, only one helical orientation is involved in the crystal structure.

Although the bulk product is racemic because the complex was derived from a spontaneous resolution of achiral components in the absence of any chiral source, each crystal has a chirality that could be exploited in functional materials. Preliminary quasi-Kurtz powder second harmonic generation experiments have shown that the complex displays a modest powder SHG efficiency approximately five times higher than that of the technically useful potassium dihydrogen phosphate (KDP). Indeed, it represents the first NLO-active bulk solid based on two-dimensional tubular coordination polymers alternatingly assembled by two types of homochiral helices with sulfide sites. Thermogravimetric analysis (TGA) shows that it has an onset temperature for decomposition above 300 °C. The stability of the complex makes it a potential candidate for practical applications.

11.5
Conclusions

In conclusion, the assembly process of silver thiolate (or thione) polymeric complexes is tunable by control of the reaction conditions, including the employment of anions with different coordination abilities, and by adjusting the rate of the transformation from soluble species formed by silver salts and N-donor thiol ligands into diverse polymers. This approach provides a means of avoiding too rapid polymerization of transition metal thiolate complexes and consequently facilitates the isolation of polymeric transition metal thiolate complexes in crystalline form. Although a great number of coordination polymers have been synthesized to date, the control or modification of their structures still remains a great challenge. Our work may provide useful information for the design of desired functional materials based on inorganic coordination polymers.

Acknowledgements

The authors gratefully acknowledge financial support from the State Key Basic Project (001CB108906), the National Natural Science Foundation of China (No. 20231020), and the Natural Science Foundation of Fujian Province.

References

1. (a) B. Moulton, M.J. Zaworotko, Chem. Rev. **2001**, *101*, 1629–1658; (b) M. Eddaoudi, D.B. Moler, H.L. Li, B.L. Chen, T.M. Reineke, M. O'Keeffe, O.M. Yaghi, Acc. Chem. Res. **2001**, *34*, 319–330; (c) B.F. Abrahams, M. Moylan, S.D. Orchard, R. Robson, Angew. Chem. Int. Ed. **2003**, *42*, 1848–1851; (d) D.B. Mitzi, Prog. Inorg. Chem. **1999**, *48*, 1–121; (e) M. Munakata, L.P. Wu, T. Kuroda-Sowa, Adv. Inorg. Chem. **1998**, *46*, 173–303.

2. (a) J.S. Seo, D. Whang, H. Lee, S.I. Jun, J. Oh, Y.J. Jeon, K. Kim, Nature **2000**, *404*, 982–986; (b) S. Noro, S. Kitagawa, M. Kondo, K. Seki, Angew. Chem. Int. Ed. **2000**, *93*, 2081–2084; (c) H. Li, M. Eddaoudi, M. O'Keeffe, O.M. Yaghi, Nature **2000**, *402*, 276–279.

3. P.J. Hagrman, D. Hagrman, J. Zubieta, Angew. Chem. Int. Ed. **1999**, *38*, 2638–2684.

4. S.I. Stupp, P.V. Braun, Science **1997**, *277*, 1242–1248.

5. D.B. Mitzi, S. Wang, C.A. Field, C.A. Chess, A.M. Guloy, Science **1995**, *267*, 1473–1476.

6. N. Matsumoto, Y. Sunatsuki, H. Miyasaka, Y. Hashimoto, D. Luneau, J.P. Tuchagues, Angew. Chem. Int. Ed. **1999**, *38*, 171–173.

7. D. Venkataraman, G.B. Gardner, S. Lee, J.S. Moore, J. Am. Chem. Soc. **1995**, *117*, 11600–11601.

8. M. Munakata, L.P. Wu, G.L. Ning, T. Kuroda-Sowa, M. Maecawa, Y. Suenaga, N. Maeno, J. Am. Chem. Soc. **1999**, *121*, 4968–4976.

9. T.L. Hennigar, D.C. MacQuarrie, P. Losier, R.D. Rogess, M.J. Zaworotko,

10 G. K. H. Shimizu, G. D. Enright, C. Ratcliffe, J. A. Ripmeester, D. D. W. Wayner, *Angew. Chem. Int. Ed.* **1998**, *37*, 1407–1409.

11 O. M. Yaghi, M. O'Keeffe, N. W. Ockwig, H. K. Chae, M. Eddaoudi, J. Kim, *Nature* **2003**, *423*, 705–714. N. L. Rosi, J. Eckert, M. Eddaoudi, D. T. Vodak, J. Kim, M. O'Keeffe, O. M. Yaghi, *Science* **2003**, *300*, 1127–1129. H. K. Chae, J. Kim, O. D. Friedrichs, M. O'Keeffe, O. M. Yaghi, *Angew. Chem. Int. Ed.* **2003**, *42*, 3907–3909. E. Q. Gao, S. Q. Bai, Z. M. Wang, C. H. Yan, *J. Am. Chem. Soc.* **2003**, *125*, 4984–4985. S. L. Zheng, J. P. Zhang, W. T. Wong, X. M. Chen, *J. Am. Chem. Soc.* **2003**, *125*, 6882–6883.

12 W. P. Su, R. Cao, M. C. Hong, W. J. Zhang, W. T. Wong, J. X. Lu, *Inorg. Chem. Commun.* **1999**, *2*, 241–243.

13 M. C. Hong, W. P. Su, R. Cao, W. J. Zhang, W. T. Wong, J. X. Lu, *Inorg. Chem.* **1999**, *38*, 600–602.

14 W. P. Su, R. Cao, R. M. C. Hong, J. T. Chen, J. X. Lu, *Chem. Commun.* **1998**, 1389–1390.

15 (a) R. V. Parish, Z. Salehi, R. G. Pritchard, *Angew. Chem. Int. Ed. Engl.* **1997**, *36*, 251–255; (b) Z. Salehi, R. V. Parish, R. G. Pritchard, *J. Chem. Soc., Dalton Trans.* **1997**, 4241–4246.

16 W. P. Su, M. C. Hong, J. B. Weng, Y. C. Liang, Y. J. Zhao, R. Cao, Y. F. Zhou, A. S. C. Chan, *Inorg. Chim. Acta* **2002**, *331*, 8–15.

17 T. L. Hennigar, D. C. MacQuarrie, P. Losier, R. D. Rogess, M. J. Zaworotko, *Angew. Chem. Int. Ed. Engl.* **1997**, *36*, 972.

18 K. Fegy, D. Luneau, T. Ohm, C. Paulsen, P. Rey, *Angew. Chem. Int. Ed.* **1998**, *37*, 1270–1273.

19 M. Monfort, I. Resino, J. Ribas, H. Stoeckli-Evans, *Angew. Chem. Int. Ed.* **2000**, *39*, 191–193.

20 W. P. Su, M. C. Hong, J. B. Weng, R. Cao, S. F. Lu, *Angew. Chem. Int. Ed.* **2000**, *39*, 2911–2914.

21 G. K. H. Shimizu, G. D. Enright, C. I. Ratcliffe, K. F. Preston, J. L. Reid, J. A. Ripmeester, *Chem. Commun.* **1999**, 1485–1487.

22 (a) C. M. Che, M. C. Tse, M. C. W. Chan, K. K. Cheung, D. L. Phillips, K. H. Leung, *J. Am. Chem. Soc.* **2000**, *122*, 2464–2468; (b) V. W. W. Yam, K. K. W. Lo, N. K. M. Fung, C. R. Wang, *Coord. Chem. Rev.* **1998**, *171*, 17–41. (c) V. W. W. Yam, *Acc. Chem. Res.* **2002**, *35*, 555–563.

23 M. C. Hong, W. P. Su, R. Cao, M. Fujita, J. X. Lu, *Chem. Eur. J.* **2000**, *6*, 427–431.

24 Y. J. Zhao, M. C. Hong, R. Cao, W. P. Su, *J. Chem. Soc., Dalton Trans.* **2002**, 1354–1357.

25 Y. J. Zhao, M. C. Hong, Y. C. Liang, R. Cao, W. J. Li, J. B. Weng, S. F. Lu, *Chem. Commun.* **2001**, 1020–1021.

26 M. C. Hong, Y. J. Zhao, W. P. Su, R. Cao, M. Fujita, Z. Y. Zhou, A. S. C. Chan, *J. Am. Chem. Soc.* **2000**, *122*, 4819–4820.

27 M. C. Hong, Y. J. Zhao, W. P. Su, R. Cao, M. Fujita, Z. Y. Zhou, A. S. C. Chan, *Angew. Chem. Int. Ed.* **2000**, *39*, 2468–2470.

28 W. Lin, O. R. Evans, R. G. Xiong, Z. Wang, *J. Am. Chem. Soc.* **1998**, *120*, 13272; H. L. Ngo, W. B. Lin, *J. Am. Chem. Soc.* **2002**, *124*, 14298–14299; Y. Cui, S. J. Lee, W. B. Lin, *J. Am. Chem. Soc.* **2003**, *125*, 6014–6015.

29 O. R. Evans, W. B. Lin, *Acc. Chem. Res.* **2002**, *35*, 511–522.

30 L. Han, M. C. Hong, R. H. Wang, J. H. Luo, Z. Z. Lin, D. Q. Yuan, *Chem. Commun.* **2003**, 2580.

12
Molybdenum(Tungsten)-Copper(Silver) Thiolates: Rationally Designed Syntheses from "Reactive" Building Blocks

Ling Chen and Xin-Tao Wu

12.1
Introduction

The exploration of the synthesis of transition metal thiolate complexes was initiated and stimulated by the chemical simulation of polynuclear centers of nitrogenase in the 1970s [1]. Since then, the area of synthetic transition metal thiolate as well as molybdenum–iron–sulfur clusters has continued to attract attention, especially after the crystal structure of the iron protein of nitrogenase was determined [2, 3]. Transition metal thiolates are an attractive subject not only in terms of understanding their chemistry and bonding, but also in view of their potential applications as catalysts [4–7] or as superconductors [8, 9].

The hetero transition metal thiolates discussed herein have been synthesized by conventional solution syntheses. These kinetically stable compounds are classical in that oxidation states can be readily assigned and they are generally characterized by an absence of bonding between the anions and cations. The structures of transition metal thiolate clusters are dictated by the bonding abilities of sulfur atoms bearing different numbers of unpaired electrons, which serve as reactive sites on the building blocks. In the formation of transition metal sulfide clusters, sulfur plays a structure-directing role.

There are six kinds of coordination patterns of sulfur, as shown in Scheme 12.1, namely term-S and μ_n-S ($n=2$–6). The reactivity of sulfur atoms decreases with decreasing number of free electron pairs according to the sequence term-S > μ_2-S > μ_3-S, μ_4-S.

Terminal M=S double bonds are observed in thiometalates, e.g. MoS_4^{2-} [10], in which each sulfur atom is coordinated to one metal atom and possesses two free electron pairs with which it can bond to other metal atoms to give rise to polynuclear structures. As the coordination number (CN) is increased up to CN=4, the sulfur atom contributes more electrons to the center metal–sulfur bonding, which diminishes its further reactivity as an electron donor. Sulfur can be encapsulated in a cage and surrounded by four, five, or six metal atoms so that no free electron pairs are left. Thus, these three kinds of sulfur atoms lose the complex reactivity while retaining the parent structure (see Scheme 12.1). In the case of μ_6-S, the empty d orbitals of sulfur are believed to

Inorganic Chemistry in Focus II. Edited by G. Meyer, D. Naumann, and L. Wesemann
Copyright © 2005 WILEY-VCH Verlag GmbH & Co. KGaA, Weinheim
ISBN: 3-527-30811-3

participate in the M–S bonding interaction, the sulfur hybrid orbitals thus being sp^3d^2 instead of the conventional sp^3.

With the exception of terminal sulfur atoms, the bridging sulfur atoms (μ_n-S, n=2–4) with their higher CN, located at the corners of many transition metal sulfide clusters, are still capable of donating their electrons to form additional M′–S bonds, thereby generating new clusters of higher nuclearity. Such reactions, classified according to nuclearity, have been summarized elsewhere [11, 12].

The WS_4^{2-} tetrahedron is a common building fragment as well as a starting material in the W/Ag/S system. There are one σ bond and three d-pπ bonds between tungsten and sulfur. The sulfur atom adopts a nearly tetrahedral sp^3 hybridization and the area between the S–M–S angle has high electron density. This accounts for the tendency to form M⟨S/S⟩M′ rhomboidal fragments through strong $S_{(donor)}$–$M'_{(acceptor)}$ bonds [13]. One-dimensional chains are common in W/Ag/S chemistry. Sharing opposite S–S edges of $[WS_4]^{2-}$ tetrahedra allows only unidirectional growth, whereas μ_3-S bridging leads to several types of cyclic arrangements or zigzag chains. The latter feature in W/Ag/S chains is based on W–Ag–W bending.

Two classes of structural motifs of hetero transition metal thiolates are discussed in this chapter, namely one-dimensional M/Ag/S polymers and M/Cu/S cage clusters (M = Mo, W).

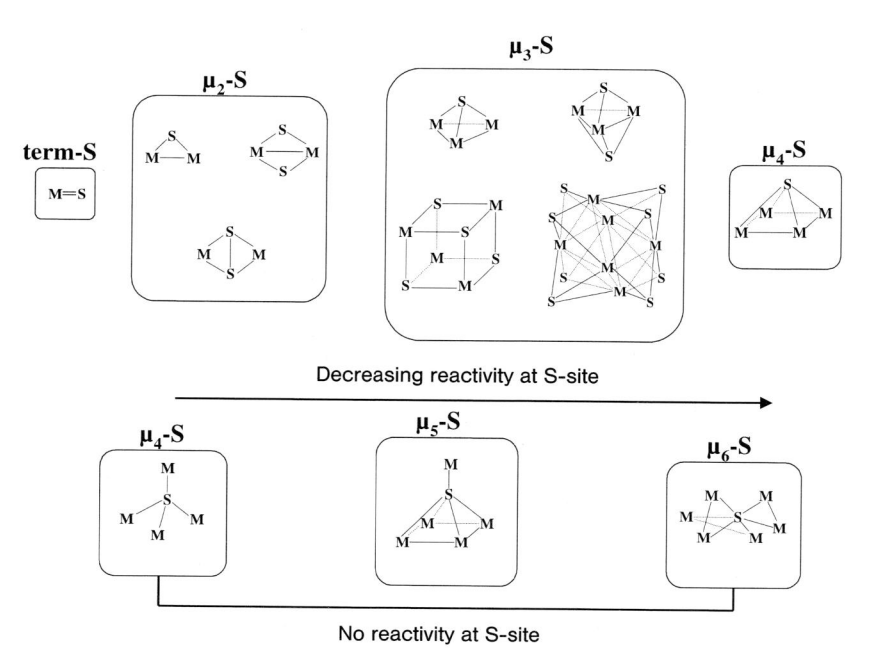

Scheme 12.1 Types of coordination of sulfur atoms and their reactivity.

12.2
One-Dimensional Thiometalate Polymers

Unconventional inorganic semiconducting materials have recently attracted much attention. One class of promising materials are one-dimensional thiometalate inorganic polymers [14]. By self-assembly reactions of $(NH_4)_2MS_4$ (M = W, Mo) and $AgNO_3$ in the presence of different cations, a series of novel W(Mo)/Ag/S chains with various 1D structures has been synthesized. The variety of motifs in these polymeric systems shows that the valence state and the size of the cation play a dominant role in determining the configuration of the anionic W(Mo)/Ag/S chains. A summary of the synthetic routes is given in Scheme 12.2 and the structural motifs of the anionic chains are listed in Tab. 12.1.

12.2.1
Linear Chains

The repeat unit of the representative compounds of this type of structure is $[(WAgS_4)^{1-}]$ (as labeled in Fig. 12.1). Compounds **1** to **4** in Tab. 12.1 belong to this class. The polymeric chain is formed through the connection of adjacent, nearly perpendicular $[-WS_2Ag-]$ rhomboidal fragments. Sharing opposite S–S edges of $[WS_4]^{2-}$ tetrahedra obviously allows only unidirectional growth. The

Tab. 12.1 Structure types of W(Mo)/Ag/S polymer chains

Comp.	Formula	Structure type	Ref.
1	$[WAgS_4NH_4]_n$	linear chain	15
2	$[(W_2Ag_2S_8)(tmenH_2)(tmen) \cdot (H_2O)]_n$	linear chain	16
3	$[WAgS_4 \cdot HNEt_3 \cdot DMF]_n$	linear chain	17
4	$[WAgS_4NH_4 \cdot NH_3C(CH_2OH)_3 \cdot 2DMF]_n$	linear chain	18
5	$[MoAgS_4(PhCH_3(C_3H_5)_3N)]_n$	wave-like chain	19
6	$[Mo_2Ag_2S_8 \cdot Ca(DMF)_6]_n$	zigzag chain	19
7	$[W_4Ag_4S_{16} \cdot 2Ca(DMSO)_6]_n$	zigzag chain	20
8	$\{[W_2Ag_2S_8] \cdot [Zn(4,4'-bipy)_2(DMF)_2(DMSO)_2]\}_n$	zigzag chain	21
9	$[W_3Ag_3S_{12} \cdot Nd(DMSO)_8]_n$	helical chain	22
10	$\{[W_3Ag_3S_{12}] \cdot [La(DMAC)_5(H_2O)_3 \cdot (DMAC)_4]\}_n$	helical chain	14
11	$\{[W_6Ag_6S_{24}] \cdot [La(DMSO)_8]_2\}_n$	helical chain	23
12	$\{[W_6Ag_6S_{24}] \cdot [Nd(DMSO)_8]_2\}_n$	helical chain	23
13	$[Mo_3Ag_3S_{12} \cdot Nd(DMSO)_8]_n$	helical chain	19
14	$[W_4Ag_5S_{16} \cdot Nd(DMF)_8]_n$	pane chain	15
15	$\{[W_4Ag_5S_{16}]_2 \cdot [La(DEF)_2(DMF)_6] \cdot [La(DEF)_4(DMF)_4]\}_n$	pane chain	24
16	$\{[W_8Ag_{10}S_{32}] \cdot [La(DEF)_8]_2\}_n$	double pane chain	14
17	$\{[W_8Ag_{10}S_{32}] \cdot [Nd(DEF)_8]_2\}_n$	double pane chain	14
18	$[WAgS_4NH_3C(CH_2OH)_3 \cdot H_2O]_n$	double chain	18
19	$\{[W_4Ag_6S_{16}] \cdot [Ca(DEAC)_6]\}_n$	hanging ladder	21

$$MS_4^{2-} + Ag^+ \quad (M=Mo,W)$$

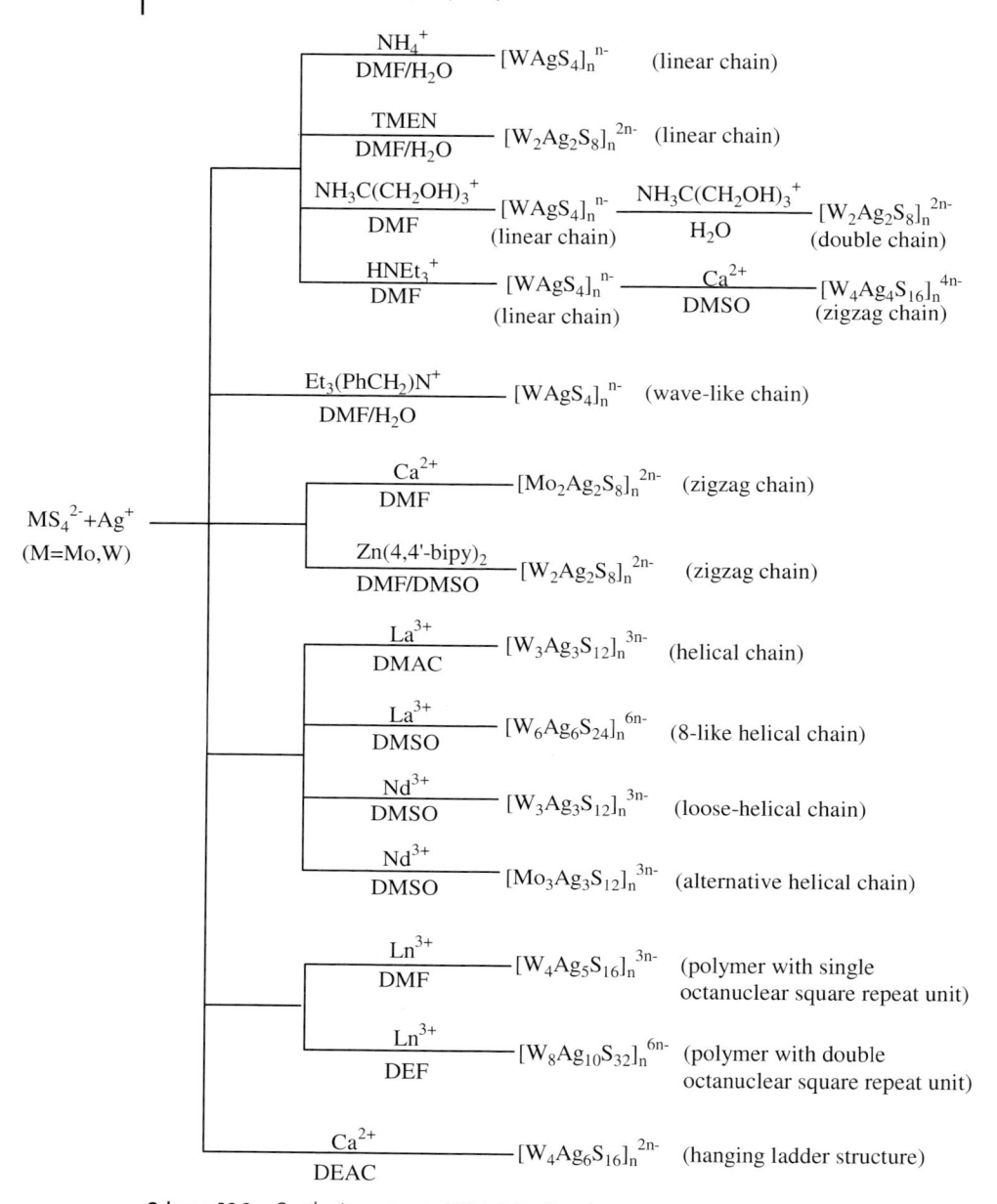

- $\dfrac{NH_4^+}{DMF/H_2O}$ — $[WAgS_4]_n^{n-}$ (linear chain)
- $\dfrac{TMEN}{DMF/H_2O}$ — $[W_2Ag_2S_8]_n^{2n-}$ (linear chain)
- $\dfrac{NH_3C(CH_2OH)_3^+}{DMF}$ — $[WAgS_4]_n^{n-}$ (linear chain) — $\dfrac{NH_3C(CH_2OH)_3^+}{H_2O}$ — $[W_2Ag_2S_8]_n^{2n-}$ (double chain)
- $\dfrac{HNEt_3^+}{DMF}$ — $[WAgS_4]_n^{n-}$ (linear chain) — $\dfrac{Ca^{2+}}{DMSO}$ — $[W_4Ag_4S_{16}]_n^{4n-}$ (zigzag chain)
- $\dfrac{Et_3(PhCH_2)N^+}{DMF/H_2O}$ — $[WAgS_4]_n^{n-}$ (wave-like chain)
- $\dfrac{Ca^{2+}}{DMF}$ — $[Mo_2Ag_2S_8]_n^{2n-}$ (zigzag chain)
- $\dfrac{Zn(4,4'\text{-bipy})_2}{DMF/DMSO}$ — $[W_2Ag_2S_8]_n^{2n-}$ (zigzag chain)
- $\dfrac{La^{3+}}{DMAC}$ — $[W_3Ag_3S_{12}]_n^{3n-}$ (helical chain)
- $\dfrac{La^{3+}}{DMSO}$ — $[W_6Ag_6S_{24}]_n^{6n-}$ (8-like helical chain)
- $\dfrac{Nd^{3+}}{DMSO}$ — $[W_3Ag_3S_{12}]_n^{3n-}$ (loose-helical chain)
- $\dfrac{Nd^{3+}}{DMSO}$ — $[Mo_3Ag_3S_{12}]_n^{3n-}$ (alternative helical chain)
- $\dfrac{Ln^{3+}}{DMF}$ — $[W_4Ag_5S_{16}]_n^{3n-}$ (polymer with single octanuclear square repeat unit)
- $\dfrac{Ln^{3+}}{DEF}$ — $[W_8Ag_{10}S_{32}]_n^{6n-}$ (polymer with double octanuclear square repeat unit)
- $\dfrac{Ca^{2+}}{DEAC}$ — $[W_4Ag_6S_{16}]_n^{2n-}$ (hanging ladder structure)

Scheme 12.2 Synthetic routes to W(Mo)/Ag/S polymers.

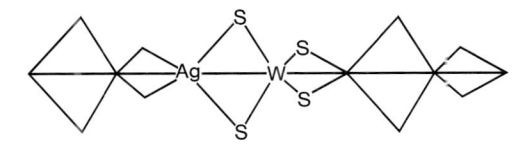

Fig. 12.1 The structure of a linear chain in the system W/Ag/S.

W–Ag–W and Ag–W–Ag angles are nearly linear, e.g. 177° for both in compound **3** [17]. In comparison, the [AgS$_4$] tetrahedra are quite distorted.

12.2.2
Wave-Like Chains

The flexibility of the M–M′–M angle allows for bending or folding of the anionic chains. Compound **5** has a wave-like anionic chain with the repeat unit [(MoAgS$_4$)$^{1-}$] (Fig. 12.2). Here, the Mo–Ag–Mo and Ag–Mo–Ag angles are 157.14(4)° and 156.47(5)°, respectively. The asymmetric geometry of the cation [(PhCH$_3$(C$_3$H$_5$)$_3$N)]$^+$ might induce the bending of the Mo–Ag–Mo angle.

12.2.3
Zigzag Chains

The maximum folding angle of the anionic chains may be around 90°, as dictated by μ_3-S. This zigzag motif is observed in compounds **6**, **7**, and **8**. The anionic skeleton of **7** is shown in Fig. 12.3, which is characterized by the repeat unit

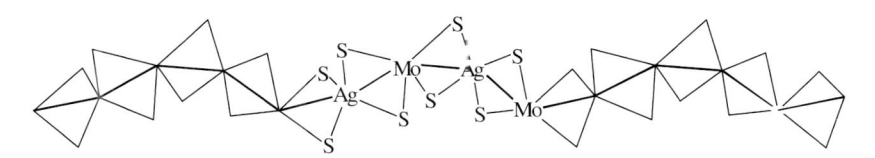

Fig. 12.2 The structure of a wave-like chain as observed in the Mo/Ag/S system.

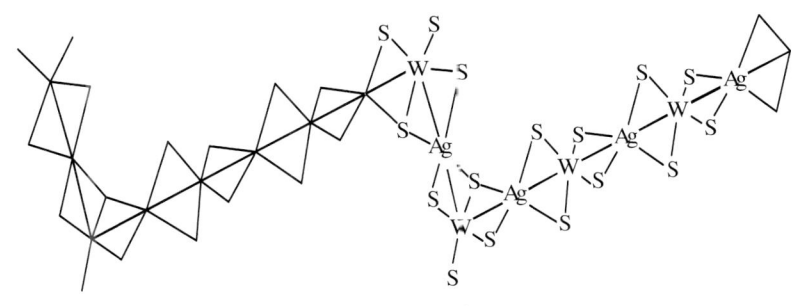

Fig. 12.3 Structure of a zigzag chain as observed in W/Ag/S systems.

$[(W_4Ag_4S_{16})^{4-}]$ built from two approximately orthogonal W–Ag fragments (indicated as WAgW and WAgWAgWAgW) with an average corner Ag–W–Ag angle of 93.7°. The presence of μ_3-S bridging at each corner redirects the growth direction of the linear chain by 90°, while all the other metal atoms (W, Ag) coordinated with four μ_2-S essentially retain their linear configurations as in the linear chain compounds **1–4**. The obvious difference between the anionic chains of these two categories (zigzag vs. linear) might be induced by the differences in the cation charges. The zigzag chain associated with the divalent cation $[Ca(DMSO)_6]^{2+}$ seems to have higher electron density in order to compensate the higher positive charge. Three compounds with anionic zigzag chains observed to date all crystallize with divalent cations [19–21].

12.2.4
Helical Chains

The flexibility of the W/Ag/S chain is again observed in compounds **9–13**, which adopt a helical motif. For example, the repeat unit of the anionic chain in **10** is $\{[W_3Ag_3S_{12}]^{3-}\}$ (see Fig. 12.4). Even though this helical chain carries a higher negative charge (judging from the counter-cation), the expected μ_3-S bridging does not appear. All metal atoms are coordinated by four μ_2-S as in compounds **1–4**. However, they do not retain the linear configuration, rather they adopt a screw pattern. The screw pitch in **10** is 1.765 nm and the helical radius is 0.888 nm. The formation of such helical chains might result from the helical linking motif of the cations as a result of hydrogen bonds. However, a cationic helical chain is not necessary. In the case of **9** and **11–13**, the trivalent cations are as discrete as usual and, as a consequence, the helical anion could also be influenced by the polarity of the coordinated solvent (see below).

12.2.5
One-Dimensional Chains with Square Units

A closed cyclic repeat unit $[(W_4Ag_4S_{16})^{4-}]$ makes up the one-dimensional chains (also called pane chains) in **14, 15, 16**, and **17**. This tetravalent anion is bridged by Ag$^+$ cations to extend in one dimension (see Fig. 12.5); four W atoms are aligned in an approximate square and the four Ag atoms nearly lie on

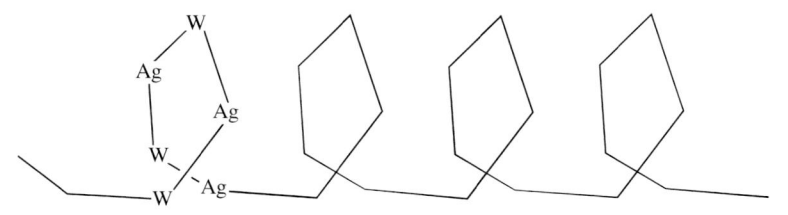

Fig. 12.4 The structure of a helical chain as observed in W/Ag/S systems.

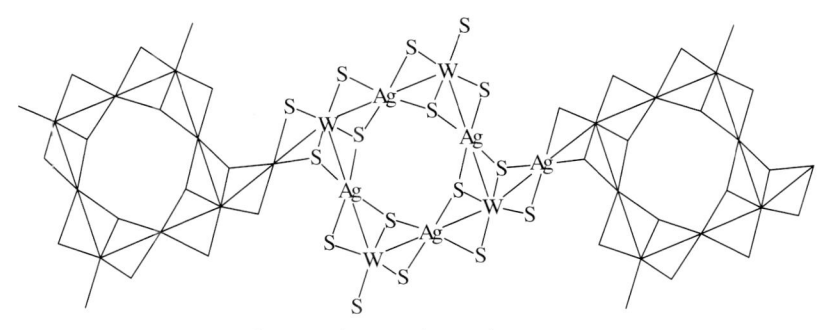

Fig. 12.5 The structure of a pane chain as observed in W/Ag/S systems.

the edges. Two neighboring squares are bridged by one Ag^+, which is located at the inversion center. A comparison of compound **9** · [Nd(DMSO)$_8$] with **14** · [Nd(DMF)$_8$] strongly suggests that the coordinated solvent plays an important role in directing the aggregation of the anion because of its size, polarity, etc.

12.2.6
Double Chains

The only example of a double chain is observed in **18** (see Fig. 12.6), which contains the monovalent repeat unit $[WAgS_4]^{1-}$. The structure can be viewed as being built from two single chains [–SWSAg–] bridged by μ_2-S atoms. Such a pattern of [WS$_2$Ag] rhombs connected by two μ_2-S is rare. This unique double chain might result from the formation of a cationic chain made up of neighboring $NH_3C(CH_2OH)_3^+$ cations linked by H_2O molecules through hydrogen bonds.

12.2.7
Hanging Ladder Chains

The anionic chain of **19** exhibits a unique hanging ladder configuration (see Fig. 12.7), in which each [AgS$_4$] shares its two opposite edges with two [WS$_4$] tetrahedra, while each [WS$_4$] shares three edges with three different [AgS$_4$], thereby generating the hanging ladder feature. Compared with **14–17**

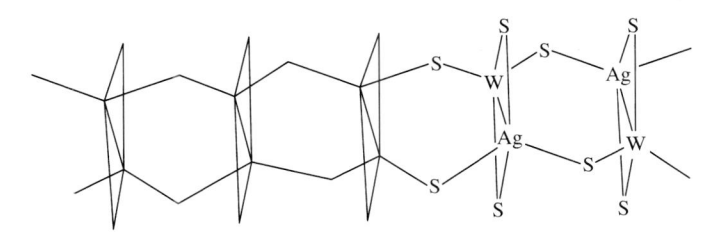

Fig. 12.6 The structure of a double chain as observed in W/Ag/S systems.

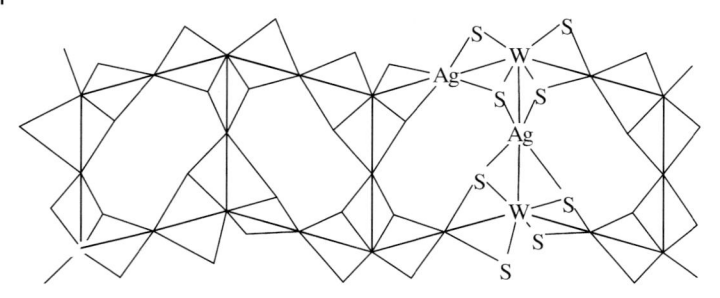

Fig. 12.7 The structure of a hanging ladder chain as observed in W/Ag/S systems.

(Fig. 12.4), no term-S atoms are present in **19**, which suggests that the term-S may be the key for stopping growth in one direction.

In summary, the diversity of counter-cations elicits numerous structure-type possibilities in W/Ag/S systems. Besides the classical role of the cations, that is, the influence of cation size and charge, shape (geometry), and polarity, aggregation patterns of cations also have distinctive effects on the assembly of the anions. On the other hand, the coordinated sulfur ligands (e.g., term-S, μ_2-S, and μ_3-S) seem to direct the assembly of [WS$_4$] and [AgS$_4$] tetrahedra. The flexibility of the sulfur bond angles allows the formation of these linear, wave, zigzag or cyclic chains.

12.2.8
Semiconductivity and Theoretical Calculations

The most intriguing feature of the above-mentioned polymeric complexes is their semiconducting properties. The electrical conductivities of **8, 10, 11, 15, 16, 17,** and **19** were measured with samples in the form of pressed pellets (two probes), and the electronic structures of the compounds were calculated using the NNEW3 program package applying extended Hückel theory. Conductivities and band gaps are listed in Tab. 12.2. The data show that all these polymers are

Tab. 12.2 Conductivities and energy gaps of some W/Ag/S polymers

Compd	Structure type	Conductivity (S·cm^{-1})	Energy gap E_g (eV)
8	zigzag chain	5.43×10^{-7}	3.87
10	helical chain	1.16×10^{-4}	3.96
11	helical chain	6.96×10^{-7}	3.81
15	pane chain	3.14×10^{-7}	3.73
16	double pane chain	2.58×10^{-4}	3.77
17	double pane chain	2.58×10^{-4}	3.77
19	hanging ladder chain	2.92×10^{-5}	3.70

semiconductors, though their conductivities are relatively low because their energy gaps E_g are somewhat large. Moreover, the properties of W/Ag/S polymers should be anisotropic in relation to their structures. The powder conductivity data may thus be taken just as a guide. Of course, it would be interesting to further study the relationship between the structures of W(Mo)/Ag/S phases and their semiconducting properties (e.g., the relationship between the dimensionalities of the W/Ag/S complexes and their conductivities).

12.3
Cage-Like Clusters

During the process of self-assembly of transition metal thiolates, the "building blocks" could become more complicated than the simple $[MS_4]^{2-}$ tetrahedron. It was found that some cluster molecules such as butterfly- and cubane-like clusters still have further reactive tendencies because of their active sulfur sites (e.g., term-S, μ_n-S, $n = 2$–4). These so-called "reactive building blocks" open the possibility of rationally designed syntheses, and have led to the discovery of numerous interesting structure types. Here, we will limit our discussion to some cage clusters.

The reaction of CuSR with the precursor butterfly cluster (so-called reactive building block) $[Cu_2S_3W(O)(PPh_3)_2]$ gave rise to the bicubane-like cluster cage in $[Cu_6S_6W_2(SCMe_3)_2(O)_2(PPh_3)_4]$ (**20**) [25] (indicated by route 1 in Scheme 12.3). This reaction is thought to involve a tetranuclear intermediate state, which undergoes dimeric assembly. Such a cage has a distant relationship to the active center of nitrogenase. A possible pathway for the reaction is shown in Scheme 12.3. The butterfly precursor could also bind to copper coordinated to the bidentate ligand (CuS$_2$COEt) and then form a cubane-like cluster as in $[MCu_3S_3(S_2COEt)(O)(PPh_3)_3]$ [25] (route 2 in Scheme 12.3).

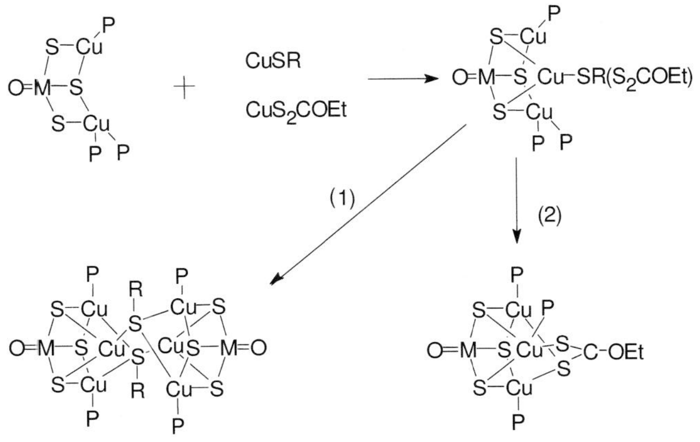

Scheme 12.3 Possible reaction pathways for the synthesis of **20**.

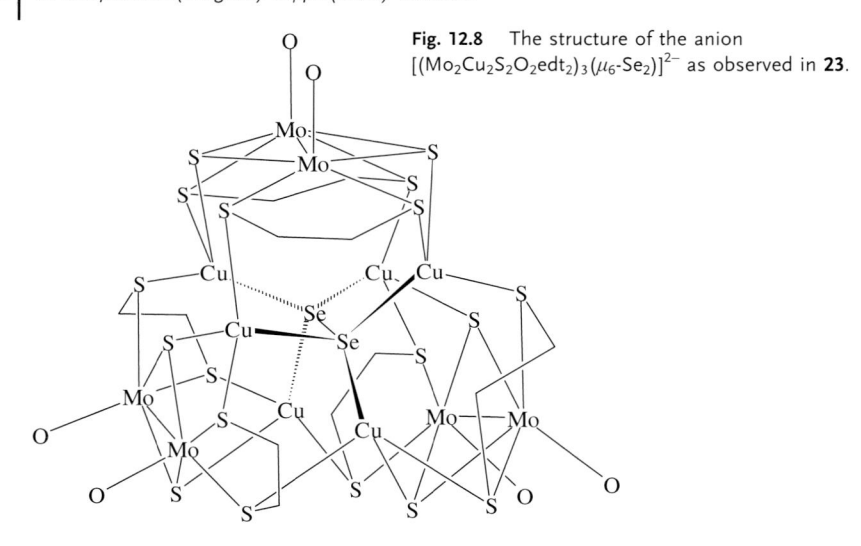

Fig. 12.8 The structure of the anion $[(Mo_2Cu_2S_2O_2edt_2)_3(\mu_6\text{-}Se_2)]^{2-}$ as observed in **23**.

Novel dodecanuclear heterobimetallic cage clusters containing μ_6-S_2 species as in $[Et_4N]_2[\{M_2Cu_2S_2O_2(edt)_2\}_3(\mu_6\text{-}S_2)]\cdot CH_2Cl_2$ (edt = $^-SCH_2CH_2S^-$; M = Mo, **21a**; M = W, **21b**) have been synthesized [26]. The anions of **21a** and **21b** are possibly derived from initially formed $[M_2O_2S_2(edt)_2]^{2-}/2Cu^+$ (M = Mo, W) fragments, which are then assembled by S_2^{2-} anions to construct cages. The Cu(I) ion can be replaced by Ag(I) to give the isostructural $[Et_4N]_2[\{M_2Ag_2S_2O_2(edt)_2\}_3(\mu_6\text{-}S_2)]$ (M = Mo, **22a**; M = W, **22b**) [27], and substitution of the central S_2^{2-} by Se_2^{2-} leads to $[Et_4N]_2[\{Mo_2Cu_2S_2O_2(edt)_2\}_3(\mu_6\text{-}Se_2)]$ (**23**) [28]. The structure of the dodecanuclear cluster cage is shown in Fig. 12.8.

Four novel tetradecanuclear clusters have been synthesized and characterized. They have the general formula $[(n\text{-}Bu)_4N]_6[M_4Cu_{10}S_{16}E_2E']\cdot H_2O$ with M = Mo, E = E' = O, **24a**; M = W, E = 1/2 O + 1/2 S, E' = O, **24b**; M = Mo, E = S, E' = 1/2 O + 1/2 S, **24c**; M = W, E = E' = S, **24d**) [29]. The cage anions consist of one incomplete cubane-like Cu_3MS_3E, one trigonal prismatic Cu_3MS_4, and two butterfly-type Cu_2MS_3E' fragments, which are bridged by two μ_3-S and one μ_4-S ligands (Fig. 12.9).

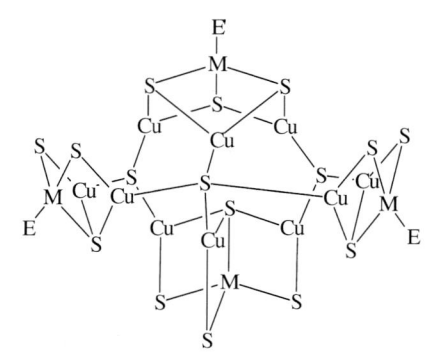

Fig. 12.9 The structure of the tetradecanuclear cluster anions as observed in $[(n\text{-}Bu)_4N]_6[M_4Cu_{10}S_{16}E_2E']\cdot H_2O$.

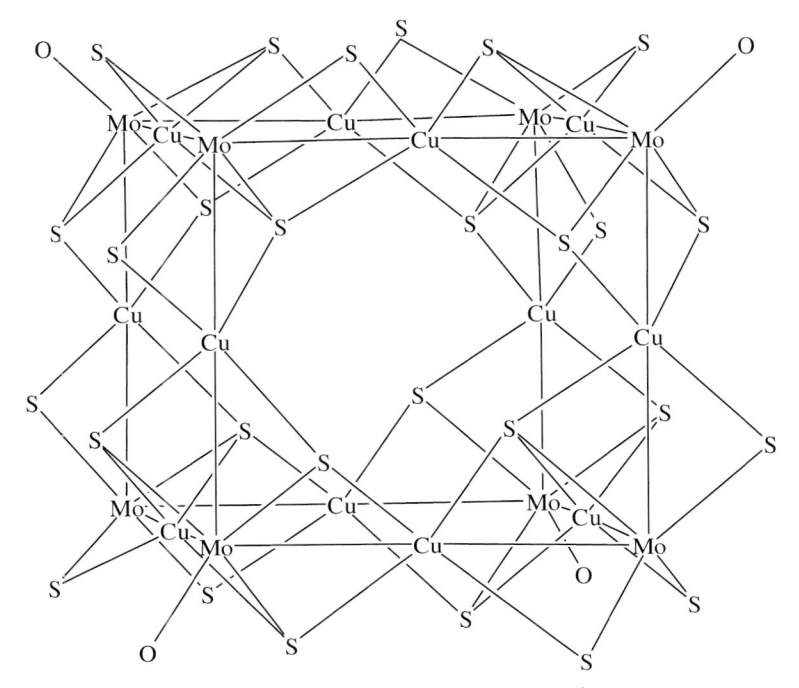

Fig. 12.10 The structure of the cluster anion $[Mo_8Cu_{12}S_{28}O_4]^{4-}$ [25a].

The bimetallic icosanuclear clusters $[Et_4N]_4[M_8Cu_{12}S_{28}O_4] \cdot DMF$ (M = Mo, **25a**; M = W, **25b**) are produced by reactions of $(NH_4)_2MS_4$ and $(NH_4)_2MO_2S_2$ (M = Mo, W) with Cu^+ ions in DMF [30]. The structures may be viewed as an approximate cubane-like array consisting of eight M atoms at the apex and twelve linearly coordinated Cu atoms lying along the edges. The anionic structure is shown in Fig. 12.10.

12.4
Conclusions

The hetero transition metal thiolates, with their highly diverse stoichiometries and structures, expand our fundamental understanding of the structural and solution coordination chemistry of transition metals. Common features of tungsten/silver thiolate systems are flexible one-dimensional anionic inorganic chains, the structures of which are to some extent controllable by the countercation. Although similar synthetic solution methods were applied to the molybdenum/copper thiolate system, the structures observed are often strikingly different compared to the W/Ag/S polymers. In particular, the Mo/Cu/S products commonly exhibit discrete high nuclearity clusters that are built up from "active building blocks" through $S_{(donor)}$–$M_{(acceptor)}$ bonding. This also opens the possibility of rationally designed syntheses. However, the solution syntheses of both systems are so complicated that the precursor species could be precipitated through a multiphase reaction–precipitation–crystallization process that is governed by kinetic rather than thermodynamic control. It would be worthwhile to explore new synthetic methods, from which numerous new structure types and interesting physical properties could be expected.

Acknowledgements

The authors gratefully acknowledge financial support from the State Key Basic Project (001CB108906), the National Natural Science Foundation of China (20333070, 20273073, 90206040), the Natural Science Foundation of Fujian Province (2002F014), and the Chinese Academy of Science.

References

1 T. E. Wolff, T. M. Berg, C. Warruck, K. O. Hodgson, R. H. Holm, *J. Am. Chem. Soc.* **1978**, *100*, 4630.
2 Y. Zhang, R. H. Holm, *J. Am. Chem. Soc.* **2003**, *125*, 3910.
3 F. Osterloh, C. Achim, R. H. Holm, *Inorg. Chem.* **2001**, *40*, 224.
4 D. Coucouvanis, A. Hadjikyriacou, M. Draganjac, M. G. Kanatzidis, O. Ileperuma, *Polyhedron* **1986**, *5*, 349.
5 M. R. DuBois, *J. Am. Chem. Soc.* **1983**, *105*, 3170.
6 M. Draganjac, T. B. Rauchfuss, *Angew. Chem. Int. Ed. Engl.* **1985**, *24*, 742.
7 M. D. Curtis, *Appl. Organomet. Chem.* **1992**, *6*, 429.
8 R. Chevrel, M. Sergent, in *Crystal Chemistry and Properties of Materials with Quasi-One-Dimensional Structures* (Ed. G. Rouxel), D. Reidel Publ. Co., Dordrecht, **1986**, 315–373.
9 T. Saito, W. Yamamoto, T. Yamagata, H. Tmoto, *J. Am. Chem. Soc.* **1988**, *110*, 1646.
10 A. Müller, C. K. Jorgensen, E. Diemann, *Z. Anorg. Allg. Chem.* **1972**, *38*, 391.
11 J.-X. Lu, *Some New Aspects of Transition-Metal Cluster Compounds*, Fujian Science Press, **1997**.
12 H.-W. Hou, X.-Q. Xin, S. Shi, *Coord. Chem. Rev.* **1996**, *153*, 25.
13 Ref. [11], Chapter 3, pp. 52.
14 L. Chen, X.-T. Wu, X.-C. Gao, W.-J. Zhang, P. Lin, *J. Chem. Soc., Dalton Trans.* **1999**, 4303–4307.
15 Q. Huang, X.-T. Wu, Q.-M. Wang, T.-L. Sheng, J.-X. Lu, *Angew. Chem. Int. Ed.* **1996**, *35*, 868–870.
16 Q. Huang, X.-T. Wu, T.-L. Sheng, Q.-M. Wang, *Acta Cryst.* **1996**, *C52*, 29–31.
17 Q. Huang, X.-T. Wu, Q.-M. Wang, T.-L. Sheng, *Acta Cryst.* **1996**, *C52*, 795–797.
18 Q. Huang, X.-T. Wu, Q.-M. Wang, T.-L. Sheng, *Inorg. Chem.* **1995**, *34*, 4931–4932.
19 H. Yu, W.-J. Zhang, X.-T. Wu, T.-L. Sheng, Q.-M. Wang, P. Lin, *Angew. Chem. Int. Ed.* **1998**, *37*, 2520–2521.
20 Q. Huang, X.-T. Wu, J.-X. Lu, *Inorg. Chem.* **1996**, *35*, 7445–7447.
21 L. Chen, H. Yu, L.-M. Wu, W.-X. Du, X.-C. Gao, P. Lin, W.-J. Zhang, C.-P. Cui, X.-T. Wu, *J. Solid State. Chem.* **2000**, *151*, 286–293.

22 Q. Huang, X.-T. Wu, J.-X. Lu, *Chem. Commun.* **1997**, 703–704.
23 L. Chen, X.-T. Wu, unpublished results.
24 L. Chen, X.-T. Wu, P. Lin, *J. Chem. Cryst.* **1999**, *29*, 629–633.
25 S. W. Du, N. Y. Zhu, P. C. Chen, X.-T. Wu, *Angew. Chem. Int. Ed. Engl.* **1992**, *31*, 1085–1087.
26 P. Lin, X.-T. Wu, Q. Huang, Q. M. Wang, T. L. Sheng, W. J. Zhang, J. Guo, J. X. Lu, *Inorg. Chem.* **1998**, *37*, 5672–5674.
27 P. Lin, X.-T. Wu, W. J. Zhang, J. Guo, T. L. Sheng, Q. M. Wang, J. X. Lu, *Chem. Commun.* **1997**, 1349–1350.
28 P. Lin, X.-T. Wu, L. Chen, L. M. Wu, W. X. Du, *Polyhedron* **2000**, *19*, 2189–2193.
29 J. Guo, X.-T. Wu, W. J. Zhang, T. L. Sheng, Q. Huang, P. Lin, Q. M. Wang, J. X. Lu, *Angew. Chem. Int. Ed. Engl.* **1997**, *36*, 2464–2466.
30 Q. Huang, X.-T. Wu, T. L. Sheng, Q. M. Wang, *Polyhedron* **1996**, *15*, 3405–3410.

13
Reactivity of Unsaturated Organic Compounds at Ruthenium(II) Centers – The Relevance of Metallacyclopentatriene Intermediates

Roland Schmid and Karl Kirchner

13.1
Introduction

Of the diverse elemental combinations, carbon–carbon bond-forming reactions have attracted great interest for their many applications in industrial and synthetic processes. This type of conversion is catalyzed mainly by transition metal compounds. The basis of the catalytic action is the remarkable ability of transition metal containing fragments to coordinate organic molecules. Beside the fascinating world of the structural chemistry thus generated, a crucial consequence is essential modification of the chemical behavior through ligation. In fact, some of the most powerful and selective methods for preparing organic molecules involve organometallic catalysis. Of specific interest to our group are C–C and C–heteroatom bond-forming reactions involving alkynes, such as the cyclotrimerization of alkynes, and in addition the cyclocotrimerization of two alkynes with unsaturated organic compounds containing C=X bonds (X=O, S, N, etc.) for obtaining carbocyclic and heterocyclic structures, as generalized in Chart 13.1.

The development of cyclization methods constitutes a continuing challenge because of the ubiquitous occurrence of cyclic structures, especially among bioactive targets. For instance, the vast majority of pharmaceuticals in use today contain either one or two five- or six-membered rings. Among the wide range of transition metal compounds that have been found to actively catalyze cyclizations, we have focused

X, Y = O, S, NR   **Chart 13.1**

Inorganic Chemistry in Focus II. Edited by G. Meyer, D. Naumann, and L. Wesemann
Copyright © 2005 WILEY-VCH Verlag GmbH & Co. KGaA, Weinheim
ISBN: 3-527-30811-3

particularly on ruthenium chemistry. Indeed, lying at the heart of the Periodic Table, ruthenium combines valuable properties of both early- and late-transition-metal relatives. Thus, a balance between the high reactivity of elements to its left and the less oxophilic and Lewis acidic nature of those to its right results in a special array of desirable properties [1]. Accordingly, ruthenium is characterized by a high capacity for multiple metal bonding on the one hand and functional group tolerance on the other. For instance, even the formation of cyclic organosulfur compounds is mediated by appropriate ruthenium complexes [2, 3], while other transition metal centers are deactivated because of strong coordination to sulfur compounds.

Elucidation of the mechanistic aspects of homogeneous catalysis has been an ambitious scientific goal from the beginning of the awareness of organometallic catalysis. Unfortunately, kinetic studies in this field of research are scarce. One of the reasons for this is the fact that often the rates of the successive steps and the stabilities of the diverse reaction intermediates are not sufficiently different to allow their separate observation and characterization. Fortunately, due to the enormous progress that has been made in computational chemistry in recent years, theoretical methods are playing an increasingly important role in identifying possible elementary reactions. These may be substitution, oxidative addition, reductive elimination, migratory insertion, hydrogen exchange, σ-hydrogen transfer, β-bond metathesis, or nucleophilic addition [4]. Ultimately, one would like to understand these fundamental transformations in order to be able to monitor and tune changes in reactivity toward an obvious synthetic purpose. Thus, it has become very useful for us to combine, whenever possible, experimental work with theoretical studies based on density functional theory (DFT) calculations.

In what follows we cover important results of our recent research in the context of the literature. Specifically, we describe the highlights of the reactions of the substitutionally labile neutral pseudo-14VE complexes Cp'Ru(COD)L (Cp' = Cp, Cp*; COD = 1,5-cyclooctadiene; L = Cl, Br) and the cationic [Cp'Ru(CH$_3$CN)$_2$L]$^+$ complexes with various triple and double bonds. The ligand L in the latter is a coligand, the nature of which has proved to be critical with regard to the outcome of such reactions, and can vary from halides to acetonitrile, tertiary phosphines, tertiary stibines, and CO (Chart 13.2). Along these lines, we will demonstrate that quite different chemistry may follow just by varying the coligand.

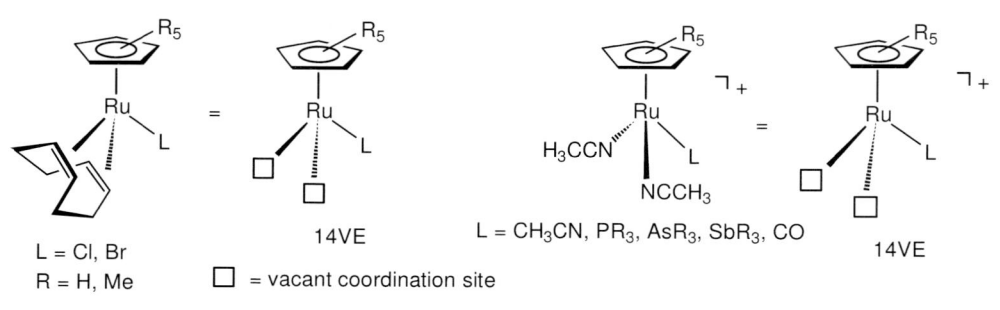

Chart 13.2

13.2
The Cyclotrimerization of Alkynes Mediated by the Cp′RuCl Fragment

The thermal cyclotrimerization of acetylene to benzene is an intriguing process. Based on experimental heats of formation of benzene and acetylene, this trimerization is extremely exothermic ($\Delta H^0 = -143$ kcal mol^{-1}) [5]. On the other hand, and contrary to the Hammond postulate, the activation energy is prohibitively high, calculated to lie in the 60–80 kcal mol^{-1} regime [6]. Such a high barrier originates from the fact that some bonding orbitals of the reactants have to become antibonding orbitals in the product and vice versa. Therefore, substantial distortions of the acetylenes have to occur so as to switch electron density from bonding regions in the reactants to bonding regions in the product [6a]. Another contribution to the barrier stems from closed-shell repulsions between the filled π orbitals (HOMO-HOMO interactions) [6b], despite some aromatic stabilization of the transition state [7].

Crucially, the barrier is diminished appreciably upon coordination of the alkynes at transition metal centers. Actually, since the pioneering work of Reppe in 1948 [8], who found that low-valent nickel complexes catalyze alkyne cyclooligomerization, the transition metal mediated [2+2+2] cycloaddition of alkynes has been developed as a very efficient method to synthesize functionalized arene systems [9]. A wide array of transition metal containing fragments has since been found to actively catalyze the reaction. Of these, derivatives of cobalt and ruthenium proved particularly efficient while tolerating the presence of varied functionalities on the alkyne. For the former, the CpCoL$_2$ (L=CO, PR$_3$, alkene) catalyst family was studied in much detail, culminating in the development of Vollhardt's benzocyclobutene synthesis [10] and Bönnemann's pyridine synthesis [11], both of which have appeared as key steps in the total synthesis of numerous natural products [9c]. In recent years, the ruthenium complexes Cp′Ru(COD)L and Ru(η^5-C$_9$H$_7$)(COD)Cl have been found to promote the cyclotrimerization of 1,6-diynes as well as the cyclocotrimerization of 1,6-diynes with terminal alkynes [12], alkenes [13], allylic ethers [14], dicyanides [15], nitriles [16], tricarbonyl compounds [17], isocyanates [2], isothiocyanates, and CS$_2$ [3].

Other successfully used metal complexes include Ni(acac)$_2$/PPh$_3$ [18] and RhCl$_2$(PPh$_3$)$_3$ [19], and metal clusters [20] as well as metal porphyrin derivatives [21] have also been reported to be effective. In addition, there are also stoichiometric transformations, in which the arene ring formed remains coordinated to the metal fragment. Pertinent examples are the alkyne coupling reactions mediated by (η^2-propene)Ti(O-i-Pr)$_2$ [22], [Cp*Fe(CH$_3$CN)$_3$]$^+$ [23], and ZrCp$_2$(C$_4$Et$_4$) [24]. Summing up, this brief survey of just the very recent literature emphasizes the ever increasing attention being paid to the development of effective cyclization strategies. In this context, one problem may be mentioned in that the reactions rarely proceed with high chemo- or regioselectivity. That is, if the method is applied to the preparation of substituted aromatic compounds from three different, unsymmetrical acetylenes, 38 homo- and cross-coupling products may conceivably be produced [25]. Thus, the assembly of such acetylenes into strictly one single aromatic compound is a formidable challenge. For this goal, knowledge of the detailed mechanism of the catalytic cycle might be very helpful. Issues to be addressed are, among others: Which intermediate controls the selectivity of the overall cyclotrimerization process? What determines the size of the rings to be formed? Actually, in the development of cyclization methods, facilitating the formation of unusual ring sizes represents a significant objective [26].

There are three recent theoretical examinations of the mechanism of acetylene cyclotrimerization. One is based on CpCo(L)$_2$ complexes [27], and the other two on the Cp'RuCl fragment [28]. The proposal derived from DFT/B3LYP calculations performed by our group is shown in Scheme 13.1, revealing a couple of uncommon intermediates. In the first step, the starting precatalyst Cp'Ru(COD)Cl undergoes a pair of ligand-substitution reactions resulting in the formation of the bis-acetylene complex **A**. Subsequent oxidative coupling of the two acetylene ligands generates the metallapentatriene complex **B**. In the next step, **B** is readily capable of coordinating a third acetylene to give the acetylene-coordinated ruthenacycle **C**. No less than three successive intermediates could be located for subsequent arene formation. The first, an unusual ruthenabicyclo[3.2.0]heptatriene **D**, rearranges to a very unsymmetrical metallaheptatetraene **E**. Ultimately, an η^2-coordinated benzene ring appears in species **F** following a reductive elimination step. Completion of the cycle is achieved by an exothermic

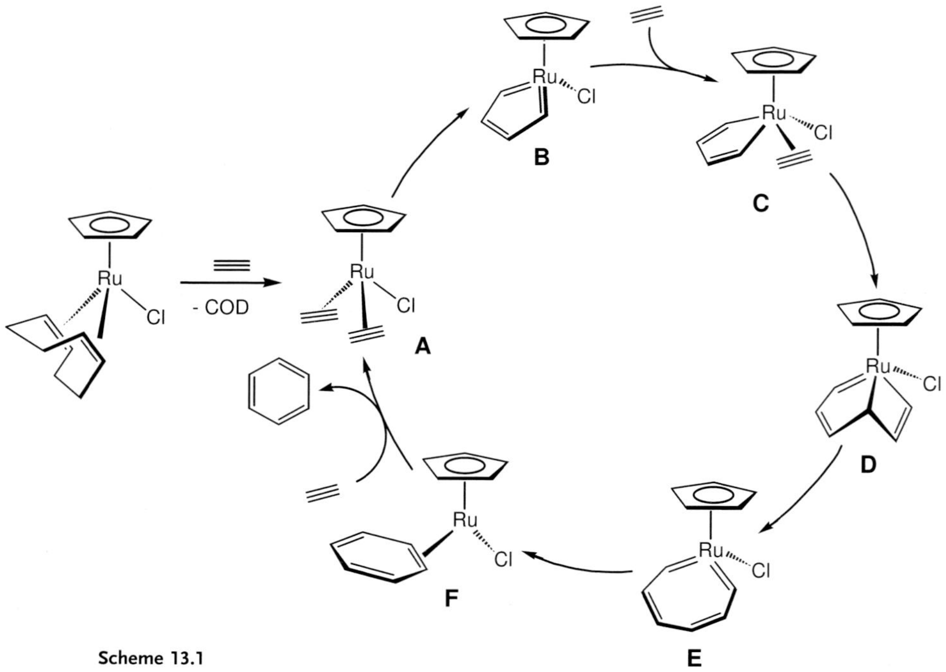

Scheme 13.1

displacement of the arene by two acetylene molecules regenerating **A**. Some remarks about the key intermediates are in order here:

B: It is now safe to assume that all cyclization processes are initiated by a metallacycle as the first key intermediate. However, the bonding modes in **B** can vary with the metal fragment. While in the similar CpCo system the respective intermediate is better formulated as a metallacyclopentadiene (type **I**), in the RuCp'Cl case **B** is clearly a metallacyclopentatriene complex (type **II**) featuring a bis-carbene functionality. In recent years, several ruthenacyclopentatriene complexes (and likewise of metals other than ruthenium) have been isolated and unequivocally characterized by NMR and X-ray crystallography [30].

In the case of the reaction of RuCp'(COD)L (L=Cl, Br) with aryl acetylenes, a stable ruthenacyclopentatriene complex is formed (Scheme 13.2) exhibiting characteristic ^{13}C NMR resonances in the range of about 270–240 ppm. The oxidative coupling takes place regioselectively in a head-to-head fashion with the substituents ending up in the a- and a'-positions. It should also be mentioned that ruthenacyclopentatrienes have been invoked as key intermediates in several catalytic reactions involving alkynes, e.g., in the double cyclopropanation of 1,6-diynes with strained cyclic alkenes [13a], in the reaction of two alkyne molecules with carboxylic acids [31], in the cycloisomerization of alkynes and propargylic alcohols [32], and in hydrative diyne cyclization [33].

The reactivity of **B** is guided by the potential for attack by another alkyne (or any other nucleophile). It has been demonstrated that these species behave as masked coordinatively unsaturated complexes and react readily with donor ligands such as tertiary phosphines, phosphites, or amines to give 'normal' metallacyclopentadiene complexes (Scheme 13.2) [30a, 34]; such reactions are not simple nucleophilic additions at the metal center but involve severe changes in the bonding mode. Addition of substrates to **B** becomes increasingly difficult in the case of a bulky coligand (see below), as well as with bulky substituents in the 1,4-positions.

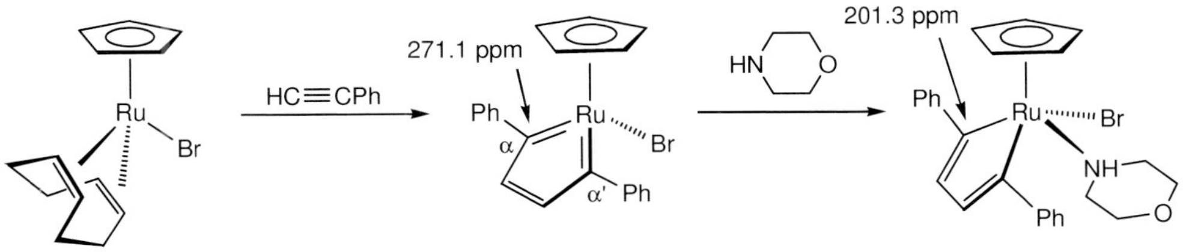

Scheme 13.2

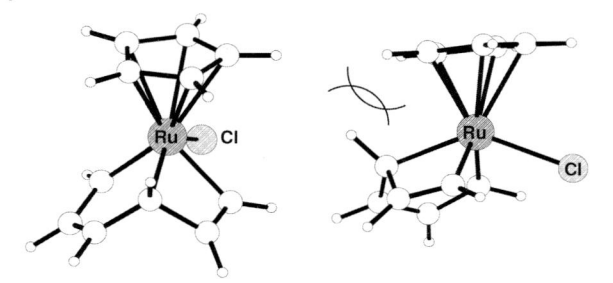

Fig. 13.1 Different perspectives of the optimized B3LYP geometry of the equilibrium structure **D** of Scheme 13.1.

D: It may be noted that very recently such a structure has indeed been isolated for the first time, viz. a stable iridabicyclo[3.2.0]hepta-1,3,6-triene, which is reversibly transformed into an iridacycloheptatriene [35]. The crucial intermediate **D** is shown in Fig. 13.1 in two different perspectives in order to emphasize the steric constraints between the hydrogen of the metallacycle (i.e., from the first two acetylene molecules) and that of the Cp ring. It can easily be seen that this intermediate controls the selectivity of the overall cyclotrimerization process. Acetylene molecules bearing two bulky substituents will therefore experience considerable repulsion at this stage. In fact, cyclotrimerization is found to work well only when parent 1,6-diynes or unsymmetrical 1,6-diynes react with terminal alkynes.

E: This intermediate is most interesting as it reveals a notable asymmetry of the Cp bonding or, in other words, ring slippage. The Cp ring can be considered as having a hapticity between η^3 and η^5, as has also been described for other systems [36]. The asymmetry of **E** removes the symmetry-forbidden character from the final step.

F: This final complex looks surprising with respect to the 18 e rule, because the benzene ring is η^2-bonded rather than η^4 as in the corresponding isoelectronic CpCo system. This becomes possible because of significant contraction of all five of the Ru–C(Cp) bonds. In this respect, the particular role played by the other co-ligand (in our case, chloride) is worth emphasizing. If it were labile, the arene would become η^6-coordinated. The resulting stable and inert sandwich complexes would deactivate the catalyst and thus quench the catalytic cycle. This is exemplified when the complexes $[RuCp(CH_3CN)_3]^+$ or $[FeCp^*(CH_3CN)_3]^+$ are used as mediators for alkyne coupling reactions [23, 37].

13.3
The Cyclocotrimerization of Alkynes with Unsaturated Organic Molecules Mediated by the Cp'RuCl Fragment

As noted above, the cyclocotrimerization of two alkynes with unsaturated organic compounds containing C=X bonds (X=O, S, N, etc.) is a synthetically useful and metal-economic process for obtaining a variety of six-membered heterocyclic systems. Here, we focus on the particularly intriguing case of coupling of diynes with ambidentate molecules RN=C=X (X=O, S). Clearly, two different kinds of ring closures are possible, leading to either pyridine-2-ones and pyridine-2-thiones or to pyrane-2-imines and thiopyrane-2-imines according to Chart 13.1. Experimentally, it was found that in the case of X=O exclusively pyridine-2-ones, that is, the nitrogen heterocycle, are formed, whereas in the case of X=S the chemoselectivity is reversed which gives thiopyrane-2-imines, that is the sulfur heterocycle. Gratifyingly, theoretical studies using DFT/B3LYP calculations performed in our laboratory reproduced well the different chemoselectivities [38]. The course of the reactions is characterized by two not so trivial intermediates, in addition to the metallacyclopentatriene **B**, viz. the bicyclic carbene **C** and the coordinatively unsaturated metallaheteronorbornene **D**, according to Scheme 13.3.

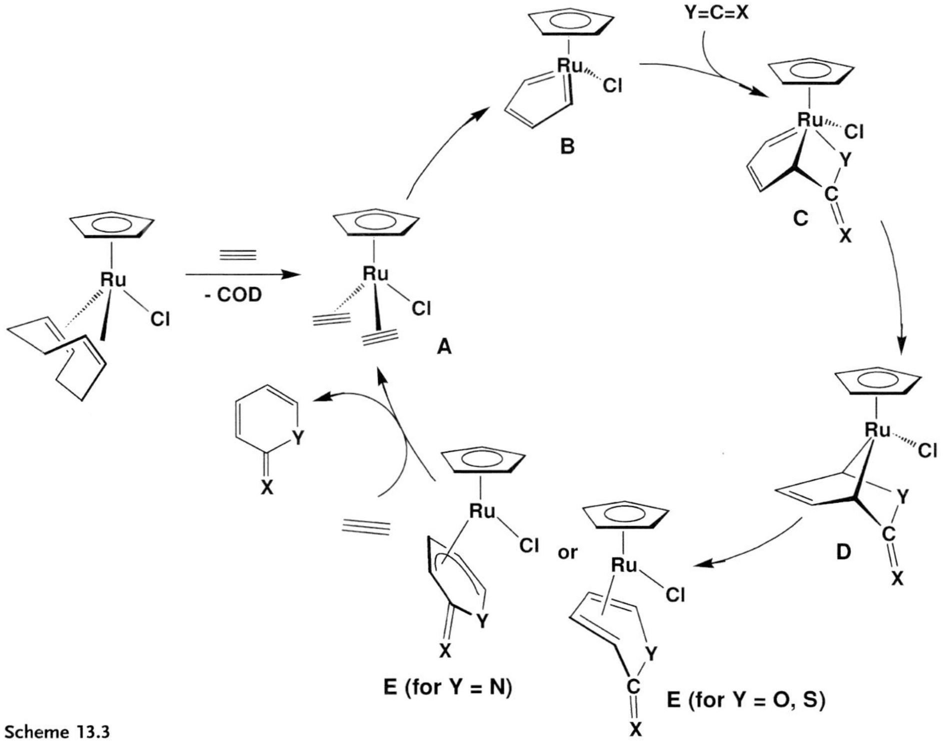

Scheme 13.3

According to the proposed mechanism, the key reaction step is addition of a double bond to the metallacyclopentatriene, giving a bicyclic carbene intermediate. This double-bond addition is initiated by η^1-attack at the ruthenium center, and it is just the attacking atom that is finally incorporated in the cycle. Thus, the chemoselectivity originates from the fact that, for HNCO, N attack is preferred over O, but for HNCS, S attack is preferred over N.

13.4
Dramatic Changes Seen Upon Replacement of Cl in Cp'RuCl with ER$_3$ (E=P, As, Sb) or CO

At first glance, the 14-electron fragment [RuCp(PR$_3$)]$^+$ would seem to be a promising candidate for mediating cyclization reactions, since it is possible to vary the properties of the phosphine ligand through varying the substituents. In this way, the regioselectivity of the coupling process might be controlled. However, although this species is catalytically active in the isomerization of allyl alcohols [39] and is a catalyst for the transfer hydrogenation of acetophenone and cyclohexanone as well as the isomerization of allyl ethers [40], the cyclotrimerization of alkynes is not initiated. Instead, a number of unusual and interesting products are obtained, depending on the structure of the alkyne and the substituent on the phosphine ligand.

The key intermediate in most reactions of the [RuCp(PR$_3$)]$^+$ fragment with alkynes is a cationic metallacyclopentatriene. This species, unlike a metallacyclopentadiene, features two highly electrophilic carbene carbon atoms. In two cases, namely the reactions of [RuCp(PCy$_3$)(CH$_3$CN)$_2$]$^+$ and [RuCp(SbPh$_3$)(CH$_3$CN)$_2$]$^+$ with 2,8-decadiyne, such a metallacyclopentatriene has been unequivocally identified by means of ^{13}C{^{1}H} NMR spectroscopy. The characteristic resonances in their ^{13}C{^{1}H} NMR spectra at δ=325.6 and 330.0 ppm, respectively, can be associated with the carbene ring carbon atoms. It is also very important to notice that the initial oxidative coupling step takes place selectively in a head-to-tail fashion rather than in the head-to-head fashion observed for the RuCp'Cl fragment. Thus, the substituents on the cationic metallacyclopentatriene intermediate end up exclusively in the α and β' positions (Scheme 13.4). The electrophilicity of the two carbene carbon atoms gives rise to a wide array of inter-ligand transformations,

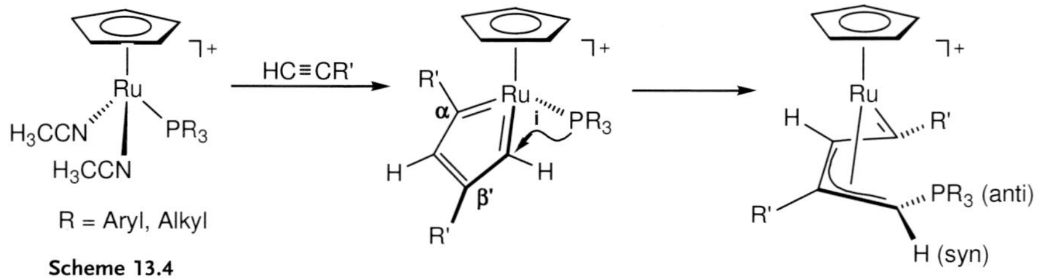

Scheme 13.4

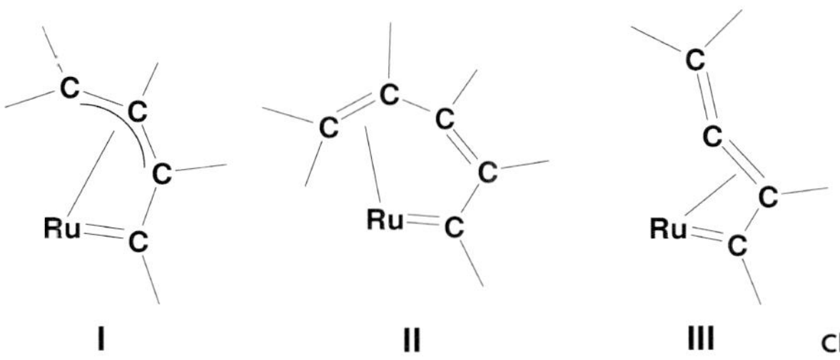

Chart 13.3

preventing the attack of a third alkyne molecule. Among other structures, ruthenium allyl carbene (**I**), butadienyl carbene (**II**), and allenyl carbene complexes (**III**) are formed (Chart 13.3) [41, 42].

(i) As shown in Schemes 13.4 and 13.5, the phosphine coligand migrates from Ru to C generating an allyl carbene. This reaction is very fast even if bulky phosphines are used and substituents are present at the a,a'-carbon atoms of the metallacycle. In the absence of kinetic data, DFT/B3LYP calculations have been performed to shed light on the mechanism of this unusual phosphine migration, with the results shown in Fig. 13.2.

The conversion of the metallacyclopentatriene **A** into the final allylcarbene complex **C** proceeds with relatively small activation barriers, the rate-limiting step being an initial distortion to produce the intermediate **B**. The reaction requires bending of the metallacycle so that the C_β carbon atoms approach the metal. Conversely, at the other side of the molecule, the Ru–C_a bond stretches and this carbon atom starts to form a new C_a–P bond. This feature is already obvious in the transition state TS_{AB} and the activation energy is 20.3 kcal mol^{-1}. TS_{AB} is much closer to **B** than to **A**. The final transformation involves complete

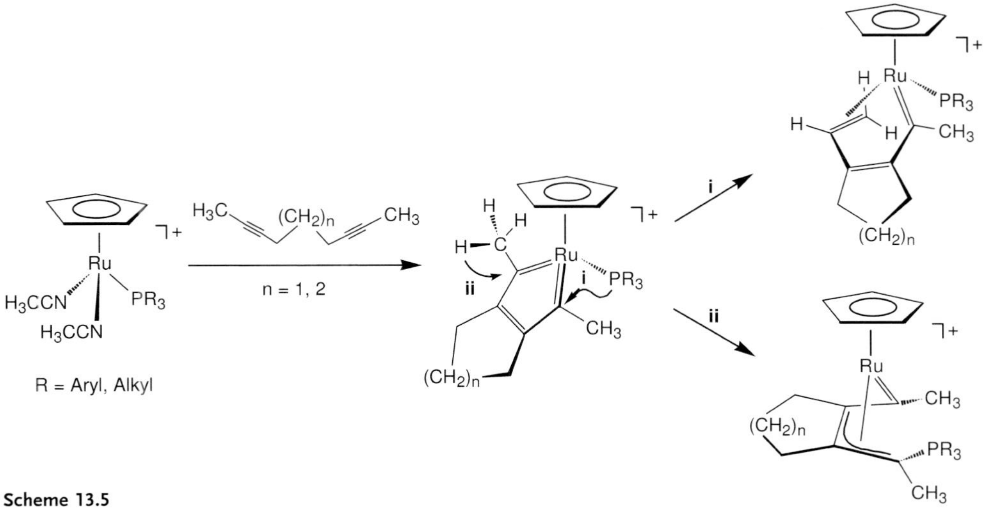

Scheme 13.5

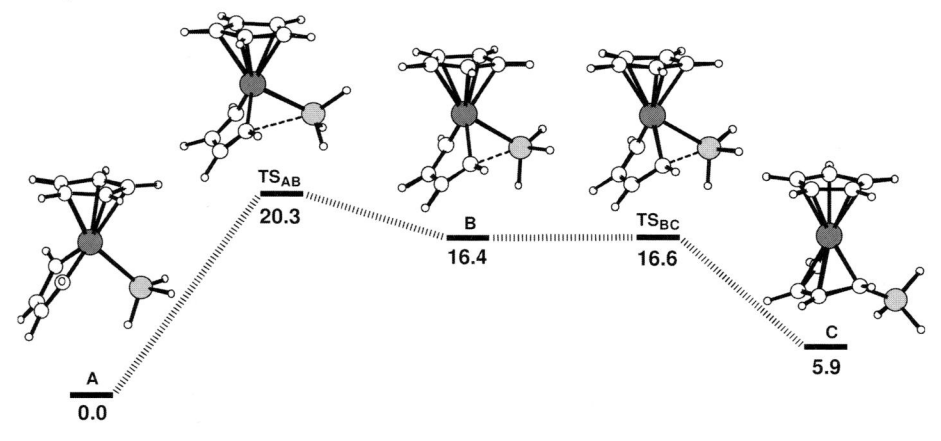

Fig. 13.2 Energy profile of the B3LYP potential energy surfaces for the conversion of the ruthenacyclopentatriene complex **A** into the allyl carbene **C** of Scheme 13.4.

Ru–P bond breaking and formation of the C_α–P bond, with simultaneous formation of the allylcarbene, and the adjustments of the carbon chain, namely formation of the Ru–C_β bonds.

(ii) If the alkyne bears an α-alkyl substituent, in some cases a competitive intramolecular 1,2-hydrogen migration results in the formation of a butadienyl carbene.

Pathway (i) is preferred over (ii) with increasing nucleophilicity of the phosphine ligand. In this respect, the mechanistic change-over from (i) to (ii) is noted when the phosphine is replaced by the other group 15 ligands AsR_3 and SbR_3 (Scheme 13.6). Thus, arsine and stibine ligands have a lesser tendency to migrate [42]. Under certain circumstances, both allyl carbene and butadienyl carbene complexes can undergo further transformations as follows.

(iii) Ruthenium allyl carbene complexes act as pseudo-16e species and are capable of activating C–H bonds of aryl and alkyl groups of the phosphine [43]. This occurs if the phosphines are sufficiently bulky, as is the case with PPh_3 or PCy_3, whereas no such reaction is observed with PMe_3. Pertinent reactions are given in Schemes 13.7 and 13.8. The initially formed allyl carbene complexes are not stable

Scheme 13.6

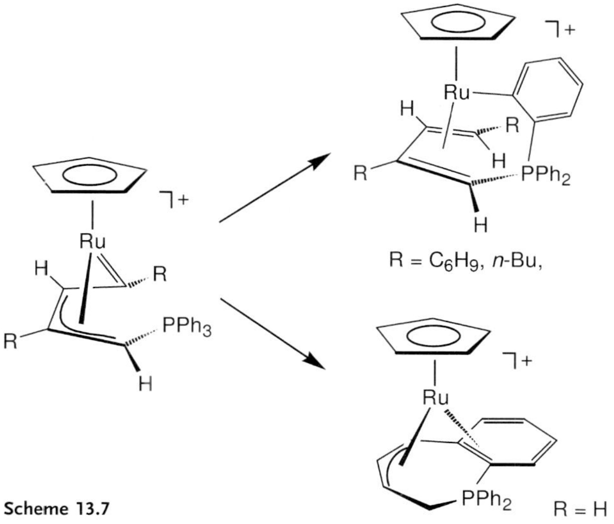

Scheme 13.7

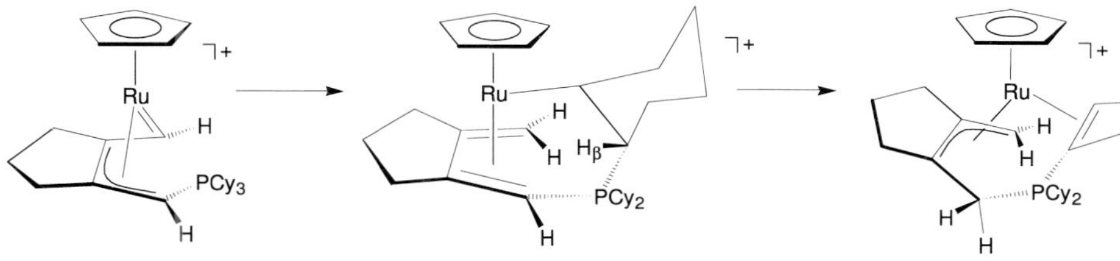

R = C$_6$H$_9$, n-Bu,

R = H

Scheme 13.8

and in some cases transform slowly at room temperature into the novel η^4-butadiene complex featuring an *ortho*-metallated arene ligand derived from PPh$_3$.

A somewhat different reaction takes place when parent acetylene is used, in that the allyl carbene intermediate eventually rearranges into [RuCp(η^3-CHCHCH-CH$_2$PPh$_2$-η^2-C$_6$H$_4$)]$^+$, as shown in Scheme 13.7. The structure and bonding features of this complex are rather unusual. First, an *ortho* C–H bond of the PPh$_3$ substituent has been activated resulting in C–C bond formation between the carbene carbon atom of the allyl carbene moiety and the *ortho* arene carbon atom. In the course of this process, the *ortho* hydrogen atom has been transferred to the terminal carbon of the allyl carbene unit with concomitant carbon sp^2 → sp^3 rehybridization. Furthermore, besides coordination to an η^3-allyl moiety, the metal center is η^2-coordinated to a double bond of a phenyl ring adjacent to the P-donor of the PPh$_3$ substituent. This bonding mode is very rare, with only a few examples having been reported recently [44].

Another example of a facile intramolecular C–H bond activation initiated by allyl carbenes is shown in Scheme 13.8. Here, one cyclohexyl substituent of the phosphine ligand is converted into an η^2-coordinated cyclohexenyl ligand. The reaction does not stop after the oxidative addition step but β-elimination takes place.

The overall transformation involves the transfer of two hydrogen atoms from a cyclohexyl ring of PCy$_3$ to the C4 chain of the allyl carbene moiety. Notably, related reactions have recently been reported, requiring in some cases the presence of extra added alkene as an external hydrogen acceptor [45].

(iv) In special cases, butadienyl carbene complexes undergo further rearrangement, such as in the reaction of [RuCp(SbR$_3$)]$^+$ with HC≡CCH$_2$Ph [42], giving an η^3-allyl-vinyl complex (Scheme 13.9).

As shown in Scheme 13.10, when HC≡CCH$_2$OH is the alkyne, an η^3-allyl-acyl complex is obtained:

(v) The formation of allenyl carbene complexes has also been observed in the reaction of [RuCp(PR$_3$)]$^+$ with a most atypical alkyne, viz. ferrocenyl and ruthenocenyl acetylene [41] (Scheme 13.11). With HC≡CSiMe$_3$, the reaction stops at the stage of a vinylidene complex.

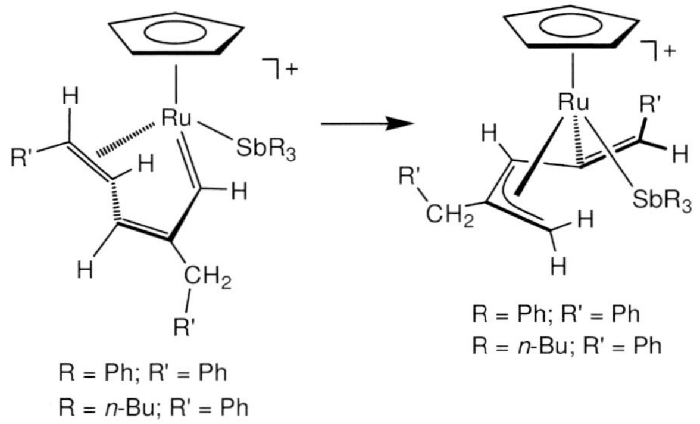

Scheme 13.9

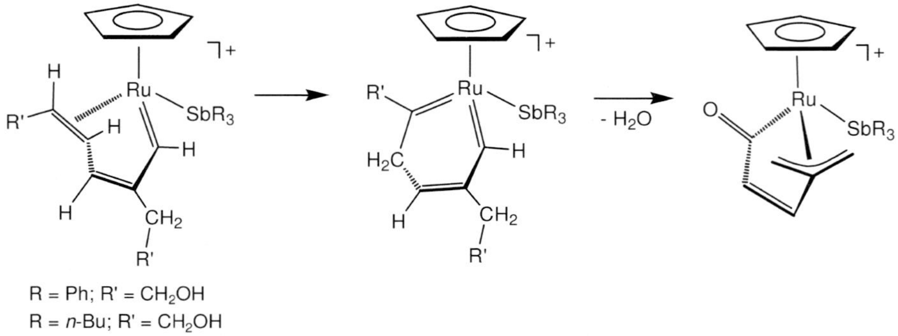

Scheme 13.10

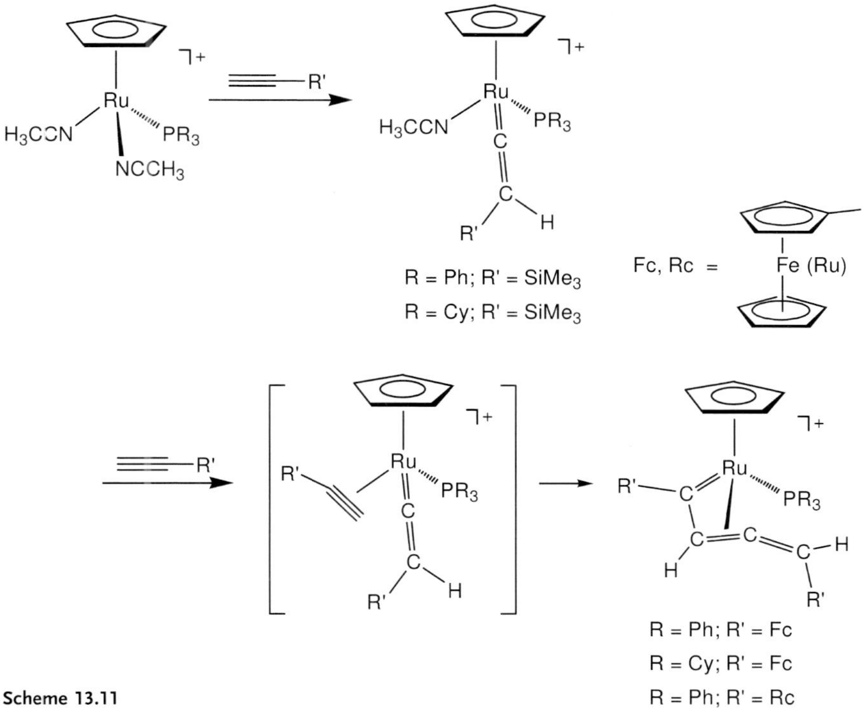

13.4 Dramatic Changes Seen Upon Replacement of Cl in Cp'RuCl

Scheme 13.11

R = Ph; R' = SiMe₃
R = Cy; R' = SiMe₃

Fc, Rc = Fe (Ru) [ferrocenyl / ruthenocenyl]

R = Ph; R' = Fc
R = Cy; R' = Fc
R = Ph; R' = Rc

This new type of C–C bond formation is effected by the strong electronic coupling of the ferrocenyl moiety with the conjugated allenyl carbene unit, which facilitates transient vinylidene formation with subsequent alkyne insertion into the Ru=C double bond. The strong electronic coupling is also apparent from the X-ray structures of the products. The alkyne-vinylidene coupling reaction is extremely facile, requiring merely 6 kcal mol^{-1} according to recent DFT/B3LYP calculations performed with the model complex [RuCp(PH$_3$)(η^2-HC≡CH) (=CCH$_2$)]$^+$ (Fig. 13.3).

(vi) Under particular circumstances, the metallacyclopentatriene can transform into an η^4-cyclobutadiene complex. This outcome has been observed in the reaction of [RuCp(PMe$_3$)]$^+$ with PhC≡CPh (Scheme 13.12).

Theoretical calculations reveal that this transformation is actually symmetry-forbidden, involving HOMO/LUMO crossing. Thus, the kinetic barrier becomes prohibitively high, albeit the reaction appears to be energetically favorable. All cyclobutadiene complexes hitherto known derive from acetylenes with bulky substituents, in particular phenyl or trimethylsilyl [30b, 46]. There is obviously a subtle balance of factors that influence the availability of the cyclobutadiene pathway. In fact, in the above example, upon switching from [RuCp(PMe$_3$)]$^+$ to [RuCp(PPh$_3$)]$^+$, a completely different reaction outcome is encountered, with

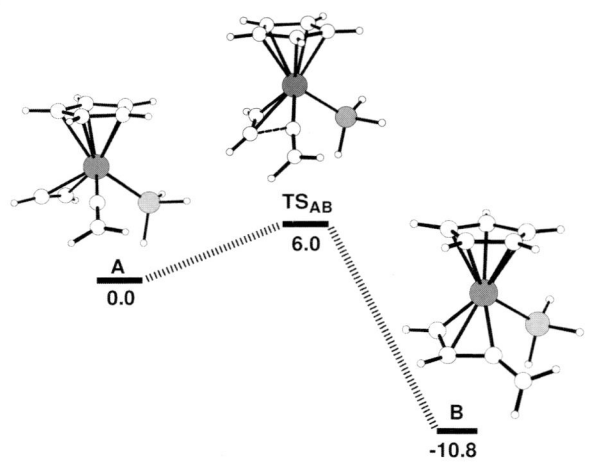

Fig. 13.3 Energy profile of the B3LYP potential energy surface for the conversion of the vinylidene–acetylene complex **A** into the allenyl carbene **B** of Scheme 13.11.

Scheme 13.12

sandwich complexes being formed. In a forthcoming paper, we will investigate the detailed mechanism of the cyclobutadiene pathway, of which to date very little is known.

(vii) Similarly to phosphine, carbon monoxide is also not a spectator ligand in the present type of conversions, but instead functions as an actor ligand. Thus, the monocarbonyl complex [RuCp'(CO)(CH$_3$CN)$_2$]$^+$ reacts with terminal and internal alkynes to afford ruthenium cyclopentadienone complexes [47] (Scheme 13.13).

This transformation is highly selective, yielding in the case of terminal alkynes only one regioisomer with the substituents exclusively in the a,a'-positions of the cyclopentadienone moiety. Again, the substituent on the alkyne is

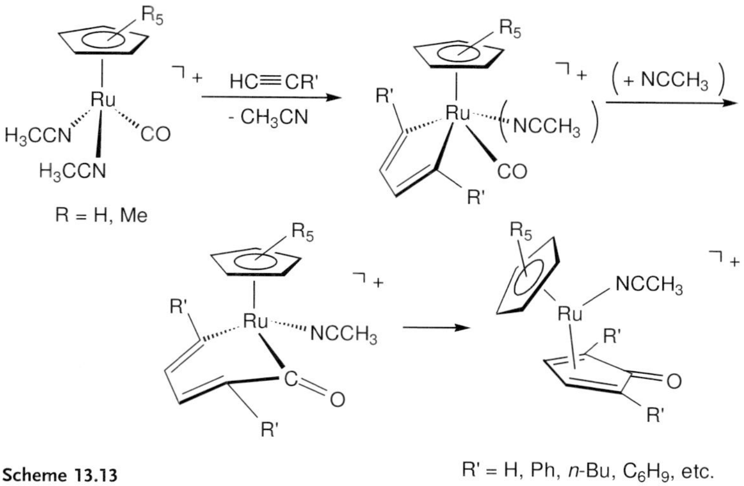

Scheme 13.13 R' = H, Ph, n-Bu, C$_6$H$_9$, etc.

of critical importance with regard to the outcome of the reaction. Specifically, the reaction with PhC≡CPh yields no cyclopentadienone complex but rather the sandwich complex [RuCp(η^6-C$_6$H$_5$PhC≡CPh)]$^+$ and a bis-carbonyl complex [RuCp(CO)$_2$(CH$_3$CN)]$^+$ in a 1:1 ratio.

Other reactions are known in which cyclopentadienone complexes are formed from carbonyls and acetylenes. The first reported cyclopentadienone complex, [Fe(η^4-C$_5$H$_4$O)(CO)$_3$], has been obtained by treating Fe(CO)$_5$ with excess HC≡CH [48]. Other examples are complexes of the types [M(η^4-C$_5$H$_4$O)(CO)$_3$] (M=Fe, Ru) [49], [MCp(η^4-C$_5$H$_4$O)] (M=Co, Rh) [50], and [V(η^4-C$_5$H$_4$O)(CO)(PMe$_3$)] [30f].

13.5
What Happens if the Phosphine Ligand is Tethered?

What is responsible for the extraordinary difference in behavior between the two fragments RuCpCl and [RuCp(PR$_3$)]$^+$? At least two reasons can be invoked. One is the charge carried by the complex. Clearly, a cationic complex will be more electrophilic than a neutral one. The other reason is the greater steric requirement of the tertiary phosphine relative to chloride. To elucidate this point further, we went out on a limb and tethered the phosphine ligand onto the Cp ring via a two-carbon linker, and performed reactions under otherwise identical conditions [51]. As a matter of fact, the arrested phosphorus moiety diverts the previous direct nucleophilic attack of the coordinated trialkylphosphine on the α carbon of the metallacyclopentatriene. Instead, a third alkyne molecule is caught up, ultimately resulting in an unusual C–C coupling process involving three alkynes and the tethered phosphine to give the cycloaddition product **H**, as shown in Scheme 13.14.

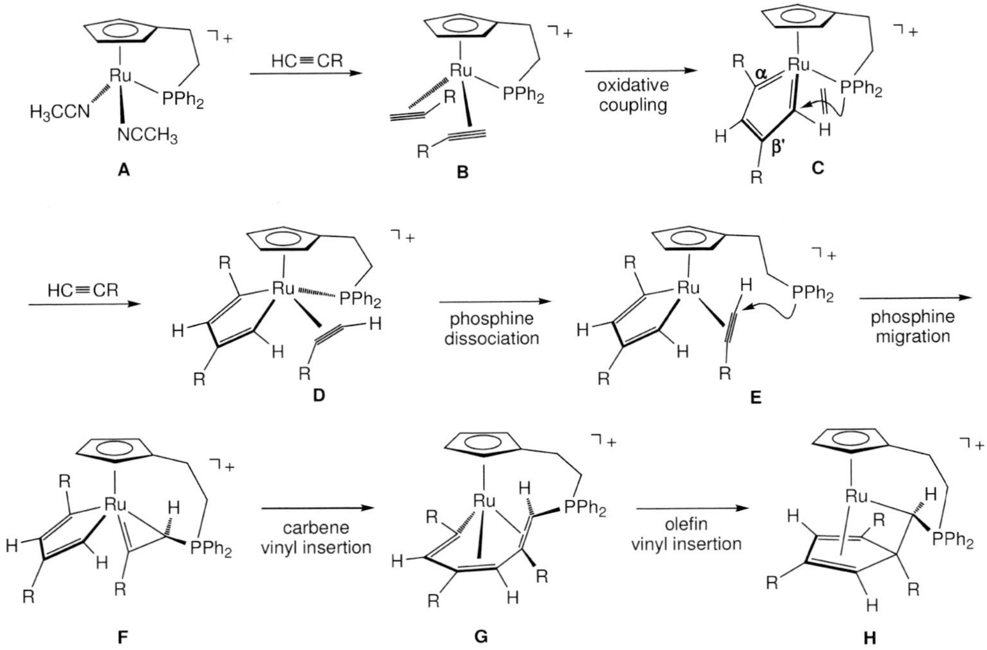

Scheme 13.14

According to DFT/B3LYP calculations, this intriguing [2+2+1] alkyne cyclotrimerization proceeds via Ru–P bond dissociation, phosphine attack at the coordinated acetylene to yield the 1-metallacyclopropene **F**, carbene vinyl insertion, and alkene vinyl insertion. Thus, in the present case, tethering can be considered as amounting to a delayed phosphine migration. That tethering of the phosphine ligand opens the interligand space between the metallacycle plane and the Ru–P vector is nicely seen in Chart 13.4. By the way, note that the corresponding angle of 103° in the tethered complex parallels that in RuCpCl, which mediates the cyclotrimerization of alkynes (*vide supra*).

Complex **C** in Scheme 13.14 is easily able to accommodate a third alkyne to afford the metallacyclopentadiene acetylene complex **D**. In the crowded construction of the latter, the Ru–P bond is labilized, giving, in an endothermic reaction, a free phosphine arm in **E**. The effect of ring strain introduced by a two-

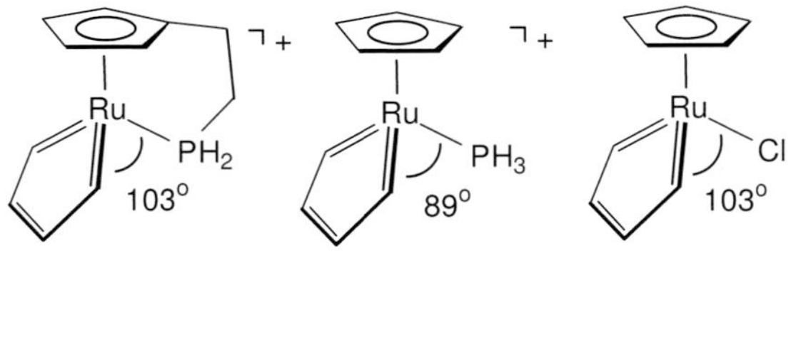

Chart 13.4

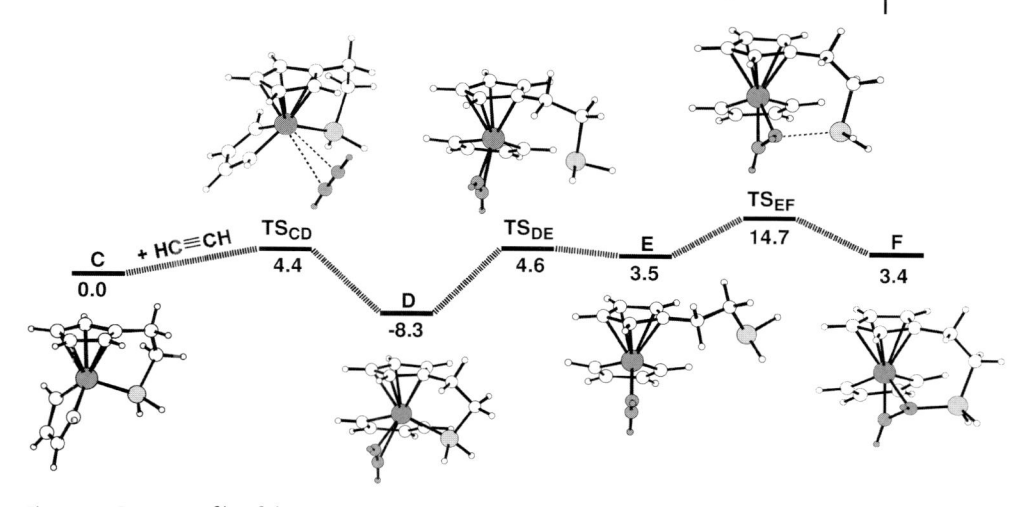

Fig. 13.4 Energy profile of the B3LYP potential energy surfaces for the conversion of the ruthenacyclopentatriene complex **C** into the 1-metallacyclopropene **F** of Scheme 13.14.

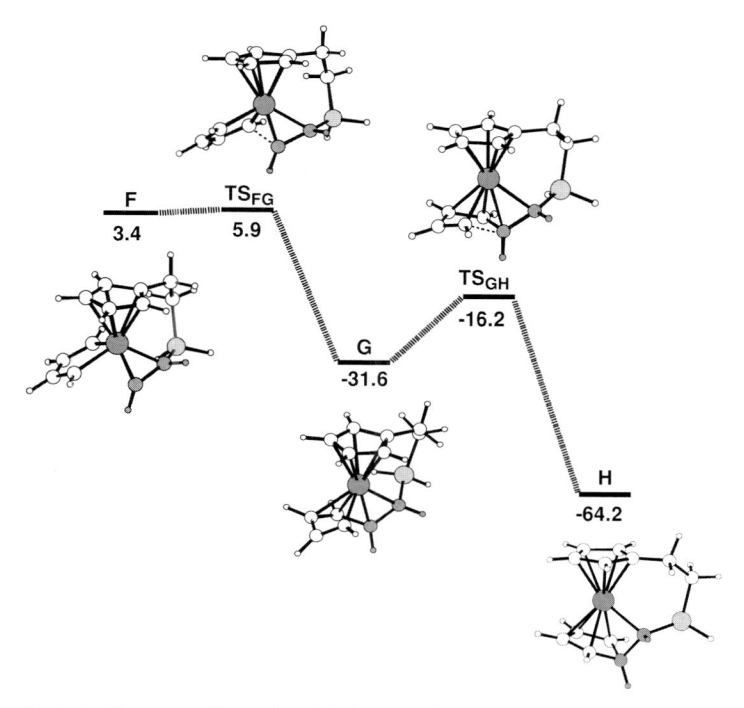

Fig. 13.5 Energy profile of the B3LYP potential energy surfaces for the conversion of the 1-metallacyclopropene **F** into the final cyclotrimerization product **H** of Scheme 13.14.

carbon tether has recently been demonstrated by Casey et al. [52]. The subsequent reaction steps are quite straightforward: nucleophilic attack of the pendant phosphine ligand at the coordinated acetylene results in the formation of the novel metallacyclopentadiene 1-metallacyclopropene complex **F**. The activation energy for this intramolecular process is 11.2 kcal mol^{-1}. It should be noted that related coupling reactions between a coordinated alkyne and a coordinated phosphine to yield 1-metallacyclopropenes and vinyl complexes have been reported in the literature [53, 54]. Intermolecular nucleophilic additions of phosphines and phosphites to alkyne ligands are also feasible [55]. Complex **F** is prone to C–C coupling between the carbene carbon atom of the 1-metallacyclopropene moiety and the unsubstituted a-carbon of the metallacyclopentadiene to yield **G**. This reaction requires an activation energy of merely 2.5 kcal mol^{-1} and is energetically very favorable releasing 35.0 kcal mol^{-1}. The final and rate-determining step is the insertion of the vinyl moiety into the η^2-alkene unit to give **H**, thus completing the [2+2+1] cyclotrimerization.

13.6
Outlook

It is highly fascinating to witness the vast and diversified varieties of rearrangements within and between molecules ligated to the ruthenium(II) center. Tertiary phosphines and carbon monoxide, which ordinarily are regarded as spectator ligands, function here as acting ligands, giving rise to a wide array of scenarios competing with the process of trimerization. This feature may be compared with the behavior of other metal centers, so as to arrive at a deeper understanding of the individuality of the various transition metals as the final goal. Unfortunately, experimental results are often difficult to compare directly because of the variable reaction conditions applied. Moreover, the corresponding reaction mechanisms put forward have often been more a facile explanation than based on particular experimental substantiation. On the other hand, recent progress in computational chemistry has shown that many important chemical and physical properties of intermediates can be predicted. In our opinion, therefore, the better choice for physically relevant comparisons should be such reliable theoretical studies.

The reactions surveyed in the present review may be compared with the theoretical study of acetylene trimerization using CpCoL$_2$ (L=CO, PR$_3$, alkene). At least two fundamental differences are obvious:

(i) For CpCoL$_2$, the coligands L are just leaving ligands that give way to the incoming acetylenes.

(ii) A larger variety of intermediates may be envisaged for the ruthenium system. This may be partly related to the atomic radii. The smaller radius of Co (1.09 vs. 1.21 Å for Ru) [56] may control interligand interactions (through-space coupling) [57] in a different way. In this respect, it might be relevant to undertake investigations on the rhodium analogue CpRh.

Admittedly, the fund of reliable theoretical treatments is still meager. However, we can be sure that in the near future an ever-increasing number of relevant data will follow, which ultimately should help us to understand chemical periodicity more fully, based on first principles.

References

1 R. H. Grubbs, C&EN **2003**, *81*, 112.
2 Y. Yamamoto, H. Takagishi, K. Itoh, *Org. Lett.* **2001**, *3*, 2117.
3 Y. Yamamoto, H. Takagishi, K. Itoh, *J. Am. Chem. Soc.* **2002**, *124*, 28.
4 S. Niu, M. B. Hall, *Chem. Rev.* **2000**, *100*, 353.
5 S. W. Benson, *Thermochemical Kinetics*, Wiley, New York, 1968.
6 (a) K. N. Houk, R. W. Gandour, R. W. Strozier, N. G. Rondan, L. A. Paquette, *J. Am. Chem. Soc.* **1979**, *101*, 6797; (b) R. D. Bach, G. J. Wolber, H. B. Schlegel, *J. Am. Chem. Soc.* **1985**, *107*, 2837.
7 H. Jiao, P. von R. Schleyer, *J. Phys. Org. Chem.* **1998**, *11*, 655.
8 (a) W. Reppe, O. Schlichting, K. Klager, T. Toepel, *Liebigs Ann. Chem.* **1948**, *560*, 3; (b) W. Reppe, W. Schweckendiek, *Liebigs Ann. Chem.* **1948**, *560*, 104.
9 (a) S. Saito, Y. Yamamoto, *Chem. Rev.* **2000**, *100*, 2901; (b) N. E. Schore, *Chem. Rev.* **1988**, *88*, 1081; (c) D. B. Grotjahn, *Transition Metal Alkyne Complexes: Transition Metal-Catalyzed Cyclotrimerization*, in *Comprehensive Organometallic Chemistry II* (Eds.: E. W. Abel, F. G. A. Stone, G. Wilkinson, L. S. Hegedus), vol. 12, Pergamon, Oxford, 1995, pp. 741–770.
10 (a) K. P. C. Vollhardt, *Acc. Chem. Res.* **1977**, *10*, 1; (b) K. P. C. Vollhardt, *Angew. Chem. Int. Ed. Engl.* **1984**, *23*, 539.
11 (a) Y. Wakatsuki, H. Yamazaki, *J. Chem. Soc., Chem. Commun.* **1973**, 280; (b) H. Bönnemann, *Angew. Chem. Int. Ed. Engl.* **1978**, *17*, 505; (c) H. Bönnemann, *Angew. Chem. Int. Ed. Engl.* **1985**, *24*, 248.
12 Y. Yamamoto, R. Ogawa, K. Itoh, *Chem. Commun.* **2000**, 549.
13 (a) Y. Yamamoto, H. Kitahara, R. Ogawa, H. Kawaguchi, K. Tatsumi, K. Itoh, *J. Am. Chem. Soc.* **2000**, *122*, 4310; (b) Y. Yamamoto, H. Kitahara, R. Hattori, K. Itoh, *Organometallics* **1998**, *17*, 1910.
14 Y. Yamamoto, H. Kitahara, R. Ogawa, K. Itoh, *J. Org. Chem.* **1998**, *63*, 9610.
15 Y. Yamamoto, R. Ogawa, K. Itoh, *J. Am. Chem. Soc.* **2001**, *123*, 6189.
16 Y. Yamamoto, S. Okuda, K. Itoh, *Chem. Commun.* **2001**, 1102.
17 Y. Yamamoto, H. Takagishi, K. Itoh, *J. Am. Chem. Soc.* **2002**, *124*, 6844.
18 (a) Y. Sato, T. Nishimata, M. Mori, *J. Org. Chem.* **1994**, *59*, 6133; (b) Y. Sato, T. Nishimata, M. Mori, *Heterocycles* **1977**, *44*, 443.
19 B. Witulski, T. Stengel, *Angew. Chem. Int. Ed. Engl.* **1999**, *38*, 2426.
20 S. Abbet, A. Sanchez, U. Heiz, W.-D. Schneider, A. M. Ferrari, G. Pacchioni, N. Rösch, *J. Am. Chem. Soc.* **2000**, *122*, 3453.
21 J. T. Fletcher, M. J. Therien, *J. Am. Chem. Soc.* **2000**, *122*, 12393.
22 D. Suzuki, H. Urabe, F. Sato, *J. Am. Chem. Soc.* **2001**, *123*, 7925.
23 K. Ferré, L. Toupet, V. Guerchais, *Organometallics* **2002**, *21*, 2578.
24 T. Takahashi, Z. Xi, A. Yamazaki, Y. Liu, K. Nakajima, M. Korora, *J. Am. Chem. Soc.* **1998**, *120*, 1672.
25 D. Suzuki, H. Urabe, F. Sato, *J. Am. Chem. Soc.* **2001**, *123*, 7925.
26 B. M. Trost, F. D. Toste, *Angew. Chem. Int. Ed. Engl.* **2001**, *40*, 1114.
27 H. Hardesty, J. B. Koerner, A. Albright, G.-Y. Lee, *J. Am. Chem. Soc.* **1999**, *121*, 6055.
28 K. Kirchner, M. J. Calhorda, R. Schmid, L. F. Veiros, *J. Am. Chem. Soc.* **2003**, *125*, 11721.
29 Y. Yamamoto, T. Arakawa, R. Ogawa, K. Itoh, *J. Am. Chem. Soc.* **2003**, *125*, 12143.

30 For metallacyclopentatriene complexes, see: (a) M.O. Albers, P.J.A. de Waal, D.C. Liles, D.J. Robinson, E. Singleton, M.B. Wiege, *J. Chem. Soc., Chem. Commun.* **1986**, 1680; (b) C. Gemel, A. LaPansée, K. Mauthner, K. Mereiter, R. Schmid, K. Kirchner, *Monatsh. Chem.* **1997**, *128*, 1189; (c) L. Pu, T. Hasegawa, S. Parkin, H. Taube, *J. Am. Chem. Soc.* **1992**, *114*, 2712; (d) W. Hirpo, M.D. Curtis, *J. Am. Chem. Soc.* **1988**, *110*, 5218; (e) J.L. Kerschner, P.E. Fanwick, I.P. Rothwell, *J. Am. Chem. Soc.* **1988**, *110*, 8235; (f) B. Hessen, A. Meetsma, F. van Bolhuis, J.H. Teuben, G. Helgesson, S. Jagner, *Organometallics* **1990**, *9*, 1925; (g) C. Ernst, O. Walter, E. Dinjus, S. Arzberger, H. Görls, *J. Prakt. Chem.* **1999**, *341*, 801; (h) Y. Yamada, J. Mizutani, M. Kurihara, H. Nishihara, *J. Organomet. Chem.* **2001**, *637–639*, 80.

31 (a) J.L. Paih, F. Monnier, S. Derien, P.H. Dixneuf, E. Clot, O. Eisenstein, *J. Am. Chem. Soc.* **2003**, *125*, 11964; (b) J.L. Paih, S. Derien, P.H. Dixneuf, *Chem. Commun.* **1999**, 1437.

32 B.M. Trost, M.T. Rudd, *J. Am. Chem. Soc.* **2002**, *124*, 4178.

33 B.M. Trost, M.T. Rudd, *J. Am. Chem. Soc.* **2003**, *125*, 11516.

34 C.S. Yi, R. Torres-Lubian, N. Liu, A.L. Rheingold, I.A. Guzei, *Organometallics* **1998**, *17*, 1257.

35 M. Paneque, M.L. Poveda, N. Rendón, K. Mereiter, *J. Am. Chem. Soc.* **2004**, *126*, 1610.

36 M.E. Stoll, P. Belanzoni, M.J. Calhorda, M.G.B. Drew, V. Félix, W.E. Geiger, C.A. Gamelas, I.S. Gonçales, C.C. Romão, L.F. Veiros, *J. Am. Chem. Soc.* **2001**, *123*, 10959.

37 E. Rüba, R. Schmid, K. Kirchner, M.J. Calhorda, *J. Organomet. Chem.* **2003**, *682*, 204.

38 R. Schmid, K. Kirchner, *J. Org. Chem.* **2003**, *68*, 8339.

39 C. Slugovc, E. Rüba, R. Schmid, K. Kirchner, *Organometallics* **1999**, *18*, 4230.

40 E. Becker, C. Slugovc, E. Rüba, C. Standfest-Hauser, K. Mereiter, R. Schmid, K. Kirchner, *J. Organomet. Chem.* **2002**, *649*, 55.

41 (a) E. Rüba, K. Mereiter, R. Schmid, V.N. Sapunov, K. Kirchner, H. Schottenberger, M.J. Calhorda, L.F. Veiros, *Chem. Eur. J.* **2002**, *8*, 3948; (b) E. Rüba, K. Mereiter, R. Schmid, K. Kirchner, *Chem. Commun.* **2001**, 1996; (c) E. Rüba, K. Mereiter, R. Schmid, K. Kirchner, H. Schottenberger, *J. Organomet. Chem.* **2001**, *637–639*, 70; (d) K. Mauthner, K.M. Soldouzi, K. Mereiter, R. Schmid, K. Kirchner, *Organometallics* **1999**, *18*, 4681.

42 (a) E. Becker, K. Mereiter, M. Puchberger, R. Schmid, K. Kirchner, *Organometallics* **2003**, *22*, 2124; (b) E. Becker, E. Rüba, K. Mereiter, R. Schmid, K. Kirchner, *Organometallics* **2001**, *20*, 3851.

43 E. Rüba, K. Mereiter, R. Schmid, K. Kirchner, E. Bustelo, M.C. Puerta, P. Valerga, *Organometallics* **2002**, *21*, 2912.

44 For related PAr$_2$(η^2-arene) complexes, see: (a) N. Feiken, P.S. Pregosin, G. Trabesinger, M. Scalone, *Organometallics* **1997**, *16*, 537; (b) N. Feiken, P.S. Pregosin, G. Trabesinger, A. Albinati, G.L. Evoli, *Organometallics* **1997**, *16*, 5756; (c) T.-Y. Cheng, D.J. Szalda, R.M. Bullock, *J. Chem. Soc., Chem. Commun.* **1999**, 1629; (d) H. Aneetha, M. Jimenez-Tenorio, M.C. Puerta, P. Valerga, K. Mereiter, *Organometallics* **2002**, *21*, 628.

45 T. Arliguie, B. Chaudret, G. Chung, F. Dahan, *Organometallics* **1991**, *10*, 2973; (b) M.L. Christ, S. Sabo-Etienne, B. Chaudrett, *Organometallics* **1995**, *14*, 1082; (c) A.F. Borowski, S. Sabo-Etienne, M.L. Christ, *Organometallics* **1996**, *15*, 1427; (d) S. Hietkamp, D.J. Hufkens, K. Vrieze, *J. Organomet. Chem.* **1978**, *152*, 347; (e) C. Six, B. Gabor, H. Görls, R. Mynott, P. Philipps, W. Leitner, *Organometallics* **1999**, *18*, 3316.

46 (a) J.R. Fritch, K.P.C. Vollhardt, *Angew. Chem. Int. Ed. Engl.* **1979**, *18*, 409; (b) G. Ville, K.P.C. Vollhardt, *J. Am. Chem. Soc.* **1981**, *103*, 5267; (c) G.

Ville, K. P. C. Vollhardt, M. J. Winter, *Organometallics* **1984**, *3*, 1177; (d) B. K. Campion, R. H. Heyn, T. Don Tilley, *Organometallics* **1990**, *9*, 1106.

47 E. Rüba, K. Mereiter, K. M. Soldouzi, C. Gemel, R. Schmid, K. Kirchner, E. Bustelo, M. C. Puerta, P. Valerga, *Organometallics* **2000**, *19*, 5384.

48 (a) W. Reppe, H. Vetter, *Justus Liebigs Ann. Chem.* **1953**, *582*, 133.

49 (a) M. L. H. Green, L. Pratt, G. Wilkinson, *J. Chem. Soc.* **1960**, 989; (b) E. Weiss, R. Merenyi, W. Huebel, *Chem. Ber.* **1962**, *95*, 1170; (c) Y. Blum, Y. Shvo, D. F. Chodosh, *Inorg. Chim. Acta* **1985**, *97*, L25; (d) A. J. Pearson, R. J. Shively, R. A. Dubbert, *Organometallics* **1992**, *11*, 4096.

50 (a) R. Markby, H. W. Sternberg, I. Wender, *Chem. Ind.* **1959**, 1381; (b) R. S. Dickson, G. R. Tailby, *Aust. J. Chem.* **1970**, *23*, 1531; (c) J. H. Bieri, A. Dreiding, T. C. C. Gartenmann, E. R. F. Gesing, R. Kunz, R. Prewo, *J. Organomet. Chem.* **1986**, *306*, 241; (d) H. Borwieck, O. Walter, E. Dinjus, J. Rebizant, *J. Organomet. Chem.* **1998**, *570*, 121; (e) E. R. F. Gesing, J. P. Tane, K. P. C. Vollhardt, *Angew. Chem.* **1980**, *92*, 1057; *Angew. Chem. Int. Ed. Engl.* **1980**, *19*, 1023.

51 E. Becker, K. Mereiter, M. Puchberger, R. Schmid, K. Kirchner, *Organometallics* **2003**, *22*, 2124.

52 For [2+2+1] cyclotrimerization reactions of alkynes, see: (a) J. M. O'Connor, A. Closson, K. Hiibner, R. Merwin, P. K. Gantzel, D. M. Roddick, *Organometallics* **2001**, *20*, 3710; (b) H. J. Kim, N. S. Choi, S. W. Lee, *J. Organomet. Chem.* **2000**, *616*, 67; (c) J. M. O'Connor, K. Hiibner, R. Merwin, P. K. Gantzel, B. S. Fong, M. Adams, A. L. Rheingold, *J. Am. Chem. Soc.* **1997**, *119*, 3631; (d) G. Moran, M. Green, A. G. Orpen, *J. Organomet. Chem.* **1983**, *250*, C15; (e) J. Moreto, K. Maruya, P. M. Bailey, P. M. Maitlis, *J. Chem. Soc., Dalton Trans.* **1982**, 1341; (f) L. S. Liebeskind, R. Chidambaram, *J. Am. Chem. Soc.* **1987**, *109*, 5025.

53 H. Ishino, S. Kuwata, Y. Ishii, M. Hidai, *Organometallics* **2001**, *20*, 13.

54 J. M. O'Connor, K. D. Bunker, *J. Organomet. Chem.* **2003**, *671*, 1.

55 (a) J. L. Davidson, *J. Chem. Soc., Dalton Trans.* **1986**, 2423; (b) J. L. Davidson, G. Vasapollo, L. Manojlovic-Muir, K. W. Muir, *J. Chem. Soc., Chem. Commun.* **1982**, 1025; (c) S. R. Allen, R. G. Beevor, M. Green, N. C. Norman, A. G. Orpen, *J. Chem. Soc., Dalton Trans.* **1985**, 435; (d) J. R. Morrow, T. L. Tonker, J. L. Templeton, W. R. Kenan, *J. Am. Chem. Soc.* **1985**, *107*, 6956.

56 C. H. Suresh, N. Koga, *J. Phys. Chem. A* **2001**, *105*, 5940.

57 V. N. Sapunov, R. Schmid, K. Kirchner, H. Nagashima, *Coord. Chem. Rev.* **2003**, *238/239*, 363.

14
Osmium(VIII) Oxide and Oxide Fluoride Chemistry

Michael Gerken and Gary J. Schrobilgen

14.1
Introduction: Oxidation State +8

The +8 oxidation state is the highest attained by any element in the Periodic Table and is limited to three elements, osmium, ruthenium, and xenon. To date, the +8 oxidation state has only been stabilized in species in which the highly electronegative ligating atoms F, O, and N are bound to the high-oxidation-state center. Among the aforementioned elements, the chemistry of Os(VIII) is the most extensive, which is largely a consequence of the high thermodynamic stability of the tetroxide, OsO_4 ($\Delta H_f^0 = -390.8 \pm 5.9$ kJ mol^{-1}) [1], the precursor for nearly all Os(VIII) compounds (see Section 14.2). The chemistry of Os(VIII) is summarized in Scheme 14.1, and illustrates its strong reliance upon OsO_4 as a synthetic precursor.

While OsO_4 reacts in aqueous alkali solution according to Eq. (1) yielding red perosmates $OsO_4(OH)_2^{2-}$ [2] (*vide infra*), RuO_4 is less stable and is readily reduced to $Ru^{VII}O_4^-$ under similar conditions (Eq. (2)) [3].

$$OsO_4 + 2OH^- \rightarrow OsO_4(OH)_2^{2-} \qquad (1)$$

$$2RuO_4 + 2OH^- \rightarrow 2RuO_4^- + \tfrac{1}{2}O_2 + H_2O \qquad (2)$$

Other than the salts of the perxenate anion, XeO_6^{4-}, all Xe(VIII) compounds, XeO_4, XeO_3F_2, and XeO_2F_4, reported to date are kinetically and thermodynamically unstable. Xenon tetroxide ($\Delta H_f^0 = 642.2$ kJ mol^{-1}) [4] is explosive even at –40 °C, rendering synthetic Xe(VIII) chemistry that starts from XeO_4 extremely challenging [5, 6].

14.2
Osmium Tetroxide and Perosmates

Osmium tetroxide is a stable, yellow solid having a melting point of 40.5 ± 0.1 °C [7] and a vapor pressure of 10 Torr at 26.0 °C [8], and can be obtained from combustion of osmium metal in a stream of oxygen above 300 °C [9] or from the reac-

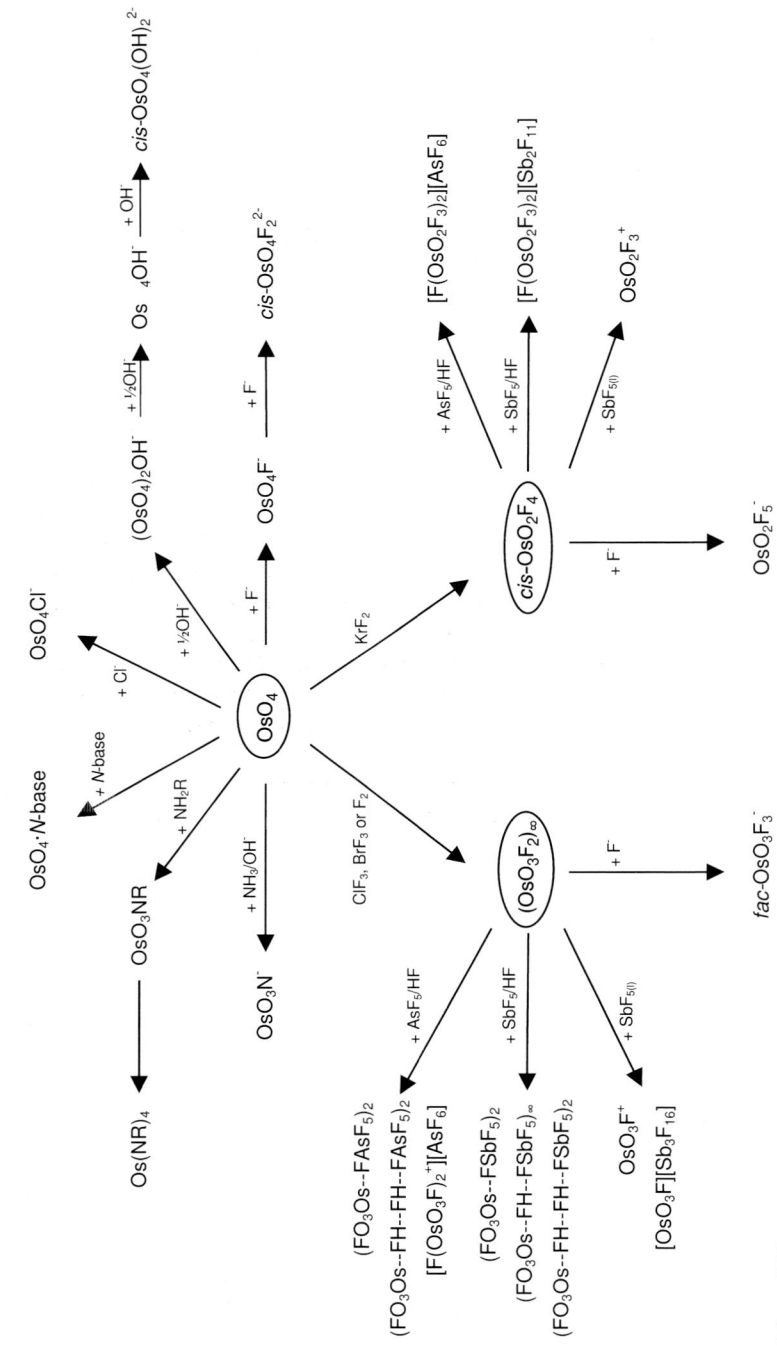

14 Osmium(VIII) Oxide and Oxide Fluoride Chemistry

$OsO_4 \cdot N\text{-base}$ ← + N-base — OsO_4 — + Cl⁻ → OsO_4Cl^-

OsO_4 — + ½OH⁻ → $(OsO_4)_2OH^-$ — + ½OH⁻ → OsO_4OH^- — + OH⁻ → $cis\text{-}OsO_4(OH)_2^{2-}$

OsO_4 — + F⁻ → OsO_4F^- — + F⁻ → $cis\text{-}OsO_4F_2^{2-}$

OsO_4 — KrF_2 → $cis\text{-}OsO_2F_4$

$cis\text{-}OsO_2F_4$ — + AsF₅/HF → $[F(OsO_2F_3)_2][AsF_6]$

$cis\text{-}OsO_2F_4$ — + SbF₅/HF → $[F(OsO_2F_3)_2][Sb_2F_{11}]$

$cis\text{-}OsO_2F_4$ — + SbF₅(l) → $OsO_2F_3^+$

$cis\text{-}OsO_2F_4$ — + F⁻ → $OsO_2F_5^-$

OsO_4 — ClF_3, BrF_3 or F_2 → $(OsO_3F_2)_\infty$

$(OsO_3F_2)_\infty$ — + F⁻ → $fac\text{-}OsO_3F_3^-$

$(OsO_3F_2)_\infty$ — + AsF₅/HF → $(FO_3Os\text{--}FAsF_5)_2$, $(FO_3Os\text{--}FH\text{--}FH\text{--}FAsF_5)_2$, $[F(OsO_3F)_2^+][AsF_6^-]$

$(OsO_3F_2)_\infty$ — + SbF₅/HF → $(FO_3Os\text{--}FSbF_5)_2$, $(FO_3Os\text{--}FH\text{--}FSbF_5)_\infty$, $(FO_3Os\text{--}FH\text{--}FH\text{--}FSbF_5)_2$

$(OsO_3F_2)_\infty$ — + SbF₅(l) → OsO_3F^+, $[OsO_3F][Sb_3F_{16}]$

OsO_4 — + NH₃/OH⁻ → OsO_3N^-

OsO_4 — + NH₂R → OsO_3NR → $Os(NR)_4$

Scheme 14.1

tion of osmium powder with a strong oxidizer, such as an aqueous solution of sodium hypochlorite [10]. Osmium tetroxide has long been used in organic chemistry as a cis-dihydroxylating agent [11, 12].

Recently, a material formulated as [Ag$_{13}$][OsO$_6$] has been synthesized and described, representing the first report of a perosmate(VIII) salt and the first homoleptic osmium(VIII) oxo-anion (Os–O distance, 190 pm) [13, 14]. The synthesis of this compound does not employ OsO$_4$ as a precursor, although OsO$_4$ may be an intermediate in the reaction. The compound was obtained in the form of olive-green crystals by the solid-state reaction of ground mixtures of silver (or silver oxide) and osmium at 300 °C and 15 MPa of O$_2$ pressure. The crystal structure of [Ag$_{13}$][OsO$_6$] may be described as [Ag@Ag$_{12}$]$^{4+}$ icosahedra and OsO$_6^{4-}$ octahedra arranged as in the body-centered cubic CsCl structure. In contrast, perxenate salts have been known for over 40 years and several salts have been structurally characterized by single-crystal X-ray diffraction [15–17].

14.2.1
Osmium Tetroxide Adducts

Osmium tetroxide has been shown to behave as a Lewis acid towards a large number of organonitrogen bases, such as quinuclidine, forming neutral adducts according to Eq. (3) [18, 19], which have been characterized by vibrational spectroscopy and single-crystal X-ray diffraction (Fig. 14.1a).

$$OsO_4 + N(CH_2CH_2)_3CH \rightarrow OsO_4 \cdot N(CH_2CH_2)_3CH \quad (3)$$

The use of chiral alkaloid derivatives that coordinate to OsO$_4$ through the nitrogen of a quinuclidine group has made efficient osmium-catalyzed asymmetric dihydroxylation possible [12].

Lewis acid-base adducts of OsO$_4$ with oxygen donor molecules have also been prepared with OH$^-$ [2, 20–28] and N-methylmorpholine N-oxide [29]. Osmium tetroxide forms anionic adducts with F$^-$ (Eq. (4)) [30] and Cl$^-$ [31], and with the pseudohalide, N$_3^-$ [31].

$$OsO_4 + [N(CH_3)_4][F] \rightarrow [N(CH_3)_4][OsO_4F] \quad (4)$$

The majority of such Lewis acid-base adducts have been characterized by elemental analysis, vibrational spectroscopy, and single-crystal X-ray diffraction (Tab. 14.1).

Osmium tetroxide can bind to one or two donor atoms. In 1:1 Lewis acid-base adducts, e.g., OsO$_4$F$^-$ [30], OsO$_4$(OH)$^-$ [2], OsO$_4$Cl$^-$ [31], and in 2:1 adducts, e.g., (OsO$_4$)$_2$OH$^-$ [2] and (OsO$_4$)$_2$ · hexamethylenetetramine (Fig. 14.1b) [19], osmium exhibits distorted trigonal-bipyramidal coordination. The O–Os–donor atom angle in most 1:1 and 2:1 adducts is near linear (Tab. 14.1), with the OsO$_4$F$^-$ anion representing a notable exception [30]. The O–Os–F angle is 156.9(5)° (Fig. 14.1c), and this has been attributed to anion–cation interactions and packing effects in the crystal lattice.

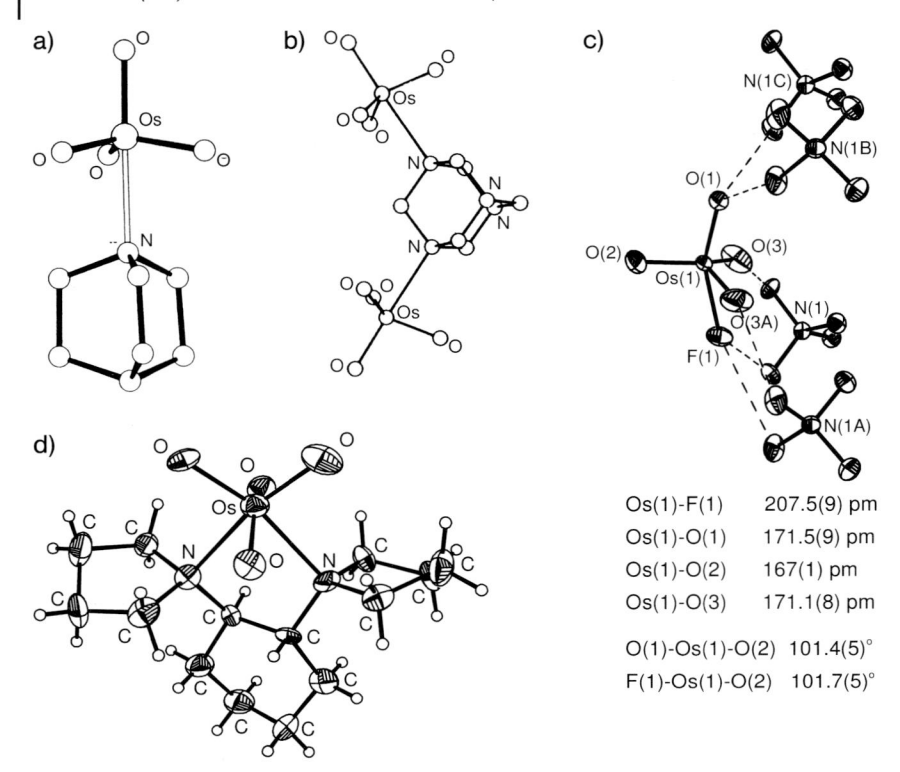

Fig. 14.1 The X-ray crystal structures of (a) the OsO$_4$·quinuclidine adduct [19], (b) the (OsO$_4$)$_2$·hexamethylenetetramine adduct [19], (c) the OsO$_4$F$^-$ anion in [N(CH$_3$)$_4$][OsO$_4$F] and its contacts with the N(CH$_3$)$_4^+$ cations (hydrogen atoms have been omitted) [30], and (d) OsO$_4$·(R,R)-trans-1,2-bis(N-pyrrolidino)cyclohexane [39].

Os(1)-F(1)	207.5(9) pm
Os(1)-O(1)	171.5(9) pm
Os(1)-O(2)	167(1) pm
Os(1)-O(3)	171.1(8) pm
O(1)-Os(1)-O(2)	101.4(5)°
F(1)-Os(1)-O(2)	101.7(5)°

Osmium in tetroxo compounds can expand its coordination number to six by coordination to two donor atoms of two monodentate bases, e.g., cis-OsO$_4$F$_2^{2-}$ [30], cis-OsO$_4$(OH)$_2^{2-}$ [24, 25], or to one bidentate chelating ligand such as in OsO$_4$ · (R,R)-trans-1,2-bis(N-pyrrolidino)cyclohexane [39]. The latter adduct, which is unstable above −40 °C and soluble in dry diethyl ether, has been characterized by low-temperature single-crystal X-ray diffraction (Fig. 14.1d). The crystal structure shows a severely distorted octahedral geometry with a cis arrangement of nitrogen donor atoms. The cis-OsO$_4$F$_2^{2-}$ anion has reportedly been obtained as the Cs$^+$ and Rb$^+$ salts upon reaction of OsO$_4$ with an excess of the corresponding alkali metal fluoride in water [21, 23, 32–34]. However, the reaction of [N(CH$_3$)$_4$][OsO$_4$F] with [N(CH$_3$)$_4$][F] (commonly referred to as "naked fluoride") in acetonitrile solvent failed to produce the cis-OsO$_4$F$_2^{2-}$ anion (Eq. (5)) in quantitative yield [30].

$$[N(CH_3)_4][F] + [N(CH_3)_4][OsO_4F] \xrightarrow{CH_3CN} [N(CH_3)_4]_2[\text{cis-OsO}_4F_2] \qquad (5)$$

Tab. 14.1 Lewis base adducts of OsO$_4$, their characterization, and selected structural data[a]

Adduct	Method of characterization	Os-donor atom bond length, pm[b]	∠O$_{ax/trans}$–Os–donor atom, deg[b]
Donor atom: Oxygen			
OsO$_4$OH$^-$	e.a. [2, 20], IR [2]		
(OsO$_4$)$_2$OH$^-$	e.a. [2, 20], IR [2], X-ray [2, 13]	221(2)/222(2) [2], 216/222 [13]	179(1)/177(1) [2]
OsO$_4$(OH)$_2^{2-}$	e.a. [2, 20, 21], IR [2, 22, 23, 28], Raman [23] X-ray [24, 25, 27, 28]	216/216 [24], 210/217 [25]	
OsO$_4$ · N-methylmorpholine N-oxide	e.a. [29], IR [29], X-ray [29]	230.5(4)	175.1(3)
Donor: Halide or Azide			
OsO$_4$F$^-$	e.a. [32], IR [30, 32] Raman [30, 32], X-ray [30]	207.5(9)	156.9(4)
OsO$_4$F$_2^{2-}$	IR [30], Raman [30][c], EXAFS [34]		
OsO$_4$Cl$^-$	e.a. [31], IR [31], X-ray [31]	276.0(2)	179.0(2)
OsO$_4$N$_3^-$	e.a. [31], IR [31]		
Donor atom: Nitrogen			
OsO$_4$ · quinuclidine	e.a. [18], IR [18], Raman, [18] X-ray [19]	237	180.0
(OsO$_4$)$_2$ · hexamethylenetetramine	e.a. [18, 19], IR [18, 19], Raman [18] ,X-ray [19]	242	n.r.
OsO$_4$ · N-methylmorpholine	e.a. [29], IR [29], X-ray [29]	244.0(7)	177.7(3)
OsO$_4$ · pyridine	e.a. [18], IR [18, 35] Raman [18, 35]		
OsO$_4$ · pyrazine	e.a. [18], IR [18]		
OsO$_4$ · triethylenediamine	e.a. [18], IR [18], Raman [18]		

Tab. 14.1 (continued)

Adduct	Method of characterization	Os-donor atom bond length, pm[b]	∠O$_{ax/trans}$–Os–donor atom, deg[b]
(OsO$_4$)$_2$ · 5-methylpyrimidine	e.a. [18], IR [18]		
OsO$_4$ · phthalazine	e.a. [18], IR [18], Raman [18]		
OsO$_4$ · isoquinoline	e.a. [18], IR [18], Raman [18]		
OsO$_4$ · 1,8-naphthyridine	Raman [36], X-ray [36]	243.8(6)	178.4(3)
OsO$_4$ · 4-pyrrolidinopyridine	X-ray [37]	231.8(13)	179.3(8)
OsO$_4$ · 4-phenylpyridine	X-ray [37]	241.9(15)	178.8(6)
OsO$_4$ · 4-cyanopyridine	X-ray [37]	245.2(8)	178.6(4)
OsO$_4$ · (dimethylcarbamoyl)dihydroquinidine	X-ray [38]	249	
(OsO$_4$) · (R,R)-trans-1,2-bis(N-pyrrolidino)cyclohexane	X-ray [39]	232.9(11)/233.4	167.3/165.8

a) Abbreviations denote: elemental analysis (e.a.), not reported (n.r.).
b) The geometric parameters are taken from the designated X-ray reference in the left-hand column unless otherwise noted.
c) The reported characterizations of OsO$_4$F$_2^{2-}$ [21, 32–34] were shown to be erroneous.

A Raman spectroscopic study of [N(CH$_3$)$_4$]$_2$[cis-OsO$_4$F$_2$] and [N(CH$_3$)$_4$][OsO$_4$F] established that in the previous report the vibrational spectrum of OsO$_4$F$^-$ had been incorrectly assigned to that of the cis-OsO$_4$F$_2^{2-}$ anion. Although the insolubility of [N(CH$_3$)$_4$][cis-OsO$_4$F$_2$] has prevented the growth of crystals for X-ray diffraction analysis, vibrational spectroscopy, in conjunction with density functional theory (DFT) calculations, has shown that the OsO$_4$F$_2^{2-}$ anion exists exclusively as the cis-dioxo isomer (Structure I) [30]. Formation of the cis isomer

$$\begin{bmatrix} & O & \\ & \| & \\ & | & ..F \\ O- & Os & -F \\ & \diagup | & \\ & O \; | & \\ & O & \end{bmatrix}^{2-}$$

Structure I

serves to minimize the number of doubly-bonded oxygen ligands that are trans to one another, whereas the cis configuration of OsO$_4$·(R,R)-trans-1,2-bis(N-pyrrolidino)cyclohexane is determined by the chelating nature of the ligand [39]. The preference for the cis-dioxo arrangement in cis-OsO$_4$F$_2^{2-}$ is electronic in origin and has been observed for other Os(VIII) species, and it is a general feature in d^0 transition metal chemistry (see Section 14.4.3).

14.3
Nitrido Osmates and Organo-Imido Osmates

The reactions of OsO$_4$ with Lewis bases such as NH$_3$ in aqueous KOH solution and with primary amines in aqueous and organic solvent media have been shown to yield the nitrido osmate anion [40] and organo-imido osmates [40–42] according to Eqs. (6) and (7), respectively.

$$OsO_4 + NH_3 + KOH \rightarrow [K][OsO_3N] + 2H_2O \tag{6}$$

$$OsO_4 + (CH_3)_3CNH_2 \rightarrow (CH_3)_3CNOsO_3 + H_2O \tag{7}$$

Like OsO$_4$, imido osmates behave as Lewis acids towards nitrogen donor molecules, such as 1,4-diazabicyclo[2.2.2]octane [43], while the OsO$_3$N$^-$ anion has been utilized as a nitrogen donor ligand in Rh, Ir, Pt, and Au complexes [44, 45]. The syntheses of several bis(imido)osmates, O$_2$Os(NR)$_2$ [41, 42, 46] tris(imido)osmates, OOs(NR)$_3$ [42, 46], and tetrakis(imido)osmates, Os(NR)$_4$ [47, 48], have also been reported.

14.4
Osmium Oxide Fluoride Chemistry

While oxygen and nitrogen are capable of stabilizing osmium in the +8 oxidation state as the homoleptic species, OsO$_4$ and Os(NR)$_4$, the more electronegative element, fluorine, may be expected, at first glance, to provide even greater stabilization of Os(VIII) compounds. However, the formal replacement of oxygen by fluorine generally leads to the destabilization of the +8 oxidation state. This is exemplified by an increase in oxidizing power in the sequence $OsO_4 < OsO_3F_2 < cis\text{-}OsO_2F_4$, and the absence of confirmed reports of $OsOF_6$ and OsF_8. Among the three possible neutral mononuclear oxide fluorides, only OsO_3F_2 [49] and cis-OsO_2F_4 [49, 50] are known. A report of the preparation of $OsOF_6$ has been shown to be erroneous (see Section 14.4.2). Although the synthesis of OsF_8 [51] has been reported, the claim was subsequently shown to be erroneous [52, 53]. The direct reaction of osmium with fluorine gives OsF_7 as the highest osmium fluoride [54]. A similar trend has been noted in Xe(VIII) chemistry, where stability decreases in the series $XeO_4 > XeO_3F_2 > XeO_4F_2$ [55].

14.4.1
Osmium Trioxide Difluoride

Osmium trioxide difluoride is a bright orange solid with a melting point of ca. 170 °C and can be prepared by fluorination of OsO_4 with F_2 [56], BrF_3 [57], or excess ClF_3 [49]. The latter synthetic route is preferred because of the ease of removal of the volatile products. Bulk OsO_3F_2 has a low vapor pressure, consistent with an oligomeric or polymeric structure in the solid state. Osmium trioxide difluoride was found to occur in three modifications: a monoclinic low-temperature phase ($\alpha\text{-}OsO_3F_2$, < 90 °C), an orthorhombic β-modification, which is stable between 90 and 130 °C, and a high-temperature γ-modification (> 130 °C) [58]. Raman spectroscopy indicates the presence of fluorine-bridged oligomers or polymers in all three modifications [58]. The structure of the monoclinic low-temperature α-modification was determined by single-crystal X-ray diffraction and contains zigzag chains of cis-fluorine-bridged fac-OsO_3F_3 moieties in which the bridging fluorine atoms are trans to the oxygen atoms (Fig. 14.2) [49]. The Raman and infrared spectra of matrix-isolated OsO_3F_2 were assigned to monomeric OsO_3F_2 of D_{3h} symmetry in which the three oxygen ligands are in equatorial positions and the fluorine ligands are axial, in accord with the predictions of the valence shell electron pair repulsion (VSEPR) model of molecular geometry [59].

More recently, OsO_3F_2 was shown to exhibit a pronounced Lewis acidity toward fluoride ion. While polymeric $\alpha\text{-}OsO_3F_2$ is insoluble in anhydrous HF, it is readily soluble in HF solutions of fluoride ion sources such as CsF and [N(CH$_3$)$_4$][F] (Eq. (8)).

$$OsO_3F_2 + [N(CH_3)_4][F] \rightarrow [N(CH_3)_4][fac\text{-}OsO_3F_3] \tag{8}$$

The reactions of OsO_3F_2 with alkali metal fluorides yield the [M][fac-OsO_3F_3] (M = Cs, Rb, K) salts. The reaction of OsO_3F_2 with liquid NOF between –78 and –60 °C yields [NO][fac-OsO_3F_3] according to Eq. (9) [30].

$$OsO_3F_2 + NOF \rightarrow [NO][fac\text{-}OsO_3F_3] \tag{9}$$

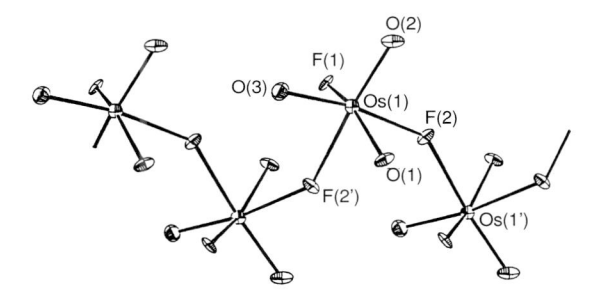

Os(1)-F(1)	187.9(1) pm	Os(1)-O(1)	172.7(1) pm
Os(1)---F(2)	212.6(1) pm	Os(1)-O(2)	168.8(1) pm
Os(1)---F(2')	210.8(1) pm	Os(1)-O(3)	167.8(1) pm

| O-Os-O | 101.3(5) - 104.2(5)° | F---Os---F | 79.6(3)° |
| F-Os---F | 77.4(4), 76.4(3)° | Os(1)---F(2)---Os(1') | 143.9(2)° |

Fig. 14.2 View of the $(OsO_3F_2)_\infty$ chain in the X-ray crystal structure of OsO_3F_2 [49].

Salts of the *fac*-$OsO_3F_3^-$ anion have been studied by vibrational spectroscopy [22, 30, 32], EXAFS [34], and ^{19}F NMR spectroscopy [30] (Tab. 14.2), and by single-crystal X-ray diffraction [30]. The crystal structure of $[N(CH_3)_4][\textit{fac-}OsO_3F_3]$ (Fig. 14.3) unambiguously established the facial geometry of the $OsO_3F_3^-$ anion that had been previously deduced from vibrational spectroscopy. The structure of $[N(CH_3)_4][\textit{fac-}OsO_3F_3]$ represents the first crystal structure of a trioxide trifluoride in a simple salt; the *fac*-$WO_3F_3^{3-}$ anion had been determined previously in the crystal structure of $Pb_5W_3O_9F_{10}$ [60].

Strong Lewis acids such as AsF_5 and SbF_5 are required to dissolve OsO_3F_2 in anhydrous HF (Eq. (10)) and yield intense yellow-orange (As) and yellow (Sb) solutions [61].

$$OsO_3F_2 + PnF_5 \xrightarrow{HF} [OsO_3F][PnF_6] \quad Pn = As, Sb \tag{10}$$

While excess AsF_5 is necessary to depolymerize $(OsO_3F_2)_\infty$ in anhydrous HF solvent, stoichiometric amounts of the stronger Lewis acid, SbF_5, are sufficient. Even under these superacidic conditions, "free" OsO_3F^+ is not present in HF solution or in the solid state. Rather, several HF solvates of the OsO_3F^+ cation have been isolated and characterized in the solid state by single-crystal X-ray diffraction. The crystal structures of orange $[OsO_3F][HF]_2[AsF_6]$ and yellow $[OsO_3F][HF][SbF_6]$ reveal $(FO_3Os-FH-FH-FAsF_5)_2$ dimers and $(FO_3Os-FH-FSbF_5)_\infty$ helical chains, respectively, with the osmium centers having coordination numbers of six and *fac*-trioxo arrangements (Fig. 14.4). The coordination number of six is achieved by fluorine-bridging to one anion and to one HF solvent molecule. With a large excess of Lewis acid, straw-yellow, unsolvated $[OsO_3F][PnF_6]$ (Pn = As, Sb) salts have been isolated and structurally characterized by X-ray crystallography. The crystal struc-

Tab. 14.2 NMR spectroscopic parameters for osmium oxide fluoride species

Os(VIII) Species	Solvent	T, °C	Counterion	$\delta(^{19}F)$, ppm [a]	$^1J(^{187}Os-^{19}F)$, Hz	$^2J(^{19}F-^{19}F)$, Hz	Ref.
OsO_3F^+	SbF_5	55	$Sb_nF_{5n+1}^-$	70.9(s)	–	–	61
$fac\text{-}OsO_3F_3^-$	CH_3CN	–20	$N(CH_3)_4^+$	–116.8(s)	32	–	30
$OsO_2F_3^+$	SbF_5	30	$Sb_nF_{5n+1}^-$	122.4(d)/129.5(t)	–	–	66
$cis\text{-}OsO_2F_4$ [b]	HF	30	–	15.8(t)/63.3(t)	35.1/59.4	138.3	50
$OsO_2F_5^-$	HF	–80	$N(CH_3)_4^+$	–6.5(qn)/–36.6(d)	–	93	67, 68

a) $\delta(^{19}F)$ with respect to $CFCl_3$; (s), (d), (t), (qn) denote singlet, doublet, triplet, and quintet, respectively.
b) $\delta(^{187}Os)$ of 1431 ± 10 ppm has been observed; $\delta(^{187}Os)$ ppm with respect to OsO_4 [50].

14.4 Osmium Oxide Fluoride Chemistry | 253

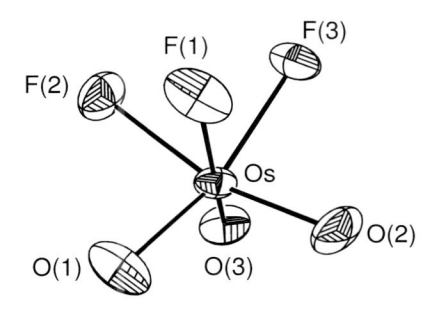

Os-F(1)	197(1) pm	Os-O(1)	170(1) pm
Os-F(2)	191(1) pm	Os-O(2)	172(1) pm
Os-F(3)	194(1) pm	Os-O(3)	173(1) pm

O-Os-O	101.2(7) - 102.8(8)°
F-Os-F	79.9(4) - 80.4(5)°
O-Os-F	86.5(6) - 89.4(5)°

Fig. 14.3 The $fac\text{-}OsO_3F_3^-$ anion in the X-ray crystal structure of $[N(CH_3)_4][fac\text{-}OsO_3F_3]$ [30].

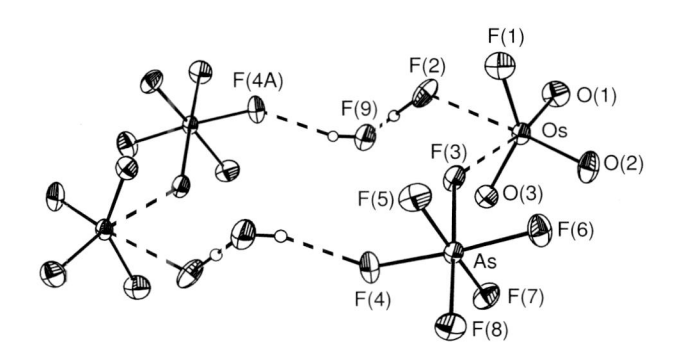

Os-F(1)	180.4(5) pm	Os-O(1)	169.4(6) pm
Os---F(2)	228.2(5) pm	Os-O(2)	166.9(6) pm
Os---F(3)	223.1(4) pm	Os-O(3)	171.9(6) pm

O-Os-O	102.4(3) - 103.5(3)°	F-Os---F	72.6(2), 74.1(2)°
O-Os-F	100.0(3) - 141.7(3)°	O-Os---F	74.1(2) - 87.5(2)°
	F(2)---Os---F(3) 82.3(2)°		

Fig. 14.4 The $(FO_3Os\text{-}FH\text{-}FH\text{-}FAsF_5)_2$ dimer in the X-ray crystal structure of $[OsO_3F][HF]_2[AsF_6]$ [61].

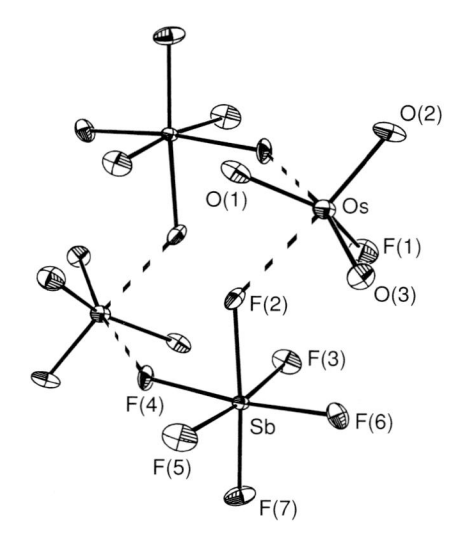

Os-F(1)	184.4(6) pm	Os-O(1)	170.8(7) pm
Os---F(2)	223.6(6) pm	Os-O(2)	168.1(8) pm
Os---F(4)	231.5(6) pm	Os-O(3)	169.6(7) pm

O-Os-O	102.2(3) - 103.9(4)°	F-Os---F	73.9(2), 75.3(2)°
O-Os-F	97.5(3) - 145.0(3)°	O-Os---F	77.8(3) - 88.9(3)°
	F---Os---F 79.2(3)°		

Fig. 14.5 The (FO$_3$Os–FSbF$_5$)$_2$ dimer in the X-ray crystal structure of [OsO$_3$F][SbF$_6$] [61].

tures of [OsO$_3$F][PnF$_6$] contain fluorine-bridged (FO$_3$Os–FPnF$_5$)$_2$ dimers in which osmium is six-coordinate with a *fac*-OsO$_3$F$_3$ arrangement (Fig. 14.5).

Addition of anhydrous HF to solid [OsO$_3$F][AsF$_6$] results in displacement of half of the AsF$_5$, yielding [μ-F(OsO$_3$F)$_2$][AsF$_6$] according to Eq. (11) [61], which

$$2[\text{OsO}_3\text{F}][\text{AsF}_6] \rightarrow [\mu\text{-F}(\text{OsO}_3\text{F})_2][\text{AsF}_6] + \text{AsF}_5 \tag{11}$$

has been characterized by Raman spectroscopy. Dissolution of OsO$_3$F$_2$ in neat SbF$_5$, a more fluoro-acidic medium, yielded a yellow solution according to Eq. (12) [61].

$$\text{OsO}_3\text{F}_2 + \text{SbF}_5 \rightarrow [\text{OsO}_3\text{F}][\text{Sb}_n\text{F}_{5n+1}] \tag{12}$$

The latter solution gave rise to a singlet in the ^{19}F NMR spectrum that was consistent with the formation of the OsO$_3$F$^+$ cation (Tab. 14.2). Below 55 °C, straw-yellow [OsO$_3$F][Sb$_3$F$_{16}$] crystallizes (Fig. 14.6). The weakly nucleophilic Sb$_3$F$_{16}^-$ anion and the absence of a donor solvent such as HF are necessary to stabilize an OsO$_3$F$^+$ cation having only weak contacts with the counter anion and tetrahedral coordination.

14.4 Osmium Oxide Fluoride Chemistry | 255

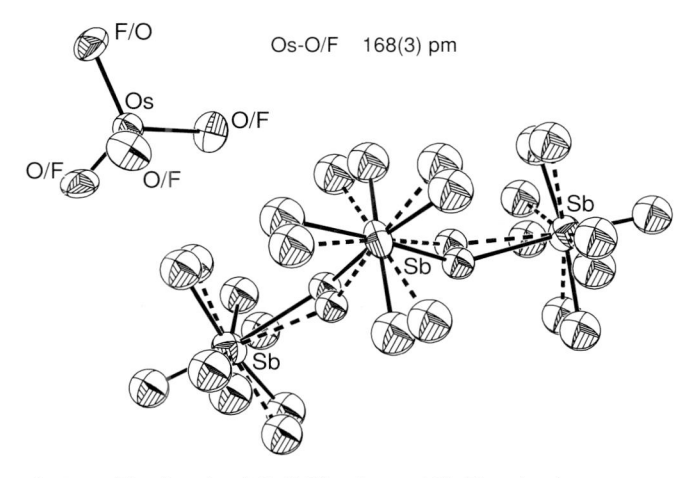

Fig. 14.6 The disordered OsO$_3$F$^+$ cation and Sb$_3$F$_{16}^-$ anion in the X-ray crystal structure of [OsO$_3$F][Sb$_3$F$_{16}$] [61].

14.4.2
Osmium Dioxide Tetrafluoride

A report of the preparation of OsOF$_6$ [62] by reaction of OsO$_4$ with KrF$_2$ in anhydrous HF was shown to be erroneous and the product that was isolated was shown to be *cis*-OsO$_2$F$_4$ [63], which is formed according to Eq. (13) [49, 50].

$$\text{OsO}_4 + 2\text{KrF}_2 \xrightarrow{\text{HF}} cis\text{-OsO}_2\text{F}_4 + 2\text{Kr} + \text{O}_2 \qquad (13)$$

Osmium dioxide tetrafluoride is a deep-magenta-colored solid with a vapor pressure of 1 Torr at room temperature [50]. The ^{19}F NMR spectrum of *cis*-OsO$_2$F$_4$ in anhydrous HF solvent shows an A$_2$X$_2$ coupling pattern (Fig. 14.7a and Tab. 14.2) [49, 50], consistent with a *cis*-dioxo isomer in solution. Vibrational spectroscopy established the *cis*-OsO$_2$F$_4$ form in the solid state. The crystal structure of *cis*-OsO$_2$F$_4$ has been reported but is severely disordered [49]. A gas-phase electron-diffraction study confirmed the presence of the *cis*-configuration in the gas phase and provided structural information [50]. The structure of *cis*-OsO$_2$F$_4$ is pseudo-octahedral with a *cis* arrangement of the two oxygen and four fluorine ligands (Fig. 14.7b).

The fluorination of RuO$_4$ has been attempted by the use of F$_2$ or KrF$_2$ as fluorinating agents in anhydrous HF solution, but only KrF$_2$ reacts with RuO$_4$ to form RuOF$_4$. The fluorination may proceed through RuO$_2$F$_4$ as an intermediate, by analogy with the reaction between KrF$_2$ and OsO$_4$ (Eq. (14)), followed by reduction to give RuOF$_4$ (Eq. (15)) [64, 65].

$$\text{RuO}_4 + 2\text{KrF}_2 \rightarrow \text{RuO}_2\text{F}_4 + 2\text{Kr} + \text{O}_2 \qquad (14)$$

$$\text{RuO}_2\text{F}_4 \rightarrow \text{RuOF}_4 + \tfrac{1}{2}\text{O}_2 \qquad (15)$$

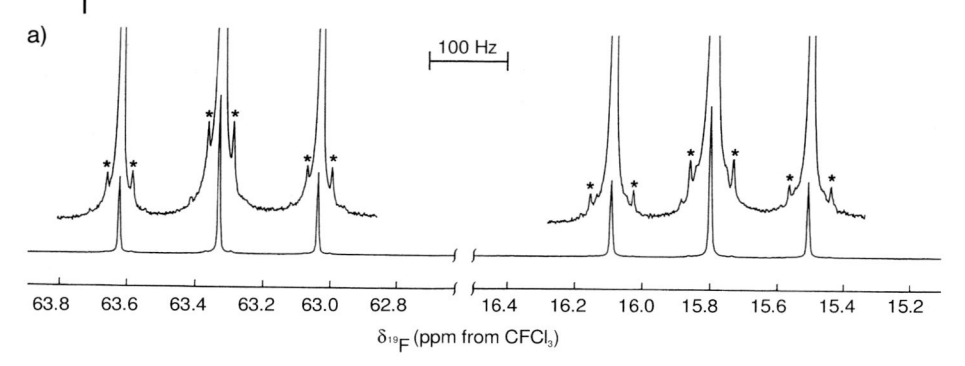

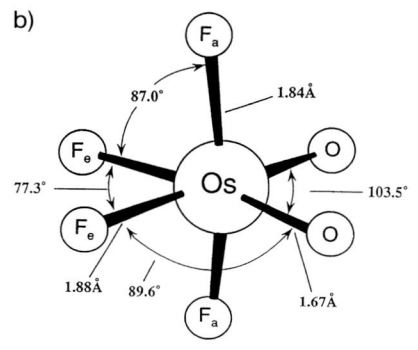

Fig. 14.7 (a) Fluorine-19 NMR spectrum of cis-OsO$_2$F$_4$ in HF at 30 °C. Asterisks denote natural abundance ^{187}Os satellites. (b) The gas-phase electron diffraction structure of cis-OsO$_2$F$_4$ [50].

The difference is attributed to the lower Ru–O bond energies compared with those of the Os–O bonds in cis-OsO$_2$F$_4$.

Dissolution of cis-OsO$_2$F$_4$ in the highly acidic medium, liquid SbF$_5$, results in yellow solutions of the OsO$_2$F$_3^+$ cation (Eq. (16)) [66].

$$\text{cis-OsO}_2\text{F}_4 + n\text{SbF}_5 \xrightarrow{\text{SbF}_s} [\text{OsO}_2\text{F}_3][\text{Sb}_n\text{F}_{5n+1}] \quad (16)$$

The ^{19}F NMR spectra of these solutions at 30 °C consist of a doublet and a triplet with relative intensities of 2:1 (Fig. 14.8a and Tab. 14.2) and are consistent with a trigonal-bipyramidal OsO$_2$F$_3^+$ cation (Structure **II**) in which one fluorine

Structure **II**

atom and both oxygen atoms are in equatorial positions and two fluorine atoms are in axial positions, as predicted by the VSEPR model of molecular geometry.

Osmium dioxide tetrafluoride also exhibits fluoride-ion donor properties towards the strong Lewis acids AsF$_5$ and SbF$_5$ in the more fluoro-basic solvent medium of anhydrous HF, forming orange solutions which are stable at room temperature [66]. Removal of the HF solvent yields [μ-F(cis-OsO$_2$F$_3$)$_2$][AsF$_6$] (Eq. (17)) and [μ-F(cis-OsO$_2$F$_3$)$_2$][Sb$_2$F$_{11}$] (Eq. (18)).

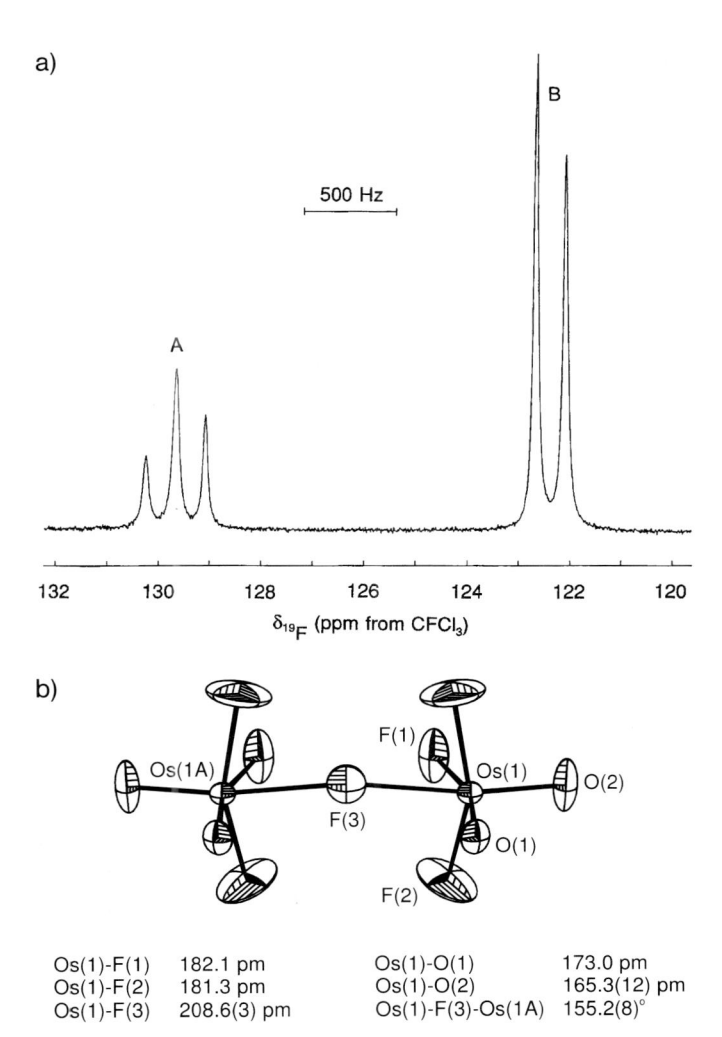

Os(1)-F(1)	182.1 pm	Os(1)-O(1)	173.0 pm
Os(1)-F(2)	181.3 pm	Os(1)-O(2)	165.3(12) pm
Os(1)-F(3)	208.6(3) pm	Os(1)-F(3)-Os(1A)	155.2(8)°

Fig. 14.8 (a) Fluorine-19 NMR spectrum of the OsO$_2$F$_3^+$ cation in neat SbF$_5$ solvent at 7 °C; A denotes the axial fluorine and B denotes the equatorial fluorine environments of the trigonal-bipyramidal cation (Structure II) and (b) the μ-F(OsO$_2$F$_3$)$_2^+$ cation in [μ-F(OsO$_2$F$_3$)$_2$][Sb$_2$F$_{11}$]; uncertainties are not provided for bond lengths that have been corrected for libration [66].

14 Osmium(VIII) Oxide and Oxide Fluoride Chemistry

$$2\ cis\text{-}OsO_2F_4 + AsF_5 \xrightarrow{HF,\ 25°C} [\mu\text{-}F(cis\text{-}OsO_2F_3)_2][AsF_6] \quad (17)$$

$$2\ cis\text{-}OsO_2F_4 + 2\ SbF_5 \xrightarrow{HF,\ 25°C} [\mu\text{-}F(cis\text{-}OsO_2F_3)_2][Sb_2F_{11}] \quad (18)$$

These salts contain the dinuclear $\mu\text{-}F(cis\text{-}OsO_2F_3)_2^+$ cation, which has been characterized in the X-ray crystal structure of $[\mu\text{-}F(cis\text{-}OsO_2F_3)_2][Sb_2F_{11}]$ (Fig. 14.8 b). Osmium in $\mu\text{-}F(cis\text{-}OsO_2F_3)_2^+$ retains hexacoordination, by fluorine-bridge formation between the two osmium centers, and the cis-dioxo arrangement observed in the neutral parent compounds.

Attempts to form the $OsOF_5^+$ cation by fluorination of $cis\text{-}OsO_2F_4$ with KrF_2 in the presence of AsF_5 in HF solvent resulted in autodecomposition of the $[KrF][AsF_6]$ that was generated *in situ* and in the formation of $[\mu\text{-}F(cis\text{-}OsO_2F_3)_2][AsF_6]$ according to Eq. (19) [66].

$$2\ cis\text{-}OsO_2F_4 + KrF_2 + AsF_5 \xrightarrow{HF,\ -20°C} [\mu\text{-}F(cis\text{-}OsO_2F_3)_2][AsF_6] + F_2 + Kr \quad (19)$$

A study of the fluoride-ion acceptor properties of $cis\text{-}OsO_2F_4$ yielded the $OsO_2F_5^-$ anion, according to Eq. (20) [67, 68].

$$cis\text{-}OsO_2F_4 + [N(CH_3)_4][F] \xrightarrow{CH_3CN} [N(CH_3)_4][OsO_2F_5] \quad (20)$$

For heptacoordinate compounds, structures based on three geometric types are possible (Fig. 14.9): (a) a monocapped octahedron, (b) a monocapped trigonal prism, and (c) a pentagonal bipyramid [69].

The ^{19}F NMR spectra of the NO^+ and $N(CH_3)_4^+$ salts of $OsO_2F_5^-$ are comprised of a doublet and a quintet with relative intensities of 4:1 (Tab. 14.2). This split-

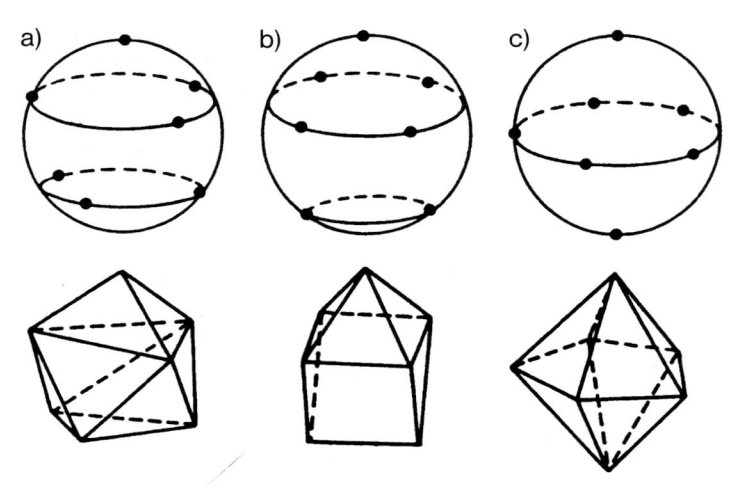

Fig. 14.9 The three possible stereochemistries for heptacoordination represented by points-on-a-sphere models and by their corresponding polyhedra: (a) monocapped octahedron, (b) monocapped trigonal prism, (c) pentagonal bipyramid [69].

ting pattern is only consistent with a monocapped trigonal-prismatic geometry having a *cis* arrangement of the two oxygen atoms and four fluorine atoms which form a square face that is, in turn, capped by a unique fluorine atom (Structure III). The $OsO_2F_5^-$ anion retains the *cis*-dioxo arrangement of its neutral precursor, *cis*-OsO_2F_4, providing the first example of a transition metal $AO_2F_5^{n-}$ system. Unlike the other known $AO_2F_5^{n-}$ species, $IO_2F_5^{2-}$ [70] and $UO_2F_5^{3-}$ [71], the geometry of $OsO_2F_5^-$ is not based on a pentagonal bipyramid, which is the preferred geometry for heptacoordinate main-group oxide fluorides [72].

Structure III

14.4.3
The *trans* Influence

The influence of multiply-bonded oxygen and nitrogen ligands on the lengths of *trans* metal–ligand single bonds has been extensively discussed in the literature [73, 74]. Besides elongation of the *trans* bonds, the most notable structural effect attributable to the *trans* influence is the exclusive existence of hexacoordinate d^0 transition metal dioxo, trioxo, and tetroxo species as the *cis*-, *fac*-, and *cis*-isomers, respectively. The dioxo, trioxo, and tetroxo species *cis*-OsO_2F_4 [49, 50], *cis*-$TcO_2F_4^-$ [75], *cis*-$ReO_2F_4^-$ [76], *cis*-$VO_2F_4^{2-}$ [77], *fac*-OsO_3F_3 [30], *fac*-$WO_3F_3^{3-}$ [60], and *cis*-$OsO_4F_2^{2-}$ [30] exemplify this generalization. This contrasts with the main-group anion $IO_2F_4^-$, which occurs as a mixture of *trans*- and *cis*-dioxo isomers that are thermodynamic and kinetic products, respectively [78, 79]. Osmium(VIII) oxide fluoride anions have presented ideal opportunities to investigate the factors that control their geometries because large numbers of oxygen and fluorine ligands do not result in high negative anion charges, making NMR spectroscopy in solution and crystal growth possible.

The *trans* influence in the dioxo isomers has been used to account for the exclusive existence of the *cis* isomer in terms of increased d_π-p_π orbital overlap when compared with the *trans*-dioxo isomer [75]. The d_{xy}, d_{xz}, and d_{yz} orbitals, the t_{2g} set of an octahedral complex, are of the correct symmetry to participate in π-bonding to π-donor ligands that are *cis* to each other, whereas in the *trans* isomer, only two of the t_{2g} orbitals have the correct symmetry for π-interaction. The facial trioxo isomer has three d orbitals, d_{xy}, d_{xz}, and d_{yz}, each of which can interact with two oxygen p_π orbitals, whereas in the meridional isomer, the p_π orbitals of three oxygen ligands compete for the single d orbital of the octahedral t_{2g} set that lies in the plane of the three oxygen ligands.

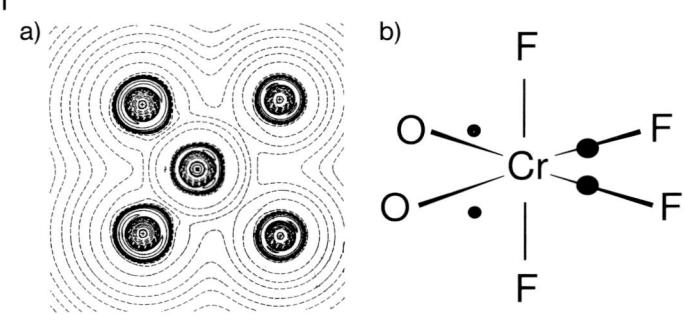

Fig. 14.10 (a) Contour map of the Laplacian, $L = -\nabla^2\rho(r)$, for cis-$CrO_2F_4^{2-}$ through the [O_2CrF_2] plane; (b) diagram showing the positions and relative sizes of the charge concentrations in the outer shell of the Cr core [80].

Gillespie and Bader [80] have shown that the oxygen–transition metal bond results in charge concentrations in the outer electron core opposite the double-bond domains that are significantly larger than that produced by a singly-bonded ligand such as fluoride. The contour map of the Laplacian of the electron density ($L=-\nabla^2\rho(r)$) for cis-$CrO_2F_4^{2-}$, which is isovalent with cis-OsO_2F_4 [49, 50], cis-$TcO_2F_4^-$ [75], and cis-$ReO_2F_4^-$ [76], is shown in Fig. 14.10 and clearly shows the local charge concentrations in the outer core of the central chromium atom. Doubly-bonded oxygens avoid these charge concentrations, so that cis-dioxo, fac-trioxo, and cis-tetroxo arrangements are favored, but trans-dioxo, mer-trioxo, and trans-tetroxo arrangements are not favored.

The trans influence also accounts for the propensity of fluorine bridges to occur trans to oxygen atoms in the infinite chain polymer $(OsO_2F_3)_\infty$ (Fig. 14.2) and in the μ-F(cis-OsO_2F_3)$_2^+$ cation (Fig. 14.8b). A further example of this behavior is provided by the $(FO_3Os–FPnF_5)_2$ dimers in the [OsO_3F][PnF_6] salts (Fig. 14.5), in which the PnF_6^- anions interact with the OsO_3F^+ cations by means of fluorine bridges. In [OsO_3F][HF]$_2$[AsF_5] (Fig. 14.4), such interactions occur between the Os(VIII) center of the cation and the fluorine atoms of HF molecules and AsF_6^- anions within the $(FO_3Os–FH–FH–FAsF_5)_2$ dimer. The X-ray crystal structures of μ-F(cis-ReO_2F_3)$_2^-$ [76], which is isoelectronic with μ-F(cis-OsO_2F_3)$_2^+$ [66], and of $Re_3O_6F_{10}$ [76], $(TcO_2F_3)_\infty$ [81], μ-F(trans-WOF_4)$_2^-$ [82], μ-F(trans-$ReOF_4$)$_2^+$ [83], μ-F(trans-$TcOF_4$)$_2^+$ [84], and $TcO_2F_3 \cdot SbF_5$ [85], reveal that the fluorine bridges of these species also occur exclusively trans to oxygen ligand atoms. The preference for fluorine bridging trans to oxygen atom ligands in these structures appears to be a consequence of the poor π-donor properties of fluorine and the fact that the bridging fluorine atoms must bear more negative charge than the terminal fluorine ligands. This charge build-up is reinforced when the strong π-donor oxygen atoms are trans to the fluorine bridges, so that a fluorine trans to oxygen must compete with the oxygen ligand for the same two $d_{t_{2g}}$ orbitals, resulting in a negligible $p(\pi) \rightarrow d(\pi)$ contribution from the bridging fluorine and a net build-up of negative charge on the fluorine ligand.

14.4.4
Computational Results and Bonding

Elements in high formal oxidation states, such as Os(VIII), represent extremely electron-poor centers having high effective electronegativities. The calculated Mulliken charges of osmium in Os(VIII) oxide fluorides are between 1.5 and 2 [30, 61, 66] and, as anticipated, are considerably less than the formal oxidation state of osmium, indicating a significant degree of covalency as a consequence of the high effective electronegativity of the Os(VIII) center. Calculated Mayer valencies for formally octavalent osmium are close to 6, illustrating that a completely covalent bonding description is also not satisfactory [30, 61, 66].

Density functional [30, 61, 66], Hartree-Fock [86], and MP2 theoretical studies [86] of osmium oxide fluorides have shown that the Os–O bonds have significantly higher covalent character when compared with the highly ionic Os–F bonds. This distinction is most apparent in the OsO_4F^- and cis-$OsO_4F_2^{2-}$ anions, for which the calculated Os–F Mayer bond orders are approximately 0.5 [30]. The ratios of calculated Os–O and Os–F bond orders in the OsO_4F^- and cis-$OsO_4F_2^{2-}$ anions are close to 3:1, which is reflected in the long Os–F bonds found in the crystal structure of $[N(CH_3)_4][OsO_4F]$ (2.075(9) Å) and calculated for the cis-$OsO_4F_2^{2-}$ anion (2.078 Å). The strongest Os(VIII)–F bond is found in the OsO_3F^+ cation, with a calculated Os–O Mayer bond order that is approximately double that of the Os–F bond, reproducing the relative bond strengths that are predicted from a simple Lewis description comprising three Os–O double bonds and an Os–F single bond [61].

Substitution of oxygen by fluorine atoms in the series OsO_4–OsO_3F_2–OsO_2F_4 leads to an increase in the Os–O bond order [86]. At the same time, the positive charge on Os increases [30]. Both findings are a consequence of electron density withdrawal from the osmium center by increasing number of the fluorine ligands over the series. Upon substituting one oxygen atom by one fluorine atom in the series, OsO_4–OsO_3F^+ and cis-$OsO_4F_2^{2-}$–$OsO_3F_3^-$, the Os–O/F bond orders and positive charge on the osmium atom increase even more dramatically, and, as a result of a decrease in the net negative charge of the species, the bond polarities are reduced. The bond orders and charges have not been calculated for cis-OsO_2F_4 at the same level of theory, but are expected to follow the trends seen in the latter series.

In a computational study of neutral Os(VIII) oxygen and fluorine compounds using ab initio methods at the Hartree-Fock and MP2 levels [86], no energy-minimized geometry was obtained for $OsOF_6$, which dissociated during the geometry optimization. Osmium octafluoride, on the other hand, has a calculated D_{2d} geometry based on a distorted Archimedean (square) antiprism [86]. Although OsF_8 has a negative calculated heat of formation, reductive elimination of F_2 from this molecule is predicted to be a facile decomposition pathway. The Os–F bonds of OsF_8 achieve maximum ionicity when compared with those in the osmium(VIII) oxide fluorides, favoring its reductive decomposition. The position of the bond critical points, in terms of the atoms-in-molecules (AIM) theory, suggests that the effective electronegativity of Os(VIII) may be higher than that of oxygen and fluorine [86].

14.4.5
NMR Spectroscopy

Fluorine-19 NMR spectroscopy has been the primary method of structural characterization for Os(VIII) compounds in solution, with $^2J(^{19}F-^{19}F)$ couplings providing the gross geometries of cis-OsO_2F_4 [49, 50], $OsO_2F_3^+$ [66], and $OsO_2F_5^-$ [67, 68] in solution (Tab. 14.2).

Osmium has two spin-active nuclei, ^{187}Os (1.64% natural abundance, $I=1/2$) and ^{189}Os (16.1% natural abundance, $I=3/2$). Osmium-189, which has a large quadrupole moment and line width factor, has therefore only proven feasible for acquiring an NMR signal in the case of highly symmetric OsO_4 [87, 88]. The spin-1/2 nuclide ^{187}Os has one of the lowest receptivities of any NMR nuclide and an extremely low resonance frequency, which dramatically limit its use as an NMR probe. Thus, ^{187}Os NMR spectroscopy has not been widely exploited for the characterization of osmium compounds and the ^{187}Os resonance of OsO_4 is the only ^{187}Os resonance that has been directly observed [88, 89]. Several studies have used inverse two-dimensional (^1H, ^{187}Os and ^{19}F, ^{187}Os) NMR experiments to obtain ^{187}Os chemical shifts and coupling constants to ^{187}Os [50, 90–93]. Among osmium oxide fluorides, this method has only been used to characterize cis-OsO_2F_4 [50] (Tab. 14.2) because of the low-frequency requirements for NMR probes. When broadening arising from exchange or solvent viscosity is minimized, ^{187}Os satellites can be observed in the ^{19}F NMR spectra of oxide fluorides such as cis-OsO_2F_4 (Fig. 14.7a) [50] and fac-$OsO_3F_3^-$ [30] (Tab. 14.2), unambiguously confirming the presence of an osmium fluoride species. The $^1J(^{187}Os-^{19}F)$ couplings in oxide fluorides are small (<60 Hz) and, therefore, difficult to observe owing to the low intensities of the ^{187}Os satellites.

The ^{19}F chemical shifts of Os(VIII) oxide fluorides range from 129.5 ppm ($OsO_2F_3^+$) to –116.8 ppm (fac-$OsO_3F_3^-$). The ^{19}F shieldings of trioxo- and dioxo-Os(VIII) species increase in the order cation < neutral oxide fluoride < anion, reflecting the increased polarity of the Os–F bond with increasing net negative charge of the species.

14.5
Outlook

The investigation of the Lewis acid properties of OsO_2F_4 and OsO_3F_2 towards neutral bases, in particular nitrogen bases that are resistant to oxidative attack, offers promise for the syntheses of new examples of neutral high-coordinate Os(VIII) species. The Lewis acid properties of cis-OsO_2F_4 and OsO_3F_2 towards CH_3CN in solution and in the solid state are currently under investigation [67]. Preliminary findings indicate a mer/fac-isomerism of the $OsO_3F_2 \cdot CH_3CN$ adduct and a structure for the cis-$OsO_2F_4 \cdot CH_3CN$ adduct that is derived from the monocapped trigonal prism of $OsO_2F_5^-$ by replacement of a non-capping fluorine by the nitrogen donor atom of CH_3CN [67].

With the availability of more sensitive NMR probes and improved pulse sequences, the ^{187}Os nuclide may become more available for use in structural elucidation in solution. Polarization-transfer techniques show the greatest promise for increasing the use of this low-receptivity nuclide. The more extensive use of solid-state NMR spectroscopy for the characterization of inorganic fluorides is currently under development [94] and should prove of general use for the characterization of Os(VIII) oxide and oxide fluoride species in the solid state.

Attempts to prepare the highest oxide fluoride, $OsOF_6$, have thus far failed [62, 63]. Advances in matrix-isolation techniques may make possible the vibrational spectroscopic identification of this elusive oxide fluoride and OsF_8. The $OsO_2F_3^+$, μ-$F(cis$-$OsO_2F_3)_2^+$, OsO_3F^+, and μ-$F(OsO_3F)_2^+$ cations represent rare examples of metal oxide fluoride cations. The syntheses of $OsOF_6$ and OsF_8 would afford a further extension of high oxidation state metal fluoride and oxide fluoride cation chemistry by providing possible routes to the $OsOF_5^+$ and OsF_7^+ cations.

The oxide fluoride chemistry of Os(VIII) is expected to continue to enhance our understanding of the factors that determine stereochemistry in high-coordinate species. Although the chemistry of Xe(VIII) is presently more limited than that of Os(VIII), the comparison of their chemistries will provide impetus to the further development of Xe(VIII) chemistry and a fuller assessment of the extent of d orbital participation in the bonding of main-group compounds. The presently known and future Os(VIII) species serve as guide posts to the syntheses of new Xe(VIII) species and it may be reasonable to speculate on the possible existence of $XeO_2F_5^-$, $XeO_3F_3^-$ [67], XeO_3F^+, $XeO_2F_3^+$, $XeOF_5^+$ [95, 96], and XeF_7^+ [97] in the light of recent developments in Os(VIII) chemistry.

Acknowledgements

We thank the donors of the Petroleum Research Fund, administered by the American Chemical Society, for support of aspects of the work reviewed in this article that have been carried out at McMaster University under ACS-PRF No. 28284-AC3. We also wish to thank Dr. Hélène P. A. Mercier for her assistance in the preparation of this manuscript.

References

1. H. V. Wartenberg, *Annalen* **1924**, *440*, 97.
2. H. C. Jewiss, W. Levason, M. Tajik, M. Webster, N. P. C. Walker, *J. Chem. Soc., Dalton Trans.* **1985**, 199.
3. R. E. Connick, C. R. Hurley, *J. Am. Chem. Soc.* **1952**, *74*, 5012.
4. S. R. Gunn, *J. Am. Chem. Soc.* **1965**, *87*, 2290.
5. H. Selig, H. H. Claassen, C. L. Chernick, J. G. Malm, J. L. Huston, *Science* **1964**, *143*, 1322.
6. M. Gerken, G. J. Schrobilgen, *Inorg. Chem.* **2002**, *41*, 198.
7. S. Aoyama, K. Watanabe, *Nippon Kagaku Zasshi* **1955**, *76*, 970; ref. (559) in *Comprehensive Coordination Chemistry* (Ed. G. Wilkinson), Vol. 4, Pergamon Press, Oxford, 1987, pp. 589.
8. D. R. Stull, *Ind. Eng. Chem.* **1947**, *39*, 540.
9. G. Brauer, *Handbook of Preparative Inorganic Chemistry*, Vol. 2, 2nd Ed., Academic Press, New York, **1965**, pp. 1603.
10. R. J. Colin, J. Jones, W. P. Griffith, *J. Chem. Soc., Dalton Trans.* **1974**, 1094.
11. M. Schröder, *Chem. Rev.* **1980**, *80*, 187.
12. H. C. Kolb, M. S. VanNieuwenhuizen, K. B. Sharpless, *Chem. Rev.* **1994**, *94*, 2483.
13. S. Ahlert, W. Klein, O. Jepsen, O. Gunnarsson, O. K. Andersen, M. Jansen, *Angew. Chem.* **2003**, *115*, 4458; *Angew. Chem. Int. Ed.* **2003**, *42*, 4322.
14. S. Ahlert, L. Diekhöner, R. Sordan, K. Kern, M. Jansen, *J. Chem. Soc., Chem. Commun.* **2004**, 462.
15. J. A. Ibers, W. C. Hamilton, D. R. Mackenzie, *Inorg. Chem.* **1964**, *3*, 1412.
16. A. Zalkin, J. D. Forrester, D. H. Templeton, *Inorg. Chem.* **1964**, *3*, 1417.
17. A. Zalkin, J. D. Forrester, D. H. Templeton, S. M. Williamson, *J. Am. Chem. Soc.* **1964**, *86*, 3569.
18. M. J. Cleare, P. C. Hydes, W. P. Griffith, M. J. Wright, *J. Chem. Soc., Dalton Trans.* **1977**, 941.
19. W. P. Griffith, A. C. Skapski, K. A. Woode, M. J. Wright, *Inorg. Chim. Acta* **1978**, *31*, L413.
20. E. Z. Fritzmann, *Anorg. Chem.* **1928**, *172*, 213.
21. F. Krauss, D. Wilken, *Z. Anorg. Allg. Chem.* **1925**, *145*, 151.
22. W. P. Griffith, *J. Chem. Soc.* **1964**, 245.
23. W. P. Griffith, *J. Chem. Soc. A* **1969**, 211.
24. N. N. Nevskii, B. N. Ivanov-Emin, N. A. Nevskaya, *Dokl. Akad. Nauk SSSR* **1982**, *266*, 628.
25. N. N. Nevskii, B. N. Ivanov-Emin, N. A. Nevskaya, N. V. Belov, *Dokl. Akad. Nauk SSSR* **1982**, *266*, 1138.
26. N. N. Nevskii, M. A. Porai-Koshits, *Dokl. Akad. Nauk SSSR* **1983**, *270*, 1392.
27. N. N. Nevskii, M. A. Porai-Koshits, *Dokl. Akad. Nauk SSSR* **1983**, *272*, 1123.
28. B. N. Ivanov-Emin, N. A. Nevskaya, B. E. Zaitsev, N. N. Nevskii, A. S. Izmailovich, *Russ. J. Inorg. Chem.* **1984**, *29*, 710; *Zh. Neorg. Khim.* **1984**, *29*, 1241.
29. A. J. Bailey, M. G. Bhowon, W. P. Griffith, A. G. F. Shoir, A. J. P. White, D. J. Williams, *J. Chem. Soc., Dalton Trans.* **1997**, 3245.
30. M. Gerken, D. A. Dixon, G. J. Schrobilgen, *Inorg. Chem.* **2000**, *39*, 4244.
31. R. Weber, K. Dehnicke, U. Müller, D. Fenske, *Z. Anorg. Allg. Chem.* **1984**, *516*, 214.
32. P. J. Jones, W. Levason, M. Tajik, *J. Fluorine Chem.* **1984**, *25*, 195.
33. B. N. Ivanov-Emin, N. A. Nevskaya, Yu. N. Medvedev, B. E. Zaitsev, I. V. Lin'ko, *Russ. J. Inorg. Chem. (Engl. Trans.)* **1986**, *31*, 1088; *Zh. Neorg. Khim.* **1986**, *31*, 1889.
34. S. A. Brewer, A. K. Brisdon, J. H. Holloway, E. G. Hope, W. Levason, J. S. Ogden, A. K. Saad, *J. Fluorine Chem.* **1993**, *60*, 13.
35. A. B. Nikol'skii, Yu. I. D'yachenko, *Russ. J. Inorg. Chem.* **1974**, *19*, 1031; *Zh. Neorg. Khim.* **1974**, *19*, 1889.
36. W. P. Griffith, T. Y. Koh, A. J. P. White, D. J. Williams, *Polyhedron* **1995**, *14*, 2019.
37. D. W. Nelson, A. Gypser, P. T. Ho, H. C. Kolb, T. Kondo, H.-L. Kwong, D. V. McGrath, A. E. Rubin, P. O. Norrby, K. P. Gable, K. B. Sharpless, *J. Am. Chem. Soc.* **1997**, *119*, 1840.

38. J. S. Svendsen, I. Marko, E. N. Jacobsen, C. P. Rao, S. Bott, K. B. Sharpless, *J. Org. Chem.* **1989**, *54*, 2264.
39. E. J. Corey, S. Sarshar, M. D. Azimioara, R. C. Newbold, M. C. Noe, *J. Am. Chem. Soc.* **1996**, *118*, 7851.
40. A. F. Clifford, C. S. Kobajashi, *Inorg. Synth.* **1960**, *6*, 204.
41. W. A. Nugent, R. L. Harlow, R. J. McKinney, *J. Am. Chem. Soc.* **1979**, *101*, 7265.
42. D. E. Wigley, *Prog. Inorg. Chem.* **1994**, *42*, 239.
43. W. P. Griffith, N. T. McManus, A. C. Skapski, A. D. White, *Inorg. Chim. Acta* **1985**, *105*, L11.
44. W. H. Leung, J. L. C. Chim, W. T. Wong, *J. Chem. Soc., Dalton Trans.* **1996**, 3153.
45. W. H. Leung, J. L. C. Chim, W. T. Wong, *J. Chem. Soc., Dalton Trans.* **1997**, 3277.
46. A. O. Chong, K. Oshima, K. B. Sharpless, *J. Am. Chem. Soc.* **1977**, *99*, 3420.
47. A. A. Danopoulos, G. Wilkinson, *Polyhedron* **1990**, *9*, 1009.
48. D. W. H. Rankin, H. E. Robertson, A. A. Danopoulos, P. D. Lyne, D. M. P. Mingos, G. Wilkinson, *J. Chem. Soc., Dalton Trans.* **1994**, 1563.
49. R. Bougon, B. Buu, K. Seppelt, *Chem. Ber.* **1993**, *126*, 1331.
50. K. O. Christe, D. A. Dixon, H. G. Mack, H. Oberhammer, A. Pagelot, J. C. P. Sanders, G. J. Schrobilgen, *J. Am. Chem. Soc.* **1993**, *115*, 11279.
51. O. Ruff, R. W. Tschirch, *Ber.* **1913**, *46*, 929.
52. B. Weinstock, E. E. Weaver, E. P. Knop, *Inorg. Chem.* **1966**, *5*, 2189.
53. B. Weinstock, J. G. Malm, *J. Am. Chem. Soc.* **1958**, *80*, 4466.
54. O. Glemser, H. W. Roesky, K. H. Helberg, H. U. Werther, *Chem. Ber.* **1966**, *99*, 2652.
55. J. Huston, *Inorg. Chem.* **1982**, *21*, 685.
56. W. A. Sunder, F. A. Stevie, *J. Fluorine Chem.* **1975**, *6*, 449.
57. M. A. Hepworth, P. L. Robinson, *J. Inorg. Nucl. Chem.* **1957**, *4*, 24.
58. M. M. Nguyen-Nghi, N. Bartlett, *C. R. Séances Acad. Sci.* **1969**, *269*, 756.
59. I. R. Beattie, H. E. Blayden, R. A. Crocombe, P. J. Jones, J. S. Ogden, *J. Raman Spectrosc.* **1976**, *4*, 313.
60. S. C. Abrahams, P. Marsh, J. Ravez, *J. Chem. Phys.* **1987**, *87*, 6012.
61. M. Gerken, D. A. Dixon, G. J. Schrobilgen, *Inorg. Chem.* **2002**, *41*, 259.
62. R. Bougon, *J. Fluorine Chem.* **1991**, *53*, 419.
63. K. O. Christe, R. Bougon, *J. Chem. Soc., Chem. Commun.* **1992**, 1056.
64. L. Meublat, M. Lance, R. Bougon, *Can. J. Chem.* **1989**, *67*, 1729.
65. R. Bougon, W. V. Cicha, J. Isabey, *J. Fluorine Chem.* **1994**, *67*, 271.
66. W. J. Casteel, Jr., D. A. Dixon, H. P. A. Mercier, G. J. Schrobilgen, *Inorg. Chem.* **1996**, *35*, 4310.
67. M. Gerken, Ph.D. Thesis, McMaster University, **2000**.
68. M. Gerken, G. J. Schrobilgen, to be published.
69. R. J. Gillespie, I. Hargittai, *The VSEPR Model of Molecular Geometry*, Allyn and Bacon, Needham Heights, Massachusetts, 1991, pp. 58–60.
70. J. A. Boatz, K. O. Christe, D. A. Dixon, B. A. Fir, M. Gerken, R. Z. Gnann, H. P. A. Mercier, G. J. Schrobilgen, *Inorg. Chem.* **2003**, *42*, 5282.
71. W. H. Zachariasen, *Acta Crystallogr.* **1954**, *7*, 783.
72. K. O. Christe, D. A. Dixon, A. R. Mahjoub, H. P. A. Mercier, J. C. P. Sanders, G. J. Schrobilgen, W. W. Wilson, *J. Am. Chem. Soc.* **1993**, *115*, 2686.
73. E. M. Shustorovich, M. A. Porai-Koshits, Y. A. Buslaev, *Coord. Chem. Rev.* **1975**, *17*, 1.
74. E. M. Shustorovich, Y. A. Buslaev, A. Yu, *Inorg. Chem.* **1976**, *15*, 1142.
75. W. J. Casteel, Jr., D. A. Dixon, N. LeBlond, H. P. A. Mercier, G. J. Schrobilgen, *Inorg. Chem.* **1998**, *37*, 340.
76. W. J. Casteel, Jr., D. A. Dixon, N. LeBlond, P. E. Lock, H. P. A. Mercier, G. J. Schrobilgen, *Inorg. Chem.* **1999**, *38*, 2340.
77. M. Leimkühler, R. Mattes, *J. Solid State Chem.* **1986**, *65*, 260.
78. A. Engelbrecht, P. Peterfy, E. Schandara, *Z. Anorg. Allg. Chem.* **1971**, *384*, 202.
79. K. O. Christe, R. D. Wilson, C. J. Schack, *Inorg. Chem.* **1981**, *20*, 2104.

80 R. J. Gillespie, I. Bytheway, T. H. Tang, R. F. W. Bader, *Inorg. Chem.* **1996**, *35*, 3954.
81 H. P. A. Mercier, G. J. Schrobilgen, *Inorg. Chem.* **1993**, *32*, 145.
82 B. F. Hoskins, A. Linden, T. O'Donnell, *Inorg. Chem.* **1987**, *26*, 2223.
83 G. J. Schrobilgen, J. H. Holloway, D. R. Russell, *J. Chem. Soc., Dalton Trans.* **1984**, 1411.
84 D. A. Dixon, N. LeBlond, H. P. A. Mercier, G. J. Schrobilgen, *Inorg. Chem.* **2000**, *39*, 4494.
85 N. LeBlond, G. J. Schrobilgen, *Inorg. Chem.* **2000**, *39*, 2473.
86 A. Veldkamp, G. Frenking, *Chem. Ber.* **1993**, *126*, 1325.
87 A. Schwenk, G. Zimmermann, *Phys. Lett. Sect. A* **1968**, *26*, 258.
88 C. Brevard, P. Granger, *Handbook of High Resolution Multinuclear NMR*, Wiley-Interscience, New York, **1981**, pp. 196–197.
89 A. Schwenk, *Z. Phys.* **1968**, *213*, 482.
90 R. Benn, E. Joussen, H. Lehmkuhl, F. L. Ortiz, A. Rufinska, *J. Am. Chem. Soc.* **1989**, *111*, 8754.
91 R. Benn, H. Brenneke, E. Joussen, H. Lehmkuhl, F. L. Ortiz, *Organometallics* **1990**, *9*, 756.
92 R. I. Michelman, G. E. Ball, R. G. Bergman, R. A. Andersen, *Organometallics* **1994**, *13*, 869.
93 A. G. Bell, W. Kózmínski, A. Linden, W. von Philipsborn, *Organometallics* **1996**, *15*, 3124.
94 M. Gerken, P. Hazendonk, J. Nieboer, G. J. Schrobilgen, *J. Fluorine Chem.* **2004**, *125*, 1163.
95 J. H. Holloway, G. J. Schrobilgen, *J. Chem. Soc., Chem. Commun.* **1975**, 623.
96 D. E. McKee, C. J. Adams, A. Zalkin, N. Bartlett, *J. Chem. Soc., Chem. Commun.* **1973**, 26.
97 K. O. Christe, D. A. Dixon, J. C. P. Sanders, G. J. Schrobilgen, W. W. Wilson, *J. Am. Chem. Soc.* **1993**, *115*, 9461.

15
Liquid-Crystalline Lanthanide Complexes

Koen Binnemans

15.1
Introduction

Liquid crystals are important materials in display technology, because they are the key components of liquid crystal displays (LCDs). Liquid crystals can be considered as liquids with the properties of an anisotropic crystal, which means that many physical properties of liquid crystals depend on the direction in which these are measured [1]. The possibility to switch liquid crystals by applying an electric field is of importance for their use in LCDs [2]. The design of new liquid crystals is still an active research field, because molecules with an unconventional molecular shape give rise to exotic liquid crystals [3]. Metal-containing liquid crystals are known as *metallomesogens*. These liquid-crystalline metal complexes combine the properties of both liquid crystals (fluidity, anisotropic physical properties, switchable in external electric and magnetic fields) and transition metals (redox behavior, unique spectroscopic and magnetic properties). Several reviews give an overview of the field of metallomesogens [4].

Metallomesogens that contain rare-earth or lanthanide ions have been much less extensively investigated than metallomesogens that contain d-group transition metals. Because some of the trivalent lanthanide ions have a large magnetic anisotropy, lanthanide-containing liquid crystals can be aligned by weaker magnetic fields than diamagnetic or even other paramagnetic liquid crystals. Lanthanides often show an intense photoluminescence and, in contrast to the fluorescence of organic molecules, the emission bands are very narrow for lanthanide compounds. Therefore, liquid-crystalline lanthanide complexes can be used as luminescent liquid crystals. It should be realized that the magnetic and spectroscopic properties of these materials strongly depend on the nature of the lanthanide ion, so that for a given application a proper choice of the lanthanide ion is necessary. Lanthanide-containing liquid crystals and surfactants have been reviewed by Binnemans and Görller-Walrand [5].

This chapter describes one of the most extensively investigated classes of liquid-crystalline lanthanide complexes, namely the Schiff's base complexes, and the magnetic properties of these complexes. Secondly, this chapter is intended

Inorganic Chemistry in Focus II. Edited by G. Meyer, D. Naumann, and L. Wesemann
Copyright © 2005 WILEY-VCH Verlag GmbH & Co. KGaA, Weinheim
ISBN: 3-527-30811-3

to give an update of the review of Binnemans and Görller-Walrand, in the sense that new developments in the field after 2001 are described.

15.2
Mesophases

The difference between crystals and liquids is that the molecules in a crystal are ordered in a three-dimensional lattice, whereas in a liquid they are not. A molecular crystal consists of a more or less rigid arrangement of the molecules, which possess both positional and orientational order. The molecules are constrained to occupy specific sites in the lattice. At low temperatures, attractive intermolecular forces in a crystal are strong enough to hold the molecules firmly in place. When a crystalline compound is heated, their thermal motion increases and eventually becomes so strong that the intermolecular forces cannot hold the molecules in place, so that the solid melts. The regular arrangement of molecules is broken down with the loss of long-range orientational and positional order to give a disordered isotropic liquid. However, this melting process, which transforms a compound from being highly ordered to being totally disordered in one step, is not universal for all types of compounds. Some phases have more order than is present in liquids but less order than typical for molecular crystals. Compounds that exhibit such phases are called *liquid crystals*, since they share properties generally associated with both liquids and crystals. A more proper name for a liquid-crystal molecule is *mesogen* and the phases it forms are known as *mesophases*. The motion of the molecules in liquid-crystalline phases is comparable with that of the molecules in a liquid, but the molecules maintain some degree of orientational order and sometimes some positional order as well (Fig. 15.1). A vector, called the *director* ($\tilde{n}$) of the liquid crystal, represents the orientation of the molecules.

A number of different types of molecules form liquid-crystalline phases. What most of them have in common is that their physical properties are anisotropic. Either the molecular shape is such that the length of one molecular axis is very different from the other two, or different parts of the molecules have very differ-

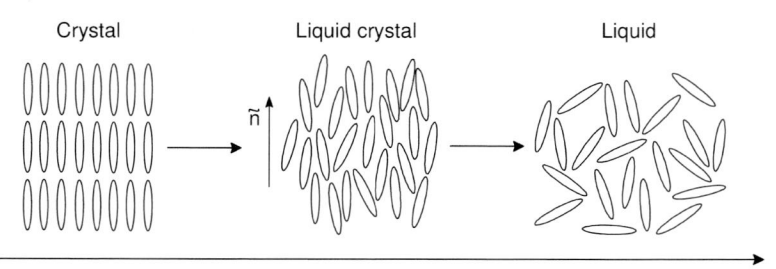

Fig. 15.1 Schematic representation of the melting behavior of a liquid crystal.

ent solubility properties (amphiphilic compounds with a hydrophobic tail and a hydrophilic head group). In the former case, a mesophase can be formed by heating and/or cooling the compounds, and these compounds are called *thermotropic liquid crystals*. In the latter case, a solvent (in most cases water) causes the formation of a mesophase. A mesophase formed in the presence of a solvent is a *lyotropic mesophase* and the compounds that form these mesophases are called *lyotropic liquid crystals*.

When a thermotropic liquid crystal is heated, it goes from the crystalline state into the liquid-crystalline state at a point called the *melting point*. By further heating, the birefringent (anisotropic) liquid crystal is transformed at the *clearing point* into an isotropic liquid: the birefringent liquid becomes clear and all molecular order is lost. When the thermotropic liquid crystal forms a mesophase in both the heating and the cooling process, the compound is called an *enantiotropic* liquid crystal. Thermodynamically unstable mesophases, which only appear in the cooling process at a temperature below the melting point, are referred to as *monotropic* liquid crystals. Thermotropic liquid crystals are generally divided into two main groups, depending on their structural features: *calamitic* mesogens (formed by rod-like molecules) and *discotic* mesogens (formed by disk-like molecules).

Rod-like molecules can form a nematic phase or one of the different types of smectic phases. The least ordered mesophase is the *nematic phase* (N). The molecules in the nematic phase possess long-range orientational order, but lack positional order. The molecules align with their long molecular axes more or less parallel to a preferred direction indicated by the *director* (ñ) (Fig. 15.2). They can move freely within the nematic phase and are able to rotate about their long molecular axes. The nematic phase is technologically the most important of the many different types of mesophases. It is used in virtually all commercially available liquid crystal displays (LCDs).

Smectic phases show a higher degree of order than the nematic phase. There is of course orientational order, but also some positional order. The molecules are not only oriented with their long molecular axes in one direction, but they are also stacked on top of one another in layers. Within the layers there can be some positional order. A number of smectic phases exist, and these phases differ from one another in the degree of order present both within and between the layers. In the *smectic A* phase (SmA), the molecules are aligned with their long molecular axes parallel to the layer normal (i.e. the director is perpendicular to the layer planes), but there is no positional order within the layer. In the

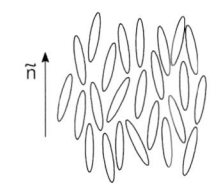

Fig. 15.2 Molecular order in the nematic phase.

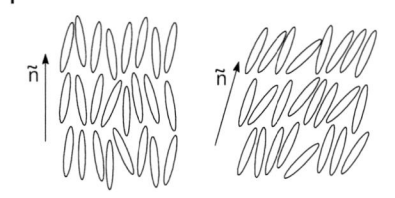

Fig. 15.3 Molecular order in the SmA phase (left) and in the SmC phase (right).

smectic C phase (SmC), the long molecular axis is tilted with respect to the normal to the layer planes. The arrangement of the molecules in the SmA and SmC phases is schematically represented in Fig. 15.3. SmA and SmC phases are the least ordered smectic phases and are also the most commonly observed. Due to the molecular mobility inherent in these phases and their relatively low viscosities, they are called *true smectic phases*. Over the years, many smectic phases with ordering within the layers have been discovered, but they are less common than SmA and SmC phases.

Most of the mesogenic molecules with a disk-like shape tend to be stacked one on top of the other into columns, and these columns are arranged side-to-side according to a two-dimensional lattice. There are several types of columnar mesophases, depending on the symmetry of the two-dimensional lattice (hexagonal, tetragonal or rectangular) and depending on the order or disorder of the molecular stacking within the columns (ordered and disordered columnar mesophases). Fig. 15.4 shows the molecular arrangement in the *hexagonal columnar phase* (Col_h). If the molecules in the columns are ordered, the symbol is further specified as Col_{ho}, disordered as Col_{hd}, and tilted as Col_{ht}. The Col_{ho} phase is best defined as a disordered crystal mesophase because of the long-range positional order of the molecules in three dimensions (thus also within the column). However, the Col_{hd} phase can be defined as being truly liquid-crystalline because of the disordered arrangement of the molecules within the columns.

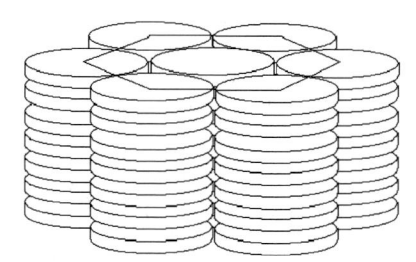

Fig. 15.4 Schematic representation of a hexagonal columnar mesophase.

15.3
Lanthanide Complexes with Schiff's Base Ligands

Because Schiff's base complexes have been used so often in the past for the design of metallomesogens, and because the trivalent lanthanides are known to form complexes with these ligands, it was expected that Schiff's bases could be used to prepare liquid-crystalline lanthanide complexes. The first calamitic lanthanide-containing liquid crystals were reported in 1991 by Galyametdinov and coworkers [6]. These authors described the synthesis and thermal behavior of lanthanide complexes with the N-alkyl salicylaldimine ligand shown in Fig. 15.5 ($R=C_7H_{15}$, $R'=C_{12}H_{25}$). The stoichiometry of the complexes was believed to be [LnL$_3$X$_2$] (X=NO$_3$ or Cl). It is noteworthy that no base was used to deprotonate the ligand. The ligand itself is mesogenic, exhibiting a nematic mesophase. The lanthanide complexes form a highly viscous smectic mesophase, which was later identified as a smectic A phase. Exploration of the physical properties of these compounds was hampered by the very high viscosity of the mesophase and by the low thermal stability of the lanthanide complexes (they decompose at the clearing point). An X-ray diffraction study of these compounds has been published by Binnemans and coworkers [7].

A major breakthrough came in 1994, when Galyametdinov and coworkers discovered lanthanide Schiff's base complexes with enhanced thermal stability and exhibiting a smectic A phase with a relatively low viscosity [8]. The ligands in these complexes are remarkable because they contain only one aromatic ring and because they do not form a mesophase themselves (Fig. 15.6); mesomorphism is induced by the lanthanide ion. Soon after the discovery of this type of metallomesogens, it became evident that they have very interesting magnetic properties, such as a huge magnetic anisotropy. Due to the high magnetic anisotropy, the mesophase formed by these compounds can be much more easily aligned in an external magnetic field than one of organic liquid crystals. The magnetic behavior of the Schiff's base complexes is discussed in detail further on.

Subsequent work performed by the research groups of Galyametdinov, Bruce, and Binnemans gave more insight into the structures of these Schiff's base complexes [9–13]. By performing the complex formation at room temperature or just above, compounds could be obtained with a stoichiometry consistent

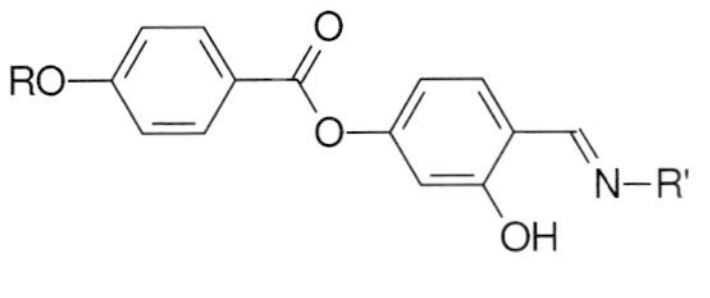

Fig. 15.5 Two-ring Schiff's base.

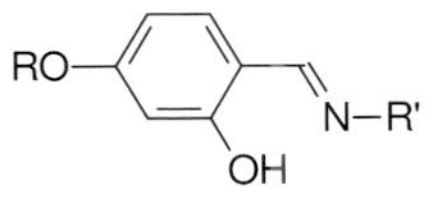

Fig. 15.6 Salicylaldimine Schiff's base ligand.

Fig. 15.7 Zwitterionic form of a salicylaldimine Schiff's base ligand.

with [Ln(*LH*)$_3$(NO$_3$)$_3$]. ^{1}H NMR studies on diamagnetic lanthanum(III) complexes showed that in the complexes the Schiff's base ligand is present in a *zwitterionic form*, i.e. the phenolic hydrogen has been transferred to the imine nitrogen (Fig. 15.7) [12]. The zwitterion is also found in the crystal structures of homologous non-mesogenic complexes, formed by ligands with short alkyl chains, so that three ligands and three nitrate groups are present for each metal ion (Fig. 15.8) [12]. The ligands coordinate to the metal ion only through the negatively charged phenolic oxygen. No binding occurs between the lanthanide ion and the imine nitrogen, and the three nitrate groups coordinate in a bidentate fashion. The coordination number of the lanthanide ion is therefore nine. The formation of a zwitterionic form can be rationalized by the tendency of the lanthanide ions to coordinate to negatively charged ligands (with a preference for O-donor ligands). By transfer of the phenolic proton to the imine nitrogen, the phenolic oxygen becomes negatively charged and can coordinate to the lanthanide ion.

All the lanthanide complexes with the one-ring salicylaldimine Schiff's base ligands exhibit a smectic A phase. Neither the alkoxy chain length nor the *N*-alkyl chain length has a substantial influence on the transition temperatures [12, 14]. However, for the nitrate series, the transition temperatures depend on the

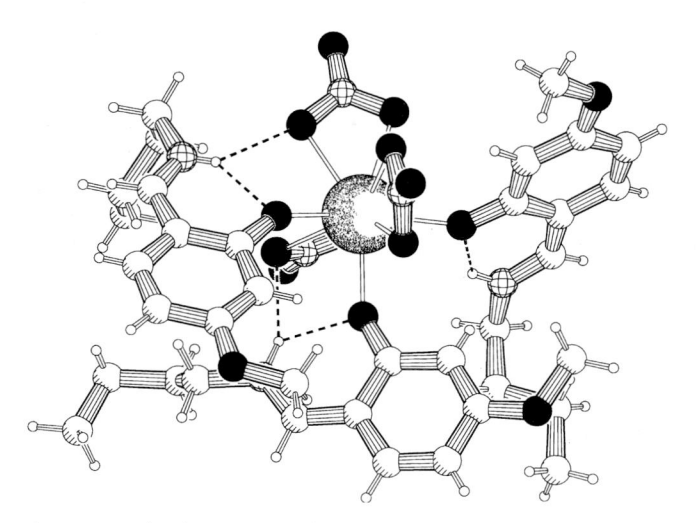

Fig. 15.8 Molecular structure of [Nd(LH)$_3$(NO$_3$)$_3$], where LH is *N*-butyl-2-hydroxy-4-methoxybenzaldimine. Atomic coordinates were taken from Ref. [12].

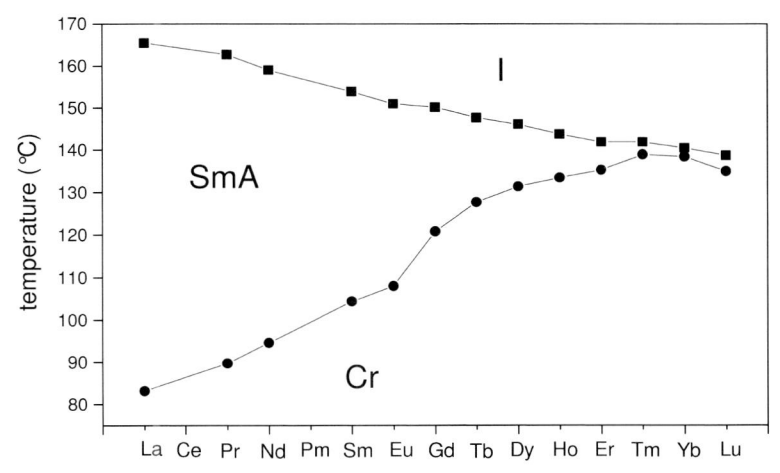

Fig. 15.9 Influence of the lanthanide ion on the transition temperatures of [Ln(LH)$_3$(NO$_3$)$_3$] complexes. LH is the Schiff's base ligand shown in Fig. 15.5, with R=C$_8$H$_{17}$ and R′=C$_{18}$H$_{37}$. Cr=crystalline solid, SmA=smectic A mesophase, I=isotropic liquid.

lanthanide (Fig. 15.9) [10]. Whereas the melting point increases over the lanthanide series, the clearing temperature decreases simultaneously, so that the overall effect is a decrease of the mesophase stability range over the lanthanide series. The chloride complexes have higher transition temperatures than the corresponding nitrate complexes. Because chloride ions act only as monodentate ligands and because the lanthanide ions prefer high coordination numbers (typically 8 or 9), complexes of the type [Ln(*LH*)$_3$Cl$_3$] are most likely dimeric or even oligomeric [15, 16].

An interesting class of compounds are the Schiff's base complexes with dodecylsulfate counter ions (or with alkylsulfate counter ions in general), because their transition temperatures are much lower than those of the nitrate or chloride complexes. Complexes with dodecylsulfate counter ions were first prepared by Binnemans et al. by means of a metathesis reaction between the corresponding chloride complex and Ag*DOS* (*DOS*=dodecylsulfate) [17]. However, Galyametdinov et al. found a more elegant synthetic route, namely the reaction of Ln(*DOS*)$_3$·*x*H$_2$O salts with the Schiff's base ligand [18]. The transition temperatures of the complexes are greatly reduced in comparison with those of the chloride complexes. These Schiff's base complexes have an opposite temperature dependence on the lanthanide ion in comparison with the nitrate complexes, in the sense that the mesophase stability range increases over the lanthanide series. Although the perfluoroalkylsulfate complexes have slightly higher transition temperatures than the alkylsulfate complexes, they have the great advantage that their mesophase has a much lower viscosity [13, 19, 20]. The low viscosity is most probably due to the immiscibility of the alkyl chains of the

Schiff's base ligand and the perfluoroalkyl chains of the counter ion. These complexes are much more easily aligned in an external magnetic field. The mesophase of the Schiff's base complexes can be supercooled into a glass state. In this way, anisotropic glasses can be obtained.

Although the crystal structures of the non-mesomorphic Schiff's base complexes have been determined, much less is known about the actual structures of the mesomorphic lanthanide complexes themselves. The presence of a smectic A phase dictates that the Schiff's base ligands have to be arranged around the central metal ion in such a way that the formation of a layered structure is possible. Spectroscopic data indicate that the first coordination sphere around the lanthanide ion is not changed by elongation of the alkyl chains. Therefore, it was anticipated that the bonding in the non-mesogenic complexes is the same as that in the mesogenic complexes. Of course, changes in the packing of the molecules can be assumed. A molecular structure has been proposed by Galyametdinov and coworkers, which can be described as a trigonal prism in which the nitrate groups occupy equatorial or axial positions [9]. Because the smectic layer thickness d (as determined by X-ray diffraction) was found to be shorter than the calculated length of the Schiff's base ligand, it was assumed that the molecules of each layer penetrate the adjacent layer to a limited depth (Fig. 15.10). Although the Schiff's

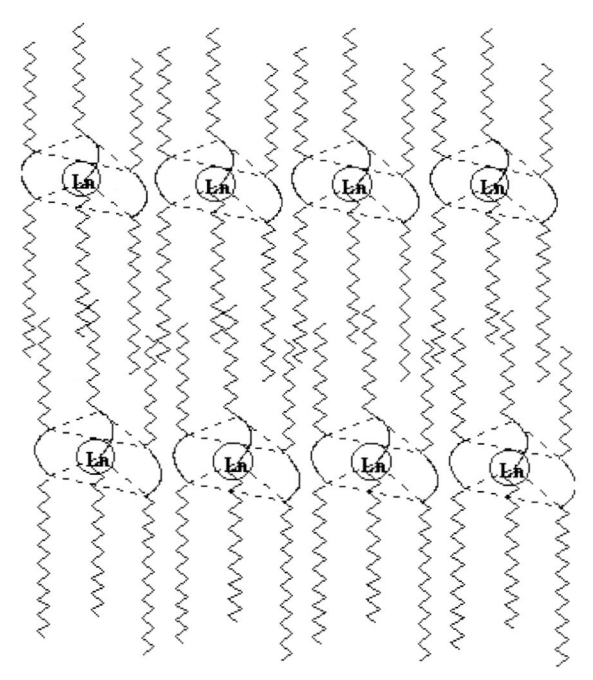

Fig. 15.10 Schematic representation of the $[Ln(LH)_3(NO_3)_3]$ Schiff's base complexes in a smectic A phase. The nitrate groups have been omitted for clarity.

bases derived from 2,4-dihydroxybenzaldehyde form liquid-crystalline lanthanide complexes, those derived from 2,5-dihydroxybenzaldehyde or from 2-hydroxybenzaldehyde (salicylaldehyde) do not [21].

15.4
Complexes of β-Diketonates

Tris(β-diketonato)lanthanide(III) complexes are well-known as NMR shift reagents [22–24]. Examples are Pr(fod)$_3$ and Eu(fod)$_3$, where fod represents the anion of 6,6,7,7,8,8,8-heptafluoro-2,2-dimethyl-3,5-octanedione. The coordination sphere of the lanthanide ion in a tris(β-diketonato)lanthanide(III) complex is unsaturated, since the coordination number of the metal ion in these complexes is only six, whereas lanthanide ions prefer to be coordinated by eight or nine donor atoms [25]. The lanthanide ion in the tris-β-diketonate complexes can expand its coordination sphere by forming adducts with neutral molecules containing oxygen or nitrogen donor atoms. Such a molecule can be water, and therefore the tris(β-diketonato)lanthanide(III) complexes can be easily transformed into dihydrates. However, polydentate ligands (e.g., 1,10-phenanthroline) can replace water in the first coordination sphere, and the resulting lanthanide complexes often show strong photoluminescence. A typical example of a luminescent lanthanide complex is [Eu(tta)$_3$(phen)], where tta is the anion of thenoyltrifluoroacetone, and phen is 1,10-phenanthroline [26–28]. Of course, it would be interesting to design lanthanide β-diketonate complexes which exhibit a liquid-crystalline phase themselves, rather than relying on host-guest systems. Although Swager and coworkers showed that β-diketones could be used to obtain eight-coordinate metallomesogens with zirconium(IV) as the central metal ion [29], attempts to synthesize liquid-crystalline lanthanide complexes with these mesogenic ligands were unsuccessful for a long time.

Binnemans and Lodewyckx showed that it is possible to obtain mesomorphic lanthanide complexes by forming Lewis base adducts of simple tris(β-diketonato)lanthanide(III) complexes with mesogenic ligands. More particularly, they studied bis-adducts of [Ln(*dbm*)$_3$] complexes (H*dbm* = 1,3-diphenyl-1,3-propanedione or dibenzoylmethane) with the one-ring Schiff's base ligands discussed above (Fig. 15.11) [30, 31]. The adducts are monotropic liquid crystals, exhibiting a smectic A phase upon cooling of the isotropic liquid. The crystal structure of [La(*dbm*)$_3$L$_2$], where L is N-butyl-2-hydroxy-4-methoxybenzaldehyde, was determined (Fig. 15.12). The temperature difference between the melting point and the clearing point of the monotropic mesophase increases over the lanthanide series. The complexes of the heavier lanthanides are not liquid-crystalline.

Galyametdinov and coworkers prepared tris-β-diketonate complexes of the lanthanides with 1,3-bis(p-tetradecyloxyphenyl)-1,3-propanedione, in which the coordination sphere is saturated by 1,10-phenanthroline, 2,2′-bipyridine or substituted 2,2′-bipyridines (Fig. 15.13) [32, 33]. This was the first example of a lanthanide complex for which the mesomorphism was induced by the β-diketo-

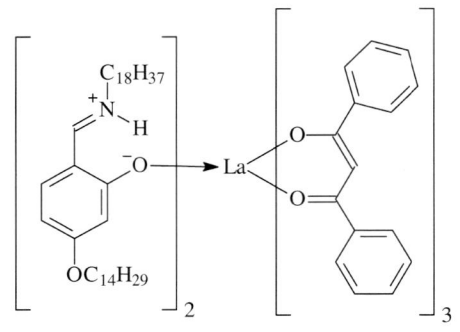

Fig. 15.11 Lewis base adduct of the tris(β-diketonato)lanthanum(III) complex [La(dbm)$_3$] and a non-mesomorphic salicylaldimine Schiff's base. The compound exhibits a monotropic smectic A phase: Cr 95 (SmA 81) I.

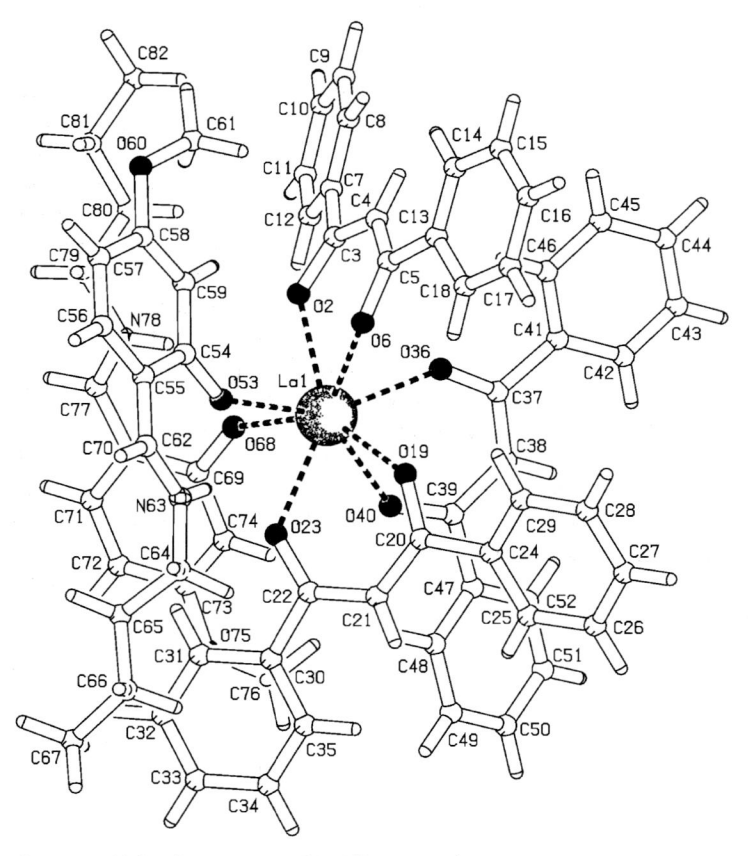

Fig. 15.12 Molecular structure of [La(dbm)$_3$L$_2$], where L is N-butyl-2-hydroxy-4-methoxybenzaldehyde.

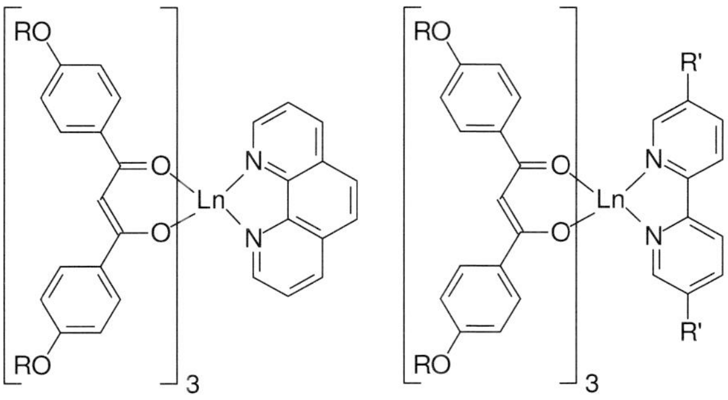

$R = C_{14}H_{29}$

1a: Ln = Tb
1b: Ln = Eu

$R = C_{14}H_{29}$

2a: Ln = Eu, R' = H
2b: Ln = Eu, R' = $CO(O)C_{11}H_{23}$
2c: Ln = Tb, R' = $CO(O)C_{11}H_{23}$

Fig. 15.13 Liquid-crystalline β-diketonate complexes.

nate ligand. The complexes form monotropic smectic A phases or monotropic highly ordered smectic phases (that were not studied in detail).

Other attempts to obtain liquid-crystalline lanthanide complexes through Lewis adduct formation to tris(β-diketonato)lanthanide(III) complexes were less successful. Hapiot and Boyaval studied adducts of [Ln(*tta*)$_3$] (H*tta*=thenoyltrifluoroacetone) and cholesteryl nonanoate and cholesteryl tetradecanoate [34, 35].

15.5 Bis(benzimidazolyl)pyridines

Piguet and coworkers investigated 5- and 6-substituted 2,6-bis(benzimidazol-2-yl)pyridines and the corresponding lanthanide complexes [Ln*L*(NO_3)$_3$] [36–38]. Conformational changes (*trans-trans* → *cis-cis*) occurred upon complexation to lanthanide ions: the I-shaped 5-substituted ligands are transformed into U-shaped lanthanide complexes, whereas the U-shaped 6-substituted ligands give I-shaped lanthanide complexes (Fig. 15.14). The 5-substituted 2,6-bis(benzimidazol-2-yl)pyridines are calamitic liquid crystals and exhibit a rich mesomorphism (SmC, SmA, and/or N). Although it was initially reported that the lanthanide complexes essentially retain the mesomorphism of the ligands, further studies showed that the lanthanide complexes have a low thermal stability and are not mesomorphic [39]. Structurally, the bent 6-substituted ligands have similarities to "banana" liquid crystals. However, no mesomorphism was detected. The corresponding lanthanide complexes have the shape of calamitic liquid crystals, but they decompose at around 180–200 °C, without evidence of mesomorphism. The absence of mesomorphism is attributed to the spatial expansion brought about by the Ln(NO_3)$_3$ core [38].

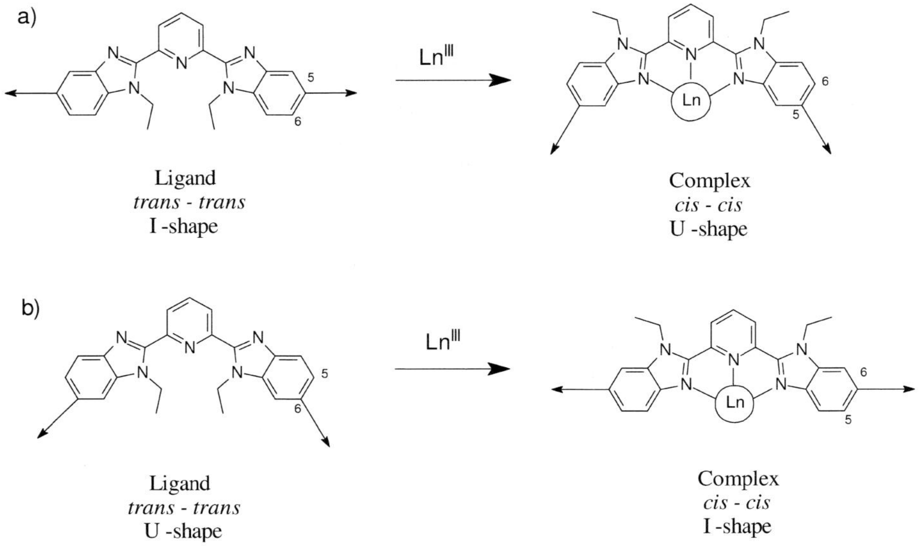

Fig. 15.14 Conformational changes (*trans-trans* → *cis-cis*) occurring upon complexation to trivalent lanthanide ions for (a) 5-substituted and (b) 6-substituted 2,6-bis(benzimidazol-2-yl)-pyridines. Linear substituents are represented schematically by arrows.

Liquid-crystalline lanthanide complexes could be obtained by increasing the number of terminal alkyl chains from two to six (Fig. 15.15) [40]. Interestingly, the type of mesophase generated depends on the lanthanide ion. For the europium(III) and dysprosium(III) complexes a bicontinuous cubic phase (Cub_v) is formed, whereas the corresponding lutetium(III) complex exhibits a hexagonal

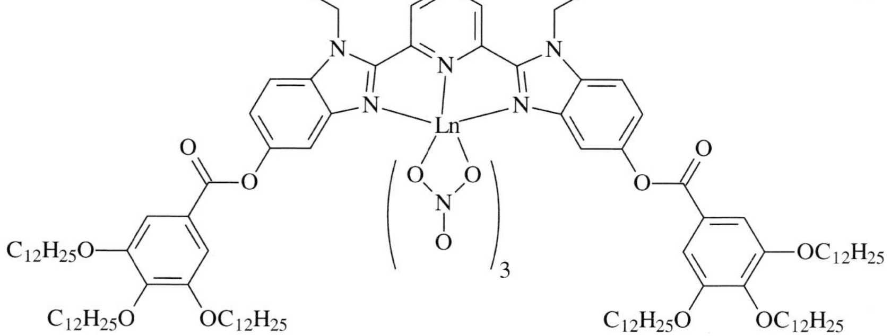

Fig. 15.15 Liquid-crystalline bis(benzimidazolyl)pyridine complexes. For Ln = Eu and Dy a cubic mesophase is formed, for Ln = Lu a hexagonal columnar mesophase is observed.

columnar mesophase (Col$_h$). Thus, a minor size difference of the lanthanide ion induces a transition between a cubic and a columnar mesophase. A bicontinuous cubic phase is a highly ordered mesophase with the appearance of a viscous isotropic phase.

15.6 Mixed f-d Metallomesogens

Studies of heteropolynuclear complexes that contain both a transition metal ion and a trivalent lanthanide ion, the so-called f-d metal complexes, have attracted a lot of interest since 1985, when Bencini et al. reported unique magnetic properties for Cu^{II}-Gd^{III} *salen* complexes [41]. The well-known *salen* ligand (*salen* = 2,2'-N,N'-bis(salicylidene)ethylenediamine) is often used to form f-d complexes by adduct formation of a transition metal *salen* complex and a lanthanide salt. Most often, Cu^{II} is chosen as the transition metal (because it is paramagnetic, d^9), but other divalent d metal ions such as Ni^{II}, Pd^{II}, Co^{II} or vanadyl can also be used to prepare f-d complexes. Recently, Sakamoto et al. reviewed these f-d complexes [42].

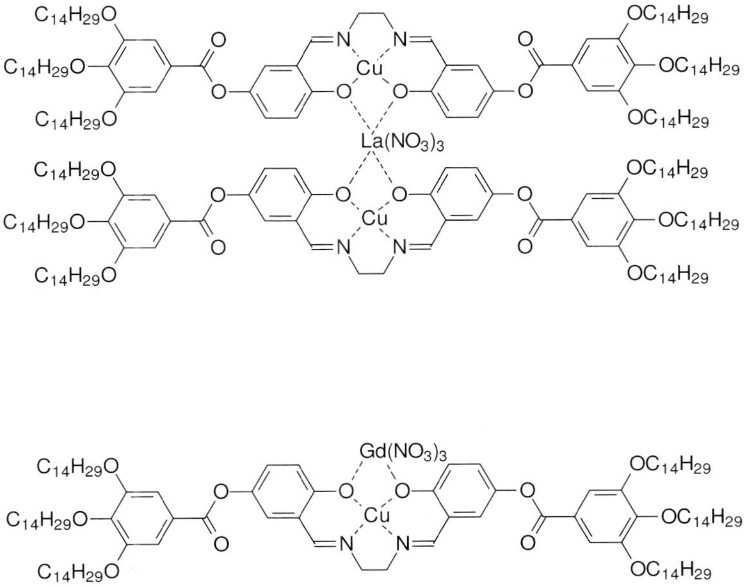

Fig. 15.16 The first examples of liquid-crystalline f-d complexes; a trinuclear copper-lanthanum-copper complex and a dinuclear copper–gadolinium complex. The ligands are functionalized salen-type ligands. The complexes exhibit a hexagonal columnar mesophase.

Binnemans and coworkers described liquid-crystalline f-d complexes, which were called mixed f-d metallomesogens [43]. Their approach involved modifying the structures of previously described non-mesomorphic f-d complexes in such a way that sufficient structural anisotropy was obtained to favor the formation of mesophases. Such a modification can be achieved by extending the aromatic parts of the central rigid core, and/or by attaching long alkyl chains to the extremities of the core complex. Kahn et al. described adducts of lanthanide(III) nitrates with the Cu(*salen*) complex [44]. For the light lanthanides, they obtained heterotrinuclear [Ln(NO$_3$)$_3${Cu(*salen*)}$_2$] compounds, whereas for gadolinium and the heavier lanthanides heterobinuclear [Ln(NO$_3$)$_3${Cu(*salen*)}] complexes were formed. The first examples of mixed f-d metallomesogens were the trinuclear copper-lanthanum complex and the dinuclear copper-gadolinium complex shown in Fig. 15.16. These complexes each form a hexagonal columnar mesophase.

In order to investigate the influence of the d metal ion, adducts were formed between a mesomorphic Ni(*salen*) complex with six terminal alkoxy chains and a lanthanide nitrate (Ln = La, Gd) [45]. Different alkoxy chain lengths were used: OC$_{12}$H$_{25}$, OC$_{14}$H$_{29}$, OC$_{16}$H$_{33}$, and OC$_{18}$H$_{37}$. Trinuclear nickel-lanthanum and nickel-gadolinium complexes [Ln(NO$_3$)$_3${Ni(*salen*)}$_2$] were obtained. The compounds exhibit a hexagonal columnar mesophase with rather low melting points.

15.7
Phthalocyanines

The bis-[2,3,9,10,16,17,23,24-octakis(alkoxymethyl)phthalocyaninato]lutetium(III) complexes [(C$_n$H$_{2n+1}$OCH$_2$)$_8$Pc]$_2$Lu (n = 8, 12, 18) described by Piechocki et al. in 1985 were the first examples of liquid-crystalline lanthanide complexes [46]. In these compounds, the lutetium(III) ion is sandwiched between two phthalocyanine macrocycles. The authors were trying to obtain one-dimensional molecular semiconductors. The conductivity behavior of the bis(phthalocyaninato)lutetium (III) sandwich complexes was known previously, but only after these studies was it realized that aligned columnar mesophases formed by such metallophthalocyanines can be considered as electrical wires at a molecular level. The mobile alkyl chains form an insulating layer. The phthalocyanine ring is one of the most stable of macrocycles. Although different substitution patterns are possible on the macrocyclic ring, the easiest compounds to obtain are the octakis-substituted phthalocyanines.

The work on liquid-crystalline bis(phthalocyaninato)lanthanide(III) complexes has been reviewed in detail by Binnemans and Görller-Walrand [5]. Typically, these compounds form hexagonal columnar mesophases, although other types of columnar mesophases have been found as well. Most studies have been limited to lutetium(III) compounds. This stems from the fact that these compounds are the easiest to prepare: the yield decreases with increasing size of the lanthanide(III) ion. In order to investigate the influence of the lanthanide ion and the chain length on the thermal properties of liquid-crystalline bis(phthalo-

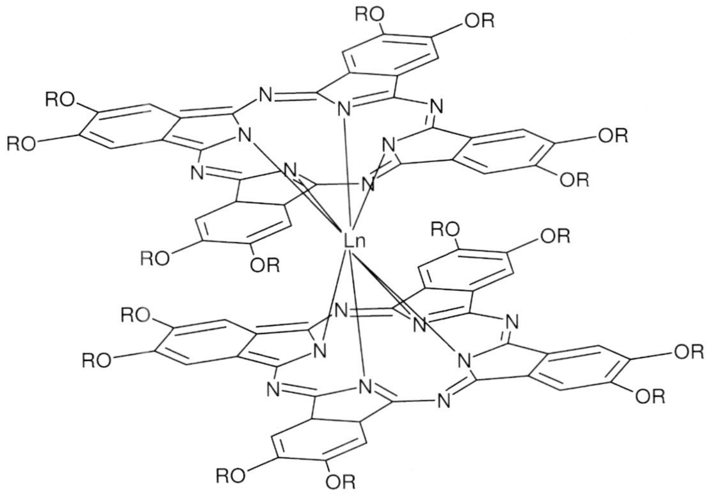

Fig. 15.17 Structure of an alkoxy-substituted bis(phthalocyaninato)lanthanide(III) complex [(RO)$_8$Pc]$_2$Ln, where R=C$_n$H$_{2n+1}$.

cyaninato)-lanthanide(III) complexes, Binnemans and coworkers prepared two series of octakis(alkoxy)-substituted compounds: a first series of double-deckers [(C$_{12}$H$_{25}$O)$_8$Pc]$_2$Ln with different lanthanide ions (Ln = Pr, Nd, Eu, Gd, Tb, Dy, Ho, Er, Tm, Yb, Lu) and a series [(C$_n$H$_{2n+1}$O)$_8$Pc]$_2$Er with different alkoxy chains (Fig. 15.17) [47]. All of these sandwich complexes exhibited a hexagonal columnar mesophase (Col$_h$) over a broad temperature range, i.e. 100–150 °C. It was found that the nature of the lanthanide ion has only a limited influence on the transition temperatures, and particularly on the melting temperature into the mesophase, whereas the transition temperatures could be tuned by appropriate choice of the alkoxy chain length. Scanning tunneling microscopy (STM) images of [(C$_{12}$H$_{25}$O)$_8$Pc]$_2$Er deposited on various substrates were recorded and compared to those of the corresponding metal-free phthalocyanine (C$_{12}$H$_{25}$O)$_8$PcH$_2$. Ohta and coworkers observed a rectangular ordered columnar mesophase (Col$_{ro}$) for [(C$_n$H$_{2n+1}$O)$_8$Pc]$_2$Lu complexes (n = 10, 12, 14, 16, 18) between the crystalline state and the hexagonal columnar mesophase [48].

15.8
Porphyrins

Porphyrins are macrocycles that are related to phthalocyanines and they can form complexes with lanthanides as well, but the mesophase behavior of these lanthanide complexes has been less extensively investigated than that of the phthalocyanine compounds (probably because it is more difficult to prepare porphyrins than phthalocyanines). Triple-decker complexes consisting of two

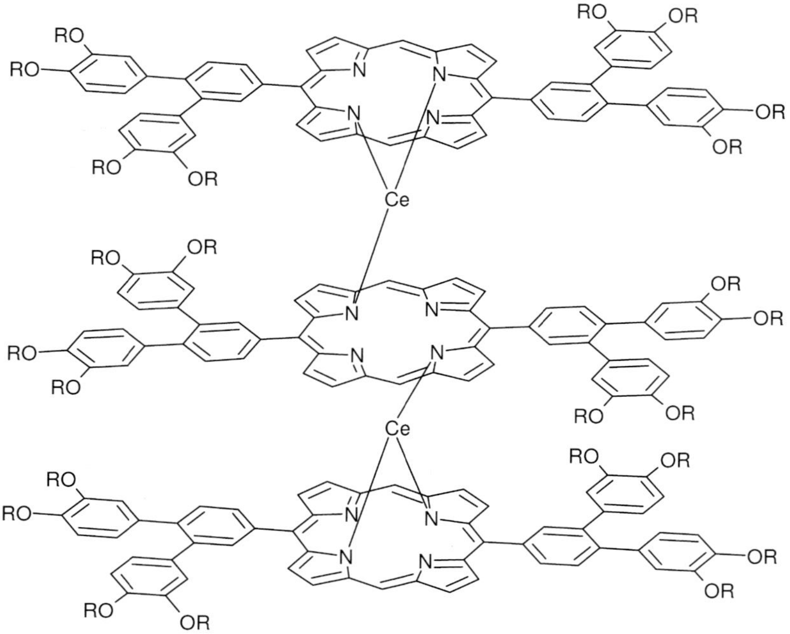

Fig. 15.18 Structural model of a porphyrinato triple-decker cerium(III) complex.

cerium(III) ions and three 5,15-diarylporphyrins were prepared by reaction between the parent porphyrins and Ce(*acac*)$_3$ · H$_2$O (H*acac*=acetylacetone) (Fig. 15.18) [49]. The work was motivated by the desire to find the border in terms of molecular structure between a discotic mesogen and a calamitic mesogen, when the molecular shape is continuously changed from disk-like to rod-like. The parent porphyrins have a strip-like shape, but they can aggregate into columnar structures. The complexes containing dialkoxyphenyl groups in the 5,15-positions of the parent porphyrin show a mesophase possessing both lamellar and columnar structures at lower temperatures (discotic lamellar columnar mesophase, D_{LC}) and a rectangular columnar mesophase (*Col*$_r$) at higher temperatures. The D_{LC} phase could only be observed in virgin samples. Complexes containing tetraalkoxylated terphenyl groups at the 5,15-positions of the parent porphyrin show only a hexagonal columnar phase (*Col*$_h$) with a single stacking distance. The differences in mesomorphism are explained by differences in the steric hindrance of the side chains for rotation along the axis connecting the center of the three porphyrins. Zhao and Liu synthesized a hydroxy{5,10,15,20-tetrakis[*p*-(decyloxy-*m*-methyloxy)phenyl]porphyrin ytterbium(III) complex [50]. This compound was found to exhibit a columnar hexagonal mesophase.

15.9
Crown Ether Complexes

Lanthanide complexes of a steroid-substituted benzo-crown ether have been synthesized (Fig. 15.19) [51]. The metal-to-ligand ratio of all such metal complexes is 1:1. The ligand 4′-[(cholesteryloxy)carbonyl]-benzo-15-crown-5 is a monotropic liquid crystal, displaying a cholesteric mesophase. The lanthanide complexes with nitrate counter ions formed a highly viscous mesophase, which decomposed at the clearing point. The complexes had a low thermal stability. The transition temperatures change as a function of the lanthanide ion, in the sense that the mesophase stability range becomes narrower as the size of the lanthanide ion is decreased.

Bünzli and coworkers observed a mesophase for europium(III) and terbium(III) complexes of a diaza-18-crown-6 with mesogenic pendant arms (Fig. 15.20) [52]. Although the mesophase was not identified unambiguously, it is very likely a columnar phase. The authors were able to detect the crystal-to-mesophase transition of the europium(III) complex by temperature-dependent luminescence measurements. To date, this is the only reported example of luminescence of a liquid-crystalline lanthanide complex in the mesophase.

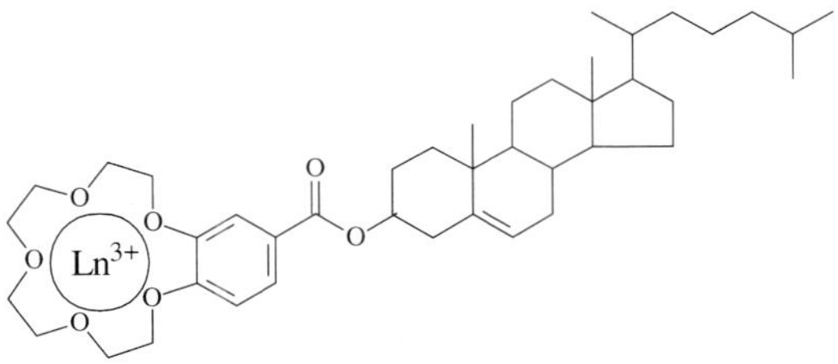

Fig. 15.19 Lanthanide complex of a steroid-substituted benzo-crown ether. The charge of the lanthanide ion is balanced by three nitrate or by three dodecylsulfate counter ions.

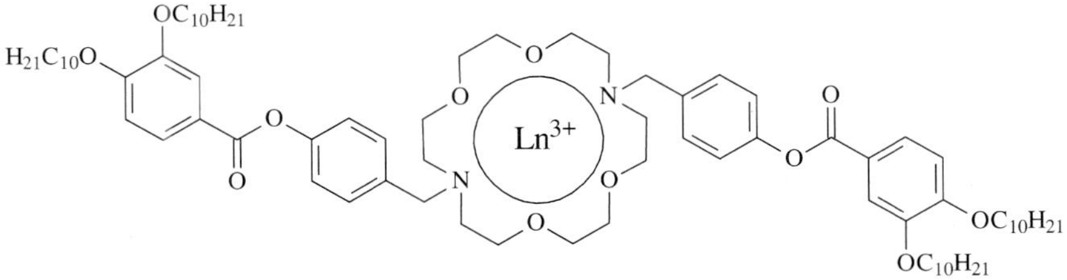

Fig. 15.20 Liquid-crystalline diaza-18-crown-6 complexes (Ln = Eu, Tb). The three nitrate counter ions have been omitted for the sake of clarity.

15.10
Magnetic Anisotropy and Alignment in a Magnetic Field

In contrast to conventional diamagnetic liquid crystals, which can be oriented only in strong magnetic fields, paramagnetic metallomesogens can be oriented in rather weak magnetic fields. For easy alignment of these metallomesogenic molecules in a magnetic field a pronounced magnetic anisotropy is required. The magnetic anisotropy $\Delta\chi$ is the difference between the magnetic susceptibility parallel ($\chi_{\parallel}$) and perpendicular ($\chi_{\perp}$) to the director $\tilde{n}$: $\Delta\chi = \chi_{\parallel} - \chi_{\perp}$. In general, the molar magnetic anisotropy of organic liquid crystals is less than 100×10^{-6} cm^{-3} mol^{-1}. This value can be increased by introducing paramagnetic metal atoms in liquid-crystalline compounds. Lanthanide-containing metallomesogens are especially attractive for these purposes as they can exhibit a very high magnetic anisotropy, often one or two orders of magnitude larger than that of organic liquid crystals.

The orientation of liquid crystals in a magnetic field depends not only on the magnitude of the magnetic anisotropy $\Delta\chi$, but also on its sign. Depending on the sign of $\Delta\chi$, the director $\tilde{n}$ of the liquid crystal is oriented either parallel or perpendicular to the external magnetic field. If $\Delta\chi$ is positive, the molecules will be oriented with their molecular long axes parallel to the magnetic field. If $\Delta\chi$ is negative, the molecules will be aligned with their molecular long axes perpendicular to the magnetic field. Whereas it is in principle possible to obtain a monodomain liquid crystal when the director aligns parallel to the applied magnetic field ($\Delta\chi > 0$), this is not the case when the director aligns perpendicular to the magnetic field ($\Delta\chi < 0$); there is only one direction parallel to the magnetic field, but if $\Delta\chi < 0$ the director can take any orientation in a plane perpendicular to the magnetic field. A polydomain is obtained, in which the different microdomains are randomly oriented, the only restriction being that the director is situated within a plane. A monodomain can be obtained by applying a second magnetic field perpendicular to the first magnetic field. If $\Delta\chi < 0$, the director will be oriented perpendicular to both magnetic field directions. Finally, it should be mentioned that a low viscosity mesophase is desirable for good alignment of the sample in a magnetic field.

The trivalent lanthanide ions (especially Tb^{3+}, Dy^{3+}, Ho^{3+}, Er^{3+}, Tm^{3+}) have very large magnetic anisotropies in comparison with other paramagnetic ions (e.g., Cu^{2+} or V=O^{2+}), so that the diamagnetic contributions to $\Delta\chi$ can be neglected in comparison to the paramagnetic contribution. An exception is the Gd^{3+} ion, which is almost magnetically isotropic due to the ground state $^{8}S_{7/2}$. The large magnetic anisotropy of liquid-crystalline lanthanide complexes was discovered soon after the first calamitic lanthanide-containing metallomesogens had been prepared [9, 53]. Because the sign of the magnetic anisotropy depends on the lanthanide ion, it is possible to obtain, with the same kind of ligand and with appropriate choice of the lanthanide ion, compounds which can either be aligned perpendicular or parallel to the magnetic field lines. Analysis of experimental magnetic susceptibility data shows that these compounds can be classified into two distinct groups, depending on the sign of $\Delta\chi$ [12, 54]. The first group contains Ce^{3+}, Pr^{3+}, Nd^{3+}, Sm^{3+}, Tb^{3+}, Dy^{3+}, and Ho^{3+} compounds, while the second group contains Eu^{3+}, Er^{3+}, Tm^{3+}, and Yb^{3+} compounds. Two compounds not belonging to the same group always have opposite sign of $\Delta\chi$. For instance, if $\Delta\chi$ is negative for the first group of lanthanide compounds, $\Delta\chi$ is positive for the second group and vice versa. All experimental results obtained to date for lanthanide-containing liquid crystals are in agreement with a negative $\Delta\chi$ value for the first group and a positive $\Delta\chi$ value for the second group. Recent theoretical work, however, shows that the reverse situation can also be expected. The highest magnetic anisotropy values are observed for the heavy lanthanides Tb^{3+}, Dy^{3+}, Ho^{3+}, Er^{3+}, and Tm^{3+} because these ions have high magnetic moments. For metallomesogens containing these ions, the absolute value of the magnetic anisotropy in the mesophase can be 5000×10^{-6} cm^{-3} mol^{-1} or even more. As shown below, the magnetic moment is related to the magnetic susceptibility and thus to the magnetic anisotropy. There is no one-to-one relationship between a high magnetic moment and a large magnetic anisotropy, but if the magnetic moment is small it is impossible to have a large magnetic anisotropy. Another requirement for a high magnetic anisotropy, besides a large magnetic moment, is that the crystal-field perturbation is strong. The $^{2S+1}L_J$ ground state splits under the influence of the crystal-field potential into a number of crystal field levels, with a maximum of $2J+1$ levels when J is an integer (even number of f electrons), and of $J + \frac{1}{2}$ levels for half-integer J (odd number of f electrons). The magnetic anisotropy arises from the fact that when the crystal-field splitting is large, not all of the crystal-field levels are statistically populated at a given temperature. The magnetic anisotropy increases when the temperature is lowered. At temperatures above room temperature, the magnetic behavior becomes more and more isotropic and approaches the predictions of the free ion approximation. However, when the crystal-field perturbation is large, it is possible to have a large magnetic anisotropy, even at temperatures at which the mesophase of lanthanide-containing liquid crystals is stable. The magnitude of the magnetic anisotropy for a series of isostructural lanthanide-containing metallomesogens can be estimated on the basis of the crystal-field splitting of the $^7F_1$ multiplet in the corresponding europium(III) compound [55]. This splitting is accessible through photoluminescence measurements. Model calculations have been used to relate the magnetic anisotropy to the type of coordination polyhedron around the central lanthanide ion [56, 57]. Another approach for explaining the magnetic anisotropy is based on the Bleaney theory, which is used to predict the lanthanide-induced shift in ^{1}H NMR spectra [58].

The main problem to cope with is the high viscosity of the mesophase, which prevents fast switching. Cooling rates of the order of one Kelvin per minute are necessary to achieve good alignment in a magnetic field when going from the isotropic liquid to the mesophase. All the Schiff's base complexes reported to date display a smectic A phase. The smectic A phase has an intrinsically higher viscosity than the nematic phase. Therefore, obtaining liquid-crystalline lanthanide complexes exhibiting a nematic phase is a major research goal. This goal is not easy to achieve, because the structure of the Schiff's base complexes tends to promote the formation of layered structures (and this means smectic mesophases). However, by appropriate choice of the counter ion, the viscosity of the mesophase can be reduced considerably. For instance, a smectic A phase with a low viscosity is found for Schiff's base complexes with perfluorinated alkylsulfates as counter ions [13].

The sign of $\Delta\chi$ cannot be determined by magnetic susceptibility measurements and has to be obtained through an independent measurement, for instance by EPR (angle dependence of the EPR signal) or by X-ray diffraction in a magnetic field (distribution of the diffraction maxima with respect to the external magnetic field). It should be realized that the maximal value for the magnetic anisotropy $\Delta\chi$ can only be measured from completely aligned samples. For real samples, alignment is never complete, so that the measured value of $\Delta\chi$ is smaller than the value that can be theoretically expected. $\chi_\parallel$ and $\chi_\perp$ can be measured directly by orientation of an aligned liquid crystal with the director parallel or perpendicular to the magnetic field. This implies that the measurements are carried out in a magnetic field that is not strong enough to reorient the liquid crystal. Alternatively, the aligned mesophase can be frozen into the glassy state. The $\chi_\parallel$ and $\chi_\perp$ values of a glassy (aligned) mesophase can be measured in the same way as one would measure these parameters for a single crystal (orienting the sample with its long molecular axis parallel or perpendicular to the magnetic field lines). The situation is quite different for a mesophase in a magnetic field above a certain threshold strength H_0 because in this case the mesophase will be aligned by the magnetic field. A nematic phase can be oriented at any temperature within the mesophase because of the low viscosity of this phase. Due to the intrinsic higher viscosity, a smectic A phase can only be oriented at a temperature close to the clearing point. The orientation at this point will be retained if the mesophase is cooled further. The magnetic anisotropy $\Delta\chi$ cannot be obtained from the relationship $\Delta\chi=\chi_\parallel-\chi_\perp$, because only one of the two components $\chi_\parallel$ or $\chi_\perp$ can be determined. This problem can be overcome by measuring the magnetic susceptibility in the isotropic phase and in the mesophase. In the isotropic phase, χ_{iso} will be measured. It is assumed that χ_{iso} is equal to the average magnetic susceptibility $\bar{\chi}$, which is defined as $\bar{\chi} = (\chi_\parallel + 2\chi_\perp)/3$. In the mesophase, the molecules will be oriented in such a way that the axis of maximal magnetic susceptibility will be parallel to the magnetic field. Because of alignment of the paramagnetic molecules, an increase in magnetic susceptibility will be observed (in comparison to $\bar{\chi}$) and the measured value is denoted as χ_{max}. If $\Delta\chi>0$, χ_{max} corresponds to $\chi_\parallel$ and $\Delta\chi = \frac{3}{2}(\chi_{max} - \bar{\chi})$. If $\Delta\chi<0$, χ_{max} corresponds to $\chi_\perp$ and $\Delta\chi = 3(\bar{\chi} - \chi_{max})$.

In Fig. 15.21, the alignment of a praseodymium(III)-containing liquid crystal is illustrated [12]. The sample exhibited Curie-Weiss behavior for the temperature dependence of its magnetic properties, when heated from the solid state through the smectic A phase until the isotropic phase was reached. When the sample was cooled from the isotropic phase to the mesophase (in the presence of a magnetic field), a drastic increase in the magnetic moment μ_{eff} was observed in the vicinity of the clearing point in comparison to the initial value re-

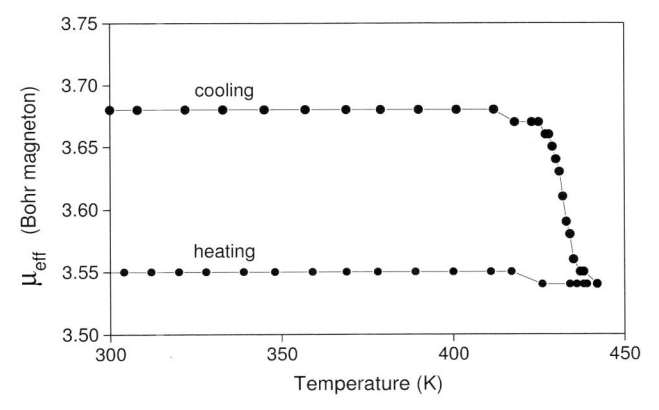

Fig. 15.21 Magnetic alignment of the praseodymium(III)-containing liquid crystal [Pr(LH)$_3$(NO$_3$)$_3$], where LH is 4-dodecyloxy-N-hexadecyl-2-hydroxybenzaldimine. Adapted from Ref. [12].

corded by heating of the sample. The increase occurred over a narrow temperature range, and, upon further cooling, the magnetic properties were found to vary according to the Curie-Weiss law, but with a different and higher μ_{eff} than in the heating cycle. This behavior can be ascribed to a magnetic-field-induced orientation in the liquid-crystalline phase of a magnetically anisotropic sample with its axis of maximal magnetic susceptibility parallel to the magnetic field. The magnetic anisotropy $\Delta\chi$ is proportional to the magnetic moment μ_{eff}. The $\Delta\chi$ values depend on the counter ion. The highest $\Delta\chi$ values have been observed for complexes with perfluorinated counter ions. Studies of such magnetic anisotropy have been largely restricted to Schiff's base complexes.

15.11
Luminescent Liquid Crystals

Although the trivalent lanthanide ions are well-known for their luminescence properties, it is not a trivial task to design luminescent liquid crystals. Most of the lanthanide-containing liquid crystals described to date have transition temperatures well above room temperature. On the other hand, radiationless deactivation becomes more and more important with increasing temperature, and often the luminescence will be totally quenched at the high temperatures necessary to melt the compound. Also, the type of ligand is of importance with regard to obtaining strongly luminescent lanthanide complexes. One has to ensure that the energy of the triplet state of the ligand (from which the excitation energy is transferred to the lanthanide ion) is not lower than the energy of the emitting level of the lanthanide ion, because otherwise no energy transfer can take place. Additionally, the presence of high-energy vibrations (such N–H and

O–H stretching vibrations) has to be avoided, since these can efficiently deactivate excited states. So far, only one study of the luminescence properties of a mesomorphic lanthanide complex in the mesophase has been reported, namely the work of Bünzli and coworkers on complexes of diaza-18-crown-6 compounds (see above) [52].

An alternative approach to obtain luminescent liquid crystals is to use the host-guest concept: a luminescent lanthanide complex is dissolved in a liquid-crystalline matrix. This approach allows the luminescence and mesomorphic properties of the liquid-crystal mixture to be optimized independently. The first to explore this concept was Larrabee, who added luminescent molecules to a nematic liquid crystal and used the resulting liquid-crystal mixture to make prototypes of luminescent LCDs [59, 60]. These luminescent molecules were excited by means of a UV source and emission was controlled electrically by the alignment of the liquid crystal. Yu and Labes suggested the use of an electric-field-induced cholesteric-to-nematic phase transition as the basis of a luminescent LCD [61, 62]. The cholesteric OFF state absorbs the excitation light more strongly than the nematic ON state, so that the emission of the luminescent additive is intense in the OFF state and is reduced in the ON state. The authors used the red-emitting europium(III) complex $Eu(tta)_3 \cdot 2H_2O$ (tta = thenoyltrifluoroacetone) as the luminescent additive and were able to achieve a contrast ratio of 9:1 (the difference in intensity between the OFF and ON states). A luminescent LCD is an emissive type of display. Advantages of such types of displays can be a high contrast and a wide viewing angle. However, the development of luminescent LCDs slowed down after the first trials. At that time, the quality of the available LCDs was sufficiently high to meet all needs. Secondly, there were no convenient UV sources available (other than the mercury discharge lamp). Presently, the situation is different. The growing market for flat computer monitors makes luminescent LCDs attractive again. The newly developed UV-LEDs have advantages as LCD backlights: light, thin, mercury-free, and noiseless. Most of the present studies on luminescent LCDs describe the use of luminescent polymer sheets placed in front of the LCD, rather than luminescent liquid-crystal mixtures.

Boyaval et al. investigated the luminescence of lanthanide complexes in a cholesteric liquid-crystal mixture [34, 63]. Binnemans and coworkers showed that a nematic liquid crystal can be an interesting host matrix to study the spectroscopic properties of luminescent lanthanide complexes [64]. These authors doped the nematic liquid crystals *N*-(4-methoxybenzylidene)-4-butylaniline (MBBA) and 4-*n*-pentyl-4′-cyanobiphenyl (5CB) with the β-diketonate complex $[Eu(tta)_3(phen)]$ (Htta = thenoyltrifluoroacetone; phen = 1,10-phenanthroline) and observed narrow-band red photoluminescence with a well-resolved crystal field splitting. Later, these authors extended their studies to lanthanide β-diketonate complexes emitting in the near-infrared (Ln = Nd, Er, Yb) [65].

15.12
Conclusions

The incorporation of trivalent lanthanide ions into liquid crystals results in a class of materials with unique spectroscopic and magnetic properties. The narrow-band emission of the lanthanides makes it possible to obtain luminescent liquid crystals that emit light of a high coloric purity. Because of the high magnetic anisotropy of paramagnetic lanthanide ions, lanthanide-containing liquid crystals can be aligned more easily in an external magnetic field than the conventional organic liquid crystals. Although several classes of liquid-crystalline lanthanide complexes have been described, this field is still dominated by the Schiff's base complexes. However, new types of mesomorphic lanthanide compounds are gradually appearing in the literature. The liquid-crystalline lanthanide complexes remain a fascinating topic in supramolecular coordination chemistry.

Acknowledgements

I thank the FWO-Flanders (Belgium) and the K.U. Leuven for financial support. I acknowledge the following persons for their continuing support or for their collaboration: Prof. Christiane Görller-Walrand (K.U. Leuven), Prof. Duncan W. Bruce (University of Exeter, UK), Prof. Yury G. Galyametdinov (Kazan Branch of the Russian Academy of Sciences, Russia), Dr. Daniel Guillon and Dr. Bertrand Donnio (ICPMS, Strasbourg, France), Prof. Wolfgang Haase (TU Darmstadt, Germany), Dr. Vladimir Mironov (Russian Academy of Sciences, Moscow, Russia), Prof. Gerd Meyer and Dr. Dirk Hinz-Hübner (Universität zu Köln, Germany). The bulk of the work on metallomesogens in my group at the K.U. Leuven has been carried out by Dr. Rik Van Deun, Dr. Liesbet Jongen, Dr. Jurgen Sleven, and Dr. Katleen Lodewyckx.

References

1. (a) D. DEMUS, J. GOODBY, G. W. GRAY, H. W. SPIESS, V. VILL (Eds.), *Handbook of Liquid Crystals*, Vols. 1–3, Wiley-VCH, Weinheim, **1998**; (b) P. J. COLLINGS, M. HIRD, *Introduction to Liquid Crystals: Chemistry and Physics*, Taylor and Francis, London, **1997**; (c) P. J. COLLINGS, *Liquid Crystals: Nature's Delicate Phase of Matter*, Princeton University Press, Princeton, NJ, **1990**.

2. (a) D. BAHADUR, *Liquid Crystals: Applications and Uses*, Vols. 1–3, World Scientific, Singapore, **1990**; (b) M. SCHADT, *Liq. Cryst.* **1993**, *5*, 57.

3. C. TSCHIERSKE, *J. Mater. Chem.* **1998**, *8*, 1485.

4. (a) A. M. GIROUD-GODQUIN, P. M. MAITLIS, *Angew. Chem. Int. Ed. Engl.* **1991**, *30*, 375; (b) S. A. HUDSON, P. M. MAITLIS, *Chem. Rev.* **1993**, *93*, 861; (c) A. P. POLISHCHUK, T. V. TIMOFEEVA, *Russian*

Chem. Rev. **1993**, *62*, 291; (d) J. L. Serrano (Ed.), *Metallomesogens, Synthesis, Properties and Applications*, VCH, Weinheim, 1996; (e) D. W. Bruce, in *Inorganic Materials*, 2nd Edition (Eds. D. W. Bruce, D. O'Hare), Wiley, Chichester, **1996**, Chapter 8, p. 429; (f) B. Donnio, D. W. Bruce, *Struct. Bonding* **1999**, *95*, 193; (g) B. Donnio, D. Guillon, R. Deschenaux, D. W. Bruce, in *Comprehensive Coordination Chemistry II* (Eds.: J. A. McCleverty, T. J. Meyer), Elsevier, Oxford, **2003**, Volume 7 (Eds.: M. Fujita, A. Powell, C. Creutz), Chapter 7.9, pp. 357–627.

5 K. Binnemans, C. Görller-Walrand, *Chem. Rev.* **2002**, *102*, 2303.

6 Yu. G. Galyametdinov, G. I. Ivanova, I. V. Ovchinnikov, *Bull. Acad. Sci. USSR, Div. Chem. Sci.* **1991**, *40*, 1109 [*Izv. Akad. Nauk SSSR, Ser. Khim.* **1991**, 1232].

7 K. Binnemans, K. Lodewyckx, R. Van Deun, Yu. G. Galyametdinov, D. Hinz, G. Meyer, *Liq. Cryst.* **2001**, *28*, 279.

8 Yu. G. Galyametdinov, G. I. Ivanova, A. V. Prosvirin, O. Kadkin, *Russian Chem. Bull.* **1994**, *43*, 938 [*Izv. Akad. Nauk., Ser. Khim.* **1994**, 1003].

9 Yu. G. Galyametdinov, M. A. Athanassopoulou, K. Griesar, O. Kharitonova, E. A. Soto Bustamante, L. Tinchurina, I. Ovchinnikov, W. Haase, *Chem. Mater.* **1996**, *8*, 922.

10 K. Binnemans, R. Van Deun, D. W. Bruce, Yu. G. Galyametdinov, *Chem. Phys. Lett.* **1999**, *300*, 509.

11 K. Binnemans, D. W. Bruce, S. R. Collinson, R. Van Deun, Yu. G. Galyametdinov, F. Martin, *Phil. Trans. R. Soc. Lond. A* **1999**, *357*, 3063.

12 K. Binnemans, Yu. G. Galyametdinov, R. Van Deun, D. W. Bruce, S. R. Collinson, A. P. Polishchuk, I. Bikchantaev, W. Haase, A. V. Prosvirin, L. Tinchurina, I. Litvinov, A. Gubajdullin, A. Rakhmatullin, K. Uytterhoeven, L. Van Meervelt, *J. Am. Chem. Soc.* **2000**, *122*, 4335.

13 Yu. G. Galyametdinov, W. Haase, L. Malykhina, A. Prosvirin, I. Bikchantaev, A. Rakhmatullin, K. Binnemans, *Chem. Eur. J.* **2001**, *7*, 99.

14 R. Van Deun, K. Binnemans, *J. Alloys Compds.* **2000**, *303*, 146.

15 R. Van Deun, K. Binnemans, *Mater. Sci. Eng. C.* **2001**, *18*, 211.

16 K. Binnemans, D. Moors, T. N. Parac-Vogt, R. Van Deun, D. Hinz-Hübner, G. Meyer, *Liq. Cryst.* **2002**, *29*, 1209.

17 K. Binnemans, Yu. G. Galyametdinov, S. R. Collinson, D. W. Bruce, *J. Mater. Chem.* **1998**, *8*, 1551.

18 Yu. G. Galyametdinov, G. I. Ivanova, I. V. Ovchinnikov, K. Binnemans, D. W. Bruce, *Russian Chem. Bull.* **1999**, *48*, 385 [*Izv. Akad. Nauk.; Ser. Khim.* **1999**, *48*, 387].

19 Yu. G. Galyametdinov, W. Wan, L. Malykhina, M. Darius, W. Haase, *Liq. Cryst.* **2002**, *29*, 1360.

20 F. K. Gainullina, L. V. Malykhina, L. M. Tinchurina, G. I. Ivanova, K. Binnemans, Yu. G. Galyametdinov, *Russian J. Coord. Chem.* **2003**, *5*, 357 [*Koord. Khim.* **2003**, *29*, 382].

21 R. Van Deun, K. Binnemans, *Liq. Cryst.* **2003**, *30*, 479.

22 C. C. Hinckley, *J. Am. Chem. Soc.* **1969**, *91*, 5160.

23 A. F. Cockerill, G. L. O. Davies, R. C. Harden, D. M. Rackman, *Chem. Rev.* **1973**, *73*, 553.

24 J. A. Peters, J. Huskens, D. J. Raber, *Prog. Nucl. Magn. Reson. Spectrosc.* **1996**, *28*, 283.

25 J. H. Forsberg, in *Gmelin Handbook of Inorganic Chemistry, Sc, Y, La–Lu Rare Earth Elements* (System No. 39), D3, Springer, Berlin, **1981**, pp. 65–251 (and references therein).

26 T. Sano, M. Fujita, T. Fujii, Y. Hamada, K. Shibata, K. Kuroki, *Jpn. J. Appl. Phys.* **1995**, *34*, 1883.

27 H. H. Li, S. Inoue, K. Machida, G. Adachi, *Chem. Mater.* **1999**, *11*, 3171.

28 K. Kuruki, Y. Koike, Y. Okamoto, *Chem. Rev.* **2002**, *102*, 2347.

29 S. T. Trzaska, H. X. Zheng, T. M. Swager, *Chem. Mater.* **1999**, *11*, 130.

30 K. Binnemans, K. Lodewyckx, *Angew. Chem.* **2001**, *113*, 248 [*Angew. Chem. Int. Ed.* **2001**, *40*, 242].

31 K. Binnemans, K. Lodewyckx, T. N. Parac-Vogt, R. Van Deun, B. Tinant, K. Van Hecke, L. Van Meervelt, *Eur. J. Inorg. Chem.* **2003**, 3028.

32 Yu. G. Galyametdinov, L. V. Malykhina, W. Haase, K. Driesen, K. Binnemans, *Liq. Cryst.* **2002**, *29*, 1581.

33 Yu. G. Galyametdinov, O. A. Turanova, V. Van, A. A. Knyazev, W. Haase, *Dokl. Chem.* **2002**, *384*, 144 [*Dokl. Akad. Nauk* **2002**, *384*, 206].

34 J. Boyaval, F. Hapiot, C. Li, N. Isaert, M. Warenghem, P. Carette, *Mol. Cryst. Liq. Cryst. A* **1999**, *330*, 1387.

35 F. Hapiot, J. Boyaval, *Magn. Reson. Chem.* **2001**, *39*, 15.

36 H. Nozary, C. Piguet, P. Tissot, G. Bernardinelli, J.-C. G. Bünzli, R. Deschenaux, D. Guillon, *J. Am. Chem. Soc.* **1998**, *120*, 12274.

37 H. Nozary, C. Piguet, J. P. Rivera, P. Tissot, G. Bernardinelli, N. Vuillermet, J. Weber, J.-C. G. Bünzli, *Inorg. Chem.* **2000**, *39*, 5286.

38 H. Nozary, C. Piguet, J. P. Rivera, P. Tissot, P. Y. Morgantini, J. Weber, G. Bernardinelli, J.-C. G. Bünzli, R. Deschenaux, B. Donnio, D. Guillon, *Chem. Mater.* **2002**, *14*, 1075.

39 H. Nozary, C. Piguet, P. Tissot, G. Bernardinelli, R. Deschenaux, M.-T. Vilches, *Chem. Commun.* **1997**, 2101 [*Chem. Commun.* **1997**, 2249 (corrigendum)].

40 E. Terazzi, J. M. Bénech, J. P. Rivera, G. Bernardinelli, B. Donnio, D. Guillon, C. Piguet, *Dalton Trans.* **2003**, 769.

41 A. Bencini, C. Benelli, A. Caneschi, R. L. Carlin, A. Dei, D. Gatteschi, *J. Am. Chem. Soc.* **1985**, *107*, 8128.

42 M. Sakamoto, K. Manseki, H. Okawa, *Coord. Chem. Rev.* **2001**, *379*, 219.

43 K. Binnemans, K. Lodewyckx, B. Donnio, D. Guillon, *Chem. Eur. J.* **2002**, *8*, 1101.

44 M. L. Kahn, T. M. Rajendiran, Y. Jeannin, C. Mathonière, O. Kahn, *C. R. Acad. Sci. II C* **2000**, *3*, 131.

45 K. Binnemans, K. Lodewyckx, *Supramol. Chem.* **2003**, *15*, 485.

46 C. Piechocki, J. Simon, J. J. André, D. Guillon, P. Petit, A. Skoulios, P. Weber, *Chem. Phys. Lett.* **1985**, *122*, 124.

47 K. Binnemans, J. Sleven, S. De Feyter, F. C. De Schryver, B. Donnio, D. Guillon, *Chem. Mater.* **2003**, *15*, 3930.

48 F. Maeda, K. Hatsusaka, K. Ohta, M. Kimura, *J. Mater. Chem.* **2003**, *13*, 243.

49 H. Miwa, N. Kobayashi, K. Ban, K. Ohta, *Bull. Chem. Soc. Jpn.* **1999**, *72*, 2719.

50 Z. X. Zhao, G. F. Liu, *Liq. Cryst.* **2002**, *10*, 1335.

51 K. Binnemans, B. Gündogan, *J. Rare Earths* **2002**, *20*, 249.

52 S. Suarez, O. Mamula, D. Imbert, C. Piguet, J.-C. G. Bünzli, *Chem. Commun.* **2003**, 1226.

53 I. V. Ovchinnikov, Yu. G. Galyametdinov, A. V. Prosvirin, *Russian Chem. Bull.* **1995**, *44*, 768.

54 V. S. Mironov, Yu. G. Galyametdinov, A. Ceulemans, K. Binnemans, *J. Chem. Phys.* **2000**, *113*, 10293.

55 K. Binnemans, L. Malykhina, V. S. Mironov, W. Haase, K. Driesen, R. Van Deun, L. Fluyt, C. Görller-Walrand, Yu. G. Galyametdinov, *ChemPhysChem* **2001**, *2*, 680.

56 V. S. Mironov, Yu. G. Galyametdinov, A. Ceulemans, C. Görller-Walrand, K. Binnemans, *Chem. Phys. Lett.* **2001**, *345*, 132.

57 V. S. Mironov, Yu. G. Galyametdinov, A. Ceulemans, C. Görller-Walrand, K. Binnemans, *J. Chem. Phys.* **2002**, *116*, 4673.

58 A. Turanov, I. Ovchinnikov, Yu. G. Galyametdinov, D. W. Bruce, *Liq. Cryst.* **2001**, *28*, 845.

59 R. D. Larrabee, *RCA Rev.* **1973**, *34*, 329.

60 R. D. Larrabee, US Patent 3,960,753 (June 1, 1976).

61 L. J. Yu, M. M. Labes, *Appl. Phys. Lett.* **1977**, *31*, 719.

62 M. M. Labes, US Patent 4,176,918 (December 4, 1979).

63 J. Boyaval, C. Li, F. Hapiot, M. Warenghem, N. Isaert, Y. Guyot, G. Boulon, P. Carette, *Mol. Cryst. Liq. Cryst.* **2001**, *359*, 337.

64 K. Binnemans, D. Moors, *J. Mater. Chem.* **2002**, *12*, 3374.

65 R. Van Deun, D. Moors, B. De Fré, K. Binnemans, *J. Mater. Chem.* **2003**, *123*, 1520.

16
Rare-Earth Borates:
An Overview from the Structural Chemistry Viewpoint

Jianhua Lin, Yingxia Wang, and Linyan Li

16.1
Introduction

Since boron has the capability of either three- or four-fold coordination with oxygen, borates often exhibit more complex crystal structures than other main group analogues and have attracted considerable attention from chemists and mineralogists over the past decades. In addition, borates have been considered as a large resource for functional materials. For example, with their complex anions and often low symmetry structures, borate materials often exhibit exceptional nonlinear optical properties. BBO (low-temperature form of BaB_2O_4) [1] and LBO-I (lithium triborate, LiB_3O_5) [2] are profoundly important nonlinear optical materials and are widely used in various technological fields. LBO-II (lithium tetraborate, $Li_2B_4O_7$) [3] exhibits piezoelectric and non-ferroelectric properties and has potential applications in high-frequency acoustic wave devices or as an effective temperature compensator.

Rare-earth(III) borates represent a special series in borate chemistry. It is well known that anhydrous polyborates are not stable with cations of high valence and small ionic radius, as expected on the basis of the Lux-Flood concept. The rare-earth(III) ions are in an intermediate situation. Simple rare-earth borates have been known for many years and exhibit high stability even at high temperatures, whereas polyborates have not been widely known and it is now established that they can only be obtained either under mild reaction conditions at ambient pressures, or by high-temperature reactions under high pressure. The rare-earth(III) borates therefore provide ideal model compounds to test the influence of ionic radius on the structures and stability. In addition, many rare-earth borates have been considered as important hosts for luminescent materials, particularly under the excitation of high-energy photons. There have been extensive studies concerning the optical properties of rare-earth borates in recent years, which, although providing abundant information about the structural properties, have often yielded conflicting results.

This chapter examines the current situation concerning rare-earth borates by concentrating on recent developments in synthetic and structural chemistry, although significant progress has also been made on the physical properties

Inorganic Chemistry in Focus II. Edited by G. Meyer, D. Naumann, and L. Wesemann
Copyright © 2005 WILEY-VCH Verlag GmbH & Co. KGaA, Weinheim
ISBN: 3-527-30811-3

and performance of the rare-earth borate materials. In addition, we limit our coverage to binary borate systems. For the specifics of ternary and more complex rare-earth borate systems, we refer the reader to an excellent review by Leskelä et al. [4].

16.2
Rare-Earth Oxyborates

The rare-earth oxyborate La_3BO_6 was identified by Levin et al. [5] in the La_2O_3/B_2O_3 phase diagram. It melts incongruently and decomposes to La_2O_3 and a liquid at about 1386 °C. This compound type was found for the entire rare-earth series, with increasing refractoriness on going from La to Lu, as represented by the increased incongruent melting points (for example, 1590 °C for Gd_3BO_6 [6]). Three different structure types of Ln_3BO_6 were proposed by Bartram et al. [7], which are all monoclinic with the space groups $P2_1/c$ for Ln = La to Nd and $C2/m$, $C2$ or Cm for Ln = Sm to Yb. Lu_3BO_6 also crystallizes in a C-centered monoclinic space group but with different lattice constants. Over the years, this tentative composition of Ln_3BO_6 had not been challenged, that is until recent synthetic studies [8] showed that single-phase products of Ln_3BO_6 could not be prepared from stoichiometric starting materials (Ln/B = 3) even if repeated annealing at high temperatures was employed. The impurity phases were $LaBO_3$ in the case of lanthanum and Y_2O_3 in the yttrium system.

A structural investigation using the single-crystal X-ray diffraction technique [8] established a rather complex composition, $La_{26}(BO_3)_8O_{27}$. In comparison with La_3BO_6, it is located slightly in the La_2O_3-rich region of the La_2O_3/B_2O_3 phase diagram. This explains the presence of $LaBO_3$ as an impurity when the stoichiometric ratio of La_3BO_6 is employed in the synthesis. The structure of $La_{26}(BO_3)_8O_{27}$ is composed of triangular borate groups and isolated oxygen anions. Most of the La ions are in cubic (CN = 8) or defect-cubic (CN = 7) polyhedra, which come together to form fluorite slabs (Fig. 16.1). To fulfil the geometric requirements of the borate groups, one of the lanthanum ions is coordinated in a square antiprism. Adopting a fluorite-related structure is not surprising, because the La/O ratio is in fact close to 2 (LaO_{2-x}, $x \approx 0.0385$), like that in fluorite compounds.

Due to increasingly high refractoriness, single-crystal growth is difficult for the later rare-earth oxyborates. As alternatives, the structural properties were studied for Y_3BO_6 and Gd_3BO_6 by using various techniques such as powder XRD, electron diffraction, and HREM [9, 10]. As shown in Fig. 16.2, the electron diffraction pattern confirmed the C-centered monoclinic cell proposed by Bartram et al. [7] and the proposed model (Cm) is composed of BO_3 and B_2O_5 anions and nine cation sites. Again, it has a complex composition, $Y_{17.33}(BO_3)_4(B_2O_5)_2O_{16}$, which, in contrast to the La case, is located slightly within the B_2O_3-rich region of the Y_2O_3/B_2O_3 phase diagram when compared with Y_3BO_6. However, this model does not account for recent ^{11}B NMR results [6], which indicate the presence of both

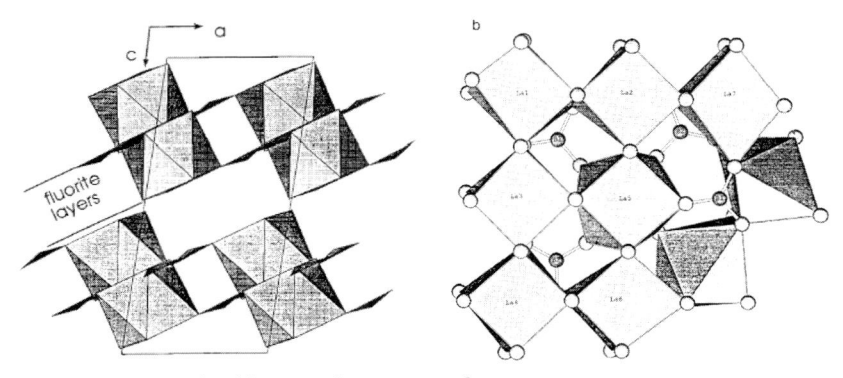

Fig. 16.1 Fluorite-related layers in the structure of $La_{26}(BO_3)_8O_{27}$, drawn along the b axis (left), and along a direction close to [101] (right) [8].

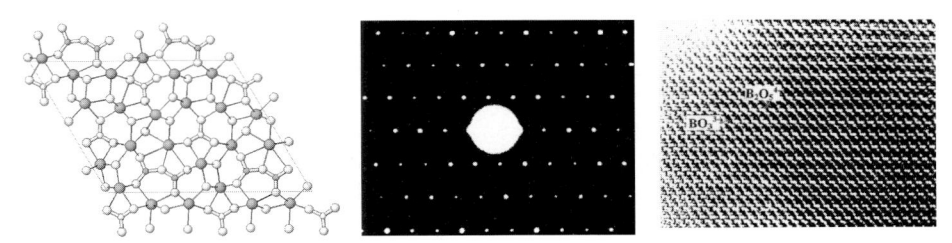

Fig. 16.2 From left to right: Structure, electron diffraction pattern, and HREM image of $Y_{17.33}(BO_3)_4(B_2O_5)_2O_{16}$ along the c axis [9].

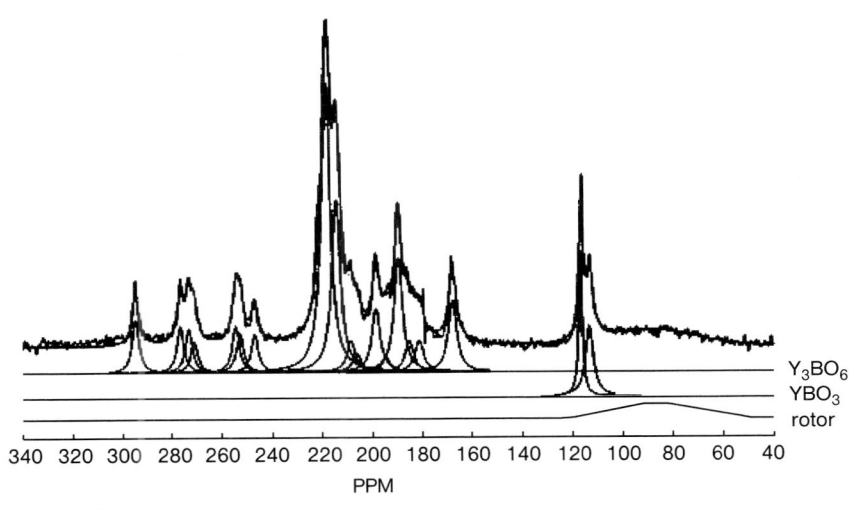

Fig. 16.3 ^{89}Y NMR spectra at 9.4 T for a sample containing Y_3BO_6 and YBO_3. For details of the simulation, see ref. [6].

BO_3 and BO_4 groups in an approximate ratio of 2:1. In addition, as reproduced in Fig. 16.3, the ^{89}Y NMR spectra [6] indicated 15 different Y sites with significantly different populations. It is a difficult puzzle to place such a large number of cation sites within the retained C-centered monoclinic cell. Resolving these problems requires careful examination of the phase relationship and more accurate structural studies, perhaps by means of neutron diffraction techniques. Finally, the structure of Lu_3BO_6, which also crystallizes in a C-centered monoclinic arrangement, has not been studied.

16.3
Rare-Earth Orthoborates

Depending on the ionic radii and the temperature, rare-earth orthoborates ($LnBO_3$) have six polymorphs [4, 11], i.e., L- [12] and $H-LaBO_3$ [13], $L-SmBO_3$ [14], L- [15, 16] and $H-YBO_3$ [16], and $L-LuBO_3$ [17]. The nomenclature of the rare-earth orthoborates follows that of the aragonite, vaterite, and calcite forms of calcium carbonate, since they exhibit similar XRD patterns. Considering the small scattering contribution of the boron atoms in X-ray diffraction, qualitative comparisons were often misleading, and incomplete for compounds that contain a large number of heavy rare-earth cations. Nevertheless, in spite of the structural details, the classification in terms of $CaCO_3$ analogues is useful to distinguish the structures of rare-earth orthoborates.

16.3.1
Aragonite and Calcite-Type Orthoborates

Although the aragonite and calcite structures are valid for the largest (La, Nd) and smallest rare-earth cations (Lu, Sc), respectively, they share a common feature in that the triangular borate groups are aligned parallel to the cation layers. The borate groups in the calcite structure adopt the positions of a close-packed arrangement and are stacked alternately with the cation layers with the sequence $AcBaCbAcBaC$ along the c axis, forming a rhombohedral structure ($R\bar{3}c$). (The capital letters in the sequence represent the positions of the borate groups and the lower-case letters represent the cation positions.) Fig. 16.4a shows the structure of a calcite layer (AcB) projected along the c axis. As the cations are all six-coordinate in calcite, only small rare-earth cations can adopt this structure. The aragonite structure, in which the triangular borate groups also form layers with a nearly close-packed arrangement, can also be described in terms of alternate stacking of cation layers and borate layers (Fig. 16.4b). In the same manner, the structure can be described as $AbA'cAbA'$, in which A and A' represent borate layers at the same positions (boron positions) but with different orientations. The structure of aragonite is orthorhombic ($Pmcn$) and the cations are nine-coordinate. Therefore, only the largest rare-earth cations, such as in $LaBO_3$ and $NdBO_3$, prefer this structure type.

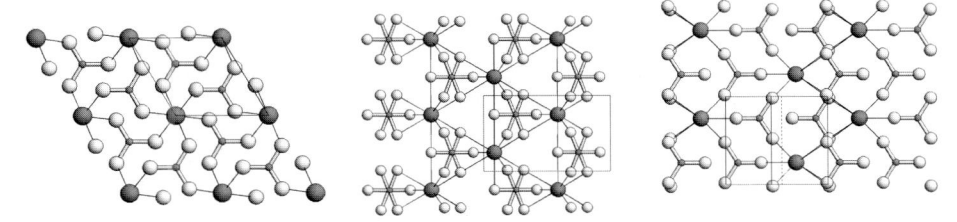

Fig. 16.4 Structures of (from left to right) L-LuBO$_3$, L-LaBO$_3$, and H-LaBO$_3$; one cation layer and two borate layers are shown in projection.

With the parallel arrangement of the borate groups as a criterion, the structure of H-LaBO$_3$ could also be assigned to this family. H-LaBO$_3$, which crystallizes in the space group $P2_1/m$ [13], is also composed of an alternate stacking of (triangular) borate layers and cation layers, although the cation layers in the structure are no longer close-packed (Fig. 16.4c). The structure contains a single cation site with nine-fold coordination and C_s point symmetry. Recently, this structural model received a challenge from optical measurements [13], which suggest that the symmetry of the cation site should be higher than C_s.

16.3.2
Vaterite-Type Orthoborates

Rare-earth orthoborates (Eu to Yb) with the vaterite-type structure constitute a unique example that attracted much attention but still left many controversies. The vaterite form of CaCO$_3$ is composed of an alternate stacking of close-packed cation layers and vertically arranged CO$_3$ layers, which, compared to aragonite and calcite, leads to a significantly smaller a/c ratio. After initial efforts [18, 19] at fitting orthoborates to the CaCO$_3$ vaterite model, spectroscopic studies [20–22] unambiguously revealed that the boron atoms all have tetrahedral rather than triangular coordination in the structure.

A structural study performed on single crystals of YBO$_3$ by Chadeyron et al. [15] demonstrated that the boron atoms are indeed tetrahedrally coordinated and connected to B$_3$O$_9$ groups. The structure was assigned the space group $P6_3/m$ with the cell constants $a = 3.776$ and $c = 8.806$ Å. In the structure, the yttrium and terminal oxygen atoms are well-defined at fully occupied specific positions, 2(b) and 4(f), respectively, which constitute a rock salt (111) layer, while the boron and the bridged oxygen atoms of the B$_3$O$_9$ groups are randomly distributed over the 6(h) positions with partial occupancies of 1/3. A subsequent electron diffraction study [16], however, revealed super-reflections (Fig. 16.5). If the super-reflections are included, a rhombohedral cell with $a = 6.6357$ Å and $c = 20.706$ Å is derived for GdBO$_3$. Thus, the hexagonal cell of Chadeyron's model [15] is a sub-cell of the rhombohedral lattice. Despite the incomplete cell setting, the hexagonal structure does provide a model to understand the vaterite-

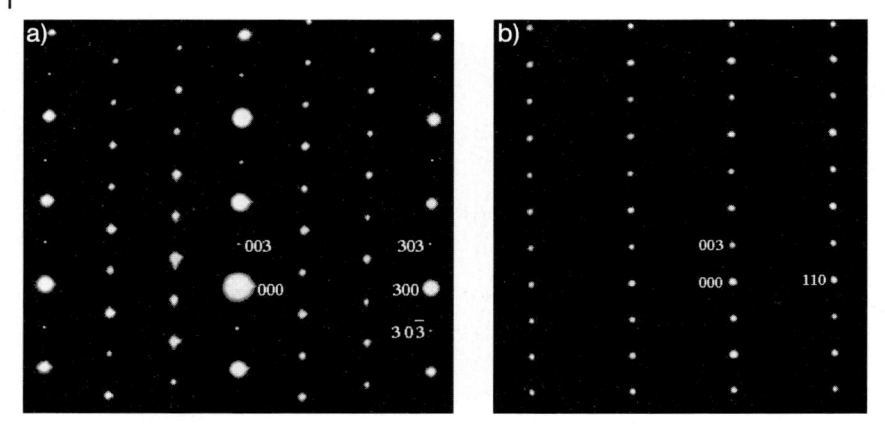

Fig. 16.5 Electron diffraction patterns of YBO$_3$ in the (left) [010] and (right) [1$\bar{1}$0] directions [16].

type orthoborates. However, one should keep in mind that the hexagonal structure is an averaged model, for which atomic parameters were not appropriately constrained and refined. Hence, great caution should be exercised when structural details are used to explain physical properties.

Based on electron diffraction, Ren et al. [16] proposed a rhombohedral structure model and refined the structure with powder X-ray diffraction data. The structure contains a single cation site of C_1 symmetry. This, however, is at variance with the high-resolution luminescence spectra, which indicated the presence of at least two Eu^{3+} sites in the doped material [23, 24]. In the rhombohedral model of GdBO$_3$, the gadolinium and terminal oxygen atoms form a rock salt (111) layer, which is exactly the same as in the other given models. Thus, the problem with this new model is again that the positions of the boron and bridged oxygen atoms are generated by high pseudo symmetry. To precisely determine the atomic positions of boron and bridged oxygen, one needs to use neutron diffraction techniques, since the coherent elastic scattering lengths of rare earths, oxygen, and boron are of the same order of magnitude for neutrons. The structure of Y$_{0.92}$Er$_{0.08}$^{11}BO$_3$ was recently revised with the aid of neutron diffraction data [25, 57]. It crystallizes in the monoclinic space group $C2/c$ (a = 11.3138, b = 6.5403, c = 9.5499 Å and β = 112.902°). This monoclinic cell is in fact a lower symmetric derivative of the rhombohedral lattice observed by electron diffraction. The major structural feature is similar to that of both the hexagonal and rhombohedral models and can be described as an alternate stacking of close-packed cation layers and B$_3$O$_9$ layers (Fig. 16.6a). Unlike in the other models, the bridged oxygen and boron atom sites are fully occupied in the B$_3$O$_9$ layer and, in addition, the two eight-coordinated cation sites, Y1 (4d) and Y2 (8f), are all well-defined in the symmetry of C_i and C_1 (Fig. 16.6b).

The luminescence spectrum of Eu^{3+} ions is sensitive to the local symmetry and this is widely used as a probe to study structural properties. This technique

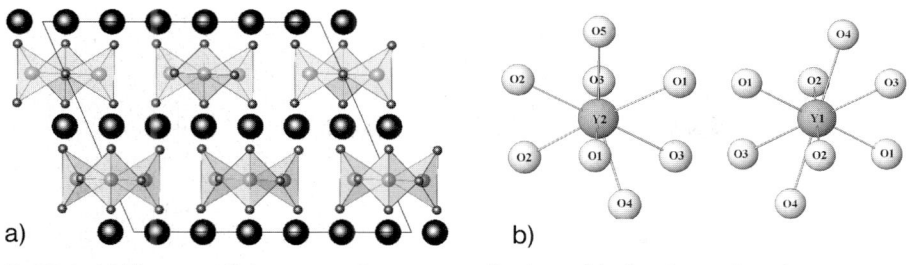

Fig. 16.6 (a) The monoclinic structure for YBO$_3$ projected along the b axis and (b) the coordination polyhedra of Y1 (4d) and Y2 (8f) atoms [25].

is of particular significance for rare-earth orthoborates, since Eu^{3+}-doped rare-earth orthoborates are themselves important luminescence materials used in plasma display panels and mercury-free lamps. The high-resolution luminescence spectrum of YBO$_3$:Eu reported by Hölsä [26] was indicative of two cation sites, assigned respective symmetries of D_{3d} and T. Chadeyron et al. developed a soft-chemistry synthetic approach [24] that can greatly enhance the emission intensity of YBO$_3$:Eu. They identified three $^5D_0$–$^7F_0$ peaks and attributed two to the intrinsic cation sites (C_3 symmetry) and the third, the intensity of which proved to be dependent on the synthesis conditions, to a perturbed cation site.

Boyer et al. [27, 28] proposed, on the other hand, three intrinsic cation sites in the YBO$_3$ structure based on the hexagonal model by inferring ordered distribution of the bridged oxygens. As indicated in Fig. 16.7, polyhedron I has an inversion center and the geometries of the polyhedra II and III are similar except for slight differences in the distances and angles. From a structural point of view, this description implies lower symmetry and an enlarged unit cell. Having the pseudo-rhombohedral lattice preserved, the symmetry could be further reduced to the triclinic system ($P\bar{1}$) by removing the twofold axis from the monoclinic cell. The cell volume of this hypothetical triclinic structure should be three times larger than that of the hexagonal cell and thereby four cation positions could be derived, of which two are at inversion centers (1(a) and 1(h)) and the other two at the general positions (2(i)) [29]. Thus, one should expect different $^5D_0$–$^7F_0$ transition peaks in the spectrum. Nevertheless, great caution should be exercised when the luminescence probe is used to interpret structural properties. The luminescence spectra do provide information on the symmetry around the probe centers, but, at the same time, they are also very sensitive to distortions induced by the probe itself or by imperfections in the crystals. A recent XAFS study [30] on YBO$_3$:Eu showed that nano-size particles are subject to considerable lattice disorder that consequently has significant influence on the luminescence spectra [31]. In addition, an ^{89}Y MAS NMR study on YBO$_3$, which accounts for the properties of the bulk material, unambiguously shows two distinct Y sites with a population ratio of about 1:2 [6], which supports the C-centered monoclinic structure model. In the light of all the available evidence, the monoclinic model is the most appropriate model for the vaterite-type rare-earth

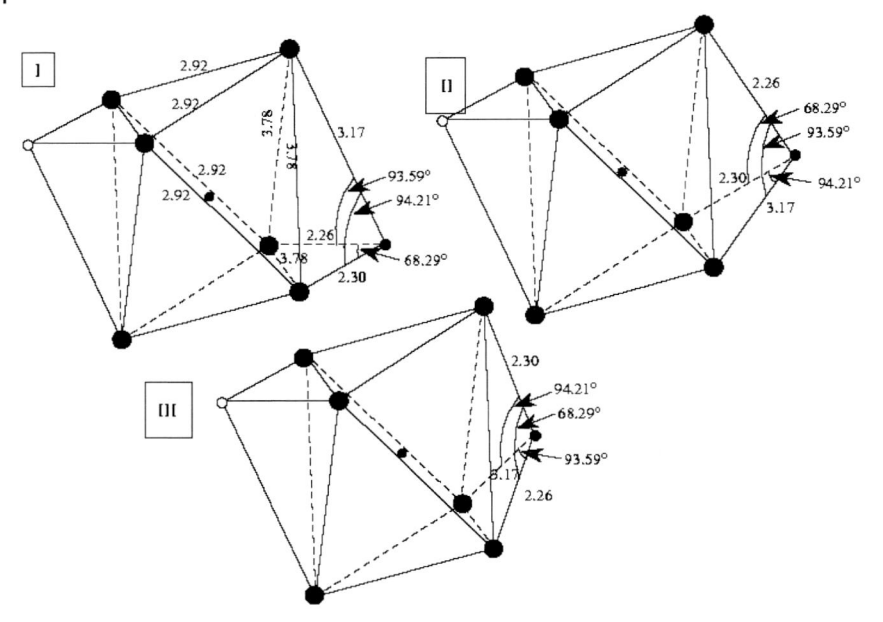

Fig. 16.7 Three kinds of YO$_8$ polyhedra in YBO$_3$ vaterite [27].

orthoborates. The other rare-earth orthoborates from Eu to Yb, as well as the high-T polymorph of LuBO$_3$, should also crystallize with this structure.

The unquenchable high-temperature phases of vaterite-type rare-earth orthoborates (H-YBO$_3$) have been known for many years [5]. Although a high-temperature ^{11}B NMR study [6] and a preliminary structural study [16] both indicated the presence of triangular BO$_3$ borate groups, an appropriate structure model had not been established prior to recent neutron diffraction studies on Y$_{0.92}$Er$_{0.08}$^{11}BO$_3$ [25]. H-YBO$_3$ also crystallizes in the monoclinic space group C2/c, with $a=12.2019$, $b=7.0671$, $c=9.3424$ Å, and $\beta=115.347°$. In comparison with L-YBO$_3$, the a-b plane is expanded whereas the (001) spacing is decreased. Fig. 16.8a displays the structure of H-YBO$_3$. It can be seen that this structure can also be described as an alternate stacking of triangular BO$_3$ borate layers and the rare-earth cation layers, just like that in the typical CaCO$_3$ vaterite. However, the orientation of the triangular borate groups is different and in the H-YBO$_3$ form they are related by a pseudo threefold axis.

Comparing the structures of L- and H-YBO$_3$, the phase transition involves the breaking and formation of three B–O bonds in the cyclic B$_3$O$_9$ groups and is accompanied by a discontinuous volume change as shown in Fig. 16.9. One can thus expect a strong influence of pressure on the phase transitions. The phase transitions under high pressure were studied many years ago by Meyer et al. [32]. They found that the vaterite orthoborate (L-YBO$_3$) is transformed into the H-LaBO$_3$ structure at 1300 °C under a pressure of 6.5 GPa. Recently, Huppertz

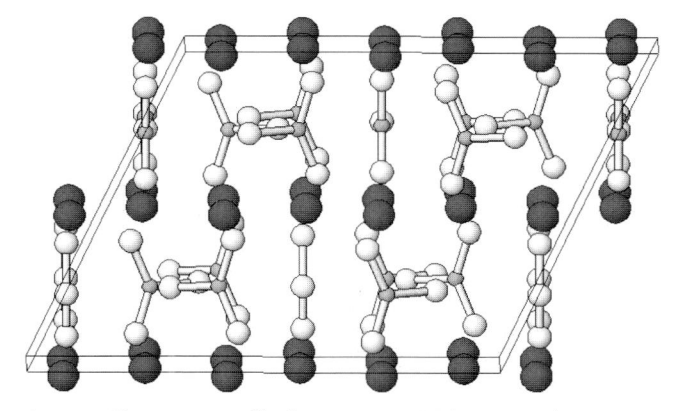

Fig. 16.8 The structure of high-temperature YBO$_3$ projected along the c-axis [25].

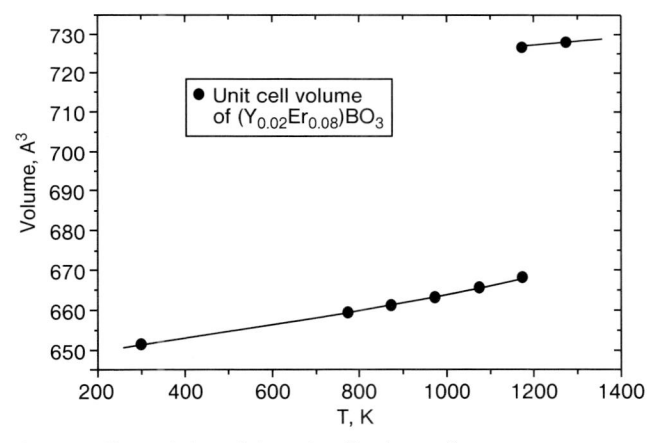

Fig. 16.9 The variation of the unit cell volume of Y$_{0.92}$Er$_{0.08}$^{11}BO$_3$ with temperature.

et al. [33] studied the phase transformation under higher pressure and found a new orthoborate polymorph (χ-DyBO$_3$). The structure of χ-DyBO$_3$ is interesting and may be considered as an intermediate phase between L-YBO$_3$ and H-YBO$_3$. The structure is triclinic and contains a non-cyclic B$_3$O$_9$ unit consisting of one BO$_3$ and two BO$_4$ groups. As shown in Fig. 16.10, the structure of χ-DyBO$_3$ can be described as an alternate stacking of borate and cation layers. As indicated in Fig. 16.11, the cyclic B$_3$O$_9$ group present in L-YBO$_3$ contains only tetrahedral borate, BO$_4$. At high temperatures, three B–O bonds in the cyclic B$_3$O$_9$ are broken resulting in isolated BO$_3$ groups in the structure of H-YBO$_3$. During the phase transition, the cations almost maintain the close-packed arrays and, furthermore, the triangular BO$_3$ groups retain pseudo threefold symmetry as in

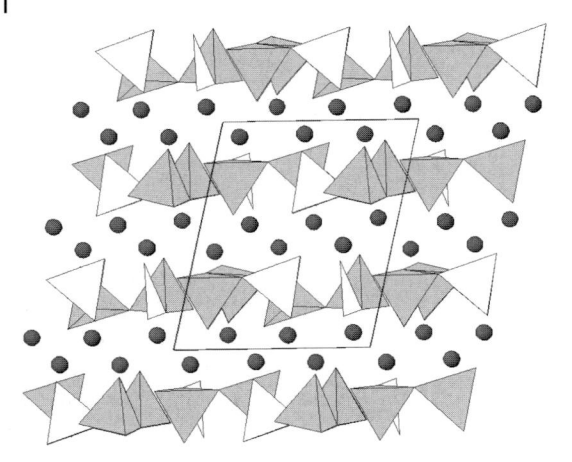

Fig. 16.10 The structure of χ-DyBO$_3$. The light polyhedra are triangular BO$_3$ and the dark ones are tetrahedral BO$_4$.

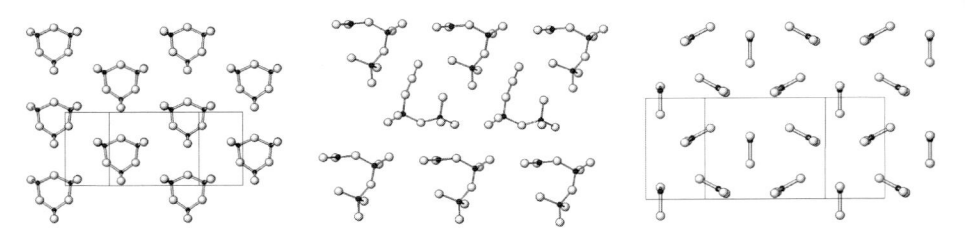

Fig. 16.11 The structures of the borate layers in (from left to right) L-YBO$_3$, χ-DyBO$_3$, and H-YBO$_3$.

the L-YBO$_3$ structure. Under high pressure, only one B–O bond per B$_3$O$_9$ group is broken, leading to a non-cyclic B$_3$O$_9$ group in the χ-DyBO$_3$ structure. In fact, χ-DyBO$_3$ has a denser structure ($\rho = 7.113$ g cm^{-3}) compared to that of L-DyBO$_3$ ($\rho = 6.68$ g cm^{-3}). In addition, an in situ X-ray diffraction study under ambient pressure shows that χ-DyBO$_3$ transforms to an unknown DyBO$_3$ phase at about 400 °C, then to the H-YBO$_3$ structure at about 920 °C, and finally to the L-YBO$_3$ structure on cooling to about 680 °C.

As far as the structure of the borate layer is concerned, L-SmBO$_3$ [4] can also be considered as a vaterite-related structure type. The nomenclature of this phase was somewhat confusing, since both L-SmBO$_3$ and H-SmBO$_3$ were used in the literature. The reason for this situation is that this phase can be prepared at either low or high temperatures, depending on the reaction conditions employed. Reactions under mild conditions, for example by means of the sol-gel method, produced this phase below 700, 800, and 900 °C for Nd, Sm, and Eu, respectively [13]. In situ X-ray diffraction measurements [23] show that L-GdBO$_3$ and L-EuBO$_3$ are stable in the temperature range below 1000 °C. Yin et al. [34]

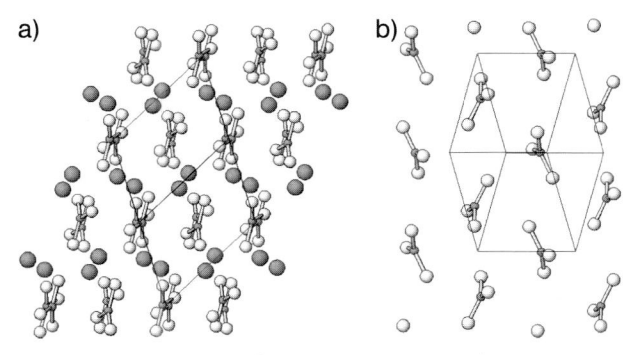

Fig. 16.12 The structure of L-EuBO$_3$ (a) projected along approximately the [112] direction and (b) the arrangement of triangular borate groups in the borate layer [14].

reported the synthesis of EuBO$_3$ and GdBO$_3$ in this structure by means of conventional solid-state reactions at 900 and 850 °C, respectively. Above 1200 °C, this structure can also be formed with large rare-earth cations such as Nd and Sm [23]. Meanwhile, in between these two temperature regions, aragonite or vaterite structures are obtained. At present, there is no evidence to suggest that the phases obtained at low and high temperatures are different in their structures. Nevertheless, all structural studies carried out to date have been conducted with crystals obtained below 900 °C. Thus, it would seem more appropriate to use "L-SmBO$_3$" for this polymorph. The structure of this phase was initially determined by Palkina et al. [35] in the triclinic space group $P1$. Recently, Corbel et al. [14] revised the structure of L-EuBO$_3$ in the centric space group $P\bar{1}$. Although the symmetry of the structure is quite low and the cations show significant deviations from an ideal close-packed layer, the structure retains the fundamental features of vaterite (Fig. 16.12) and can also be described as an alternate stacking of rare-earth cation layers and triangular borate layers. In addition, the borate groups are aligned vertically with respect to the cation layers just as in the typical vaterite. The rare-earth cations are all eight-coordinated in triangulated dodecahedra [14].

Tab. 16.1 lists structural data for rare-earth orthoborates and CaCO$_3$. In addition to the real lattice parameters, reduced hexagonal cell constants are also included in this table. The reduced cell is defined as a pseudo primary hexagonal cell that contains only one LnBO$_3$ formula unit. This definition allows one to compare geometric differences between structures of different symmetries. It can be seen that the a_r/c_r ratio can be used to classify the structure types of rare-earth orthoborates. The calcite and aragonite structures normally have greatly reduced a_r/c_r ratios since the triangular anions adopt a nearly parallel arrangement within the a-b plane. The difference between the calcite and aragonite structures lies in the position of the triangular anions, which is controlled mainly by the cation radii and the required coordination number, i.e., six-coordinate in calcite and nine-coordinate in aragonite, respectively. On the other hand,

Tab. 16.1 Lattice parameters (Å, deg.) of selected rare-earth orthoborates and $CaCO_3$

Structure type	Compound	a	b	c	α, β, γ	S.G.	C.N.	a'	c'	a'/c'
Aragonite	$LaBO_3$	5.104	8.252	5.872		$Pmcn$	9	4.851	2.936	1.652
	$CaCO_3$	4.961	7.967	5.741		$Pmnc$	9	4.693	2.870	1.635
	$H\text{-}LaBO_3$	6.3481	5.0841	4.1861	107.891	$P2_1/m$	9	5.08	2.87	1.77
Calcite	$LuBO_3$	4.915		16.212		$R\bar{3}c$	6	4.915	2.702	1.819
	$CaCO_3$	4.989		17.053		$R\bar{3}c$	6	4.989	2.842	1.755
Vaterite	$CaCO_3$	4.13	7.15	8.48		$Pbnm$	8	4.134	4.235	0.975
	$L\text{-}EuBO_3$	6.468	6.477	6.220	107.8, 108.0, 93.1	$P\bar{1}$	8	4.08	4.23	0.965
	$H\text{-}YBO_3$	12.205	7.067	9.344	115.4	$C2/c$	6	4.067	4.224	0.963
	$L\text{-}YBO_3$	11.314	6.540	9.550	112.9	$C2/c$	8	3.771	4.399	0.857
	$H\text{-}LuBO_3$	3.727		8.722		$P6_3/mmc$	8	3.727	4.361	0.855

the triangular anions are vertically arranged in the vaterite structure leading to a less reduced a_r/c_r ratio. In addition, the vaterite structures may be further classified into two groups. The low-temperature orthoborate structure exhibits an even smaller a_r/c_r ratio than the high-temperature orthoborate and the vaterite $CaCO_3$. Finally, it is noteworthy that the structures of most of the rare-earth orthoborates are different from the corresponding $CaCO_3$ structures. Nevertheless, classification in terms of the $CaCO_3$ structure types, i.e., aragonite, calcite, and vaterite, does convey the fundamental features of the rare-earth orthoborate structures and hence can be used to describe the structural properties of these phases.

16.4
Anhydrous Rare-Earth Polyborates

Borate groups, BO_3 and BO_4, may share corners to form infinite chains, layers or three-dimensional frameworks in polyborates. The tendency to form anhydrous polyborates decreases for cations with high positive charges and small ionic radii [36, 37]. In practice, cations of low valence and large radius often form a large number of anhydrous polyborates, whereas borates with cations of high valence are rare. This tendency is clearly manifested in the rare-earth borate systems. Eu(II) has a number of anhydrous polyborates [38, 39], isostructural with the alkaline earth polyborates, whereas Ce(IV) borates have not yet been observed. Ln(III) is in an intermediate situation, that is to say, the polyborates of Ln(III) may exist but their stability is low, particularly at high temperatures. Moreover, one would expect the stabilities of the polyborates to decrease on going from La to Lu in the rare-earth series.

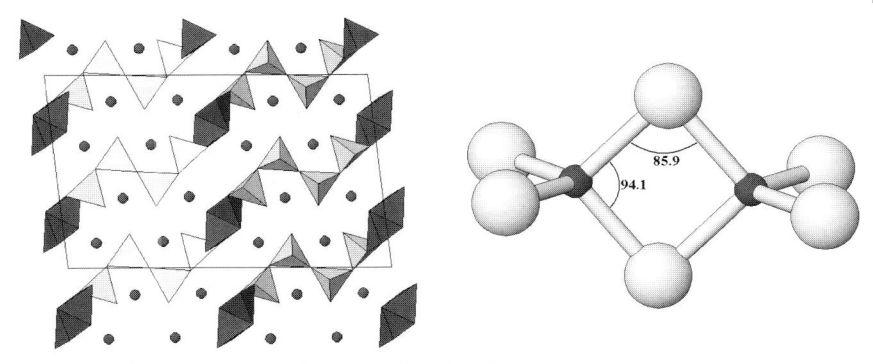

Fig. 16.13 The structure of $Dy_4B_6O_{15}$ and the edge-sharing tetrahedral borate [40].

16.4.1
$Ln_4B_6O_{15}$

The rare-earth borate $Dy_4B_6O_{15}$ was obtained at 1000 °C under a pressure of 8 GPa [40]. The structure of $Dy_4B_6O_{15}$ is built up from corrugated layers of linked BO_4 tetrahedra, between which the rare-earth atoms are located (Fig. 16.13a). Within the borate layer, the BO_4 groups are linked via shared corners and edges. It has been well established that oxide tetrahedra normally link to each other through sharing corners. This is, in fact, a general rule that is well accepted for oxide systems. The edge-sharing BO_4 identified in this compound is, therefore, remarkable since it provides an exceptional example which may stimulate reconsideration of the structure rules. The bond angles of the edge-sharing borate (Fig. 16.13b) are rather small in comparison with typical tetrahedral geometry. This means that the borate group is subject to considerable tension and thus may only be stable under high pressures.

16.4.2
$Ln_2B_4O_9$

The rare-earth borates with ratios of $Ln_2O_3:B_2O_3=1:2$ were also obtained under high pressure at high temperatures. Two polymorphs have been identified so far, which are α-$Ln_2B_4O_9$ (Ln = Eu, Gd, Tb, and Dy) [41, 42] and β-$Dy_2B_4O_9$ [43], respectively. The α polymorph, obtained at 1150 °C under a pressure of 10 GPa, crystallizes in a monoclinic structure in the space group $C2/c$. Fig. 16.14a shows the crystal structure of α-$Ln_2B_4O_9$ projected along the [010] direction. The structure of α-$Ln_2B_4O_9$ exhibits a complex network of linked BO_4 tetrahedra. It also contains edge-sharing tetrahedra, as shown by the dark polyhedra in the figure. The fundamental building block in the tetrahedral framework (Fig. 16.14b) consists of two centered edge-sharing tetrahedra surrounded by eighteen corner-sharing tetrahedra. The coordination polyhedra of the rare-earth atoms are irreg-

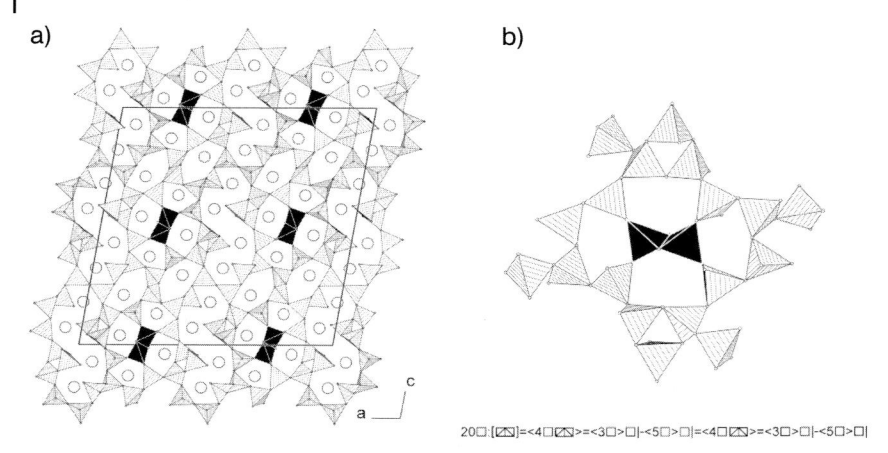

Fig. 16.14 (a) The crystal structure of α-Ln$_2$B$_4$O$_9$ projected along the [010] direction and (b) the fundamental building block in the structure [41].

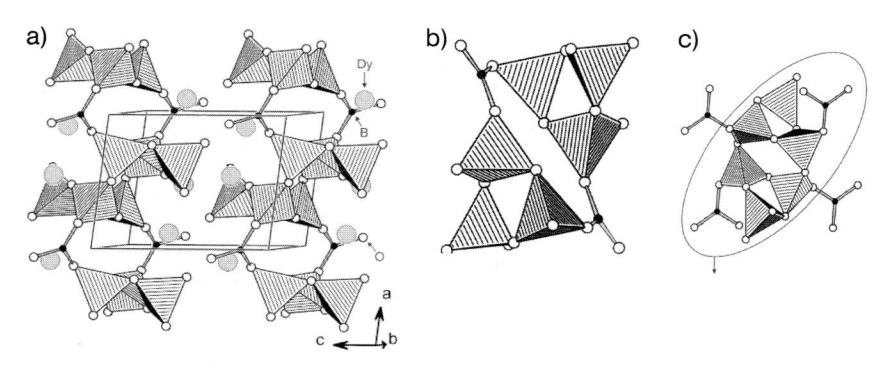

Fig. 16.15 (a) The crystal structure of β-Dy$_2$B$_4$O$_9$ and the linkage of (b) six-membered rings and (c) four-membered rings in the structure [43].

ular and the coordination number ranges from 8 to 11 in this structure. The β polymorph was obtained at 1000 °C under a pressure of 8 GPa. It crystallizes in a triclinic structure in the space group $P\bar{1}$. The borate framework in β-Dy$_2$B$_4$O$_9$ is that of a layered structure consisting of cyclic tetrahedral B$_3$O$_9$ groups and triangular BO$_3$ groups (Fig. 16.15a). The B$_3$O$_9$ group is a common borate unit which has been discussed in detail for the vaterite rare-earth borates. In the β-Dy$_2$B$_4$O$_9$ structure, the B$_3$O$_9$ groups are linked to the next B$_3$O$_9$ ring via two triangular BO$_3$ forming a six-membered ring (Fig. 16.15b). Additionally, the B$_3$O$_9$ ring is connected to the next B$_3$O$_9$ ring via two corners forming a four-membered ring unit (Fig. 16.15c). The Dy atoms in the structure are nine- and ten-coordinate, respectively.

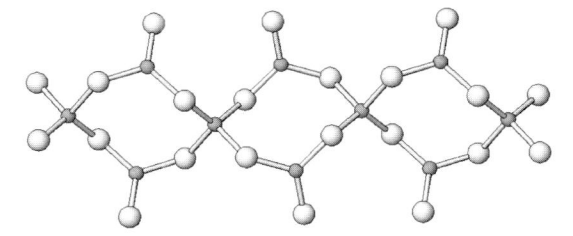

Fig. 16.16 The borate chain as observed in metaborate structures.

16.4.3
Rare-Earth Metaborates

Rare-earth metaborates, LnB_3O_6, have been known for many years [7] and can be readily synthesized by conventional solid-state reactions at high temperatures [44]. Bambauer et al. [45, 46] reported that LnB_3O_6 (Ln = La to Tb) are all isostructural and crystallize in a monoclinic structure and feature an infinite borate chain formed by BO_4 and BO_3 groups [47], as shown in Fig. 16.16. The chains are parallel and bound to rare-earth ions with tenfold coordination. Pakhomov et al. [48] indicated that TbB_3O_6 crystallizes in an orthorhombic rather than monoclinic structure and hence is not isostructural with the other LnB_3O_6. In addition, Tananaev et al. [49] indicated the presence of metaborates for the other heavy rare-earth ions and claimed that there are two other structure types for Ln = Dy to Tm and Ln = Yb and Lu, respectively. These polymorphs, however, have not yet been confirmed by other studies and, even if they do exist, could only be prepared under mild reaction conditions because of the low stability expected on the basis of the Lux-Flood acid-base concept. Recently, Sieke et al. [50] showed that $Pr(BO_2)_3$ synthesized at 850 °C has triclinic symmetry ($P\bar{1}$) and contains identical borate chains. It is also noteworthy that recent studies have shown that rare-earth metaborates have promising luminescence properties, particularly under excitation by VUV light, which may offer potential applications as luminescent materials in PDP display panels.

16.4.4
Rare-Earth Pentaborates

To obtain more borate-rich anhydrous polyborate phases, Li et al. [51–53] proposed a new synthetic approach under mild conditions. They first carried out reactions in a flux of boric acid at low temperatures (200 °C) in a closed system and then annealed the new hydrated rare-earth polyborates thus obtained at moderate temperatures. With this approach, they were able to isolate two polymorphs of rare-earth pentaborates, LnB_5O_9.

The pentaborates α-LnB_5O_9 (Ln = Pr to Eu) crystallize in a tetragonal structure in the space group $I4_1/acd$, with lattice constants of $a = 8.23813(6)$ Å and $c = 33.6377(4)$ Å for GdB_5O_9 [51, 52]. The fundamental building blocks in the α-LnB_5O_9 structure are double-three-membered rings of B_4O_9 and BO_3 units. The

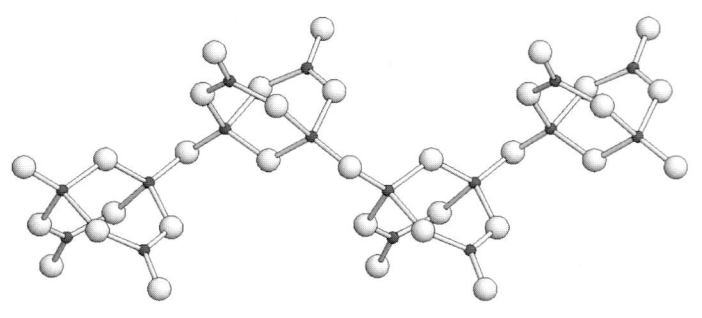

Fig. 16.17 A borate chain as observed in α-LnB$_5$O$_9$ formed by connection of B$_4$O$_9$ units [53].

B$_4$O$_9$ units are linked via terminal oxygen atoms (on BO$_4$) forming one-dimensional chains (Fig. 16.17), which are further connected to an additional BO$_3$ group forming a three-dimensional framework (Fig. 16.18). The Gd atoms occupy the cavities of the borate framework, within a distorted tricapped trigonal-prismatic polyhedron.

The anhydrous pentaborates β-LnB$_5$O$_9$ are only stable with large rare-earth cations (Ln = La and Ce) [53] and crystallize in the monoclinic space group $P2_1/c$ with lattice constants $a = 6.4418(1)$, $b = 11.6888(3)$, $c = 8.1706(2)$ Å, and $\beta = 105.167(1)°$ for LaB$_5$O$_9$. Fig. 16.19a shows the structure projected along the [100] direction. The fundamental borate building block is a [B$_5$O$_{12}$] (2T+3Δ) unit consisting of a three-membered ring (2T+Δ) and two BO$_3$ (Δ) groups. The three-membered ring units and BO$_3$ groups are linked forming a nine-membered ring borate sheet (Fig. 16.19b). The three-dimensional borate framework is constructed by linking the nine-membered ring sheets through additional BO$_3$ groups (Fig. 16.19a). The rare-earth cations are located in the cavities of the framework within nine-coordinate polyhedra.

The pentaborates LnB$_5$O$_9$ are not stable and decompose to other simple borates at high temperatures. Fig. 16.20 presents a phase diagram for the rare-earth borate system. The decomposition temperatures of hydrated hexaborates are almost constant at about 600 °C across the rare-earth series, whereas the stability ranges of the pentaborates depend strongly on the ionic radii of rare-earth cations. Lanthanum pentaborate (β-LaB$_5$O$_9$) is quite stable and no decomposition is observed up to 1000 °C. With decreasing ionic radius, the stability of the pentaborates decreases, and, for the smallest rare-earth cations (Ln = Yb and Lu), the pentaborates do not exist. The decomposition products of the pentaborates also depend on the radii of the rare-earth cations; thus, metaborates are formed with La to Tb and orthoborates with Dy to Tm. This observation provides evidence that the metaborates might not exist for the heavy rare-earth cations from Dy to Lu. Finally, the LnB$_5$O$_9$ borates are well-defined anhydrous borates and should appear in the Ln$_2$O$_3$/B$_2$O$_3$ phase diagram. The absence of this phase from the Ln$_2$O$_3$/B$_2$O$_3$ phase diagram [6] may be attributed to low decomposition temperatures. To summarize the structural information, Tab. 16.2 lists the lattice constants and the space groups for all of the anhydrous pentaborates.

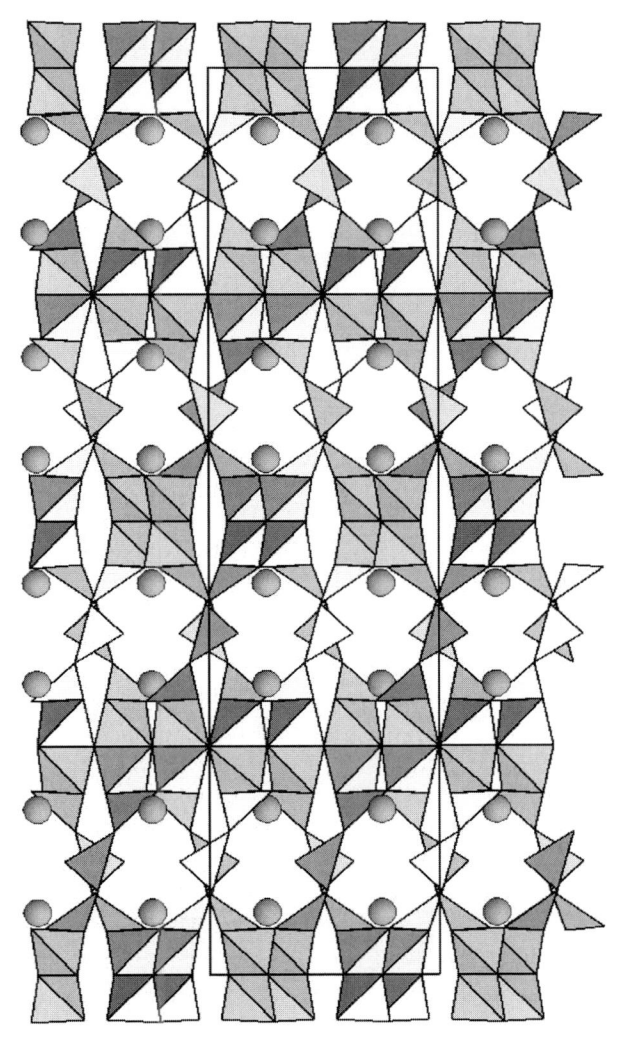

Fig. 16.18 The structure of α-LnB$_5$O$_9$; the borate groups BO$_4$ and BO$_3$ are represented by triangles and tetrahedra, respectively [53].

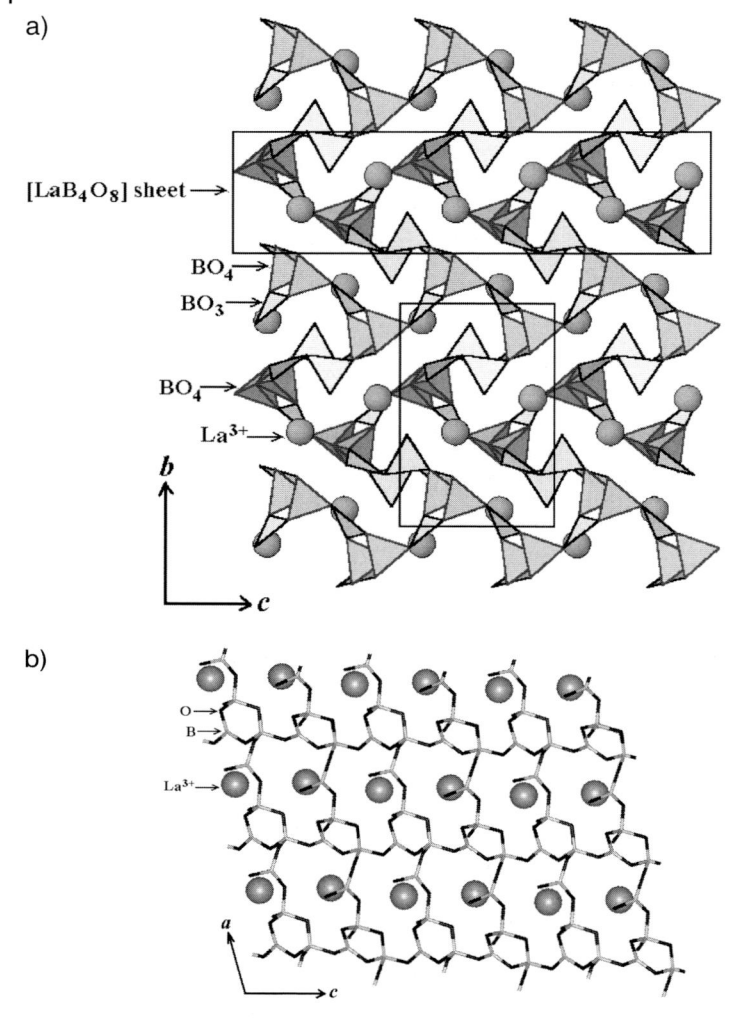

Fig. 16.19 (a) The structure of β-LnB$_5$O$_9$ projected along [100] direction, and (b) the structure of the buckled borate layer [52].

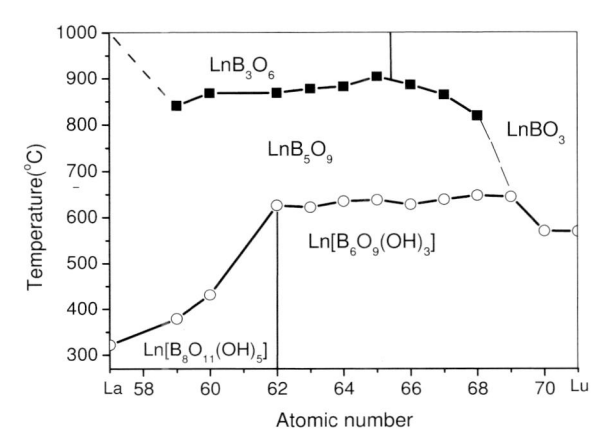

Fig. 16.20 "Phase diagram" for rare-earth polyborates. The two lines represent the dehydration temperature of the hydrated rare-earth polyborates and the decomposition temperatures of LnB_5O_9 to $Ln(BO_2)_3$ or $LnBO_3$ [51].

Tab. 16.2 Lattice constants (Å, deg.) of rare-earth pentaborates

Compound	S.G.	a (Å)	b (Å)	c (Å)	$\beta(°)$
β-LaB$_5$O$_9$	$P2_1/c$	6.4418(1)	11.6888(3)	8.1706(2)	105.167(1)
a-PrB$_5$O$_9$	$I4_1/acd$	8.3912(2)		34.0541(16)	
a-NdB$_5$O$_9$	$I4_1/acd$	8.3120(1)		33.7962(12)	
a-SmB$_5$O$_9$	$I4_1/acd$	8.2789(4)		33.7320(16)	
a-EuB$_5$O$_9$	$I4_1/acd$	8.2556(2)		33.6625(13)	
a-GdB$_5$O$_9$	$I4_1/acd$	8.2381(6)		33.6377(4)	
a-TbB$_5$O$_9$	$I41/acd$	8.2178(2)		33.5759(11)	
a-DyB$_5$O$_9$	$I4_1/acd$	8.2033(2)		33.5165(12)	
a-HoB$_5$O$_9$	$I4_1/acd$	8.1769(2)		33.4525(16)	
a-ErB$_5$O$_9$	$I4_1/acd$	8.1546(8)		33.4176(63)	

16.5
Hydrated Rare Earth Polyborates

Hydrated polyborates are common for alkali and alkaline earth cations but have been less frequently encountered for rare-earth systems due to the strong tendency for hydrolysis of Ln(III) cations. Hydrated rare-earth polyborates can only be synthesized by using molten boric acid as a medium. It is well known that boric acid may be dehydrated in a stepwise manner and polymerized to metaboric acid and then to boron oxide [54].

$$H_3BO_3 \xrightarrow{175°C} HBO_2 \xrightarrow{230°C} B_2O_3 \quad (1)$$

This is a condensation process and, in a closed system, all these boric acid species reach an equilibrium with water. Obviously, introducing rare-earth oxides into the reaction system may enhance the polymerization process, and consequently leads to the formation of hydrated rare-earth polyborates in almost quantitative yields. Four different types of hydrated rare-earth polyborates have been obtained by this method [51–53], the formation reactions of which can be summarized as follows:

$$16H_3BO_3 + Ln_2O_3 \xrightarrow{240\,°C} 2Ln[B_8O_{11}(OH)_5] + 19H_2O \quad Ln = La - Nd \quad (2)$$

$$18H_3BO_3 + Ln_2O_3 \xrightarrow{240\,°C} 2Ln[B_9O_{13}(OH)_4] \cdot H_2O + 21H_2O \quad Ln = Pr - Eu \quad (3)$$

$$10H_3BO_3 + 2Ln(NO_3)_3 \xrightarrow{240\,°C} 2Ln[B_5O_8(OH)]NO_3 \cdot 3H_2O + 4NO_2 + O_2$$
$$+ 8H_2O \quad Ln = La, Ce \quad (4)$$

$$12H_3BO_3 + Ln_2O_3 \xrightarrow{240\,°C} 2H_3LnB_6O_{12} + 15H_2O \quad (5)$$

The structures of these four hydrated polyborates are different, but they can be classified into three groups: The first group contains the octaborates $Ln[B_8O_{11}(OH)_5]$ (Ln=La–Nd) and nonaborates $Ln[B_9O_{13}(OH)_4]\cdot H_2O$ (Ln=Pr–Eu), which have structures related to tunellite, $Sr[B_6O_9(OH)_2]\cdot 3H_2O$ [55]. The fundamental borate building block in the tunellite structure is an $[MB_6O_{11}]$ layer (B_6-layer), as shown in Fig. 16.21a, which contains nine-membered borate rings in which the metal cations are located. The tetrahedral boron labeled as T' in the figure is an active site that can further link to other borate groups to form polyborates

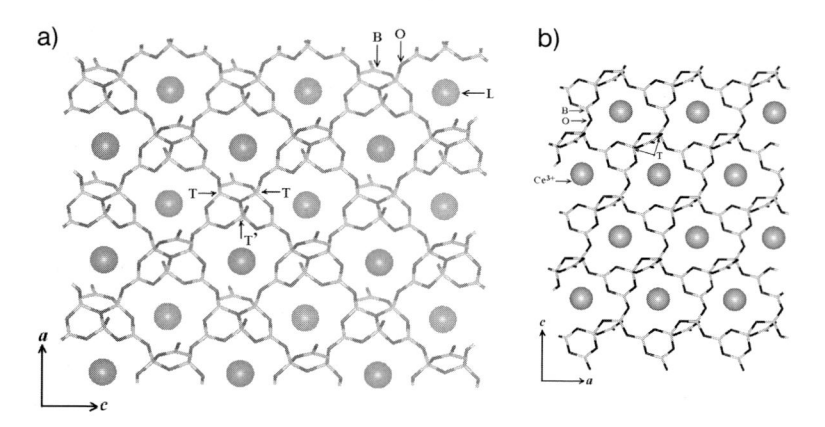

Fig. 16.21 The structures of the fundamental borate layers in (a) hydrated octa- and nona-borate and (b) hydrated cerium pentaborate [52].

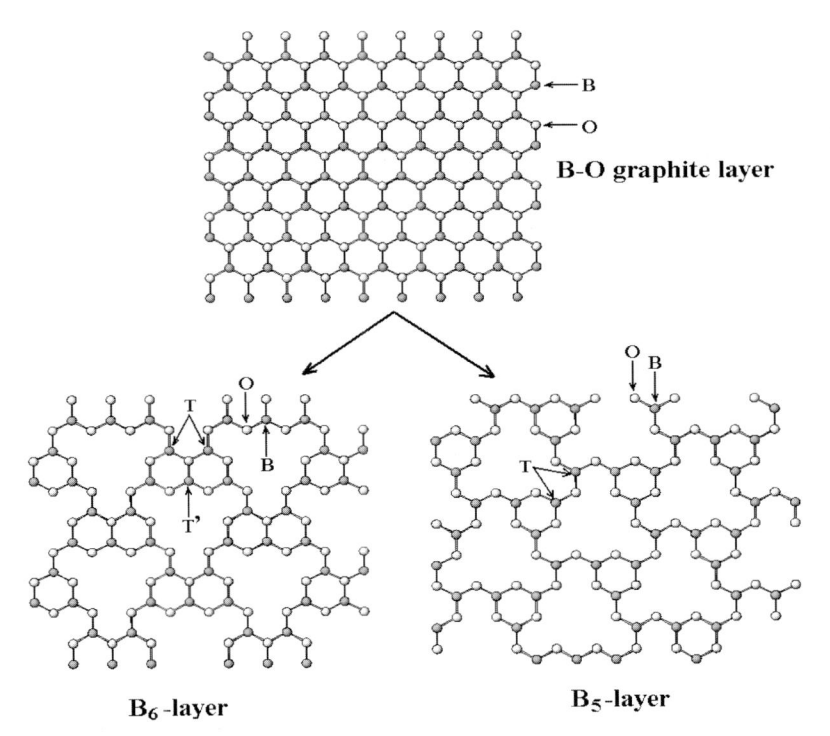

Fig. 16.22 A hypothetical graphite-like [BO] sheet and its two variants, i.e. B_6 and B_5 layers, respectively. The B_6 and B_5 layers are formed by removing one B_3O unit from every 8[BO] unit and every 7[BO] unit, respectively. The dark balls represent boron atoms, and the gray balls represent oxygen atoms [52].

with different compositions. $Ln[B_8O_{11}(OH)_5]$ is formed by adding a diborate $[B_2O_6]$ group as a side chain. In $Ln[B_9O_{13}(OH)_4] \cdot H_2O$, the side chain is a four-membered ring unit $[B_6O_{14}]$. The second group comprises the pentaborates $Ln[B_5O_8(OH)]\text{-}NO_3 \cdot 3H_2O$ (Ln = La, Ce). The borate layer also contains nine-membered rings (B_5-layer) (Fig. 16.21 b), but consists of different fundamental building units ($[B_5O_{11}]$). These two borate layers can be related by a hypothetical graphite net (BO) as displayed in Fig. 16.22. The B_6 layer can be obtained by removing B_3O from every 8BO unit in an ordered way, while removing B_3O from every 7BO unit results in the B_5 layer. The borate frameworks in these two groups are formed by stacking the aforementioned borate layers on top of one another, and, in fact, the layers are linked by strong ionic Ln–O bonds. Therefore, the structure could be considered as three-dimensional as shown in Fig. 16.23.

The third group of hydrated polyborates is $LnB_6O_9(OH)_3$ (Ln = Sm–Lu), which contains a three-dimensional borate framework with a six-membered ring (B_6O_{15}) as the fundamental borate unit (Fig. 16.24 a). The structure of $LnB_6O_9(OH)_3$ can be considered as a stacking of the six-membered ring units in a cubic close-packed fashion along the c axis. The B_6O_{15} units are directly

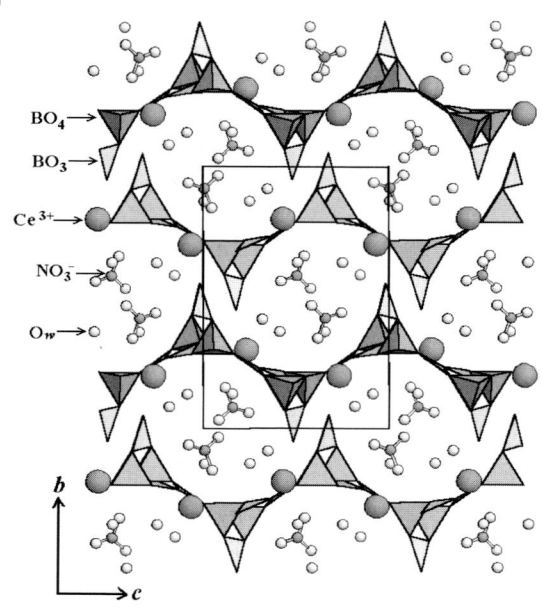

Fig. 16.23 The structure of hydrated cerium pentaborate projected along the [100] direction [52].

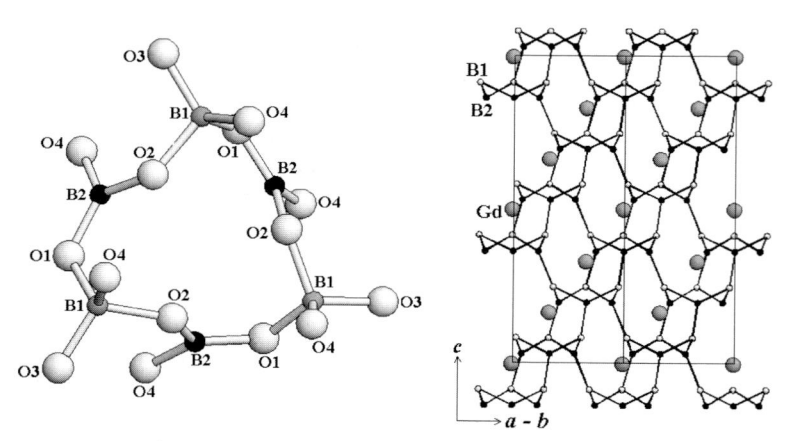

Fig. 16.24 (a) The structure of the six-membered ring unit and (b) the borate framework in hydrated hexaborates [51].

bound to the six neighbors in the adjacent layers forming a 3D framework. Fig. 16.24b gives an overview of the framework and, to emphasize the overall connectivity of the structure, only the boron and gadolinium atoms are illustrated. Rare-earth cations occupy the cavities in the framework and are coordinated by nine oxygen atoms forming a tricapped trigonal-prismatic polyhedron.

Tab. 16.3 Lattice constants (Å, deg.) of hydrated rare-earth polyborates

Compound	S.G.	a (Å)	b (Å)	c (Å)	$\beta(°)$
$La[B_8O_{11}(OH)_5]$	$P2_1/n$	9.8994(3)	14.2678(3)	7.9934(2)	90.061(4)
$La[B_5O_8(OH)]NO_3$	$P2_1/n$	6.466(1)	15.618(3)	10.780(2)	90.41(3)
$Ce[B_5O_8(OH)]NO_3$	$P2_1/n$	6.465(1)	15.571(3)	10.656(2)	90.24(3)
$Ce[B_8O_{11}(OH)_5]$	$P2_1/n$	9.8726(9)	14.1875(4)	7.9610(8)	90.064(15)
$Pr[B_8O_{11}(OH)_5]$	$P2_1/n$	9.8461(9)	14.1621(4)	7.9270(8)	90.065(17)
$Pr[B_9O_{13}(OH)_4]$	$P2_1/n$	9.8947(17)	16.7533(4)	7.7336(12)	90.035(33)
$Nd[B_8O_{11}(OH)_5]$	$P2_1/n$	9.8256(20)	14.2054(6)	7.9140(17)	90.140(37)
$Nd[B_9O_{13}(OH)_4]$	$P2_1/n$	9.8806(11)	16.6995(5)	7.7190(8)	90.087(2)
$Sm[B_9O_{13}(OH)_4]$	$P2_1/n$	9.8538(9)	16.5877(4)	7.6887(7)	90.08(16)
$Sm[B_6O_9(OH)_3]$	$R3c$	8.4336(2)		20.8003(4)	
$Eu[B_9O_{13}(OH)_4]$	$P2_1/n$	9.8589(8)	16.5673(5)	7.6917(6)	90.065(1)
$Eu[B_6O_9(OH)_3]$	$R3c$	8.4191(1)		20.7745(2)	
$Gd[B_6O_9(OH)_3]$	$R3c$	8.4081(1)		20.7426(2)	
$Tb[B_6O_9(OH)_3]$	$R3c$	8.3952(1)		20.7094(3)	
$Dy[B_6O_9(OH)_3]$	$R3c$	8.3895(1)		20.6687(4)	
$Ho[B_6O_9(OH)_3]$	$R3c$	8.3825(1)		20.6276(2)	
$Er[B_6O_9(OH)_3]$	$R3c$	8.3698(1)		20.5848(6)	
$Tm[B_6O_9(OH)_3]$	$R3c$	8.3605(2)		20.5404(7)	
$Yb[B_6O_9(OH)_3]$	$R3c$	8.3580(8)		20.5089(33)	
$Lu[B_6O_9(OH)_3]$	$R3c$	8.3506(10)		20.4756(46)	

Tab. 16.3 lists the lattice parameters of the hydrated rare earth polyborates. It can be seen that the large rare-earth cations prefer the octaborate, whereas the small ones prefer the hexaborate. The nonaborates are formed with the intermediate rare-earth ions.

16.6
Conclusions and Further Studies

Although significant progress has been made in recent years, there are still a number of unsolved problems in the structural chemistry of rare-earth borates, which have been discussed in this chapter. In addition to these known phases, an interesting question is whether there are any other unidentified phases in the rare-earth borate system. As far as the phase diagrams are concerned, at the two end-regions, i.e., B_2O_3-rich and Ln_2O_3-rich, it may well be possible to have new rare-earth borates. In the B_2O_3-rich region, Lu [56] is in fact observed as a trace of impurity phase during the decomposition of the hydrated hexaborate, which could be attributed to a more B_2O_3-rich phase. At the other end of the phase diagram, i.e., in the Ln_2O_3-rich region, one could consider the possibility of intergrowth structures being formed by, for example, the orthoborates and rare-earth oxides. As mentioned above, the vaterite structure may be considered as an alternate stacking of rock salt (111) layers and vertically aligned borate layers. The A-type rare-earth oxide, on the other hand, could be described as an alternate stacking of fluorite and rock salt layers. Therefore, it will be interesting to ascertain whether stacking of A-type Ln_2O_3 layers and vaterite borate layers can produce new rare-earth oxoborates with intergrowth structures.

Remarkable progress in the field of rare-earth borates has been achieved through high-pressure syntheses. Unlike the other main group elements, boron can be coordinated in either triangular or tetrahedral geometries. In general, tetrahedral coordination is favored under high pressures. Of course, other factors, such as packing density, may also influence the stability of the borate coordination geometry in the structures. This pressure-sensitive feature of borate structures may provide a new field in borate chemistry.

Acknowledgements

We are grateful for financial support from the NSFC and the State Key Basic Research Program of China.

References

1 D. Eimerl, L. Davis, S. Velsko, E. K. Graham, A. Zalkin, *J. Appl. Phys.* **1987**, *62*, 1968.
2 C. Chen, Y. Wu, A. Jiang, B. Wu, G. You, R. Li, S. Lin, *J. Opt. Soc. Am.* **1989**, *B6*, 616.
3 M. Adachi, T. Shiosaki, H. Kobayasi, O. Ohnishi, K. Kawabata, *Proc. IEEE Ultrasonic Symp.* **1987**, 228.
4 M. Leskelä, L. Niinistö, *Handbook on the Physics and Chemistry of Rare Earths*, (Eds.: K. A. Gschneidner, Jr., L. Eyring), North-Holland Publishing Company Amsterdam, New York, Oxford, **1986**, Vol. 8, pp. 203.
5 E. M. Levin, C. R. Robbins, J. L. Waring, *J. Am. Ceram. Soc.* **1961**, *44*, 87.
6 M. Th. Cohen-Adad, O. Aloui-Lebbou, C. Goutaudier, G. Panczer, C. Dujardin, C. Pedrini, P. Florian, D. Massiot, F. Gerard, Ch. Kappenstein, *J. Solid State Chem.* **2000**, *154*, 204.
7 S. F. Bartram, *Proc. 3rd Conf. Rare Earth Res.*, Clearwater, FL, **1963**, N.S.A. 12, No. 32035, pp. 165, 1964.
8 J. H. Lin, M. Z. Su, K. Wurst, E. Schweda, *J. Solid State Chem.* **1996**, *126*, 287.
9 J. H. Lin, S. Zhou, L. Q. Yang, G. Q. Yao, M. Z. Su, *J. Solid State Chem.* **1997**, *134*, 158.
10 J. H. Lin, L. P. You, G. L. Lu, L. Q. Yang, M. Z. Su, *J. Mater. Chem.* **1998**, *8*, 1051.
11 E. M. Levin, R. S. Roth, J. B. Martin, *Am. Miner.* **1961**, *46*, 1030.
12 E. Antic-Fidancev, J. Aride, J. Chaminade, M. Lemaître-Blaise, P. Porcher, *J. Solid State Chem.* **1992**, *97*, 74.
13 S. Lemanceau, G. B. Chadeyron, R. Mahiou, M. El-Ghozzi, J. C. Cousseins, P. Conflant, R. N. Vannier, *J. Solid State Chem.* **1999**, *148*, 229.
14 G. Corbel, M. Leblanc, E. Antic-Fidancev, M. Lemaître-Blaise, J. C. Krupa, *J. Alloys Compds.* **1999**, *287*, 71.
15 G. Chadeyron, M. El-Ghozzi, R. Mahiou, A. Arbus, J. C. Cousseins, *J. Solid State Chem.* **1997**, *128*, 261.
16 M. Ren, J. H. Lin, Y. Dong, L. Q. Yang, M. Z. Su, *Chem. Mater.* **1999**, *11*, 1576.
17 D. A. Keszler, H. Sun, *Acta Cryst.* **1988**, *C44*, 1505.
18 R. E. Newnham, M. J. Redman, R. P. Santoro, *J. Am. Ceram. Soc.* **1963**, *46*, 253.

19 W. F. Bradley, D. L. Graf, R. S. Roth, Acta Crystallogr. **1966**, *20*, 283.
20 J. P. Laperches, P. Tarte, Spectrochim. Acta **1966**, *22*, 1201.
21 H. M. Kriz, P. J. Bray, J. Chem. Phys. **1969**, *51*, 3642.
22 J. H. Denning, S. D. Ross, Spectrochim. Acta **1972**, Part A28, 1775.
23 G. B. Chadeyron, M. El-Ghozzi, D. Boyer, R. Mahiou, J. C. Cousseins, J. Alloys Compds. **2001**, *317–318*, 183.
24 G. Chadeyron, R. Mahiou, M. El-Ghozzi, A. Arbus, D. Zambon, J. C. Cousseins, J. Lumin. **1997**, *72–74*, 564.
25 J. H. Lin, D. Cheptiakov, Y. X. Wang, P. Allenspach, Chem. Mater. **2004**, *16*, 2418.
26 J. Hölsa, Inorg. Chim. Acta **1987**, *139*, 257.
27 D. Boyer, G. Bertrand, R. Mahiou, J. Luminesc. **2003**, *104*, 229.
28 D. Boyer, G. B. Chadeyron, R. Mahiou, C. Caperaa, J. C. Cousseins, J. Mater. Chem. **1999**, *9*, 211.
29 Ren Min, Ph. D. Thesis, **2000**, Peking University.
30 Z. G. Wei, L. D. Sun, C. S. Liao, J. L. Yin, X. C. Jiang, C. H. Yan, J. Phys. Chem. **200**, *B 106*, 10610.
31 Z. G. Wei, L. D. Sun, X. C. Jiang, C. S. Liao, C. H. Yan, Chem. Mater. **2003**, *15*, 3011.
32 H. J. Meyer, Naturwissenschaften **1969**, *56*, 458.
33 H. Huppertz, B. von der Eltz, R. D. Hoffmann, H. Piotrowski, J. Solid State Chem. **2002**, *166*, 203.
34 M. Yin, G. Corbel, M. Leblance, E. Antic-Fidancev, J. C. Krupa, J. Alloys Compds. **2000**, *302*, 12.
35 K. K. Palkina, V. G. Kuznetsov, L. A. Butman, B. F. Dzhurinskii, Acad. Sci. URSS **1976**, *2*, 286.
36 H. Flood, T. Förland, Acta Chem. Scand. **1947**, *1*, 592.
37 N. I. Leonyuk, J. Cryst. Growth **1997**, *174*, 301.
38 K. I. Machida, G. Y. Adachi, J. Shiokawa, Acta Cryst. **1980**, *B36*, 2008.
39 K. I. Machida, G. Y. Adachi, J. Shiokawa, M. Shimada, M. Koizumi, Inorg. Chem. **1980**, *19*, 983.
40 H. Huppertz, B. von der Eltz, J. Am. Chem. Soc. **2002**, *124*, 9376.
41 H. Emme, H. Huppertz, Chem. Eur. J. **2003**, *9*, 3633.
42 H. Emme, H. Huppertz, Z. Anorg. Allg. Chem. **2002**, *628*, 2165.
43 H. Huppertz, S. Altmannshofer, G. Heymann, J. Solid State Chem. **2003**, *170*, 320.
44 J. Weidelt, Z. Anorg. Allg. Chem. **1970**, *374*, 26.
45 H. V. Bambouer, J. Weidelt, J. S. Ysker, Naturwissenschaften **1968**, *55*, 81.
46 H. V. Bambouer, J. Weidelt, J. S. Ysker, Z. Krist. **1969**, *130*, 207.
47 G. K. Abdullaev, Kh. S. Manedov, G. G. Dzhafarov, Sov. Phys. Crystallogr. **1975**, *20*, 161.
48 V. I. Pakhomov, G. B. Silnitskaya, B. F. Dzhurinskii, Inorg. Mater. **1971**, *7*, 539.
49 I. V. Tananaev, B. F. Dzhurinskii, V. I. Chistova, Inorg. Mater. **1975**, *11*, 69.
50 C. Sieke, T. Nikelski, T. Schleid, Z. Anorg. Allg. Chem. **2002**, *628*, 819.
51 L. Y. Li, P. C. Lu, Y. Y. Wang, X. L. Jin, G. B. Li, Y. X. Wang, L. P. You, J. H. Lin, Chem. Mater. **2002**, *14*, 4963.
52 L. Y. Li, X. L. Jin, G. B. Li, Y. X. Wang, F. H. Liao, G. Q. Yao, J. H. Lin, Chem. Mater. **2003**, *15*, 2253.
53 P. C. Lu, Y. X. Wang, L. P. You, J. H. Lin, Chem. Commun. **2001**, 1178.
54 A. F. Wells, Structural Inorganic Chemistry, 5th Edition, Oxford University Press, **1984**, 1045.
55 J. R. Clark, Am. Mineral. **1964**, *49*, 1549.
56 P. C. Lu, M.Sc. Thesis, **1999**, Peking University.
57 We were aware in the press of this paper that Morgan had proposed a pseudowollastonite model for YBO_3 many years ago (P. E. D. Morgan, P. J. Carroll and F. F. Lange, Mat. Res. Bull. **1977**, *12*, 251). The detail structure characterization on $CaSiO_3$ shows that the pseudowollastonite polymorph of $CaSiO_3$ has the same monoclinic structure as that of YBO_3 (H. Yang and C. T. Prewitt, Am. Miner. **1999**, *84*, 929; ibid **1999**, *84*, 1902).

17
Ordered Siliceous Mesostructured Materials: Synthesis and Morphology Control

Pegie Cool, E. F. Vansant, and O. Collart

17.1
Introduction

Since the first development of microporous (<2 nm) aluminosilicate zeolites as catalysts/catalyst supports in the early 1950s, interest in their industrial utilization has increased substantially. Their open structure consists of cavities and micropores created by a particular arrangement of atoms. The shape and size of these pores can influence the selectivity of a catalytic reaction by permitting only certain molecular configurations to be synthesized. This fact has motivated researchers not only to perform chemical modifications but also to model the pores. However, even after extensive modeling, the dimensions of the zeolite pores were restricted to the sub-nanometer scale. This limited the application of these pore systems to small molecular entities. Therefore, a constant challenge during the last decades has been to vary the physical and chemical properties of these materials in order to increase the number of potential applications. One of the major concerns is the accessibility of the core of the porous structures to large molecules (e.g. large metal complexes, enzymes). In 1992, ExxonMobil patented the first generation of mesoporous materials (M41S) with large, uniform, tunable pores arranged in a well-defined matter. The pore diameter of these new materials was almost three times that of the largest pores in zeolites. Materials and catalysis scientists rapidly recognized the potential of the new mesoporous materials, which now offer many applications in the fields of catalysis, optics, photonics, sensors, separation, drug delivery, sorption, and acoustic or electrical insulation.

In this chapter, first the chemical tools that are necessary to construct siliceous mesoporous materials, such as MCM-41, MCM-48, and SBA-15, are discussed. Their syntheses are based on the classic parameters of zeolite synthesis, but make use of surfactants to organize the pore structures. Much emphasis is placed on the particular interaction between the silicate building blocks and the organic surfactant molecules. In the second part, it is shown how the mesophase and the morphology of mesoporous MCM materials can be controlled by the addition of alcohols.

Inorganic Chemistry in Focus II. Edited by G. Meyer, D. Naumann, and L. Wesemann
Copyright © 2005 WILEY-VCH Verlag GmbH & Co. KGaA, Weinheim
ISBN: 3-527-30811-3

17.2
Mesoporous Siliceous Materials: State-of-the-Art

As mentioned above, a continuous interest in developing materials with ever larger pores finally led to the development of materials with mesopores.

The first major breakthrough in the synthesis of these new mesoporous materials – structures possessing pores between 2 and 50 nm in diameter – came with the publication of patents by ExxonMobil in 1992 [1, 6] and the corresponding papers on mesoporous materials in *Nature* [7] and *JACS* [8]. These were the first publications to identify the newly synthesized materials as being ordered mesoporous structures. The authors classified their different structures as belonging to the 'M41S' class materials (family name given by ExxonMobil). These structures were, in the initial stages of the work, purely siliceous materials. Soon thereafter, however, the first successful attempts at aluminum incorporation into the framework of these structures were published [9]. This was done in order to obtain materials comparable to zeolites.

One of the fundamental aspects that distinguishes the newly developed mesoporous materials from classical zeolites is the concept of ordering. In zeolites, each atom has a well-defined position in the structure. It is this specific position and its repetitive character in the material framework that creates the final structure. In the case of mesoporous materials, a new concept has to be introduced. On the one hand, the atoms constituting the walls have no specific position. The pore walls therefore have to be considered as amorphous. On the other hand, the pores within the structures are disposed in a well-determined and repetitive way, creating an ordered pattern on a large scale. In other words, it is not the atoms that determine the ordering of the final structure, but the pore arrangements.

By changing the synthesis conditions (pH, time, temperature, presence of surfactant, etc.), it is possible to influence the ordering of the pores and thus to create new types of structures. The synthesis conditions also have a certain effect on the stability of the pore walls. This means that, essentially, the physical characteristics of the mesomaterials can be regulated. The chemical characteristics, on the other hand, largely depend on the type of heteroatoms present, either on the surface of or within the silicate pore walls.

17.2.1
Mesoporous Structures

The determination and classification of a structure is essentially based on the way in which the pores are ordered within it. Other classification considerations are the synthesis conditions (pH and material source) and the family of organic surfactant used to synthesize the structure. Thus, contrary to the zeolites, the atom ordering is not considered when identifying and classifying a mesoporous material.

ExxonMobil regrouped their mesoporous structures under the common family name M41S. Three important structures are contained in this family, namely the hexagonal MCM-41 [8], the cubic MCM-48 [8, 10–13], and the lamellar MCM-50. MCM is the acronym of **M**obil **C**omposition of **M**atter (the number has no link to any physico-chemical characteristics of the material). Hexagonal, cubic, and lamellar refer to the symmetry found in the materials. All these structures are synthesized in a high-pH solution.

Stucky successfully performed syntheses of mesoporous structures in acidic media. He identified and named the different structures as the cubic SBA-1 [14], SBA-11 [15], and SBA-16 [15, 16]; the hexagonal SBA-3 [14, 17] and SBA-15 [18]; the 3D hexagonal SBA-2 [17] and SBA-12 [15]; and the rectangular SBA-8 [19] (SBA stands for **S**anta **B**arbara **A**cid; the number follows the internal agreements of the Santa Barbara laboratory).

Inagaki, for his part, synthesized a mesoporous structure by folding kanemite layers and merging them together to form a hexagonal structure. He named his structure FSM-16 [20], the acronym standing for **F**olded **S**heet **M**echanism.

Other structures, such as HSM, KIT-1, and the MSU series, have also been identified. These have the particularity of possessing mesopores that are more randomly disposed compared to those previously summarized. Some authors describe their porosity as 'wormholes'. These structures will not be further discussed in this overview. Di Renzo has enumerated almost all of the mesoporous structures known to date [21].

In Tab. 17.1, the most commonly known mesoporous structures have been regrouped and classified according to their crystal system (created by the pore ordering). The space group of each structure is also listed in order to further differentiate its configuration.

Tab. 17.1 Mesoporous structures with their crystal system and space group

Structure	Crystal system	Space group	Ref.
MCM-41	2D hexagonal	$p6m$	8
FSM-16	2D hexagonal	$p6m$	20
SBA-3	2D hexagonal	$p6m$	14
SBA-15	2D hexagonal	$p6m$	18
MCM-48	3D cubic	$Ia3d$	8
SBA-1	3D cubic	$Pm3n$	14
SBA-11	3D cubic	$Pm3m$	15
SBA-16	3D cubic	$Im3m$	15
SBA-2	3D hexagonal	$P6_3/mmc$	17
SBA-12	3D hexagonal	$P6_3/mmc$	15
SBA-8	2D rectangular	cmm	19

From this table, it is apparent that although a wide variety of structures exist, they can easily be classified according to the symmetry element in their crystal system. Closer study shows that two symmetry elements dominate, namely the hexagonal and the cubic unit cell. Furthermore, this table also indicates that a lot of materials, while differentiated by their synthesis conditions or sources, ultimately possess the same structure and can therefore be considered as identical on a structural basis.

In order to provide a better insight into the structural compositions of these mesoporous materials, the two dominant unit cells, the hexagonal and cubic ones, are briefly discussed in the following subsections.

17.2.1.1 Hexagonal Mesoporous Structures MCM-41, SBA-15

The hexagonal mesoporous materials consist of an amorphous (alumino, metallo)-silicate framework, forming hexagonal pores. These are unidirectional and arranged in a honeycomb structure as shown by the TEM photograph in Fig. 17.1 A.

The image in Fig. 17.1 B represents a 3D magnification of a few hexagonal pores, emphasizing their parallel positioning. The hexagonal shape of the pore creates a heterogeneous pore surface. In the corners of the pore, an enhanced presence of siloxane bridges has been detected. These bridges help to create hydrophobic regions on the pore surface [22, 23], whereas the flat areas of the pore surface are hydrophilic due to the presence of hydroxyl groups, hence the slightly heterogeneous pore surfaces. In the classic MCM-41 synthesis, an average pore diameter of 3.6 nm is obtained. The amorphous pore walls are, in most cases, no thicker than 1 nm.

MCM-41 is probably one of the most discussed and characterized structures in the literature. Selvam recently presented a review [24] discussing the major advances in relation to the MCM-41 structure.

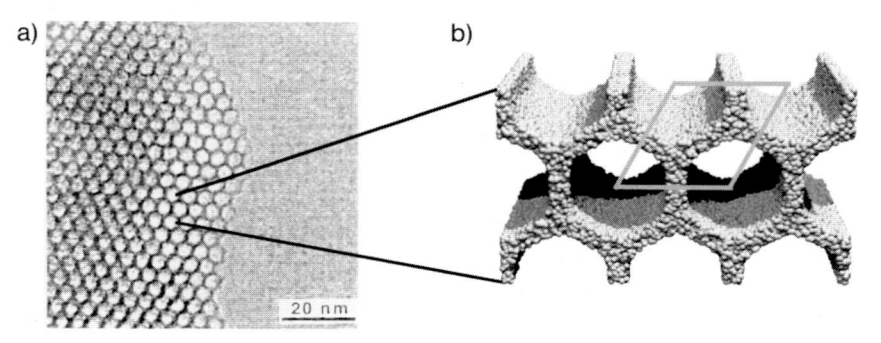

Fig. 17.1 (a) TEM image of the honeycomb structure of MCM-41. (b) Representation of some hexagonal-shaped unidimensional pores with the unit cell delineated by the gray parallelogram.

SBA-15 basically has the same structure as MCM-41, namely a unidirectional hexagonal pore system. The major difference resides in the dimensions of the pores and pore walls. A classic SBA-15 synthesis produces a structure with an average pore diameter of 6 nm and amorphous pore walls of 3 nm thickness. Another interesting feature of these SBA-15 materials is the presence of micropores within the pore walls.

17.2.1.2 Cubic Mesoporous Structure MCM-48

MCM-48 possesses a much more complex structure than MCM-41 or SBA-15. The unit cell has cubic symmetry (Fig. 17.2 A). Inside this cube, the pore walls (amorphous) are arranged following the so-called "minimal surface". This "minimal surface" is a mathematically constructed surface (Fig. 17.2 B), first described by Schoen [25]. Each point on the surface is a saddle point and presents an average curvature of zero. No physical consideration should be looked for behind these descriptions.

The minimal surface defining MCM-48 has been identified as a gyroid G or G-surface (Figs. 17.2 A and 17.2 B are comparable). Further mathematical aspects can be found in the literature, e.g. in the articles by Ravikovitch [26] and Anderson [27]. This minimal surface divides the cube into two identical but separate compartments, creating two independent 3D pore systems. Fig. 17.2 A shows such a unit cell with the two micelle systems (light- and dark-gray rods) following the pore system. The two independent pore systems are interlocked and extend along the [111] and [100] directions, but never cross or join each other. This is clarified in Fig. 17.2 C, which shows 2 × 4 unit cells without the pore walls. The pore systems are represented by micelle rods, which progress in spirals around each other towards the [100] direction.

The pore diameter of MCM-48 varies with the synthesis conditions, and lies between 2.7 and 4.6 nm. The average pore wall thickness is about 0.9 nm, which is comparable to the thickness found for MCM-41. No micropores have been detected within the pore walls. More detailed characterizations of the MCM-48 structure can be found in the literature [26–31].

17.2.2
Synthesis of Siliceous Mesoporous Materials

Mesoporous solids are synthesized by polymerizing a silica source around an organic template in a predominately aqueous solvent. The organic template is then removed through calcination in order to vacate the pores. However, the interaction between the organic template molecules and the silicate framework is influenced by the synthesis parameters. It is therefore essential to understand and control these parameters.

Since the first publication on the M41S mesoporous materials, a remarkable evolution of the synthesis parameters of mesoporous materials prepared using quaternary ammonium surfactants has been observed. Scientists have put a lot

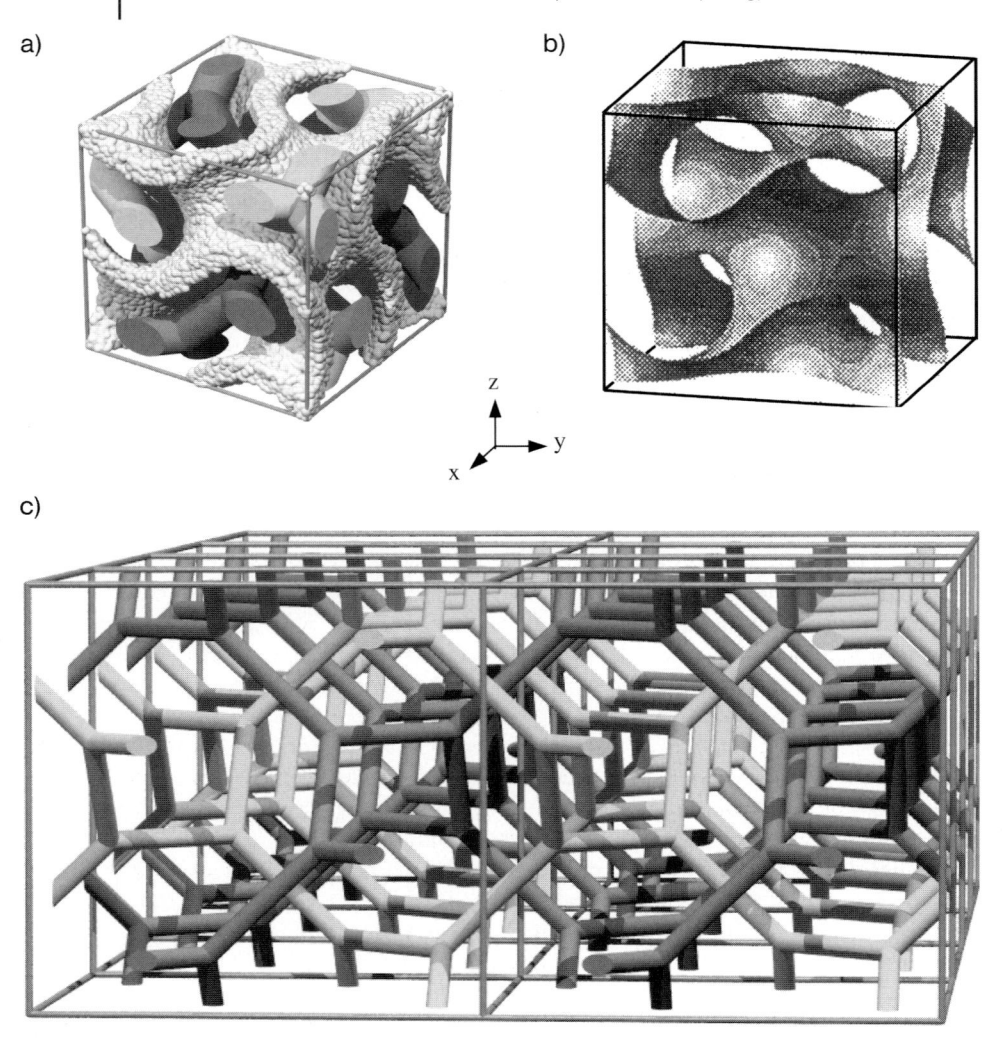

Fig. 17.2 (a) Cubic unit cell of an MCM-48 with two independent micelle systems (dark- and light-gray rods) separated by the pore wall. (b) Mathematical representation of a G gyroid minimal surface. (c) Representation of 2 × 4 cubic unit cells without the pore walls. The rods represent two independent micelle systems (dark- and light-gray) moving towards the [100] direction.

of effort into unraveling the synthesis mechanism in order to gain better control over the final characteristics of the materials [32]. Although they have mainly worked on MCM-41 materials, the presented mechanisms are valid for all mesoporous materials synthesized with quaternary ammonium surfactants (MCM-41, MCM-48, MCM-50). Patarin [33] has briefly reviewed the possible synthesis mechanism for quaternary ammonium surfactants.

The first mechanistic proposal was made by Beck [8] and relied on observations made in surfactant science. He reported that microscopy and XRD results presented for the MCM-41 material quite closely resembled those obtained for surfactant/water liquid crystals or micellar phases [34–36]. Beck therefore proposed a liquid-crystal templating (LCT) mechanism. He advanced the theory that the structure is defined by the organization of quaternary ammonium surfactant molecules into liquid crystals, which serve as templates for the formation of the mesoporous structure. In other words, the first step in the synthesis would correspond to the formation of a micelle rod around the surfactant micelle, which, in a second step, produces a hexagonal array of rods; this is then followed by the incorporation of an inorganic array (silica, silica/alumina) around the rod-like structures (Fig. 17.3).

However, considering that the liquid-crystal structures formed in surfactant solutions are highly sensitive to the overall characteristics of the solution, the authors also take into account the possibility that the addition of the silicate results in the ordering of the subsequent silicate-encaged surfactant micelles. When they synthesized materials with short-tailed surfactant molecules (C_6–C_8), they did not obtain an MCM-41 structure. Such short-tailed quaternary ammoniums are much more soluble and therefore the formation of aggregated structures is not necessary to minimize the hydrophobic interactions. It was thus concluded that the M41S materials are formed by way of a mechanism in which aggregates of cationic surfactant molecules in combination with anionic silicate species form a supramolecular structure.

Davis, by carrying out in situ ^{14}N NMR spectroscopic measurements, concluded that the liquid-crystalline phase is not present in the synthesis medium during the formation of MCM-41, and consequently that this phase cannot be the mesoporous material, in agreement with the already proposed mechanism through route 2. Thus, the randomly ordered rod-like organic micelles interact

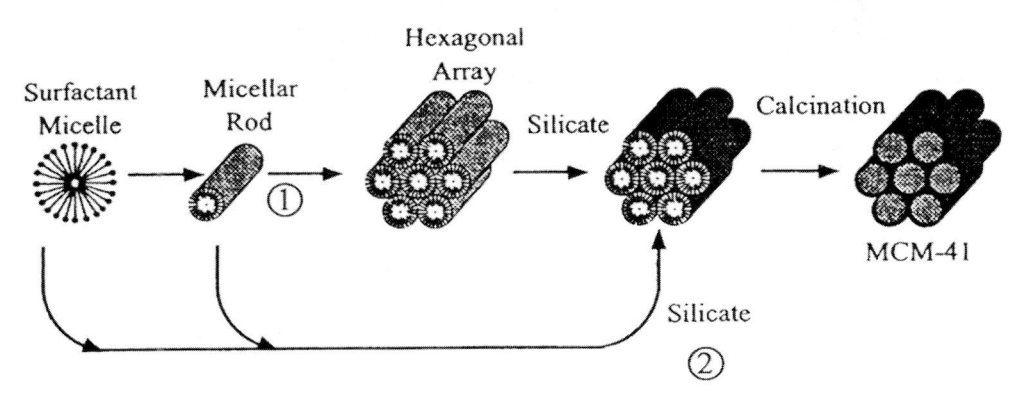

Fig. 17.3 Possible mechanistic pathways proposed by Beck for the formation of a mesoporous structure: (1) liquid crystal initiated, and (2) silicate anion initiated [8].

with the silicate species to yield two or three monolayers of silica around the external surface of the micelles. Subsequently, these composite species spontaneously form the long-range order characteristics of the mesoporous materials (Fig. 17.4).

If the surfactant is removed just at the point when the long-range order is achieved, the material will collapse due to the large number of uncondensed silicate species.

Finally, Stucky [10, 14, 17, 37, 38] developed a model that makes use of the cooperative organization of inorganic and organic molecular species into three-dimensionally structured arrays. He divided the overall process into three reaction steps: (1) multidentate binding of the silicate oligomers to the cationic surfactant, (2) preferential silicate polymerization in the interface region, and (3) charge density matching between the surfactant and the silicate. Furthermore, he stated that in this model the properties and structure of a particular system were not determined by the organic arrays that have long-range, pre-organized order, but by the dynamic interplay among inorganic and organic ion-pair species, so that different phases can be readily obtained through small variations of controllable synthesis parameters, including mixture composition and temperature. He confirmed this by examining the structures at various stages during their synthesis, by means of the small-angle neutron scattering (SANS) technique [39].

Other authors confirmed this mechanism by means of *in situ* experiments. Galarneau [40, 41] used EPR-sensitive surfactant molecules in order to investigate *in situ*: (a) the kinetics of formation of different structures and ordered mesoporous materials, (b) the interactions between the micelle surfaces and the solid silica that was formed and structured at the interface, and (c) the modification of the micelle structure over time, due to solid formation, at various distances from the micellar surface.

She confirmed the model presented by Stucky by identifying a first step in which, in the first minutes of the synthesis, the silicate anions replace the counterions of the cationic surfactant. At the same moment, a poorly ordered mesophase is formed in the solution (cf. steps 1 and 2 of Stucky). In a second step, she observed the occurrence of strong interactions between the surfactant molecules and the condensed silica. During this step, a lengthening and condensation of the silica-coated micelles takes place (cf. step 3 of Stucky). This led to the synthetic scheme represented in Fig. 17.5.

Calabro also observed, by means of an *in situ* ATR/FTIR study [42], the formation of micelles through interactions of the silicate oligomers with surfactant molecules. Others have used fluorescence techniques [43] to follow the micelle formation *in situ*.

The charge density matching between silica and surfactant not only initiates the micelle formation but also has a subsequent influence on the formation of the mesophase and thus on the type of final structure formed. Tolbert [38] proved by means of time-resolved X-ray diffraction that this mechanism has a decisive influence on the kinetics of phase transformation. She surmised that lamellar-to-hex-

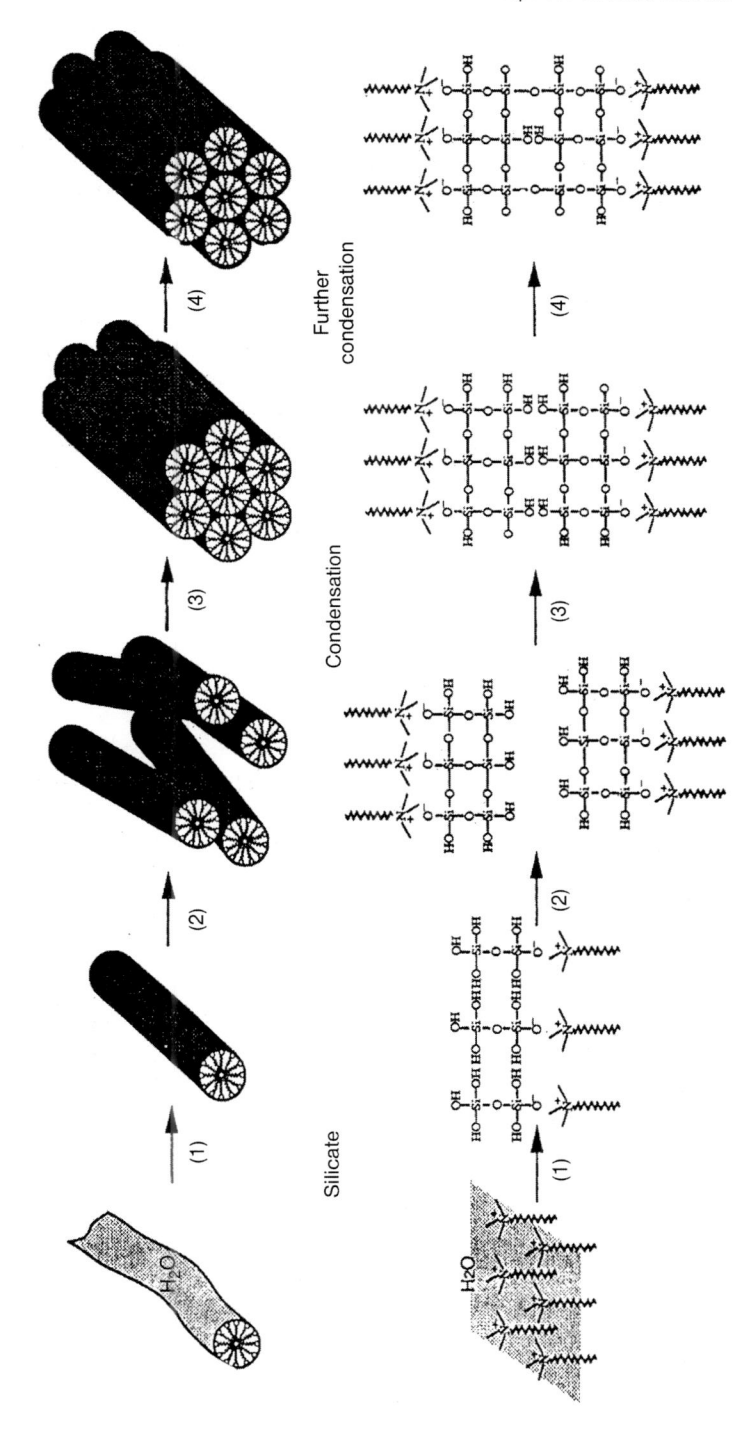

Fig. 17.4 Mechanism for the formation of mesoporous materials proposed by Davis.

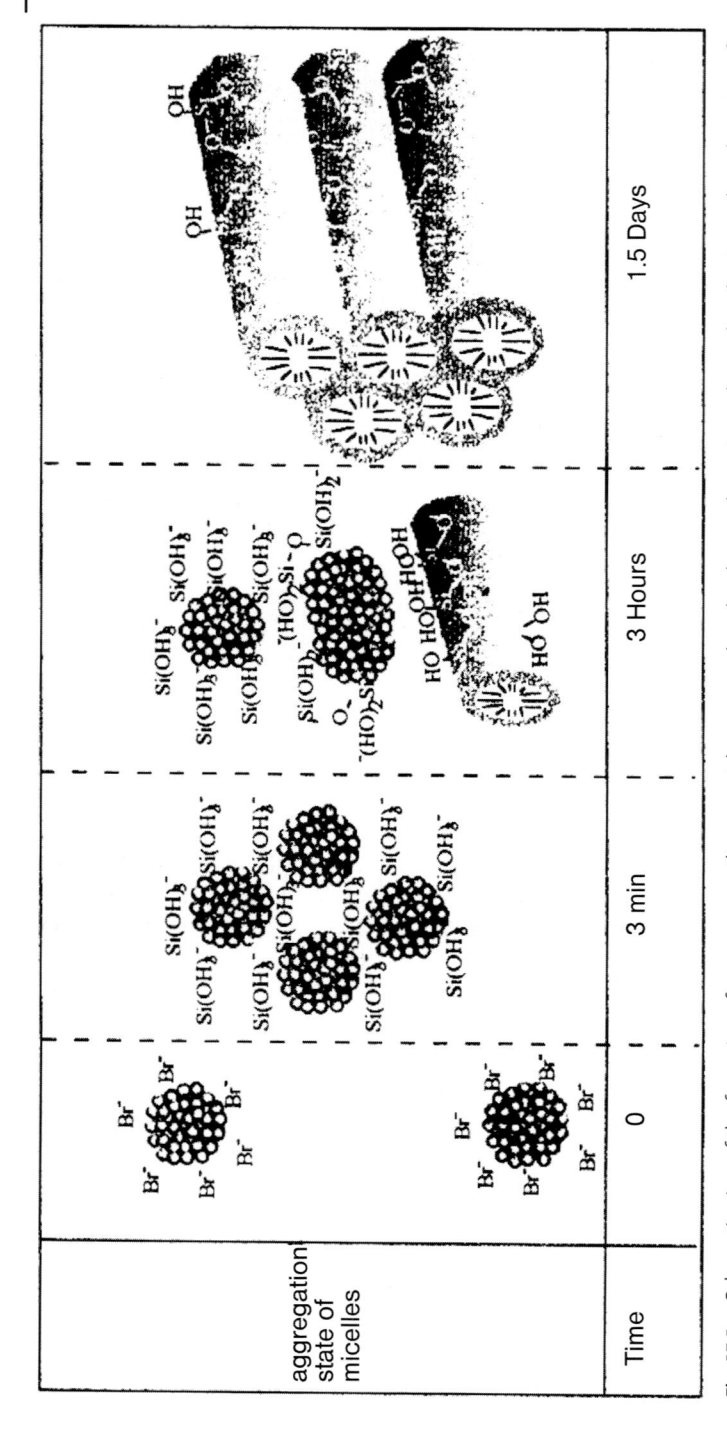

Fig. 17.5 Schematic view of the formation of mesoporous silica materials at 323 K. First step: in the first few minutes of the synthesis, silicate anions replace bromide as the counterion of the cationic surfactant, forming a poorly ordered mesophase. Second step: lengthening and condensation of silica-coated micelles, ultimately creating an ordered mesophase [40].

agonal phase transformation is favored by a continuous charge density matching, whereas a hexagonal-to-lamellar transformation would have to go against these charge density changes and is therefore kinetically less favored.

17.2.2.1 Key Steps in the Synthesis

The key steps that can be identified in the synthesis reaction of zeolites are gel ageing, ion transportation and silicate polymerization, and finally crystallization.

In the case of mesoporous materials, elucidation of the synthesis mechanism revealed that the main step appears to be the specific interaction between the silicate particles and the surfactant molecules. This interaction is the driving force that initiates the surfactant molecules to organize themselves into micellar rods, which, in turn, order themselves in a particular way.

The interaction between the silicate particles and the surfactant molecules depends mainly on the latter. When charged surfactant molecules with quaternary ammonium headgroups are used, a charge-matching mechanism between the cationic headgroup and the anionic silicate particles occurs. Such interplay is favored under synthesis conditions where the pH is above the point of zero charge (PZC) (at pH 2) of silica. Above this point, the silicic acid ($Si(OH)_4$) species will lose a proton and become negatively charged [44], making interaction with the cationic surfactant possible and simultaneously inducing the necessary charge compensation to create the micelle (Fig. 17.6 A). The close interaction between the cationic headgroup and the silica surface has been demonstrated by Simonutti [45] by means of ^{13}C and ^{29}Si solid-state NMR.

In more acidic synthesis media, another interaction mechanism occurs. At low pH (pH < 2), the silicate species are fully protonated, making direct interaction with the surfactant molecules impossible. Therefore, a counter ion must necessarily participate in the interaction mechanism by taking its place between the cationic surfactant molecule and the protonated silicic acid (Fig. 17.6 B).

When neutral tri-block copolymers (*vide infra*) are used (SBA-15, SBA-16), no direct charge-matching mechanism can be applied. However, the acidic conditions of the synthesis will provide protonated water, which can interact with the

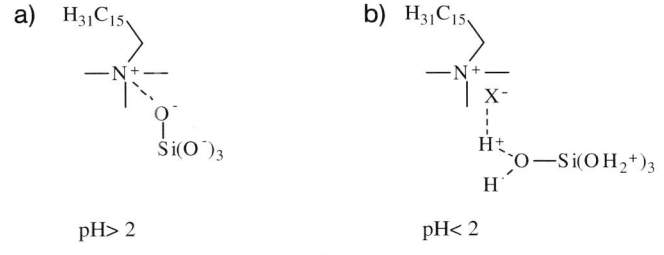

Fig. 17.6 Interaction mechanism between a quaternary ammonium surfactant molecule and a silicate species, (a) above pH 2 and (b) below pH 2 (X⁻: Cl⁻ or Br⁻).

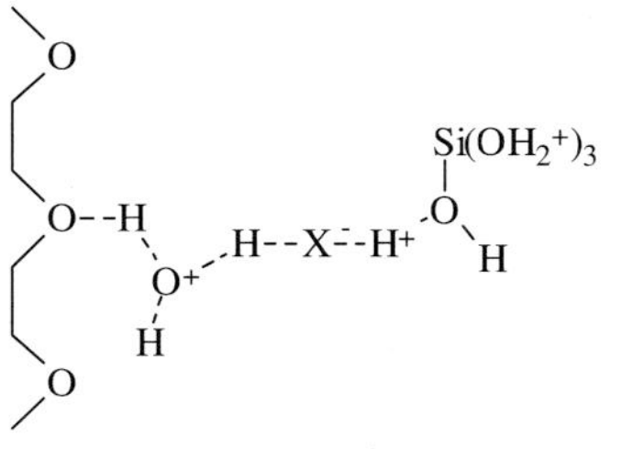

Fig. 17.7 Interaction between neutral tri-block copolymers and silicate species in an acidic medium.

copolymer through a hydrogen bridge. The incorporation of an anion between the protonated water and silicate species then initiates the interaction between the surfactant molecule and the silicate species (Fig. 17.7).

After the formation of the micelle rods as a result of the interaction with the silicate particles, a crystallization step is necessary in order to stabilize the structure formed by further polymerization of the framework.

17.2.2.2 Key Parameters in the Syntheses: Structure-Directing Agents

The organic surfactant molecules represent the key parameter in the synthesis of mesoporous materials. Their long organic tails give them the ability to create micelles, around which the (alumino)silicate species can polymerize to create the mesoporous structure. The symmetry of the resulting structure depends on the phase formation of the micelles. This phase formation essentially depends on the surface curvature of the micelles. The surface curvature is, in turn, influenced by such parameters as the size, shape, and surface charge density, or the presence of co-surfactants. It is therefore useful to have a good understanding of these effects, in order to be able to control the design of mesophases [46].

Since ExxonMobil synthesized its first mesoporous materials in 1992, using cationic quaternary ammonium surfactants, a wide variety of surface-active agents have been proposed, each creating materials with different characteristics and used under different synthesis conditions. Tab. 17.2 gives a short list of the most common surfactants used, together with the mesophase formed and the conditions under which they are used.

When charged surfactants (quaternary ammonium headgroup) are used, a fairly good approach to predicting the final mesophase can be found in the classical and contemporary molecular description of surfactant organization in amphiphilic liquid-crystal arrays [47].

When surfactants are dissolved in aqueous solution, a balance between three forces determines their packing or shape. The first force is the tendency of the alkyl chains to minimize their water contact and to maximize their inter-organic interactions, the second is the coulombic interaction between charged head groups, and the third stems from energies of solvation. The hydrophilic-hydrophobic balance, involving the first two of these interactions, is measured in terms of the local effective surfactant packing parameter g, expressed as [48, 49]:

Tab. 17.2 Common surfactants used, together with the mesophase formed and synthesis conditions

	Surfactant	Mesophase	Synth. conditions
Quaternary ammonium	cetyltrimethylammonium (CTA$^+$Cl$^-$/Br$^-$)	MCM-41, MCM-48, MCM-50	basic aq. solution
		SBA-1, SBA-3	acidic aq. solution
	Gemini	MCM-41, MCM-48	basic aq. solution
Non-ionic surfactant	amines	HMS	alcoholic solutions
	tri-block copolymers Pluronic P$_{123}$	SBA-15	acidic aq. solution
	tri-block copolymers Pluronic P$_{127}$	SBA-16	acidic aq. solution

$$g = \frac{V}{a_0 l} \tag{1}$$

where V represents the total volume of the surfactant chains plus any co-solvent organic molecules between the chains, a_0 is the effective head group area at the surface of the micelles, and l is the kinetic surfactant tail length or the curvature elastic energy [50] (Fig. 17.8). The value of g increases as a_0 decreases, V increases, or l decreases.

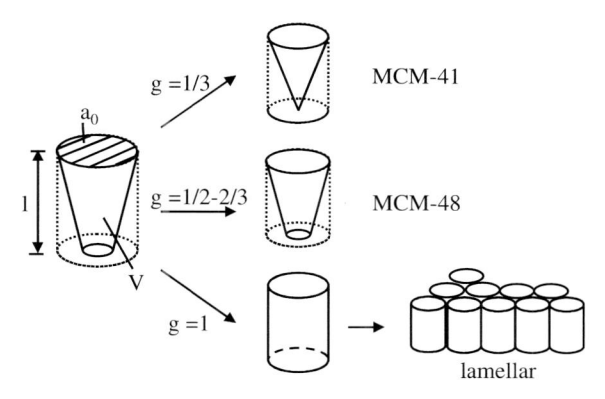

Fig. 17.8 Representation of the g packing parameter and the effect of different values thereof on the mesophase formation.

In classical micelle chemistry, as the value of g increases above critical values, mesophase transitions occur. The expected mesophase sequence as a function of the packing parameter is summarized in Tab. 17.3.

These transitions reflect a decrease of the surface curvature from cubic through viscular to lamellar. For surfactants to associate in a spherical structure, the surface area occupied by their polar head groups should be large. If, on the other hand, the head groups are permitted to pack tightly, the aggregation number will increase, and rod or lamellar packing will prevail.

As mentioned above, the value of g will increase accordingly with an increase in the volume fraction V. This volume includes the volume occupied by co-surfactants. The most common surfactants used are alcohols. The partially hydrophobic character of the alcohol will drive the molecules located at the organic–inorganic interface further inside the organic (surfactant) phase upon heating. This will indisputably cause an increase in the volume fraction V and thus also an increase in the surfactant packing parameter g. The larger the alcohol, the greater the increase in the surfactant packing.

The alcohol can originate from the alkoxide source as its hydrolysis product. Upon heating in a closed system, the alcohol will diffuse into the micelles causing a mesophase transition [51, 52].

As previously discussed, the synthesis conditions (pH, temperature) should also be carefully taken into account. Both parameters can have a determining effect on the charges of the different particles in the solution. A change in the charge density of the surfactant micelles can also have an effect on the surfactant packing. Micelle rods have a lower surface curvature and a higher charge density than globular micelles. Due to the charge-matching mechanism, the number of charged particles in the solution (depending on pH and temperature) can influence the surface curvature of the micelles and thus the final mesophase. At high pH, the ionic atmosphere surrounding the ionic head groups will decrease (due to greater compensation by the silicate anions), causing a smaller electrical repulsion between the head groups in the micelle [53], and hence leading to a closer packing of the surfactant molecules. This closer packing means that there is an increase in the g packing parameter and the possibility of obtaining other phases.

By the addition of a co-surfactant or a change of the synthesis medium, other phases or phase transitions can be induced. It is therefore important to carefully

Tab. 17.3 Mesophase sequence as a function of the packing parameter

g	Mesophase
1/3	cubic (*Pm3n*)
1/2	hexagonal (*p6m*)
1/2–2/3	cubic (*Ia3d*)
1	lamellar

Silica Mesoporous Phases

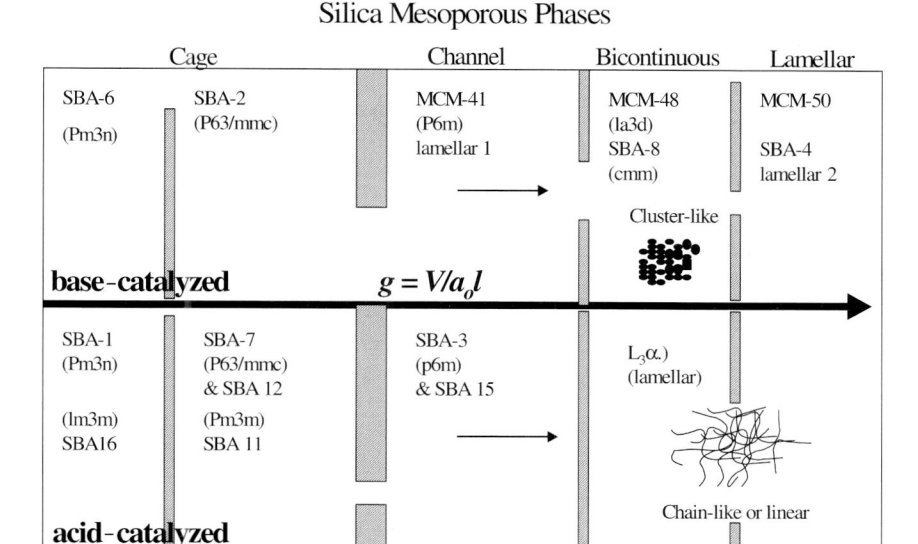

Cage		Channel	Bicontinuous	Lamellar
SBA-6 (Pm3n)	SBA-2 (P63/mmc)	MCM-41 (P6m) lamellar 1	MCM-48 (Ia3d) SBA-8 (cmm)	MCM-50 SBA-4 lamellar 2
			Cluster-like	

base-catalyzed $g = V/a_o l$

SBA-1 (Pm3n) (Im3m) SBA16	SBA-7 (P63/mmc) & SBA 12 (Pm3m) SBA 11	SBA-3 (p6m) & SBA 15	$L_3 \alpha$.) (lamellar)	
			Chain-like or linear	

acid-catalyzed

Fig. 17.9 Succession of synthesized mesophases following the g packing parameter as a function of the pH of the synthesis media.

select the sources as well as the synthesis conditions in order to obtain the correct mesophase.

The newest generation of surfactants, namely non-ionic tri-block copolymers, also follow the g packing theory, but the characteristic volumes of the hydrophilic headgroups and hydrophobic portions of each surfactant have to be taken into account. These surfactants consist of two polyethylene oxide chains (PEO) on either side of a polypropylene oxide chain (PPO) and are denoted as PEO-PPO-PEO.

As previously mentioned, the temperature can greatly influence the hydrophilicity of the surfactant, which can have an important influence on the formation of the micelles.

A general overview of the succession of synthesized mesophases as a function of the g packing factor is given in Fig. 17.9. The horizontal division of Fig. 17.9 corresponds to the phase successions in either a basic synthesis medium (upper part) or in an acidic synthesis solution (lower part).

In summary, when synthesizing a mesophase, it is not only important to consider the type of surfactant used, but also to carefully study its behavior under various synthesis conditions. Temperature, pH, co-solvents, salts, etc., can all have a certain effect on the surface charge of the micelles and thereby influence the charge-matching mechanism that occurs during the mesophase formation.

17.3
Case Study: Control of the Morphology and the Structural Ordering of MCM Materials by the Addition of Alcohols

The influence of the amount of short-chain alcohols [$C_nH_{2n+1}OH$, $n=1, 2, 3$] in TEOS/C_{16}TMABr/ammonia/alcohol/water mixtures at room temperature on the structure of mesoporous MCM solids has been investigated. The aim of the study was to investigate the possibility of controlling the morphology and structure of the mesoporous particles by systematic addition of alcohols. The different phase transitions up to high alcohol concentrations were studied by means of X-ray diffraction, N_2 sorption, SEM, and TEM analyses. The results reported here are mainly focused on the influence of ethanol addition, although the effects of the smaller methanol and the larger propanol were also investigated [54].

The basic protocol involved dissolving n-hexadecyltrimethylammonium bromide ($C_{16}TMA^+Br^-$) in a certain volume of de-ionized water before the addition of aqueous ammonia. A known volume of ethanol was added to this solution, which was then stirred for 15 minutes, and then the TEOS was added. Finally, a gel was obtained with the following composition: 1 TEOS : 0.3 C_{16}TMABr : 11 NH_3 : x alcohol : 144 H_2O.

Accordingly, a sample containing x moles (per mole TEOS) of ethanol was denoted as Ex. This solution was stirred for 2 h, and then the resulting white precipitate was collected by filtration and washed with distilled water. Subsequently, the dried sample was calcined by heating to 823 K at a heating rate of 1 K min^{-1}, and kept at this temperature for 6 h in order to remove the surfactant.

17.3.1
Structure Determination

Fig. 17.10 depicts the X-ray diffraction patterns for the different samples with increasing addition of ethanol to the mother solution.

The X-ray diffractogram of E20 clearly demonstrates that by adding a specific amount of ethanol an MCM-48 can be synthesized. The cubic unit cell can be identified in the diffractogram after calcination by the two distinct (211) and (220) diffraction peaks at 2.79° 2θ and 3.03° 2θ, respectively, and a broader reflection at 5.09° 2θ (Fig. 17.10; E20).

On the other hand, when no alcohol is added, typical MCM-41 with good crystallinity is obtained, as indicated by a strong (100) diffraction peak at 2.69° 2θ in the diffractogram of the calcined sample and two other distinguishable peaks of the (200) and (210) reflections between 4° and 6° 2θ. The additional peaks detected in the X-ray diffractogram before calcination correspond to the surfactant diffraction patterns.

When the amount of added ethanol is doubled, i.e. in the case of E40, a comparison between the X-ray diffractograms before and after calcination (Fig. 17.10; E40) shows that part of the broad XRD signal disappears upon calci-

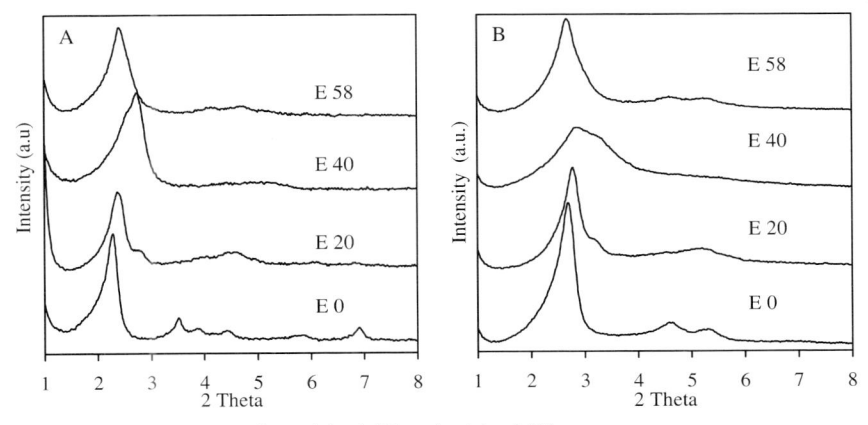

Fig. 17.10 XRD patterns of uncalcined (A) and calcined (B) mesoporous materials synthesized with an increasing amount of ethanol. The Si/alcohol molar ratio is indicated by the number of the labels.

nation. This suggests that E40 consists of mixed mesophases, in which a lamellar mesophase is responsible for the pore collapse during the removal of the organic surfactant. The remaining signal points towards some ordered structure on a local scale; no clear structure identification is possible though.

Further increase of the amount of ethanol to a molar ratio of 58 (sample E58) again produces an MCM-41-like diffraction pattern with a clear (100) reflection at 2.69° 2θ and two broader peaks in the region between 4° and 6° 2θ, attributed to the (200) and (210) reflections (Fig. 17.10; E58). Although there is a clear diffraction pattern, the peaks are broader compared to those of the E0 sample, indicating that this hexagonal-like E58 mesophase is less ordered.

The isotherms of the different mesophases (Fig. 17.11) are of type IV in the IUPAC classification, showing a capillary condensation at a relative pressure P/P_0 of between 0.2 and 0.3. The surface area, the pore volume, and the pore diameter of the E0 sample (MCM-41) are 1129 $m^2 g^{-1}$, 0.707 mL g^{-1}, and 3.2 nm, respectively. With the change of mesophase from hexagonal (MCM-41) to cubic (MCM-48), the sample E20 exhibits a small decrease in pore size to 3.1 nm, but significant increases in the surface area and pore volume, which attain values of 1379 $m^2 g^{-1}$ and 0.819 mL g^{-1}, respectively. Upon further increasing the alcohol content to a molar ratio of 40, the isotherm presents a much less well defined shape, although the low-pressure region to $P/P_0=0.2$ is identical to that of the E20 sample (cubic MCM-48), strongly suggesting that the remaining mesophase possesses cubic symmetry. The isotherm data of the sample with the highest ethanol addition, E58, indicate an identical total pore volume for both E0 and E58, and a slightly smaller pore diameter for E58. The surface area, pore volume, and pore diameter of E58 are 1211 $m^2 g^{-1}$, 0.707 mL g^{-1}, and 3 nm, respectively. All physical data are summarized in Tab. 17.4.

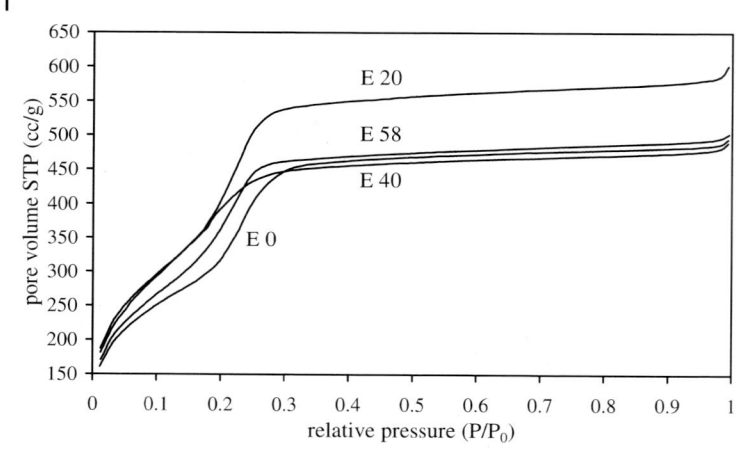

Fig. 17.11 N$_2$ sorption isotherms of calcined mesoporous materials synthesized with increasing amounts of ethanol.

Tab. 17.4 Physical data of the mesoporous silicate structure as a function of the ethanol concentration

	Phase	Pore diameter	Tot. Pore volume	Surface area
E0	MCM-41	3.2 nm	0.7 mL g^{-1}	1129 m^2 g^{-1}
E20	MCM-48	3.1 nm	0.8 mL g^{-1}	1379 m^2 g^{-1}
E58	spherical particles	3.0 nm	0.7 mL g^{-1}	1211 m^2 g^{-1}

Scanning electron microscopy (SEM) permits the study of the morphology of the different mesophase particles (Fig. 17.12). The MCM-41 particles in Fig. 17.12 A have an irregular shape with a particle size of about 1 µm.

The same irregular morphology is found for the MCM-48 mesophase particles in E20 (Fig. 17.12 B). When the amount of ethanol is further increased, as in sample E40, a significant change in the SEM images can be observed (Fig. 17.13 A). The mixed mesophase E40 consists mainly of spheres, although there is still a large contribution from particles of undefined shape. The dimension of the particles has now decreased to 0.5–1 µm. Finally, the SEM image of E58 (Fig. 17.13 B), prepared with a large amount of ethanol, reveals small regular particles of perfectly spherical shape, with an average size of about 0.5 µm.

To further identify the order and the change in the pore arrangements of the different mesophases upon addition of ethanol, a TEM study was performed.

The HREM image of E0 taken along the [001]$_H$ zone axis is shown in Fig. 17.14. This HREM image reveals a hexagonal ordered and parallel arrangement of the pores, which is confirmed by the FT pattern.

The HREM image of E20 also shows a nicely ordered pore system (Fig. 17.15). However, the corresponding FT pattern reveals a cubic arrangement

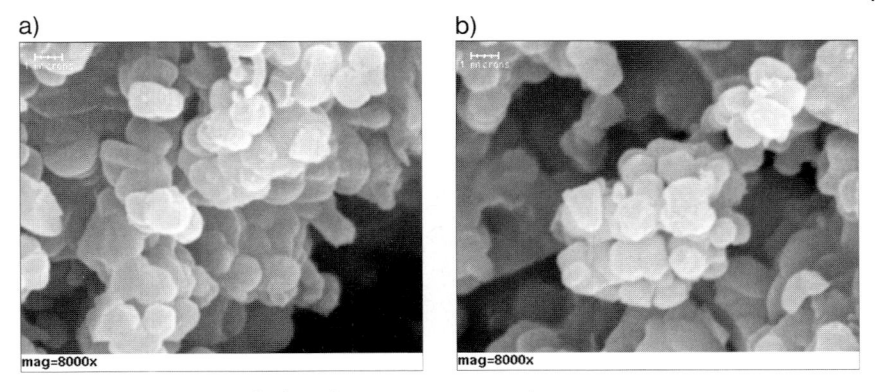

17.3 Case Study: Control of the Morphology and the Structural Ordering of MCM Materials | 337

Fig. 17.12 SEM images of calcined mesoporous materials synthesized with ethanol/Si ratios of 0 (a) and 20 (b).

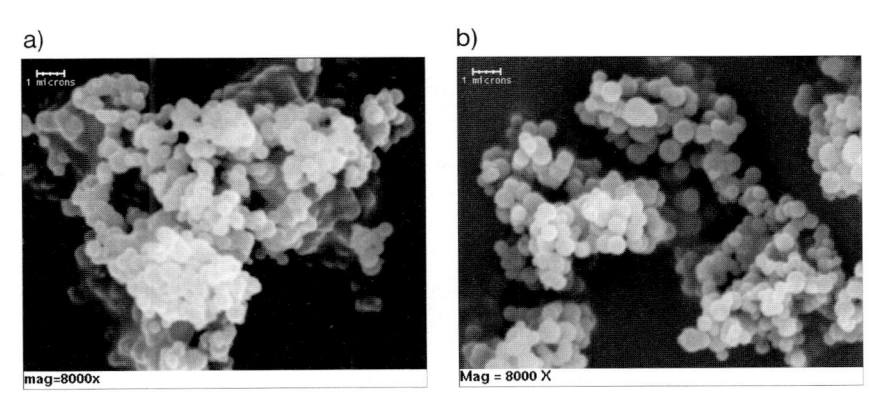

Fig. 17.13 SEM images of calcined mesoporous materials synthesized with ethanol/Si ratios of 40 (a) and 58 (b).

of the pores along the $[011]_C$ direction, exhibiting two identical d-spacings corresponding $\langle 111 \rangle$ planes and an angle of around 71° in between.

The E58 sample exhibits a pronounced spherical morphology of the particles (Fig. 17.16). The image shows a radial arrangement of the pores. At the edges of the spherical particles, this radial arrangement tends to disappear and a more wormhole-like structure is seen.

Such an observation is less clear in smaller spherical particles, where the radial arrangement is predominant. The enlargement of the HRTEM image in the inset is taken around the exact center part of the spherical particle (marked by a white square). The image shows hexagonal pore packing (being the closest packing possible) around the core of the particle. An additional proof of the hexagonal local structure is illustrated in the FT pattern taken over the central part of the particle (Fig. 17.16d). The weakness of the spots in the FT pattern indi-

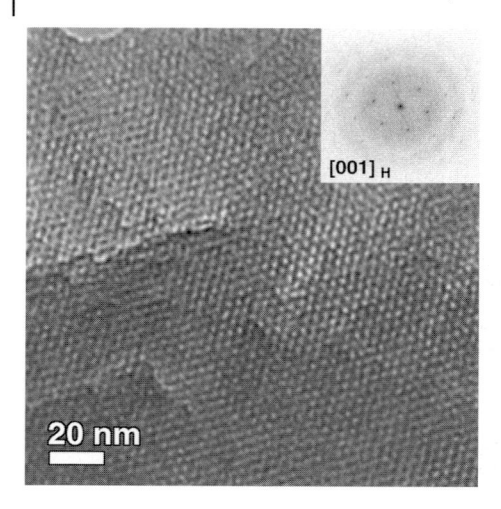

Fig. 17.14 HREM image of the MCM-41 hexagonal structure taken along the [001]$_H$ zone axis from the E0 sample. Inset: Fourier transform pattern.

cates that the hexagonal arrangement only occurs on a local scale, around the center part of the spherical particle. The neighboring pores around the central pore are distorted from the hexagonal symmetry due to the spherical shape of the particle. Closer inspection of the radial disposition of the pores reveals that they are grouped in a few parallel pores organized in a radial manner. Deviations from the hexagonal symmetry are bound to occur because of the spherical shape of the particles.

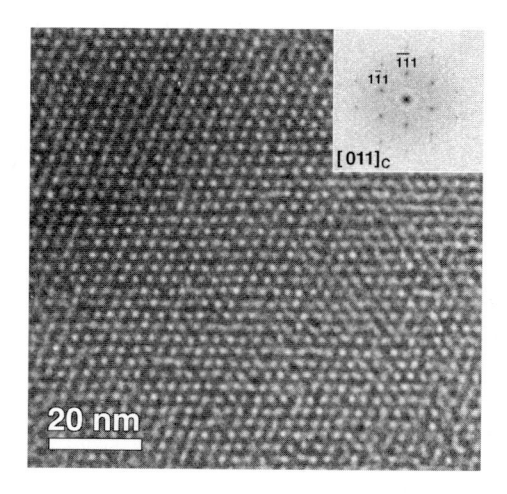

Fig. 17.15 HREM image of the MCM-48 cubic structure taken along the [011]$_C$ zone axis from the E20 sample. Inset: indexed Fourier transform pattern.

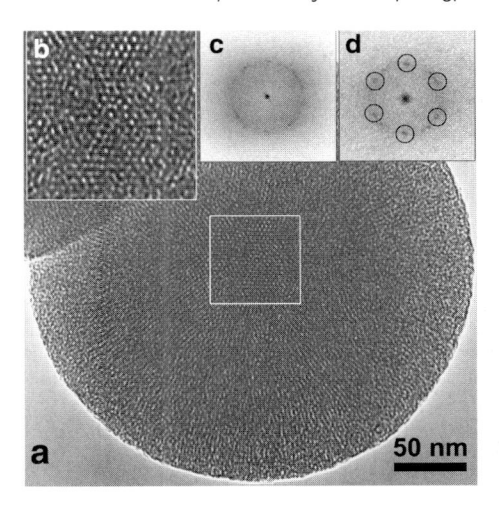

17.3 Case Study: Control of the Morphology and the Structural Ordering of MCM Materials | 339

Fig. 17.16 HREM image of a spherical particle (E58) with a well-resolved core. Enlargement of the core (marked with a white square frame), shown as an inset in (b), and the corresponding FT pattern (d), clearly show the hexagonal symmetry. The 10-10 and 11-20 peaks of a hexagonal pattern (circled) are clearly visible; (c) represents the FT pattern of the whole particle.

17.3.2
Effect of the Alcohol Concentration

In the room temperature TEOS/surfactant/ammonia/water/ethanol system, the X-ray diffractograms (Fig. 17.10) as well as the TEM observations (Figs. 17.14 and 17.15) clearly demonstrate that ethanol needs to be added to the synthesis solution in order to transform the MCM-41 phase (E0) into an MCM-48 cubic phase (E20). This result confirms that increasing the amount of ethanol causes a change in the surfactant packing parameter $g = V/a_0^* l$ (see Subsection 17.2.2.2).

At lower alcohol concentrations, driven by the hydrophobic forces, the alcohol molecules penetrate into the surfactant micelles. This results in an increase in the true volume of the surfactant (V) and therefore in an increase in the g value, causing high-to-low curvature phase transitions. Consequently, mesophases with decreasing curvature are sequentially formed, going from hexagonal to cubic. This has been widely proven in many cases and is referred to as the *co-surfactant* effect of the alcohol [38, 46, 55–63].

By doubling the ethanol concentration compared to that used to obtain the MCM-48 phase, more ethanol will penetrate inside the micelles, further decreasing the surface curvature. This will inevitably lead to the formation of a lamellar phase. This can clearly be seen on comparing the X-ray diffraction patterns before and after calcination (Fig. 17.10; E40). After calcination, part of the diffraction signal disappears. The remaining peak after calcination indicates that some ordering other than a lamellar phase must exist. TEM analysis could not reveal a straightforward phase ordering, suggesting that the structured arrangement must be either on a very local scale or else very unstable under the applied electron beam. Important information in this respect is provided by the isotherm of this phase (Fig. 17.11; E40). The low relative pressure data would seem indicative of the same shape as the MCM-48 sample. This suggests the presence of some MCM-48-like unit cells. This strongly supports the notion that ethanol cannot homogeneously decrease the surface curvature to give rise to a logical phase transition from a cubic MCM-48 phase to a lamellar structure. Ethanol present inside the micelles will only disturb the formation and ordering of an identifiable phase, leaving MCM-48-like unit cells behind.

Upon further increasing the ethanol content, spherical particles are observed, both in the SEM as the TEM images. The latter (Fig. 17.16) also reveals a remarkable phase regression phenomenon by showing pores organized in a hexagonal close-packing phase. This is confirmed by the bulk crystallographic analysis by X-ray diffraction (Fig. 17.10; E58). It shows a hexagonal diffraction pattern with a (100), (110) and a (200) diffraction signal. A tentative explanation for this phase regression can be given as follows. First, consider the changes at the level of micelle formation. At a high concentration, more and more alcohol molecules reside on the outer boundaries of the surfactant micelles, so the ratio of alcohol in the aqueous solution to that in the micelles can be expected to start increasing. This means that the role of the alcohol is shifting from a co-surfactant to a co-solvent function. This co-solvent will have a diluting effect on the eventual micelles, making them less packed, which automatically favors high surface curvatures. Secondly, the high concentration of alcohol in the solution will slow down the hydrolysis reaction of TEOS to a large extent, due to preferential solvation in the alcohol phase [64]. A suppressed hydrolysis of TEOS decreases the number of charged silicate particles in the solution. Such a reduction in the amount of charge-compensating solution enhances the repulsive forces between the adjacent head groups of the surfactant molecules. Consequently, the effective headgroup surface area a_0 of the surfactant molecules becomes larger, and the g value decreases in favor of the formation of a hexagonal phase or, at high surface curvature, even of some globular micelles. This preferential dissolution of TEOS in the alcohol will also aid homogenization of the mother solution and thereby influence the type of silicate species formed during the hydrolysis process. Once the negatively charged hydrolysis products are formed, they will only slowly replace the bromides as counter ions of the cationic surfactants in order to electrostatically neutralize them, since bromide is known to be a strong binding anion [65]. Moreover, the chosen reaction conditions of room temperature and the use of ammonia instead of sodium hydroxide as the base also promote a slowing down of the hydrolysis of the TEOS molecules.

The slow hydrolysis of TEOS, forming a variety of silicate species in the solution, together with the disturbed mechanism of charge matching between the silicate species and the CTAB surfactant as well as the potential delayed condensation of silicate-surfactant aggregates, all due to the massive presence of alcohol at the inorganic–organic interface, are part of the complex mechanism that leads to the growth of rather straight micelles on the surfaces of the particle cores. This growth will be fed by the addition of globular or straight silica-coated micelles [40, 63]. The polymerization of the silicate walls will force these long flexible mi-

17.3 Case Study: Control of the Morphology and the Structural Ordering of MCM Materials

celles to adopt close-packing and thus a hexagonal ordering. When the micelle length is further increased, the high flexibility of long micelles could induce a decrease in order, resulting in a more wormhole-like structure, as shown near the outer limit region of the particle (Fig. 17.16). The more prevalent ordering in smaller spherical particles is supportive of this. However, measurement artifacts as well as interference cannot be excluded as being responsible for the absence of ordering in the outer region of the spherical particles.

The morphology of a growing particle (spherical or of more irregular form) depends on the balance between the rate of polymerization of the negatively charged silicate micelles and the rate of mesostructure formation [66]. In the case of a slow polymerizing silicate seed at high alcohol concentrations, as in the Stöber silica synthesis [67], the growth is driven by global surface tension forces, which cause it to adopt a spherical shape in order to minimize its surface free energy, as observed in the SEM and TEM images. In contrast, at lower alcohol concentrations, the polymerization will be relatively faster and the morphology will be controlled by the deposition of silicate-micellar species onto specific regions of the growing seed crystal, resulting in non-uniform agglomerated blocks, as observed in the relevant SEM images.

While the Stöber synthesis can account for the fact that a particle adopts a spherical shape, it does not explain why the pores are arranged in a preferentially radial geometry. In the logical order of the g packing parameter, it could be that during the nucleus formation the micelles are ordered in a cubic symmetry. Further homogeneous micelle growth on the different planes of a cube will favor the radial arrangement of the micelles.

In the present concept of a seed growth mechanism, the core of the particles should contain a seed, around which the particle has grown. Evidence of this could provide verification of such a mechanism for particle growth. In order to determine the structure of this SSP core, it should be oriented either along $[111]_C$ or $[100]_C$ in the case of a cubic structure or along $[001]_H$ in the case of a hexagonal structure. Only in these directions would a continuous projection of the pores be obtained. Therefore, in order to confirm the eventual presence of a cubic phase structure, we produced direct images of the structure along $[100]_C$ and along $[111]_C$ (Fig. 17.17).

Fig. 17.17 shows a HREM image of a spherical particle with a square pore arrangement in the central part. The corresponding FT pattern exhibits an unambiguous square pattern of sharp spots superimposed on a diffuse ring. When the particle is rotated about one axis through, e.g., 35° (Fig. 17.17b), the square dot arrangement and the corresponding spots in the FT pattern disappear. Simultaneously, two distinctive regions are detected inside the particle: an apparently crystalline inner part and radial bundles of cylindrical pores on the side. The tilting does not affect the particle microstructure: the positions of the radial pores and the crystalline part remain unaltered; only the central symmetry has changed. Following a similar filtering procedure as described above, a Bragg mask was applied to image the inner and outer regions of the particle (Fig. 17.18).

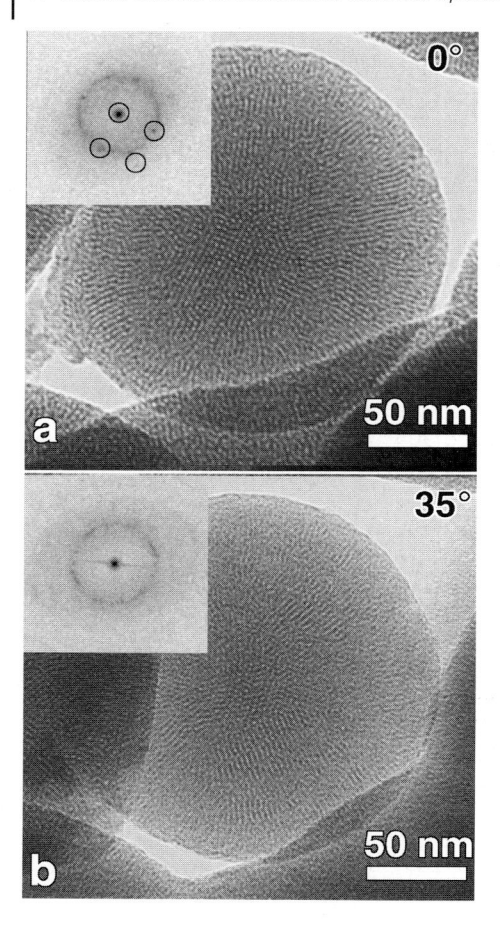

Fig. 17.17 (a) HREM image of a small spherical particle exhibiting square pore ordering in the central part. The corresponding FT pattern given in the inset clearly shows a square pattern of spots (marked by black circles), corresponding to the core structure marked by black circles. (b) HREM image of the same particle tilted by 35° along one axis. Note that the square pore ordering and corresponding pattern in the FT have disappeared.

The cubic symmetry is very pronounced in the central part (Fig. 17.18a); the cylindrical pores start more or less from the inner crystalline part and extend radially to the surface of the particle (Fig. 17.18b).

The FT pattern and the filtered images clearly demonstrate two features of interest. The cylindrical pores are not continuous from the center to the border of the particle and shift over half of the channel similar to antiphase boundaries (Fig. 17.18c). The bundles of parallel cylindrical pores are aligned close to the crystallographic directions of the projected core structure. For instance, when the core is imaged along $[001]_C$, they are parallel to the $\langle 100 \rangle_C$ direction (Fig. 17.18c), for a hexagonal core projection, and they are parallel to $\langle 110 \rangle_C$ when referring to a cubic structure.

The TEM results strongly suggest that the spherical particle consists of a core possessing a cubic symmetry and bundles of pores extending in a radial manner from the core surface to the edges of the particle.

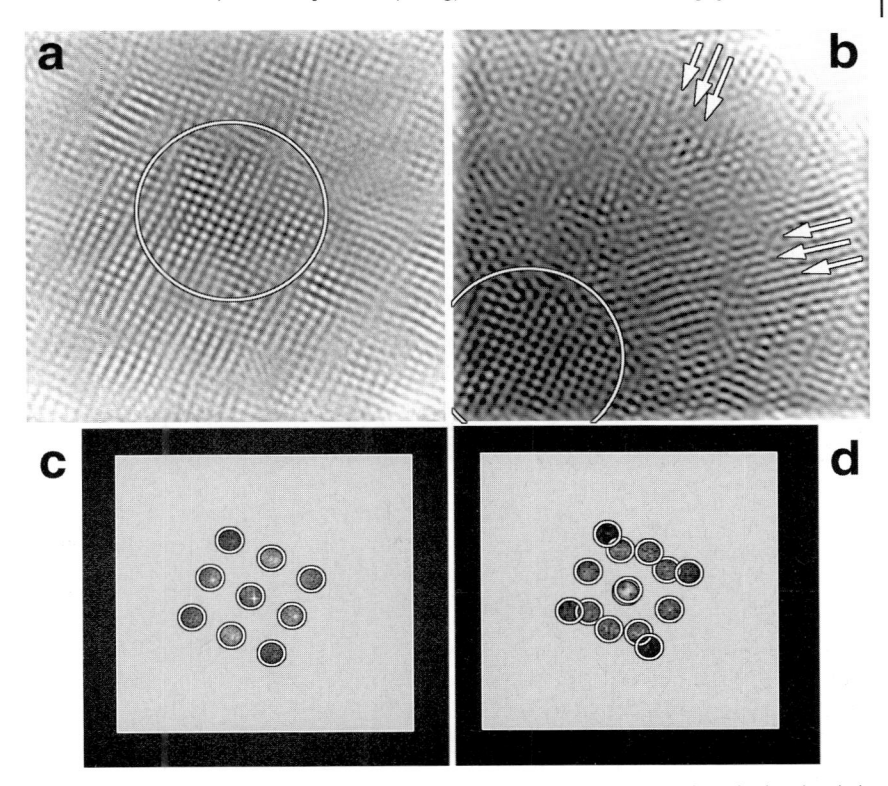

17.3 Case Study: Control of the Morphology and the Structural Ordering of MCM Materials | 343

Fig. 17.18 Enlarged HREM images of central (a) and side regions (b) of Fig. 17.17. A Bragg mask filter was used to select spots responsible for the hexagonal pore ordering (c), and spots produced by cylindrical radial pores (indicated by white arrows) (d). The core structure is marked by a white circle.

The reconstruction of the exact crystallographic shape of the core of the spherical particle from HREM observations is rather difficult. It is nevertheless reasonable to suggest that the cubic core possesses the 3-D "crystal" shape of the cubic MCM-48. This would then be a polyhedron in the form of a truncated octahedron with cubic symmetry. Different research groups have indeed shown that MCM-48 crystals can adopt this shape depending on the growth conditions [57, 68]. Anderson [69, 70] has shown that the truncated octahedral shape of the MCM-48 particles is a reflection of the underlying symmetry of the gyroid surface combined with the growth mechanism. In the ideal case, the external surface can be bound by six {100} and eight {111} planes. The crystal morphology will adopt the equilibrium form determined by minimal surface energy. In the case of MCM-48, the surface energy of the {111} face is 0.77 times that of the {100} face, and all other faces have a higher surface energy [70]. The presence of pores distributed in a radial fashion over the particle around a cubic core strongly supports the seeded growth mechanism theory. In this case, an MCM-48 crystal, with the shape of a truncated octahedron, will function as the seed on which pores will slowly be grown through the deposition of micellar aggregates. The facets will play the role of the substrate, on which the cylindrical pores grow in an epitaxial manner perpendicular to the core surface and, accordingly, with the symmetry of the surface plane. These pores will have a parallel arrangement and will eventually adopt a hexagonal close-packing (as in MCM-41).

The above results suggest a two-step formation of the particle. The first step involves a very rapid formation of mesoporous particles in a cooperative process with micelles. However, the added alcohol will have partially migrated into the micelles creating surfactant rods with low surface curvatures. This favors the synthesis of MCM-48, which is in full agreement with the phase sequence of the g packing parameter. After a certain period of time, the co-solvent effect of the alcohol will prevail and create other synthesis conditions favoring the formation of straight micelles, which will organize themselves into a stable hexagonal packing.

17.4 Conclusions

In the first part of this chapter, a summary of the most recent developments in material research has been given, together with a full description of the most important hexagonal and cubic siliceous mesoporous structures. Subsequently, the synthesis mechanisms for the formation of different mesoporous materials have been discussed, highlighting their key steps. Particular emphasis has been placed on structure-directing agents (templates), which certainly exert a decisive effect in controlling the mesophase formation.

Secondly, it has been demonstrated how the morphology and the mesophase formation within MCM materials can be controlled by the addition of an alcohol. Increasing the amount of alcohol present in a room temperature synthesis induces a shift in the mesophase formation from the hexagonal MCM-41 to the cubic MCM-48. A further increase of the alcohol content in the synthesis mixture has a significant effect on both the inner structure and the outer morphology of the MCM mesoparticles, resulting in spheres containing both cubic and hexagonal symmetry.

References

1 C.T. KRESGE, M.E. LEONOWICZ, W.J. ROTH, J.C. VARTULI, U.S. Patent 5 098 684, 1992.
2 J.S. BECK, C.T. CHU, I.D. JOHNSON, C.T. KRESGE, M.E. LEONOWICZ, W.J. ROTH, J.C. VARTULI, U.S. Patent 5 108 725, 1992.
3 J.S. BECK, D.C. CALABRO, S.B. McCULLEN, B.P. PELRINE, K.D. SCHMITT, J.C. VARTULI, U.S. Patent 5 145 816, 1992.
4 J.S. BECK, C.T. KRESGE, M.E. LEONOWICZ, W.J. ROTH, J.C. VARTULI, U.S. Patent 5 264 203, 1993.
5 J.S. BECK, K.D. SMITH, J.C. VARTULI, U.S. Patent 5 334 368, 1994.

6 J.S. Beck, C.T. Kresge, S.B. McCullen, W.J. Roth, J.C. Vartuli, U.S. Patent 5 370 785, 1994.

7 C.T. Kresge, M.E. Leonowicz, W.J. Roth, J.C. Vartuli, J.C. Beck, *Nature* **1992**, *359*, 710.

8 J.S. Beck, J.C. Vartuli, W.J. Roth, M.E. Leonowicz, C.T. Kresge, K.D. Schmitt, C.T.-W Chu, D.H. Olson, E.W. Sheppard, S.B. McCullen, J.B. Higgins, J.L. Schlenker, *J. Am. Chem. Soc.* **1992**, *14*, 10834.

9 C.Y. Chen, S.L. Burkett, H.X. Li, M.E. Davis, *Microp. Mater.* **1993**, *2*, 27.

10 A. Monnier, F. Schüth, Q. Huo, D. Kumar, D. Margolese, R.S. Maxwell, G.D. Stucky, M. Krishnamurty, P. Petroff, A. Firouzi, M. Janicke, B.F. Chemlka, *Science* **1993**, *261*, 1299.

11 J.C. Vartuli, K.D. Schmitt, C.T. Kresge, W.J. Roth, M.E. Leonowicz, S.B. McCullen, S.D. Hellring, J.S. Beck, J.L. Schlenker, D.H. Oslon, E.W. Sheppard, *Chem. Mater.* **1994**, *6*, 2317.

12 P. Van Der Voort, M. Mathieu, F. Mees, E.F. Vansant, *J. Phys. Chem. B* **1998**, *102*, 8847.

13 O. Collart, P. Van Der Voort, E.F. Vansant, D. Desplantier, A. Galarneau, F. Di Renzo, F. Fajula, *J. Phys. Chem. B* **2001**, *105*, 12771.

14 Q. Huo, D.I. Margolese, U. Ciesla, P. Feng, T.E. Gier, P. Sieger, R. Leon, P.M. Petroff, F. Schüth, G.D. Stucky, *Nature* **1994**, *368*, 317.

15 D. Zhao, Q. Huo, J. Feng, B.F. Chmelka, G.D. Stucky, *J. Am. Chem. Soc.* **1998**, *120*, 6024.

16 D. Zhao, P. Yang, N. Nelosh, J. Feng, B.F. Chmelka, G.D. Stucky, *Adv. Mater.* **1998**, *10*, 1380.

17 Q. Huo, R. Leon, P.M. Petroff, G.D. Stucky, *Science* **1995**, *268*, 1324.

18 G.D. Stucky, D. Zhao, P. Yang, W. Lukens, N. Nelosh, B.F. Chmelka, *Stud. Surf. Sci. Catal.* **1998**, *117*, 1.

19 D. Zhao, Q. Huo, J. Feng, J. Kim, Y. Han, G.D. Stucky, *Chem. Mater.* **1999**, *11*, 2668.

20 S. Inagaki, Y. Fukushima, K. Kuroda, *J. Chem. Soc., Chem. Commun.* **1993**, 680.

21 F. Di Renzo, A. Galarneau, P. Trens, F. Fajula, in *Handbook of Porous Materials* (Eds.: F. Schüth, K.S. W. Sing, J. Weitkamp), Wiley-VCH, Weinheim, 2002, p. 1383.

22 A. Cauvel, D. Brunel, F. Di Renzo, E. Garronne, B. Fubini, *Langmuir* **1997**, *13*, 2773.

23 A. Galarneau, D. Desplantier-Giscard, F. Di Renzo, F. Fajula, *Catal. Today* **2001**, *68*, 191.

24 P. Selvam, S.K. Bhatia, C.G. Sonwane, *Ind. Eng. Chem. Res.* **2001**, *40*, 3237.

25 A.H. Schoen, NASA Technical Note D-5541, Washington DC, 1970.

26 P.I. Ravikovitch, A.V. Neimark, *Langmuir* **2000**, *16*, 2419.

27 M.W. Anderson, *Zeolites* **1997**, *19*, 220.

28 V. Alfredsson, M.W. Anderson, T. Ohsuna, O. Terasaki, J. Jacob, M. Bojrup, *Chem. Mater.* **1997**, *9*, 2066.

29 A. Carlsson, M. Kaneda, Y. Sakamoto, O. Terasaki, R. Ryoo, S.H. Joo, *J. Electr. Microsc.* **1999**, *48*, 795.

30 M. Kaneda, T. Tsubakiyama, A. Carlsson, Y. Sakamoto, T. Ohsuna, O. Terasaki, S.H. Joo, R. Ryoo, *J. Phys. Chem. B* **2002**, *106*, 1256.

31 M.S. Morey, A. Davidson, G.D. Stucky, *J. Porous Mater.* **1998**, *5*, 195.

32 J. Patarin, B. Lebeau, R. Zana, *Curr. Opin. Coll. & Interf. Sci.* **2002**, *7*, 107.

33 G.J. de A.A. Soler-Illia, C. Sanchez, B. Lebeau, J. Patarin, *Chem. Rev.* **2002**, *102*, 4093.

34 P.A. Winsor, *Chem. Rev.* **1968**, *68*, 1.

35 P. Ekwall, in *Advances in Liquid Crystals* (Ed.: G.H. Brown), Academic Press, New York, 1971, p. 1.

36 V. Luzzati, R. Vargas, P. Mariani, A. Gulik, H. Delacroix, *J. Mol. Biol.* **1993**, *229*, 540.

37 G.D. Stucky, A. Monnier, F. Schüth, Q. Huo, D.I. Margolese, D. Kumar, M. Krishnamurty, P. Petroff, A. Firouzi, M. Janicke, B.F. Chmelka, *Mol. Cryst. Liq. Cryst.* **1994**, *240*, 187.

38 S.H. Tolbert, C.C. Landry, G.D. Stucky, B.F. Chmelka, P. Norby, J.C. Hanson, A. Monnier, *Chem. Mater.* **2001**, *13*, 2247.

39 C. J. Glinka, J. M. Nicol, G. D. Stucky, E. Ramli, D. I. Margolese, Q. Huo, in *Advances in Porous Materials; Mater. Res. Proc.* (Eds.: S. Komarneni, D. M. Smith, J. S. Beck), Material Research Society, Pittsburgh, PA, 1995, Vol. 371, p. 47.

40 A. Galarneau, F. Di Renzo, F. Fajula, L. Mollo, B. Fubini, M. F. Ottaviani, *J. Colloid Interf. Sci.* **1998**, *201*, 105.

41 M. F. Ottaviani, A. Galarneau, D. Desplantier-Giscard, F. Di Renzo, F. Fajula, *Micropor. Mesopor. Mater.* **2001**, *44/45*, 1.

42 D. C. Calabro, E. W. Valyocsik, F. X. Ryan, *Micropor. Mater.* **1996**, *7*, 234.

43 L. Sicard, J. Frasch, M. Soulard, B. Lebeau, J. Patarin, T. Davey, R. Zana, F. Kolenda, *Micropor. Mesopor. Mater.* **2001**, *44/45*, 25.

44 R. K. Iler, *The Chemistry of Silica: Solubility, Polymerization, Colloid and Surface Properties*, Wiley-Interscience Publications, New York, 1979, pp. 186.

45 R. Simonutti, A. Comotti, S. Bracco, P. Sozzani, *Chem. Mater.* **2001**, *13*, 771.

46 Q. Huo, D. I. Margolese, G. D. Stucky, *Chem. Mater.* **1996**, *8*, 1147.

47 P. Mariani, V. Luzzati, H. Delacroix, *J. Mol. Biol.* **1988**, *204*, 165.

48 J. N. Israelachvili, D. J. Mitchell, B. W. Ninham, *J. Chem. Soc., Faraday Trans. 2*, **1976**, *72*, 185.

49 J. N. Israelachvili, D. J. Mitchell, B. W. Ninham, *Biochim. Biophys. Acta* **1977**, *470*, 185.

50 S. M. Gruner, *J. Phys. Chem.* **1989**, *93*, 7562.

51 K. W. Gallis, C. C. Landry, *Chem. Mater.* **1997**, *9*, 2035.

52 C. C. Landry, S. H. Tolbert, K. W. Gallis, A. Monnier, G. D. Stucky, P. Norby, J. C. Hanson, *Chem. Mater.* **2001**, *13*, 1601.

53 M. J. Rosen, *J. Surf. Interfac. Phenom.*, John Wiley, New York, 1989, pp. 108.

54 S. Liu, P. Cool, O. Collart, P. Van Der Voort, E. F. Vansant, O. I. Lebedev, G. Van Tendeloo, M. Jiang, *J. Phys. Chem. B* **2003**, *107*, 10405.

55 K. W. Gallis, C. C. Landry, *Chem. Mater.* **1997**, *9*, 2035.

56 C. C. Landry, S. H. Tolbert, K. W. Gallis, A. Monnier, G. D. Stucky, P. Norby, J. C. Hanson, *Chem. Mater.* **2001**, *13*, 1600.

57 J. M. Kim, S. K. Kim, R. Ryoo, *J. Chem. Soc., Chem. Comm.* **1998**, 259.

58 J.-H. Sun, M.-O. Coppens, *J. Mater. Chem.* **2002**, *12*, 1.

59 Y. S. Lee, D. Surjadi, J. F. Rathman, *Langmuir* **2000**, *16*, 195.

60 R. Ryoo, S. H. Joo, J. M. Kim, *J. Phys. Chem. B* **1999**, *103*, 7435.

61 M. Kruk, M. Jaroniec, R. Ryoo, S. H. Joo, *Chem. Mater.* **2000**, *12*, 1414.

62 Y. Liu, A. Karkamkar, T. J. Pinnavaia, *J. Chem. Soc., Chem. Comm.* **2001**, 1822.

63 P. Ågren, M. Linden, J. B. Rosenholm, R. Schwarzenbacher, M. Kriechbaum, H. Amenitsch, P. Laggner, J. Blanchard, F. Schüth, *J. Phys. Chem. B* **1999**, *103*, 5943.

64 C. Brinker, G. Scherer, *Sol-Gel Science: The Physics and Chemistry of Sol-Gel Processing*, Academic Press, New York, 1990.

65 H. Yang, G. Vovk, N. Coombs, I. Sololov, G. A. Ozin, *J. Mater. Chem.* **1998**, *8*, 743.

66 H. B. S. Chan, P. M. Budd, T. V. Naylor, *J. Mater. Chem.* **2001**, *11*, 951–957.

67 C. A. P. Leite, E. F. de Souza, F. Galembeck, *J. Braz. Chem. Soc.* **2001**, *12(4)*, 519–525.

68 M. Kaneda, T. Tsubakiyama, A. Carlsson, Y. Sakamoto, T. Ohsuna, O. Terasaki, S. H. Joo, R. Ryoo, *J. Phys. Chem. B* **2002**, *106*, 1256.

69 M. W. Anderson, *Zeolites* **1997**, *19*, 220.

70 V. Alfredsson, M. W. Anderson, *Chem. Mater.* **1996**, *8*, 1141.

18
Local Crystal Chemistry, Structured Diffuse Scattering, and Inherently Flexible Framework Structures

Ray L. Withers and Yun Liu

18.1
Introduction

Displacively flexible, corner-connected polyhedral framework structures represent a large, technologically important and ever-growing family of crystalline phases (zeotypic, microporous molecular sieve materials [1–3], nanoporous molecular framework structures [4, 5], the silica polymorphs and their aluminophosphate analogues along with the numerous stuffed or framework doped variants thereof [6–8], the widespread ReO_3 and perovskite-related family of phases [9–11], metaphosphates [12], hexacelsians [13], fresnoites [14], etc.) of fundamental importance to solid-state chemistry, mineralogy, and materials science. They exhibit a wide range of useful physico-chemical properties, including low or even negative thermal expansion [15–19], selective adsorption and/or diffusion of gas molecules or alkali metal ions through the channels of the framework [20–22], shape-selective catalysis enhanced by the displacive flexibility of the host framework structure [1, 21, 22], triboluminescence, piezoelectricity, pyroelectricity, and ferroelectricity [14, 23, 24], amongst a host of others. Their structures all share common features: they are built from essentially rigid molecules and/or cation-centered polyhedra (octahedral, tetrahedral, square-pyramidal, etc.) that are linked through sharing corners (see, e.g., Fig. 18.1 below).

Despite the rigidity of the individual constituent polyhedral building blocks, as well as the fixed nature of their topological connectedness, many such framework structures nonetheless retain considerable inherent displacive flexibility [9, 25–27]. This structural flexibility arises from the surprisingly numerous different ways in which it is possible to change the relative orientation and positioning of neighboring polyhedral units without distorting either the shape or size of the individual polyhedral units themselves. When the net effect of the excitation of many such polyhedral rotation modes is relatively small, conventional crystallographic refinement still gives a good idea of local crystal structure. When the cumulative effect of these polyhedral rotation modes gives rise to large amplitude rotations on the local scale, however, conventional average structure refinements invariably give chemically unsound structures with short metal–oxygen distances within constituent polyhedra, chemically unreasonable

Inorganic Chemistry in Focus II. Edited by G. Meyer, D. Naumann, and L. Wesemann
Copyright © 2005 WILEY-VCH Verlag GmbH & Co. KGaA, Weinheim
ISBN: 3-527-30811-3

Fig. 18.1 Average tetrahedral framework structure of the hexagonal microporous AlPO$_4$-5 phase projected down its c direction. The light and dark tetrahedra correspond to AlO$_4$ and PO$_4$ tetrahedral building blocks, respectively.

180° metal–oxygen–metal angles, along with large and/or strongly anisotropic temperature factors [27–30]. Crystal chemical problems of this sort have long plagued average structure refinements of framework structures and often lead to a great deal of controversy in the literature as regards 'the correct' local crystal structure for many such materials [29–32].

A recent new approach to the resolution of crystallographic/crystal chemical inconsistencies of this sort and to the structural understanding of flexible framework structures in general is based upon taking proper account of the inherent, large amplitude, dynamical disorder of such systems. Until this inherent orientational disorder is properly taken into consideration, the local crystal chemistry of such materials can never be fully understood. The likely incidence of dynamical disorder shows up in calculations of crystalline lattice motion as modes of essentially zero frequency and in scattering from such solids as structured diffuse scattering [29, 30, 33–36]. Inherent flexibility of this sort is not a nuisance that can be ignored but a fundamental structural characteristic with important physico-chemical consequences. Large amplitude, nanoscale orientational flexibility of this type, for example, significantly distorts both the instantaneous local sizes and shapes of channel windows in zeotypic systems and hence plays an important role in physico-chemical properties such as the dynamical response of zeotypic framework structures to the adsorption and/or diffusion of guest molecules [21, 22, 37, 38]. It also often leads to polymorphism and a strong susceptibility to phase transformation upon alteration of composition through doping or of external conditions such as temperature, pressure, etc. [8, 34, 39–42]. It is therefore of fundamental importance to be able to quantitatively characterize this inherent displacive flexibility.

18.2
Recent Methodological Developments

While the basic notion of essentially rigid polyhedral building blocks corner-linking together to form flexible framework structures has been known for many years, it had until relatively recently been rather difficult to detect experimentally let alone systematically quantify the nature and extent of the displacive flexibility of most framework structures. (One exception is the ReO_3 and perovskite-related family of phases, where inherent displacive flexibility and its diffraction as well as real-space structural consequences have long been understood and used to relate and classify perovskite-related phases [9, 10].)

The use of temperature-dependent electron diffraction [29, 30, 33–36] in conjunction with the relatively recent introduction of the RUM (or Rigid Unit Mode) approach, particularly the development of the lattice dynamic program CRUSH [41], has now begun to change this and, in principle, enables the soft, or zero frequency, RUM modes of vibration of any framework structure to be mapped out as a function of wave-vector in reciprocal space. (RUM modes of vibration are co-operative modes of vibration of a framework structure which involve rotation and translation, but not distortion, of the constituent rigid polyhedral units.) While the appropriate idealized parent structure with respect to which such lattice dynamical calculations should be carried out is not always obvious, such calculations have nonetheless shown themselves to be extremely useful and quite predictive in specific cases [29, 30, 39, 40, 43, 44]. Lattice dynamical calculations of this sort can now be relatively routinely used to theoretically search for "soft" RUM modes, often giving what at first sight appear to be surprising results [21, 22, 39, 43].

Direct experimental proof of the large amplitude excitation of such "soft" RUM modes of distortion can most immediately be found, however, in the form of characteristic and highly structured diffuse intensity distributions in temperature-dependent diffraction studies, particularly electron diffraction studies as a result of the sensitivity of electron diffraction to weak features of reciprocal space [29, 30, 33–36] (see, for example, the characteristic diffuse intensity distributions in Fig. 18.2).

The observed characteristic diffuse intensity distributions arise directly from the thermal excitation of the various low energy RUM phonon modes [26, 27, 39, 40] and provide a direct reciprocal space map of potential soft phonon modes that may (and often do) condense, or freeze out, e.g. upon lowering of temperature. The simultaneous, large amplitude excitation of such "soft" RUM phonon modes gives rise to a dynamically disordered, high temperature structure. It is the failure to take into account the simultaneous co-existence of a multitude of such large amplitude RUM modes of distortion that leads to chemically unsound 'average' crystal structures for many framework structures [31, 32].

The soft wave-vectors traced out in such diffuse distributions have been found in many cases to be rather surprising (in particular, those that fall on continuous hyperbolic surfaces) and by no means immediately intuitively obvious [29,

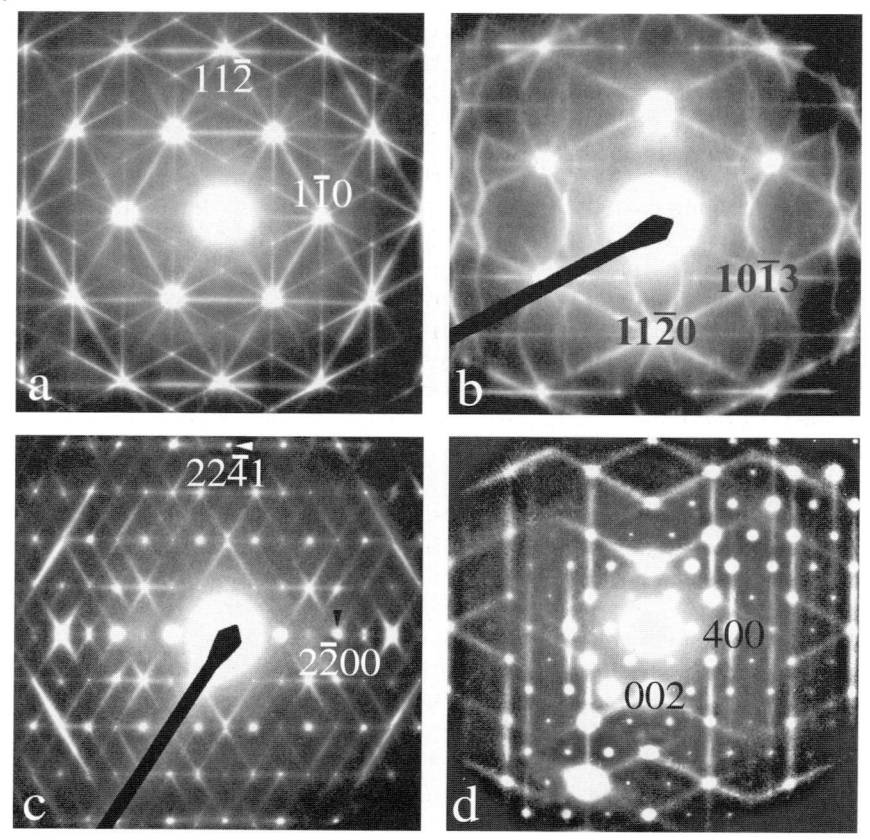

Fig. 18.2 (a) [111] Zone axis electron diffraction pattern (EDP) typical of the highest temperature polymorph of SiO$_2$ cristobalite [33], (b) a [−3,3,1] zone EDP typical of the highest temperature polymorphs of SiO$_2$ tridymite [34], (c) a [1,1,−4] zone axis EDP of the high temperature polymorph of Ba-hexacelsian [36], and (d) a close to [010] zone axis EDP of AlPO$_4$-11 at room temperature [30]. The sharp, continuous diffuse intensity distributions observed in such EDPs map out the essentially zero frequency RUM phonon modes of distortion of these various tetrahedral framework structures.

30, 34]. The factors that constrain RUM modes to such very specific modulation wave-vectors, and that determine the associated displacement eigenvectors and hence atomic displacement patterns, need to be analytically understood [36, 43, 44]. They are the same factors that control, for example, the dynamical responsiveness of flexible framework structures to external stimuli at higher temperatures and/or determine the structural relationship of lower temperature polymorphic forms to their higher temperature parent structures in the event of a structural phase transition upon, e.g., lowering of temperature. At high enough temperature, all such "soft" RUM phonon modes are significantly excited. Upon lowering of temperature, however, one such RUM mode or combination of

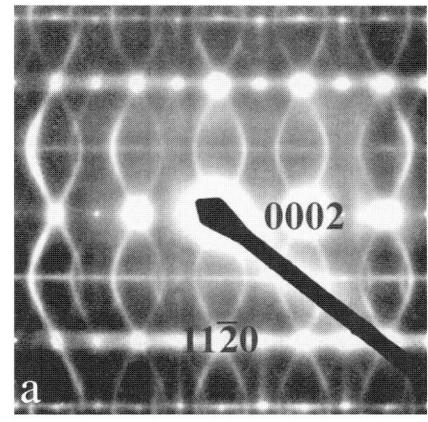

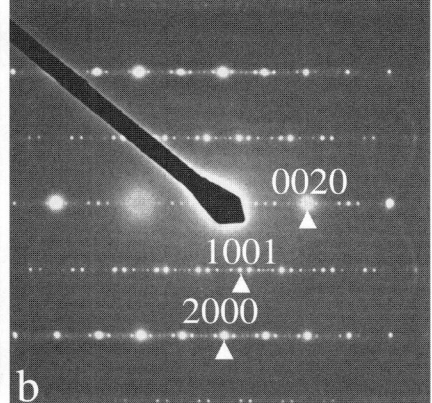

Fig. 18.3 A $\langle -1,1,0 \rangle$ zone axis EDP of (a) a dynamically disordered high temperature polymorph of SiO_2-tridymite [34] juxtaposed with (b) the equivalent zone axis EDP of a room temperature incommensurately modulated polymorph [34]. Note that the strongest additional satellite reflections in the latter phase have condensed out at positions which fall onto the higher temperature diffuse distribution.

RUM modes may condense out triggering structural change and leading to a cascade of lower temperature derivative structures [45]. The difference in free energy between these lower temperature derivative structures is then expected to be relatively small, and their structures therefore susceptible to alteration through external influences such as temperature, pressure, loading of adsorbed molecules or isomorphous doping.

In reciprocal space, such polymorphic phase transitions are indicated by the sudden appearance of additional satellite reflections, the strongest of which condense out at positions on the higher temperature diffuse distribution. Temperature-dependent electron diffraction is an excellent probe of such phase transition behavior (see, e.g., Fig. 18.3), while the Rigid Unit Mode (RUM) approach provides a robust theoretical framework for describing such phase transition behavior in terms of the condensation of one or more RUM modes of an appropriately chosen underlying parent structure [39, 40, 43, 45].

At higher temperature, where extensive orientational disorder is found to occur, the recently developed total neutron scattering technique (involving fitting high quality powder diffraction data simultaneously to both Bragg *and* diffuse scattering) can be used in conjunction with Reverse Monte Carlo (RMC) simulation to quantitatively model the disorder [27, 28, 45–48]. Appropriate crystal chemical constraints sometimes need to be incorporated into the RMC simulation process to ensure that realistic real-space pictures of the local crystal structures with all the flexible modes of distortion of the framework simultaneously included are obtained. In this context, knowledge of the location in reciprocal space of the "soft" RUM phonon modes obtained by electron diffraction is very useful. While total scattering methods have long been the only means of obtaining structural information on glasses or liquids, it has only been over the last few years that this powerful technique has begun to be systematically applied to disordered crystalline framework structures. This new methodology is very much in its development phase, but is beginning to revolutionize our understanding of orientational disorder in flexible framework structures by giving a realistic view of instantaneous local crystal structure and its capacity to respond to external stimuli [21, 22, 27, 28, 45–48].

18.3
Applications

In this section, we review in somewhat more detail than above the evidence for the existence of zero-frequency RUM modes of distortion and their relationship to lower temperature polymorphism in: (1) microporous molecular sieve materials (in particular, zeotypic phases exhibiting one-dimensional channels the average structures of which are indicative of displacively flexible framework structures) [49–61], (2) the piezoelectric and pyroelectric $A_2BM_2O_8$ (A = Sr, Ba; B = Ti, V; M = Si, Ge, ...) family of layered fresnoites [14, 35, 36, 62–64], and (3) the family of silica polymorphs and the various stuffed analogues thereof [6–8, 26–28, 33, 34, 39–41].

18.3.1
Microporous Molecular Sieve Materials

Ever since the first synthesis of an aluminosilicate zeolite in the late 1940s, the number, the chemical diversity, as well as the importance of such microporous solids have continued to increase. The extensive aluminophosphate ($AlPO_4$) and metal-doped aluminophosphate family of molecular sieve materials was first discovered in the early 1980s [49] and has since been studied intensively because of applications in ion exchange, catalysis, and as molecular sieves for the separation of large molecules [38]. Transition metal doped $AlPO_4$-11 materials, for example, have been reported to be active, selective, and stable catalysts for the skeletal isomerization of n-butene. Their structural and chemical properties are closely related to those of the aluminosilicate zeolites. Indeed, many $AlPO_4$ molecular sieve materials are isostructural with known zeolites. As for zeolites, molecular dynamics simulations have shown that the inherent displacive flexibility of the $AlPO_4$ tetrahedral framework plays an important role in determining their properties and behavior [37, 38]. They are synthesized hydrothermally using organic template molecules as structure-directing agents, which are subsequently removed by calcination at 500–600 °C [50].

The reported highest symmetry average crystal structures of each of the four aluminophosphate framework structures, $AlPO_4$-5 [51–54], $AlPO_4$-8 [55], $AlPO_4$-11 [56], and $AlPO_4$-54 (also known as $AlPO_4$:VPI-5) [57–59], all show some abnormally large thermal parameters and contain crystal chemically unreasonable Al–

O–P angles close to 180°, indicating the probable large amplitude excitation of RUM modes. In addition, there is considerable confusion as to the appropriate unit cells and space group symmetries of several of these phases. The room temperature structure of AlPO$_4$-5, for example, has been refined in space groups P6cc [51], P6/mcc [52], Pcc2 [53], Pnn2 [54], and P6 [31]. In addition, theoretical calculations have led to proposals of yet further possible unit cells and space group symmetries, such as P6 or Cc [32, 60, 61]. The initially reported [51] (and highest possible Al/P ordered) space group symmetry for AlPO$_4$-5 was P6cc. A major problem with P6cc, and the prime reason behind the search for larger unit cells and/or alternative lower symmetry space groups for AlPO$_4$-5, is that such a unit cell and space group symmetry necessitates the existence of crystal chemically unreasonable 180° Al–O–P angles as well as short Al–O and P–O bond distances [29]. Given that Al–O–P angles are invariably $\sim$140–150° when no such displacive disorder is present [27], it is clear that large amplitude ($\sim$15–20°) rotational modes must be occurring dynamically on the local scale that are simply ignored by a conventional crystallographic 'average structure' approach. Space group and unit cell ambiguities of this type are indicative of displacively flexible framework structures for which there are many potential RUM modes that could condense, or freeze out, upon lowering the temperature.

Such a situation is strongly reminiscent of the high temperature, high symmetry polymorphs of SiO$_2$-cristobalite and SiO$_2$-tridymite, for which similar problems were encountered, in which case unit cell expansion and lowering of space group symmetry (associated with specific condensed RUM modes of various sorts) was also proposed in order to resolve such complications [65, 66]. It has subsequently been shown that the resolution of this apparent problem did not lie in symmetry lowering of the underlying average structure but rather in the realization that a multitude of zero-frequency RUM modes were always simultaneously excited, thereby giving rise to a dynamically disordered structure in which the instantaneous crystal structure is always crystal chemically reasonable [26–28, 33, 40, 45–48]. One characteristic signature of such a situation is the existence of a highly structured, diffuse intensity distribution demonstrating the large amplitude excitation of a multitude of zero-frequency RUM modes. In the case of AlPO$_4$-5, the existence of just such a highly structured characteristic diffuse intensity distribution [29] (see, e.g., Fig. 18.4) confirms the presence of large amplitude RUM modes.

That the observed diffuse distribution in Fig. 18.4 is not localized in the vicinity of the $\langle-1,1,0\rangle$ and $\langle110\rangle$ zone axis orientations, but is part of an essentially continuous hyperbolic diffuse intensity distribution, has been demonstrated, as has the fact that each point on the observed diffuse distribution arises from an essentially zero-frequency RUM mode of distortion of the ideal AlPO$_4$-5 tetrahedral framework structure [29]. The atomic displacement patterns associated with the various predicted RUM modes never entail rotation about the parent c axis, but only about in-plane rotation axes running through the basal planes of the tetrahedra themselves, coupled with an associated rigid body shift along the c_p axis [29, 41].

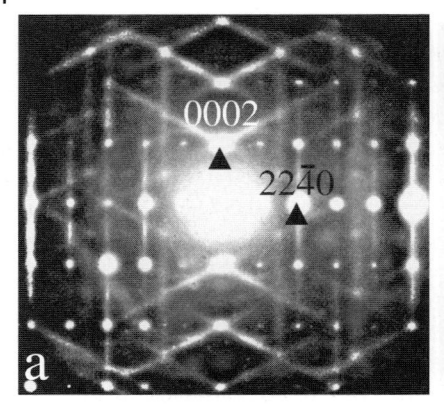

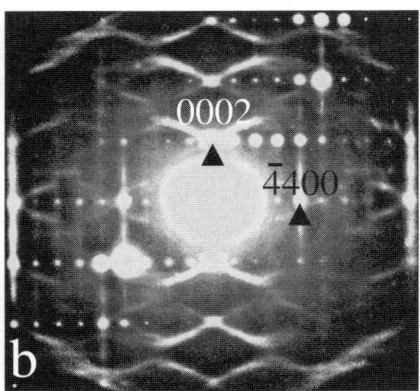

Fig. 18.4 (a) A close to $\langle -1,1,0 \rangle$ and (b) a close to $\langle 110 \rangle$ zone axis EDP of the hexagonal microporous AlPO$_4$-5 phase [29]. Notice the presence of strong (transverse polarized) diffuse streaking along c^* in both (a) and (b), as well as a spectacular curved diffuse distribution tracing out some of the zero-frequency RUM modes of this tetrahedral framework structure.

Each point on the observed diffuse distribution then represents a potential soft mode that may freeze in upon lowering the temperature or some other change in the external conditions. It is interesting to note that just such a condensed RUM mode actually does freeze out in the case of the all-silicate analogue of AlPO$_4$-5, SSZ-24 [67]. The incommensurate primary modulation wave-vector characteristic of this polymorph occurs at $\mathbf{q} = 0.304\,\mathbf{c}^*$ and falls on the observed diffuse distribution of Fig. 18.4b, as does the second harmonic modulation wave-vector thereof.

The discovery that AlPO$_4$-5 is characterized by numerous, significantly excited, RUM modes at room temperature is a significant discovery and changes the way in which we should think of this material and other related microporous molecular sieve materials [21, 22, 29, 30]. Instead of searching for one particular condensed RUM mode giving rise to one particular lower symmetry space group allowing non-180° Al–O–P angles along the c direction, it is clear that there are a multitude of such RUM modes, which are simultaneously excited, giving rise to a dynamically disordered structure.

18.3.2
The A$_2$BM$_2$O$_8$ (A = Sr, Ba; B = Ti, V; M = Si, Ge, ...) Family of Layered Fresnoites

The mineral fresnoite, Ba$_2$TiSi$_2$O$_8$ (hereafter written as BTS), is a useful piezoelectric and pyroelectric crystal with large electromechanical coupling factors and small temperature coefficients for shear mode resonance frequency [14]. It has been studied primarily for its polar properties, with much of the research having been centered on the possibility of using BTS in surface acoustic wave devices. The Ge analogue, Ba$_2$TiGe$_2$O$_8$ (or BTG), has also been investigated for

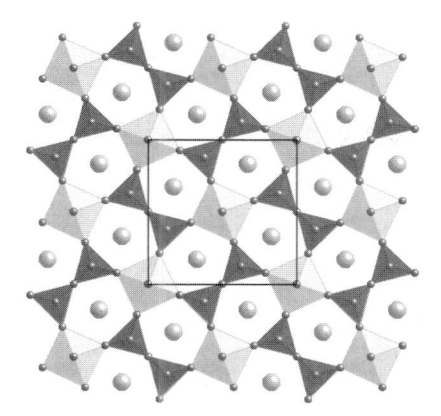

Fig. 18.5 (a) An [001] projection of the ideal *P4bm*, Ba$_2$TiGe$_2$O$_8$ fresnoite-type framework structure with its characteristic layers of essentially rigid, corner-connected TO$_4$ (T = Si or Ge) tetrahedra and TiO$_5$ square-pyramidal polyhedral units, and (b) A [−1,0,1]$_p$ ('p' for parent) "zone axis" EDP of the (3+1)-d incommensurately modulated room temperature phase of the fresnoite, Ba$_2$TiGe$_2$O$_8$ [43]. Indexation in (b) is with respect to the four basis vectors {**a*** = 1/2[1,−1,0]$_p$*, **b*** = 1/2 [110]$_p$*, **c*** = 1/2 [001]$_p$*, **q** = 0.63**b*** ≡ 0.31[110]$_p$*}.

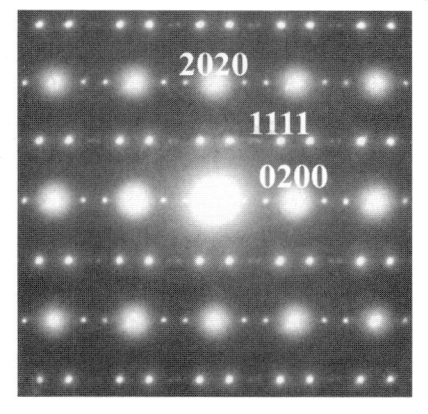

its polar properties and for insight into its high temperature paraelastic to ferroelastic phase transition [62, 63] and low-temperature re-orientable ferroelectric phase [24].

The polyhedral building blocks of the ideal *P4bm* fresnoite parent structure type are MO$_4$ tetrahedral units and BO$_5$ square-pyramidal units, each individually corner-connected to form a two-dimensional framework structure perpendicular to *c* with these individual (001) layers linked together by Ba or Sr ions [64] (see Fig. 18.5a).

Both types of polyhedral building block can reasonably be expected to be essentially rigid on general crystal chemical grounds, while their topological connectivity makes the parent fresnoite structure type an ideal candidate for the existence of low frequency RUM modes of distortion.

Electron diffraction investigations of four separate fresnoites, BTS, BTG, Sr$_2$TiSi$_2$O$_8$ (STS), and Ba$_2$VSi$_2$O$_8$ (BVS), have indicated that all four are incommensurately modulated at room temperature [43, 62, 68, 69] (see, e.g., Fig. 18.5b), with essentially the same in-plane primary modulation wave-vector given by **q** ∼ 0.30⟨110⟩$_p$*. BTG is found to be modulated along only one of the two initially symmetry-equivalent ⟨110⟩$_p$* directions and hence is ferroelastic, while BTS and BVS are modulated along both ⟨110⟩$_p$* directions simultaneously and hence are not ferroelastic. A lattice dynamical investigation of the ideal *P4bm* fresnoite parent structure type using CRUSH shows that there exists a soft in-plane RUM mode of distortion at the observed modulation wave-vector positions and that the modulation(s) are due to condensed RUM modes involving rotation of the constituent polyhedra about *c* combined with rigid body shifts in the basal plane [43].

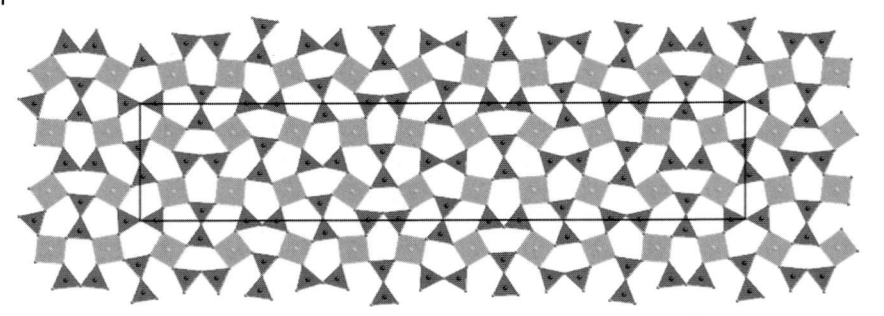

Fig. 18.6 Predicted resultant RUM pattern of distortion of one layer of the fresnoite Ba$_2$TiGe$_2$O$_8$ (BTG) projected along **c** [43] (cf. Fig. 1.5a). The Ba ions have been omitted for the sake of simplicity. The modulation wave-vector has been taken as 3/10 exactly, giving rise to a five-fold repeat along the parent [110]$_p$ direction (horizontal). Note how the condensed RUM mode distorts the cavities of the ideal fresnoite framework structure.

Indeed, a comparison of the recently refined (3+1)-d incommensurately modulated structure of BTG [70] with that predicted by the RUM approach [43] (see Fig. 18.6) shows excellent agreement.

The lattice dynamical calculations also show that the observed in-plane modulation wave-vector at $\sim 0.30\langle 110\rangle_p^*$ is the only possible soft RUM mode wave-vector involving polyhedral rotation about c. The reason for this is structural frustration – there is no simple solution to the problem of correlated polyhedral rotation when the parent structure contains rings made up of an odd number of polyhedra [43].

Intriguingly, such lattice dynamical calculations also show that there exist a further six zero-frequency RUM modes of distortion involving polyhedral rotation about in-plane axes that are soft at any modulation wave-vector. A similar situation (in the sense of there existing at least one soft phonon mode for any particular modulation wave-vector) has also been shown to occur in the case of hexacelsian [36], as well as for numerous zeolites such as sodalite, zeolite A, zeolite LTA, paulingite, faujasite [21, 22], and zeolite rho [71]. As shown by Hammonds et al. [21, 22], a soft RUM mode at any modulation wave-vector implies the existence of localized modes that only involve atoms moving within a localized region of real space. In the case of fresnoites, it is not difficult to see that such localized modes can and do indeed exist (see Fig. 18.7).

Hammonds et al. [21, 22], in the case of the zeolites, have discussed in some detail the importance of the existence of such 'localized RUM modes' along with the consequences for associated physico-chemical properties such as the adaptability of channel windows to the shapes of diffusing molecules and in enabling catalytically active transition metal ions to attach to the framework.

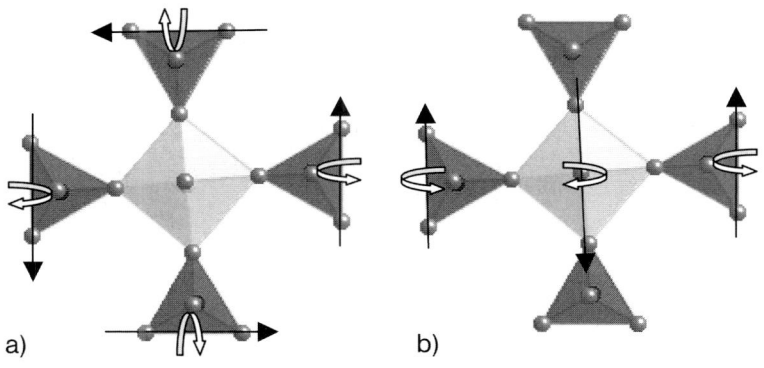

Fig. 18.7 Schematic diagrams (in projection along c_p) illustrating the six zero-frequency RUM modes of distortion which can exist for any particular modulation wave-vector of the ideal fresnoite framework structure. The tetrahedra are understood to be rotating in a clockwise sense about the basal plane rotation axes represented by the arrowed line in each case. The sense of polyhedral rotation is also indicated by the curved arrows. Note that there are two four-fold related variants of the mode shown in (b) and two localized polyhedral groupings (of the type shown in (a)) per parent unit cell (see Fig. 18.5 a), giving $2 \times (1+2) = 6$ zero-frequency RUM modes. Note also that such modes of distortion can be completely localized and not propagate from one unit cell to the next (cf. Fig. 18.5 a), i.e. they can be zero-frequency RUM modes of distortion whatever the choice of wave-vector **q**.

18.3.3
The Silica Polymorphs

The silica polymorphs (quartz, cristobalite, and tridymite), along with the various displacive variants and stuffed analogues thereof [6–8, 26–28, 33, 34, 39–41], have long played, and continue to play, an important role in mineralogy, solid-state chemistry, and materials science. Clinkers, for example, produced by hazardous waste incinerators, have only recently been shown to contain unknown variant phases based upon the tridymite structure type. A prerequisite for the development of new methods of waste recovery and recycling of this industrial waste ash will necessarily require an understanding of the solid-state chemistry of these stuffed tridymite-related phases. Apart from their usefulness and the numerous members of the family, they represent in many ways the simplest (archetypal) displacively flexible tetrahedral framework structures and were really the materials that first stimulated the development of the RUM approach [40, 72–75].

As discussed briefly above, the average structures of the higher temperature polymorphs of quartz, cristobalite, and tridymite each suffer from the usual crystal chemical problems indicative of large amplitude displacive flexibility, i.e. 180° Si–O–Si angles, SiO$_4$ tetrahedra that are too small in size [26–28, 33, 34, 40, 65, 66], and so on. The significant excitation of essentially zero-frequency RUM modes in the high temperature silica polymorphs is again signalled by the observation of characteristic diffuse intensity distributions [26, 33, 34, 76, 77] in both X-ray- as well as electron-diffraction patterns (see Figs. 18.2 a, b and 18.3 a). In the case of β-quartz, the soft phonon branch associated with RUMs has also been directly observed by means of single-crystal inelastic neutron scattering [73–75], while indirect evidence for the existence of RUMs in β-cristobalite comes from inelastic neutron scattering from polycrystalline samples [78].

Dove and colleagues [27, 28, 40, 41, 45–48] have recently investigated the various high temperature dynamically disordered polymorphs of each of the silica polymorphs in great detail by means of total neutron scattering (involving the simultaneous fitting of high quality neutron powder diffraction data to both Bragg *and* diffuse scattering) and constrained RMC simulation. Instantaneous snapshots obtained from the resultant RMC simulations (see, for example, Fig. 11 of ref. [27] or Fig. 16 of ref. [46]) clearly show large local rotations away from the average, high symmetry tetrahedral framework structures without the formation of any recognizable domain structures [65, 66]. The amplitudes of these local rotations are sufficiently large that the SiO_4 tetrahedra become essentially regular in size, and, for example, the Si–O–Si bond angle distribution functions likewise become homogenized, e.g. the average Si–O–Si angle obtained from these RMC simulations is ~ 140–$150°$ for both β-cristobalite and the highest temperature (HP) polymorph of tridymite (see Fig. 18.8). Clearly, regularly sized SiO_4 tetrahedra are able to co-operatively rotate without significant distortions in order to avoid $180°$ Si–O–Si angles. By taking into account the inherent dynamical displacive flexibility of these silica polymorphs, it has been possible to show that the local crystal chemistry of each of these high temperature silica polymorphs is always entirely reasonable on the local scale. It is precisely these sorts of studies that are needed for a wide range of other displacively flexible framework structures.

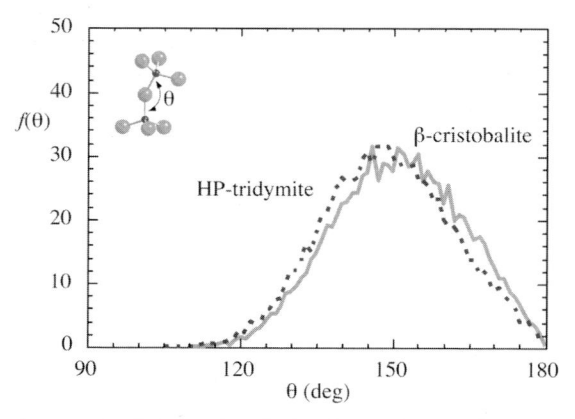

Fig. 18.8 Si–O–Si bond angle distribution function for the highest temperature polymorphs of tridymite and cristobalite obtained from analysis of RMC configurations [27]. Figure courtesy of Martin Dove.

Given such dynamically disordered high temperature parent structures, lower temperature polymorphic phase transitions are only to be expected and indeed lower temperature polymorphic phase transitions arising from condensed RUM modes occur in all three silica polymorphs. By far the most complicated sequence of such polymorphic phase transition sequences occurs in the case of SiO_2-tridymite, for which three distinct room temperature phases and up to five further higher temperature polymorphic phases have been reported [27, 79, 80]. The relationship between the reciprocal lattices of one of these reported lower temperature polymorphic forms and a higher temperature dynamically disordered form has already been illustrated in Fig. 18.3. (In this particular case, the primitive incommensurate primary modulation wave-vector **q** characteristic of this particular incommensurate polymorphic form is given by $\mathbf{q}=1/2[1,1,-2,0]^* + 0.3918\,\mathbf{c}^*$ when indexed with respect to the reciprocal lattice of the hexagonal, highest temperature polymorph.) Such complex polymorphism indicates a very delicate balance between distinct possible distortions of the ideal tridymite tetrahedral framework structure.

The constraint equations controlling the existence of soft RUM modes of distortion of the ideal tridymite tetrahedral framework structure have recently been derived [44] and used to obtain an analytical solution for the potential soft phonon mode wave-vectors and their associated displacement patterns. One of two types of RUM known to exist is found to be characterized by modulation wave-vectors perpendicular to $\langle 110 \rangle$ and atomic displacement patterns involving tetrahedral edge rotation, while a second type of RUM is found to be soft on a hyperbolic reciprocal space distribution (see, e.g., Figs. 18.2b and 18.3a) described by the equation $\sin^2 \pi l = 8/9\{1 - \cos\pi h \cos\pi k \cos\pi[h+k]\}$ and to involve tetrahedral rotation about in-plane (i.e. perpendicular to c) rotation axes [44].

With analytical expressions of this type for both the soft modulation wave-vectors and the associated displacement eigenvectors, it becomes feasible to attempt to predict the structure of tridymite-related polymorphs (not just of SiO_2 type, but also of stuffed tridymite analogues such as kalsilite, nepheline, etc.) in terms of condensed RUM modes, as has been done for the simpler case of cristobalite-related derivative structures [8]. (Intriguingly, while only one polymorphic phase transition involving a condensed RUM mode is known for pure SiO_2-cristobalite, as opposed to the large number of polymorphic phase transitions reported for pure SiO_2-tridymite, the reverse seems to be the case as far as stuffed derivatives of cristobalite and tridymite are concerned (see, for example, Tab. 1 in ref. [8]). Fig. 18.9, for example, shows the atomic dispacement pattern resulting from a condensed type II RUM mode of modulation wave-vector $\mathbf{q}= 1/2[1,1,-2,0]^* + 0.3918\,\mathbf{c}^*$ (see Fig. 18.3b) in projection along the parent **a+b** direction.

Note how this particular condensed RUM mode significantly shifts some of the initially 180° local Si–O–Si angles towards much more crystal chemically reasonable values, as well as significantly distorting the shapes of the cavities within this tetrahedral framework structure.

Whether Fig. 18.9 represents a reasonable approximation of the actual crystal structure of this particular incommensurate polymorph, however, is not so clear

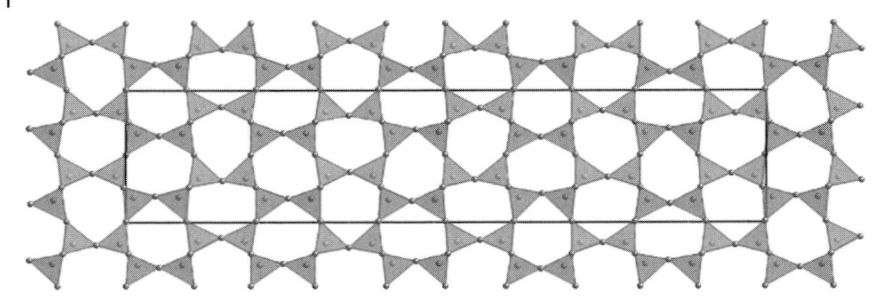

Fig. 18.9 The resultant atomic displacement pattern arising from a condensed type II RUM mode of modulation wave-vector $\mathbf{q} = 1/2 [1,1,-2,0]^* + 0.3918 \mathbf{c}^*$ in projection along $\mathbf{a} + \mathbf{b}$ with $\mathbf{c}$ horizontal. The incommensurate $\mathbf{c}^*$ component of the modulation wave-vector, 0.3918, has been moved very slightly to the rational fraction 2/5 for ease of representation, thus giving rise to a five-fold repeat along $\mathbf{c}$. For details, see Ref. [44].

as it is highly likely that there may well be more than one such RUM mode simultaneously condensed out within this polymorphic form [27, 80–83].

Despite extensive investigation over many years, further detailed structural work is clearly still needed to characterize each of the known polymorphs of tridymite [80–83] and their structural relationships.

18.4
Future Directions and Conclusions

The development of advanced flexible framework materials (such as microporous materials for applications in ion exchange, catalysis, and as molecular sieves for the separation of large molecules; ferroelectric, piezoelectric, and pyroelectric fresnoite materials for use as actuators and sensors; stuffed tridymites and their role in the design of systems to control industrial waste ash, etc.) requires quantitative understanding of local nanoscale structure as well as of their inherent ability to adapt to external influences. The new methodological approach for obtaining such information outlined in this contribution is beginning to revolutionize our understanding of such flexible framework structures by giving a realistic view of instantaneous local crystal structure as well as its capacity to respond to external stimuli.

In the case of microporous zeotypic materials, for example, such an approach provides quantitative insight into the inherent flexibility of channel windows within molecular sieve materials and the resultant consequences for adsorption, docking, and diffusion of gas molecules [21, 22]. Likewise, a quantitative understanding of the inherent displacive flexibility of the fresnoite framework structure has provided a quantitative insight into the polymorphism and temperature-dependent ferroic properties of these materials [43].

The anticipated outcomes of such research include: (i) realistic real-space pictures of local crystal structure with all flexible modes of distortion simultaneously included, (ii) the resolution of long-standing controversies regarding the local, as opposed to the (often chemically unsound) average, crystal structures of a range of zeotypic, microporous molecular sieve materials, (iii) consequent quantitative insight into how the shapes of channel windows in microporous molecular sieve materials can adapt to the shapes of diffusing molecules, as well as into how catalytically active cations might be attached to the framework, (iv) quantitative understanding of the crystal chemistry underlying piezoelectric and pyroelectric phase transitions in the $A_2BM_2O_8$ family of layered fresnoites, and (v) a systematic investigation of the crystal chemistry of stuffed tridymites commonly found in industrial waste incinerator clinker with potential applications to solid waste recovery and recycling technologies.

Acknowledgements

The authors gratefully acknowledge the contributions of colleagues, in particular Martin Dove and Thomas Höche, to various aspects of the work presented in this chapter.

References

1 V. M. HARTMANN, L. KEVAN, *Chem. Rev.* **1999**, *99*, 635–663.
2 J. M. THOMAS, *Angew. Chem. Int. Ed.* **1999**, *38*, 3588–3628.
3 J. PLÉVERT, T. M. GENTZ, A. LAINE, H. LI, V. G. YOUNG, O. M. YAGHI, M. O'KEEFFE, *J. Am. Chem. Soc.* **2001**, *123*, 12706–12707.
4 C. J. KEPERT, M. J. ROSSEINSKY, *Chem. Commun.* **1999**, 375–376.
5 G. J. HALDER, C. J. KEPERT, B. MOUBARAKI, K. S. MURRAY, J. D. CASHION, *Science* **2002**, 1762–1763.
6 A. PUTNIS, *An Introduction to Mineral Sciences*, Cambridge Univ. Press, Cambridge, UK, 1992.
7 P. J. HEANEY, *Reviews in Mineralogy and Geochemistry*, **1994**, *29*, 1–40.
8 J. G. THOMPSON, R. L. WITHERS, S. R. PALETHORPE, A. MELNITCHENKO, *J. Solid State Chem.* **1998**, *141*, 29–49.
9 A. M. GLAZER, *Acta Cryst. B* **1972**, *28*, 3384–3392.
10 M. T. ANDERSON, K. B. GREENWOOD, G. A. TAYLOR, K. R. POEPPELMEIER, *Prog. Solid State Chem.* **1993**, 197–233.
11 R. E. SCHAAK, T. E. MALLOUK, *Chem. Mater.* **2002**, *14*, 1455–1471.
12 H. IMOTO, H. FUKUOKA, S. TSUNESAWA, H. HORIUCHI, T. AMEMIYA, N. KOGA, *Inorg. Chem.* **1997**, *36*, 4172–4181.
13 W. F. MULLER, *Phys. Chem. Minerals* **1977**, *1*, 71–82.
14 T. ASAHI, T. OSAKA, J. KOBAYASHI, S. C. ABRAHAMS, S. NANAMATSU, M. KIMURA, *Phys. Rev. B* **2001**, *63*, 1–13.
15 V. KORTHUIS, N. KHOSROVANI, A. W. SLEIGHT, N. ROBERTS, R. DUPREE, W. W. WARREN, *Chem. Mater.* **1995**, *7*, 412–417.
16 J. S. O. EVANS, T. A. MARY, A. W. SLEIGHT, *J. Solid State Chem.* **1997**, *133*, 580–583.
17 R. L. WITHERS, J. S. O. EVANS, J. HANSON, A. W. SLEIGHT, *J. Solid State Chem.* **1998**, *137*, 161–167.
18 V. HEINE, P. R. WELCHE, M. T. DOVE, *J. Am. Ceram. Soc.* **1999**, *82*, 1793–1802.
19 J. Z. TAO, A. W. SLEIGHT, *J. Solid State Chem.* **2003**, *173*, 442–448.
20 M. GREGORKIEWITZ, *Solid State Ionics* **1986**, *18/19*, 534–538.

21 K. D. Hammonds, V. Heine, M. T. Dove, *J. Phys. Chem. B* **1998**, *102*, 1759–1767.
22 K. D. Hammonds, H. Deng, V. Heine, M. T. Dove, *Phys. Rev. Lett.* **1997**, *78*, 3701–3704.
23 C. N. Xu, T. Watanabe, M. Akiyama, X. G. Zheng, *Appl. Phys. Lett.* **1999**, *74*, 2414–2416.
24 A. Halliyal, A. S. Bhalla, L. E. Cross, *Ferroelectrics* **1985**, *62*, 3–9.
25 K. D. Hammonds, M. T. Dove, A. P. Giddy, V. Heine, B. Winkler, *Amer. Miner.* **1996**, *81*, 1057–1079.
26 R. L. Withers, J. G. Thompson, Chapter 13 in *In Situ Microscopy in Materials Research* (Ed.: P. L. Gai), Kluwer, Boston, **1997**, 301–330.
27 M. T. Dove, A. K. A. Pryde, D. A. Keen, *Miner. Mag.* **2000**, *64*, 267–283.
28 M. G. Tucker, M. T. Dove, D. A. Keen, Total scattering and Reverse Monte Carlo modelling of disordered crystalline materials, in *From Semiconductors to Proteins: Beyond the Average Structure* (Eds.: S. J. L. Billinge, M. F. Thorpe), Kluwer Academic/Plenum Publishers, New York, 2002, pp 85–103.
29 Y. Liu, R. L. Withers, L. Norén, *Solid State Sciences* **2003**, *5*, 427–434.
30 Y. Liu, R. L. Withers, *J. Solid State Chem.* **2003**, *172*, 431–437.
31 G. J. Klap, H. van Koningsveld, H. Graafsma, A. M. M. Schreurs, *Micropor. Mesopor. Mater.* **2000**, *38*, 403–408.
32 J. Gúlín-González, C. de las Pozas del Río, *Chem. Mater.* **2002**, *14*, 2817–2825.
33 R. L. Withers, J. G. Thompson, T. R. Welberry, *Phys. Chem. Minerals* **1989**, *16*, 517–523.
34 R. L. Withers, J. G. Thompson, Y. Xiao, R. J. Kirkpatrick, *Phys. Chem. Minerals* **1994**, *21*, 421–433.
35 Y. Tabira, R. L. Withers, Y. Takeuchi, F. Marumo, *Phys. Chem. Minerals* **2000**, *27*, 194–202.
36 R. L. Withers, Y. Tabira, J. A. Valgoma, M. Aroyo, M. T. Dove, *Phys. Chem. Minerals* **2000**, *27*, 747–756.
37 T. M. Nenoff, J. B. Parise, G. A. Jones, L. G. Galya, D. R. Corbin, G. D. Stucky, *J. Phys. Chem.-US* **1996**, *100*, 14256–14264.
38 P. Demontis, J. Gúlín-González, G. B. Suffritti, A. Tilocca, C. de las Pozas, *Micropor. Mesopor. Mater.* **2001**, *42*, 103–111.
39 M. T. Dove, K. D. Hammonds, V. Heine, R. L. Withers, Y. Xiao, R. J. Kirkpatrick, *Phys. Chem. Minerals* **1996**, *23*, 56–62.
40 M. T. Dove, V. Heine, K. D. Hammonds, M. Gambhir, A. K. A. Pryde, Short range disorder and long range order: implications of the Rigid Unit Mode model, in *Local Structure from Diffraction* (Eds.: S. J. L. Billinge, M. F. Thorpe), Plenum Press, New York, 1998, pp 253–271.
41 K. D. Hammonds, M. T. Dove, A. P. Giddy, V. Heine, *Amer. Miner.* **1994**, *79*, 1207–1209.
42 N. L. Ross, *Reviews in Mineralogy and Geochemistry*, **1994**, *41*, 257–287.
43 R. L. Withers, Y. Tabira, Y. Liu, T. Höche, *Phys. Chem. Minerals* **2002**, *29*, 624–632.
44 R. L. Withers, *Solid State Sciences* **2003**, *5*, 115–123.
45 D. A. Keen, M. T. Dove, *Miner. Mag.* **2000**, *64*, 447–457.
46 M. T. Dove, M. G. Tucker, D. A. Keen, *Eur. J. Miner.* **2002**, *14*, 331–348.
47 M. G. Tucker, M. T. Dove, D. A. Keen, *J. Appl. Cryst.* **2001**, *34*, 630–638.
48 S. A. Wells, M. T. Dove, M. G. Tucker, K. Trachenko, *J. Phys.: Cond. Matt.* **2002**, *14*, 4645–4657.
49 S. T. Wilson, B. M. Lok, C. A. Messina, T. R. Cannan, E. M. Flanigen, *J. Am. Chem. Soc.* **1982**, *104*, 1146–1147.
50 S. C. Popescu, S. Thomson, R. F. Howe, *Phys. Chem. Chem. Phys.* **2001**, *3*, 111–118.
51 J. M. Bennett, J. P. Cohen, E. M. Flanigen, J. J. Pluth, J. V. Smith, Intrazeolite chemistry in *ACS Symp. Ser.*, American Chemical Society, Washington D.C., 1983, Vol. 218, p 109.
52 J. W. Richardson, J. J. Pluth, J. V. Smith, *Acta Cryst. C* **1987**, *43*, 1469–1472.
53 A. J. Mora, A. N. Fitch, M. Cole, R. Goyal, R. H. Jones, H. Jobic, S. W. Carr, *J. Mater. Chem.* **1996**, *6*, 1831–1835.

54 T. Ikeda, K. Miyazawa, F. Iizumi, Q. Huang, A. Santoro, *J. Phys. Chem. Solids* **1999**, *60*, 1532–1535.
55 J. W. Richardson, T. Vogt, *Zeolites* **1992**, *12*, 13–19.
56 J. W. Richardson, J. J. Pluth, J. V. Smith, *Acta Cryst. B* **1988**, *44*, 367–373.
57 J. W. Richardson, J. V. Smith, J. J. Pluth, *J. Phys. Chem.* **1989**, *93*, 8212–8219.
58 C. E. Crowder, J. M. Garces, M. E. Davis, *Advances in X-ray Analysis* **1990**, *32*, 507–514.
59 D. M. Poojary, J. O. Perez, A. M. Clearfield, *J. Phys. Chem.* **1992**, *96*, 7709–7714.
60 N. J. Henson, A. K. Cheetham, J. D. Gale, *Chem. Mater.* **1996**, *8*, 664–670.
61 A. R. Ruiz-Salvador, G. Sastre, D. W. Lewis, C. R. A. Catlow, *J. Mater. Chem.* **1996**, *6*, 1837–1842.
62 T. Höche, C. Rüssel, W. Neumann, *Solid State Commun.* **1999**, *110*, 651–656.
63 S. A. Markgraf, A. S. Bhalla, *Phase Transitions* **1989**, *18*, 55–76.
64 P. B. Moore, J. Louisnathan, *Z. Kristallogr.* **1969**, *130*, 438–448.
65 A. F. Wright, A. J. Leadbetter, *Phil. Mag.* **1975**, *31*, 1391–1401.
66 D. M. Hatch, S. Ghose, *Phys. Chem. Minerals* **1991**, *17*, 554–562.
67 Z. Liu, N. Fujita, O. Terasaki, T. Ohsuna, K. Hiraga, M. A. Camblor, M.-J. Díaz-Cabanas, A. K. Cheetham, *Chem. Eur. J.* **2002**, *8*, 4549–4556.
68 T. Höche, W. Neumann, S. Esmaeilzadeh, R. Uecker, M. Lentzen, C. Rüssel, *J. Solid State Chem.* **2002**, *166*, 15–23.
69 T. Höche, S. Esmaeilzadeh, R. L. Withers, H. Schirmer, *Z. Kristallogr.*, to be published.
70 T. Höche, S. Esmaeilzadeh, R. Uecker, S. Lidin, W. Neumann, *Acta Crystallogr. B* **2002**, *59*, 209–216.
71 A. Bieniok, K. D. Hammonds, *Micropor. Mesopor. Mater.* **1998**, *25*, 193–200.
72 H. Grimm, B. Dorner, *J. Phys. Chem. Solids* **1975**, *36*, 407–413.
73 M. Vallade, B. Berge, G. Dolino, *J. Physique I* **1992**, *2*, 1481–1495.
74 H. Boysen, B. Dorner, F. Frey, H. Grimm, *J. Phys. C* **1980**, *13*, 6127–6146.
75 G. Dolino, B. Berge, M. Vallade, F. Moussa, *Physica B* **1989**, *156/157*, 15–16.
76 H. Arnold, *Z. Kristallogr.* **1965**, *121*, 145–157.
77 G. L. Hua, T. R. Welberry, R. L. Withers, J. G. Thompson, *J. Appl. Cryst.* **1988**, *21*, 458–465.
78 I. P. Swainson, M. T. Dove, *Phys. Rev. Lett.* **1993**, *71*, 193–196.
79 H. Graetsch, O. W. Flörke, *Z. Kristallogr.* **1991**, *195*, 31–48.
80 A. K. A. Pryde, M. T. Dove, *Phys. Chem. Minerals* **1998**, *26*, 171–179.
81 W. A. Dollase, *Acta Crystallogr.* **1967**, *23*, 617–623.
82 H. Graetsch, *Am. Miner.* **1998**, *83*, 872–880.
83 H. Graetsch, *Phys. Chem. Minerals* **2001**, *28*, 313–321.

19
Non-Oxide Optical Glasses: Properties, Structures, and Applications

Bruno Bureau and Jean Luc Adam

19.1
Introduction

Non-oxide glasses are materials of high interest for optical applications because they show unique transmission properties in the infrared. They are usually classified into two families: halide glasses and chalcogenide glasses. The former group is essentially represented by fluoride glasses based on heavy metals such as Zr, In or Ga as glass formers, associated with barium or lead fluoride, the Ba^{2+} and Pb^{2+} ions acting as glass modifiers. The latter group includes glasses based on S, Se, Te associated with neighboring elements of the Periodic Table, As and Sb, as well as Ge.

In this chapter, we review the most relevant non-oxide glass compositions together with their basic physical properties. Special attention is paid to applications of non-oxide glasses in fiber lasers, optical amplifiers, infrared optics, and fiber-optic sensors.

19.2
Glass Compositions and Properties

19.2.1
Fluoride Glasses

A vitreous phase is found in the ZrF_4/BaF_2 binary system if fast quenching is used to cool the mixture. In this system, large Ba^{2+} ions introduce disorder into the ZrF_n framework. Lanthanum and aluminum are strong stabilizers of the vitreous state in fluorozirconate glasses, allowing glass formation in the Zr/Ba/La/Al (ZBLA) system by gentle cooling. Further addition of a second glass-network modifier (Na^+) reduces the tendency of fluorozirconates to crystallize, which makes ZBLAN the most stable of fluoride glasses, suitable for fiber drawing. The chemical composition of ZBLAN glass is given in Tab. 19.1.

A second group of fluoride glasses are the fluoroindate glasses based on barium, indium, and gallium fluorides, in association with zinc fluoride. BIG glass, shown

Inorganic Chemistry in Focus II. Edited by G. Meyer, D. Naumann, and L. Wesemann
Copyright © 2005 WILEY-VCH Verlag GmbH & Co. KGaA, Weinheim
ISBN: 3-527-30811-3

Tab. 19.1 Chemical compositions and physical properties of fluoride glasses

Glass	Composition (mol%)	T_g (±2 °C)	T_x (±2 °C)	T_m (±3 °C)	d (±0.02) (g cm^{-3})	n_D (±0.002)	α (±6×10^{-7} K^{-1})
ZBLAN	53ZrF$_4$/20BaF$_2$/4LaF$_3$/3AlF$_3$/20NaF	262	352	455	4.34	1.498	200
ZBLA	57ZrF$_4$/34BaF$_2$/5LaF$_3$/4AlF$_3$	305	388	545	4.58	1.512	187
BIG	30BaF$_2$/18InF$_3$/12GaF$_3$/20ZnF$_2$/10YbF$_3$/6ThF$_4$/4ZrF$_4$	332	460	576	5.44	1.505	171
PZG	35.3PbF$_2$/23.5ZnF$_2$/34.3GaF$_3$/4.9YF$_3$/2AlF$_3$	270	325	547	≈5.5	≈1.59	170

in Tab. 19.1, is representative of this family, in which the glass formers are trivalent cations such as In^{3+} and Ga^{3+} [1]. One should note the relatively high concentrations of rare-earth fluorides, which are typically of the order of 15%. In addition to being essential for glass formation, such high concentrations allow more flexibility in rare-earth combinations for spectroscopic studies, especially for up-conversion processes due to energy transfers between lanthanide ions. PZG is a special fluoride glass characterized by a high lead content, which gives it a high refractive index. Rare-earth-doped PZG glasses were studied essentially because they can be formed into active films and waveguides by vapor-phase deposition.

The glass transition temperatures, T_g, of fluoride glasses are of the order of 300 °C, and melting occurs at temperatures T_m between 450 and 600 °C, depending on the glass composition. An estimation of the glass stability is given by the Hruby factor, $H_r = (T_x - T_g / T_m - T_x)$, where T_x is the crystallization temperature, which must be as large as possible. On this basis, ZBLAN appears to be very stable with respect to devitrification. In addition, thermal analysis curves show a weak crystallization peak for this glass. These favorable features are confirmed experimentally by an improved fiber-drawing ability for ZBLAN as compared to other fluoride glasses. The densities (d), refractive indices at the sodium D-line (n_D), and linear expansion coefficients (α) are also given in Tab. 19.1.

Most fluoride glasses show refractive indices of about 1.5 at the sodium D-line. However, this parameter has to be accurately controlled for the production of optical fibers, the core index of which must be significantly larger than that of the cladding glass. Usually, the increase is obtained by addition of heavy, highly polarizable ions such as Pb^{2+}. Substitution of 4 mol% of BaF_2 by PbF_2 in both the BIG and ZBLAN glasses is sufficient to increase the refractive index by 10^{-2}, the minimum value required to ensure light guiding in the fiber [2]. Some applications, such as optical amplification at 1.3 µm, necessitate the use of optical fibers with high numerical aperture, that is, with high refractive index difference. This is achieved by the addition of a higher content of lead fluoride [3]. However, this is detrimental to the thermal stability of the glass towards devitrification.

For sodium-containing glasses such as ZBLAN or BIG-Na, another common way to increase the refractive index is to substitute Na^+ by Li^+ ions, despite the

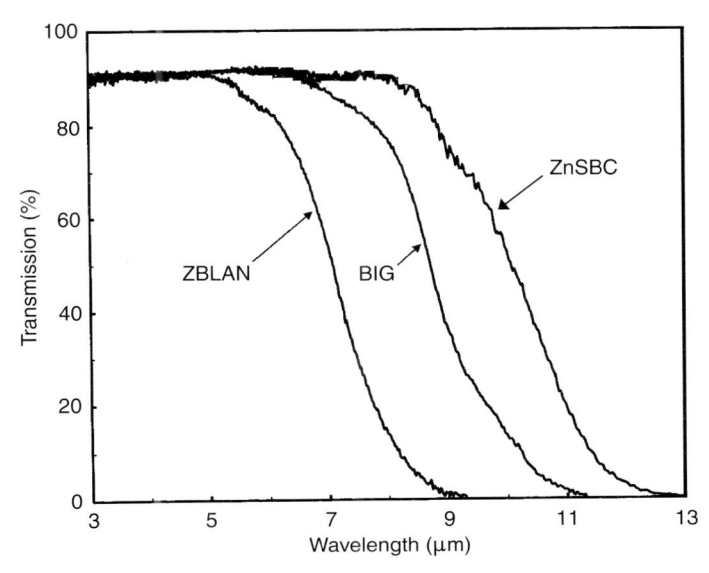

Fig. 19.1 Infrared transmission edges of fluoride glasses.

lower polarizability of the small lithium ions. Actually, Li$^+$ ions are so small that they induce a local collapse of the glassy network, resulting in densification and an increase in the refractive index [4–6].

Optical transmission is an essential physical characteristic of halide glasses in the sense that, with the exception of beryllium-based fluoride glass, this property readily differentiates halide glasses from traditional oxide glasses. Indeed, it is because of their broad transmission in the mid-infrared that heavy metal fluoride glasses were first investigated.

Fluoride glasses usually start transmitting at around 250 nm in the UV, which corresponds to an energy gap of ~ 5 eV. For comparison, pure silica shows a better UV transparency with an energy gap of ~ 8 eV ($\lambda \approx 160$ nm).

Infrared transmission spectra are shown in Fig. 19.1 for typical tetravalent-, trivalent-, and divalent-based fluoride glasses. Basically, glasses with large, heavy, and low-charged elements will possess lower phonon energies and consequently an extended infrared transparency.

Thus, for ZBLAN, the IR edge is in the 5–8 µm range with a 50% transmission located at ~ 7.1 µm (1400 cm^{-1}) for a 2 mm thick sample. This is to be compared with silica glass, the transparency of which starts to decrease at ~ 3 µm. The IR edge is due to multi-phonon absorption processes related to the fundamental vibration frequencies of the host. For ZBLAN, Zr–F vibrations at around 580 cm^{-1} account for the multi-phonon absorption, while for silica the position of the IR edge is due to Si–O vibrations at 1100 cm^{-1}.

When zirconium is substituted by hafnium, the greater atomic weight of the latter results in fundamental vibrations with lower frequencies and, consequently, the IR edge is shifted towards longer wavelengths. The IR edges of several ZrF_4, HfF_4, and ThF_4-based fluoride glasses can be found in refs. [7–10].

Fluoride glasses based on trivalent metal ions belong essentially to two families with distinct optical properties: InF_3/GaF_3-based glasses and AlF_3-based glasses. IR transmission for a typical InF_3/GaF_3 glass, referred to as BIG, is illustrated in Fig. 19.1. A shift by more than 1.5 μm in the IR is observed as compared to standard ZBLAN glass [11–13]. This is due to the lower vibrational frequency of In–F bonds (510 cm^{-1}) compared to Zr–F bonds. The detrimental effect of aluminum shifts the IR transparency of AlF_3-based glasses down to 6–7 μm [14, 15], as compared with 8 μm for an Al-free BIG glass.

With the exception of BeF_2, glasses based on divalent fluorides show a broader IR transparency than glasses based on indium or zirconium, as shown for $ZnF_2/SrF_2/BaF_2/CdF_2$ glass (ZnSBC) in Fig. 19.1, which exhibits full transparency up to 9 μm [4].

BeF_2 glass shows a special behavior, as far as optical transmission is concerned. Although it is a pure divalent glass, the light, small beryllium cation induces a narrow IR transmission with a cut-off wavelength in the 4–5 μm range, comparable to most silica glasses [16]. Ohno et al. [17] have studied the influence of BeF_2 content on AlF_3-based glasses. The IR edge shifts dramatically from ~ 7 μm (50% T) with 0% BeF_2 down to ~ 5 μm with just 25 mol% BeF_2.

19.2.2
Chalcogenide Glasses

Depending on the main constituent, sulfur, selenium or tellurium, chalcogenide glasses show specific optical transmissions, as depicted in Fig. 19.2. Thus, the IR edge is shifted towards longer wavelengths in the series S < Se < Te. This is due to the presence of elements with increasing mass, which results in reduced phonon energies of the glass matrix.

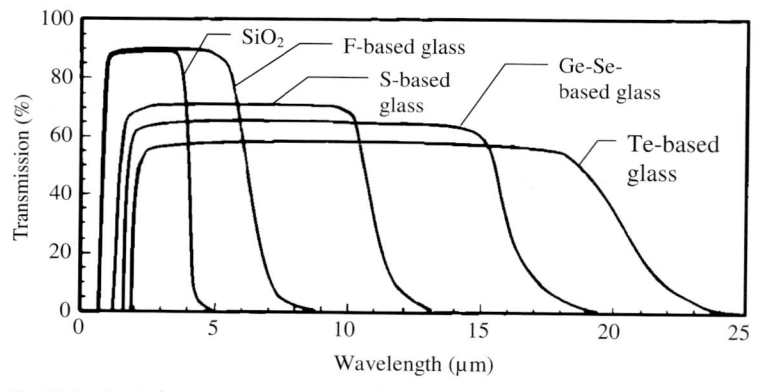

Fig. 19.2 Optical transmission spectra of silica and non-oxide glasses.

Tab. 19.2 Chemical compositions and physical properties of chalcogenide glasses. Refractive indices measured at 589 nm [a] or 632.8 nm [b]

Glass	Composition (mol%)	T_g (±2 °C)	T_x (±2 °C)	T_m (±3 °C)	n (±0.002)	a (±6×10^{-7} K^{-1})
GLS	70 Ga_2S_3/30 La_2S_3	556	681	830	≈ 2.48 [a]	90
GNS	66 Ga_2S_3/34 Na_2S	484	599	720	–	145
GGSbS	$Ge_{20}Ga_5Sb_{10}S_{65}$	305	494	–	2.3657 [b]	162
GGSbSe	$Ge_{25}Ga_5Sb_{10}Se_{60}$	283	–	–	~2.6	155
TAS	$Te_2As_3Se_5$	137	–	–	~2.8	~200

Besides a broad transparency window up to 10 µm in the infrared, sulfide glasses are of interest because most of them can be doped with rare-earth ions. The most famous lanthanide-containing sulfide glass is made of gallium, lanthanum, and sulfur and is referred to as GaLaS or GLS in the literature. Its formulation is given in Tab. 19.2. With 30% lanthanum sulfide in the glass composition, GLS can be doped with large amounts of active rare-earth ions. Other chalcogenide glasses, such as Ga/Na/S and Ge/Ga/Sb/S glasses, can only tolerate low concentrations of rare-earth ions, typically of the order of 1%.

Selenium-based glasses are also amenable to the incorporation of rare-earth ions. Stable glass compositions are found in the Ge/Ga/Se ternary system [18]. Partial substitution of Ge by As leads to glasses suitable for fiber drawing [19]. Up to 1 cat. % lanthanide ions can be incorporated into selenium-based glasses. Overall, selenide glasses are unrivalled for their high transparency in the mid-infrared, which can extend from 2 to 22 µm. However, the price one has to pay to obtain such broad infrared transmission is a low T_g and relative brittleness of the glasses. Nevertheless, two systems present an interesting compromise between acceptable mechanical behavior and large transmission windows: Ge-GaSbSe and TAS [20, 21]. In view of their properties (see Fig. 19.1 and Tab. 19.2), these glasses are promising for applications such as remote spectroscopy with optical fibers, as described in Section 19.4.4.

Chalcogenide glasses are usually synthesized from the individual chemical elements. The final optical quality of the material depends strongly on the absence of any impurities, especially hydrogen and oxygen-based entities. It is therefore of prime importance that the starting materials are of ultra-high chemical purity, 5N at least. Additional treatments such as the distillation of sulfur and de-oxidation of the glass are needed for the preparation of low-loss optical fibers required for applications [22].

19.3
Structural Aspects

19.3.1
Fluoride Glasses

19.3.1.1 Fluorozirconate Glass: ZBLAN

Due to their technological importance, stemming from their broad optical transparency that extends from the near-UV to the mid-IR, the fluorozirconate glass structures have been extensively studied during the past decades. A large number of works have dealt with the high conductivity and diffusion of fluoride ion [23–27], or the fragile behavior in terms of viscosity–temperature relationships as compared to conventional networks such as silicates [28–30]. However, papers concerned with the direct characterization of the vitreous structure of these glasses are not so common [31–34]. These few works were carried out by X-ray scattering and absorption, Raman and NMR spectroscopies. They have shown that the vitreous framework of ZB glasses (ZrF_4 + BaF_2) is built up of ZrF_n polyhedra with $7 \leq n \leq 8$, with bridging by shared F atoms. The introduction of glass modifier fluorides such as LaF_3, AlF_3, and NaF into the initial composition (so-called ZBLA and ZBLAN glasses) disrupts the connectivity of the framework and unshared F atoms appear. Experimentally, the breaking of Zr–F–Zr linkages has also been observed upon increasing the temperature of the glass above T_g [30]. From a theoretical point of view, this behavior is usually invoked to explain the fragility of the glass-forming liquids. Nevertheless, definitive precise structural data as well as an explanation of the F atom mobility have not yet been presented. To achieve this goal, the modern NMR spectrometer is undoubtedly the tool of choice, provided that it is equipped with probes that permit the sample to be spun at 35 kHz spinning speed. Such equipment enables a definite reduction of the strong homonuclear dipolar interaction present in solid-state fluorine compounds, and hence an improvement on the poor resolution of the ^{19}F static spectra presented in the past [23, 25]. Generally speaking, solid-state NMR spectroscopy is established as an element-selective and quantitative method, well-suited and indeed unrivalled for the study of vitreous, amorphous, or complex systems [36, 37]. Very recently, this approach has been successfully applied to fluorozirconates by Sen and Youngman [35], and ongoing complementary works will soon be published to propose assignments of the three ^{19}F lines seen in these spectra.

19.3.1.2 Transition Metal Fluoride Glasses: PZG

Structural investigations based on solid-state NMR have been carried out for transition metal fluoride glasses (TMFG) containing lead, i.e. $PbF_2/M^{II}F_2/M^{III}F_3$ systems [38–45]. The PZG glass ($PbF_2/ZnF_2/GaF_3$) is certainly the most famous within this fluoride glass family. Earlier studies have provided an abundance of information concerning TMFG, e.g. from UV/visible spectroscopy [46], EXAFS

[47], neutron scattering [48], static NMR [49], and Raman spectroscopy [50], coupled with molecular dynamics (MD) calculations [51]. Based on these works, TMFG are usually described as mixed packings of F$^-$ ions and Pb^{2+} cations due to the closeness of their ionic radii, with M^{II} and M^{III} being divalent and trivalent transition metal cations in octahedral holes. It is also common to describe the framework as being built up of chains of fluorine corner-sharing octahedra, (ZnF$_6$)$^{4-}$ and (GaF$_6$)$^{3-}$ for PZG, with Pb^{2+} ions in the interstitial sites.

In the 1990s, EPR studies were undertaken to investigate more precisely the octahedral network of PZG glasses [52–55]. The main conclusion obtained by this technique was that the octahedra are slightly distorted. The distributions of the radial and angular distortions were evaluated by calculating the EPR spectra in correlation with MD data.

Because of the presence of interesting nuclei amenable to NMR investigation in the different sites of the glass networks, NMR spectroscopy appeared to be a very appropriate technique to quantify the short- and medium-range orders in these materials. In this framework, a series of structural investigations using ^{19}F [38, 40, 42], ^{69,71}Ga [39, 44], and ^{207}Pb [41], nuclei as local probes was developed.

High-resolution solid-state ^{19}F MAS-NMR experiments on PZG exhibit three resolved lines corresponding to three different types of fluorine in the network. As many structural features are believed to be similar in the glasses and in the crystalline phases of close compositions, some ^{19}F MAS-NMR experiments were performed on selected crystalline compounds of interest, either because they are initial constituents (PbF$_2$, ZnF$_2$, GaF$_3$) or recrystallization compounds (Pb$_2$ZnF$_6$, PbGaF$_5$, Pb$_3$Ga$_2$F$_{12}$, Pb$_9$Ga$_2$F$_{24}$) of the glassy phases, or because they are specific of some particular connectivity of the fluorine octahedral network (CsZnGaF$_6$). This complete work permitted a division of the ^{19}F chemical shift scale into three distinct parts, each of the three lines of the PZG glasses falling within one of these zones as depicted in Fig. 19.3. Then, the 'a' lines were assigned to the so-called free fluorine sites (F$^-$ ions not involved in octahedra and surrounding Pb^{2+} ions), the 'b' lines to the unshared fluorine sites (belonging to only one octahedron), and the 'c' lines to the shared fluorine sites (F$^-$ ions connecting two octahedra).

Furthermore, the ratios of the so-called free, shared, and unshared fluorines were quantitatively determined from the spectra recorded for PZG glasses of various compositions. Thus, it was possible to show that the degree of cross-linking of these octahedra is strongly dependent on the glass composition, and may be varied from nearly 1D to nearly 3D connected octahedra. For instance, it may be inferred that for the usual composition of PZG glass generally used in optical applications, the mean number of shared fluorines per octahedron is about five, which is intermediate between 2D- and 3D-connected octahedra.

Similar results were obtained by ^{19}F NMR for other TMFG networks, such as PbF$_2$/ZnF$_2$/InF$_3$ (PZI), PbF$_2$/BaF$_2$/InF$_3$ (PBI), PbF$_2$/LiF/GaF$_3$ (PLG), PbF$_2$/NaF/GaF$_3$ (PNG), and PbF$_2$/KF/GaF$_3$ (PKG). Once again, the initial glass compositions were chosen in order to check their influence on the octahedral network connectivity [40].

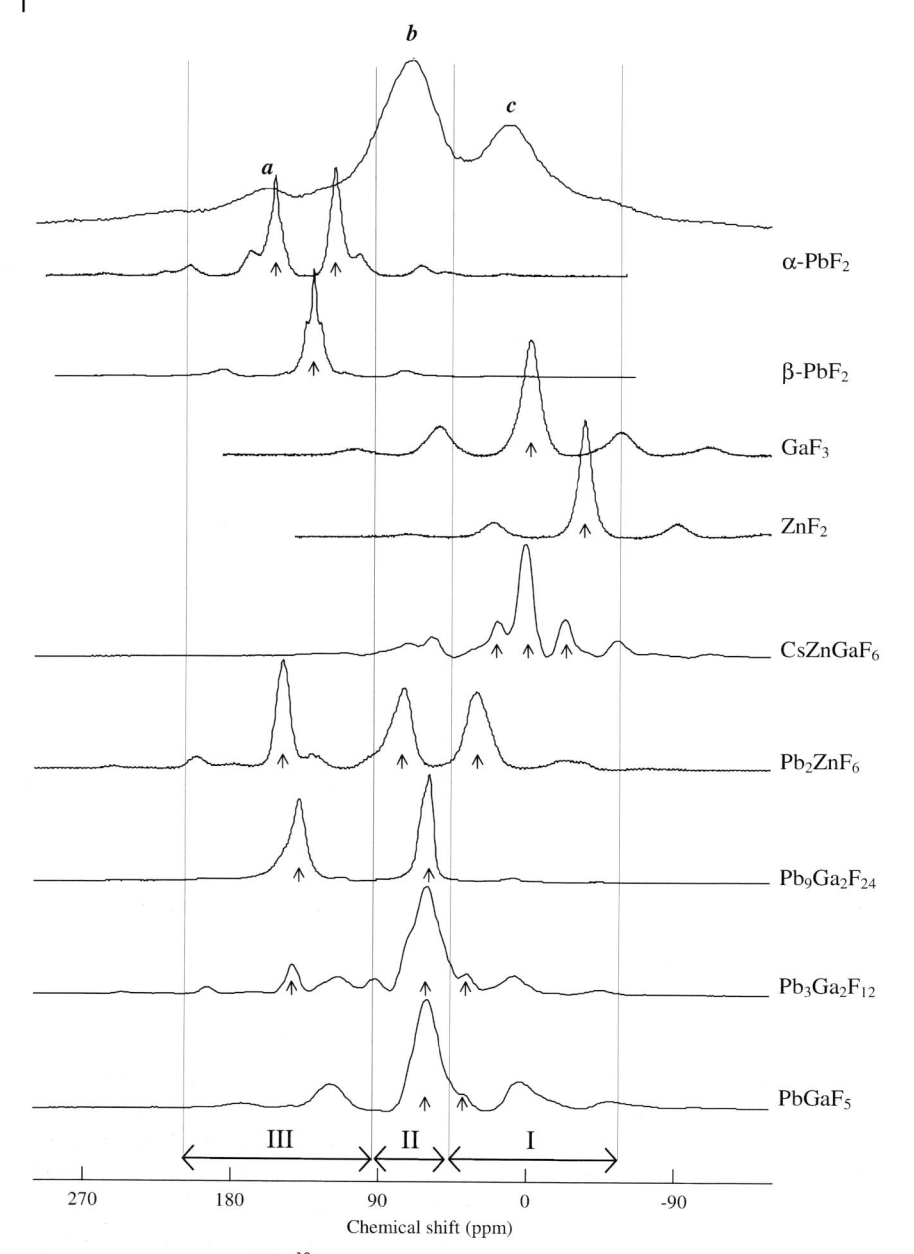

Fig. 19.3 Comparison of the ^{19}F NMR spectra for a PZG glass, the starting constituents, and the recrystallization compounds. The arrows denote the central lines of each site of these MAS spectra. The division of the chemical shift range into three distinct zones enables assignment of the a, b, and c lines of the glass as explained in the text.

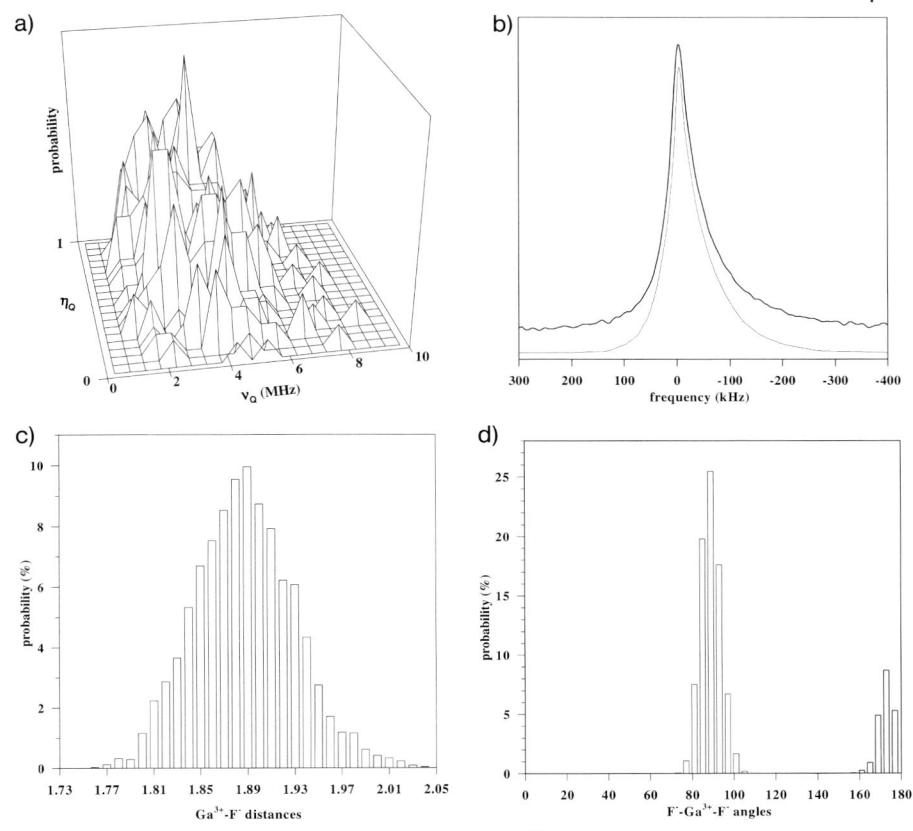

Fig. 19.4 (a) ^{71}Ga quadrupolar parameter distribution calculated by polarizable point charge model from ionic position data generated by molecular dynamic calculation. (b) Comparison of the experimental and the calculated ^{71}Ga spectra. (c) Related Ga–F distance and (d) F–Ga–F angle distributions in GaF$_6$ octahedra obtained by molecular dynamics calculations.

Finally, it has been shown that high-resolution solid-state ^{19}F NMR is a powerful tool to investigate the connectivity of the glass by probing the anion site. It has also been demonstrated that the Pb^{2+} ions play a fundamental role in differentiating the chemical shifts of fluorine sites [42]. For example, in fluoride glasses containing BaF$_2$ instead of PbF$_2$, the ^{19}F NMR spectra are not so well resolved, even at high spinning speeds [56].

For PZG glasses, ^{69,71}Ga NMR spectra also provided valuable information on both radial and angular distortion of the fluorine octahedra, by virtue of the quadrupole interaction between the quadrupole moments of the nuclei ($I = 3/2$ for both isotope) and the electric field gradients (EFG) originating from the distribution of electric charges around the nuclei (Fig. 19.4). Using the shifted echo and VOCS techniques [57], broad ^{69}Ga and ^{71}Ga spectra were recorded in amorphous GaF_3, which may be considered as a model compound for disordered fluorides with fluorine corner-sharing octahedra. These spectra were found to be identical to those obtained from PZG glasses and were simulated with continuous quadrupolar parameter Czjzek's distributions [58, 59] and a unique Ga^{3+} chemical shift value. From this value, it was inferred that in the disordered GaF_3 phase the Ga^{3+} ions are at the center of fluorine octahedra, $(GaF_6)^{3-}$, as previously found in several crystalline compounds [39].

In order to quantify the disorder, electric field gradient calculations were undertaken using a polarizable point-charge model. Atomic position sets generated by molecular dynamics were shown to give quadrupolar parameter distributions that look like Czjzek's, irrespective of the approximations made in the electric field gradient calculations. Allowing for adjustment of polarizabilities, these distributions allow reconstruction of the experimental NMR spectra, which correspond to slightly distorted $(GaF_6)^{3-}$ octahedra in agreement with previous experimental studies by EXAFS [47], Raman [50], and EPR spectroscopies [52–55]. Note that it was shown that only the six F^- nearest neighbors contribute significantly to the quadrupolar parameter distributions in simulating the experimental spectra. Thus, it was concluded that the $^{69,71}Ga$ NMR spectra are only sensitive to short-range order, preventing us from determining any topological disorder or medium-range order. From this standpoint, $^{69,71}Ga$ experiments perfectly complement the ^{19}F results.

To complete previous work, ^{207}Pb NMR experiments were also carried out on PZG glasses and parent crystalline phases [41]. Using ^{207}Pb cross-polarization magic angle spinning (CP-MAS) NMR with ^{19}F decoupling, it was shown that the lead isotropic chemical shift varies on a large scale and is very sensitive to the next nearest neighbors of the Pb^{2+} ions. In this way, it was demonstrated that the ^{207}Pb chemical shift is an interesting probe to investigate medium-range order in glassy fluoride systems.

19.3.2
Chalcogenide Glasses

Generally speaking, the structural organization of the non-oxide chalcogenide glasses reflects the thermodynamic and mechanical properties of the materials, and a convenient way to classify these numerous glass compositions is as follows: Glasses such as the vitreous selenium or the TeX glasses (X=Cl, Br, I), such as Te_3I_2, are built-up from chains with a one-dimensional (1D) framework, which leads to poor network rigidity and consequently to a low glass transition temperature, T_g, usually below 100 °C; for instance, T_g for Se is 40 °C. The second family, represented by As_2S_3 and As_2Se_3, originates from the two-dimensional (2D) connection of $AsSe_3$ pyramids and, due to the better degree of reticulation, these exhibit more rigidity and consequently higher T_g values; for example $T_g = 185$ °C for As_2Se_3. The third group, represented by GeS_2 or $GeSe_2$, are the 3D glasses, which result from the connection of tetrahedra sharing corners, and these exhibit the highest T_g values.

Over the past decades, numerous works have been published on the physical properties of these glasses based on chalcogens (S, Se, and Te) and combinations thereof, often in relation to their important potential applications as semiconductors or infrared optical compounds. Interesting papers were devoted to investigation of the short-range order by means of X-ray and neutron scattering, EXAFS, and Raman techniques [60–72]. More recently, solid-state NMR spectroscopy has proven to be a powerful technique for determining the local environment and the medium-range order of phosphorus chalcogenide glasses using ^{31}P and, above all, ^{77}Se as local probes [73]. A ^{77}Se liquid-state NMR study was also carried out on a molten g-As_xSe_{1-x} sample [74]. This background led us to perform solid-state ^{77}Se NMR measurements to obtain a deeper insight into the network of the chalcogenide glasses at room temperature. The low natural abundance of ^{77}Se (7.58%), its low relative sensitivity (6.93×10^{-3}), together with its long longitudinal relaxation time, make this nucleus very rarely studied in solid-state materials. It is a challenge to obtain informative spectra under such circumstances. Nevertheless, it has been shown that such experiments can provide evidence for different Se environments in two binary systems: As_xSe_{1-x} and Ge_xSe_{1-x}.

Two theories directly related to the way in which the $AsSe_3$ pyramids (or $GeSe_4$ tetrahedra, respectively) extend over the network have been proposed to describe the medium-range order for the Se-enriched chalcogenide glasses. In the first model, known as the "chains crossing model", the As (or Ge) atoms are considered as being homogeneously distributed throughout the network. The Se chains then cross-link the $AsSe_3$ pyramids (or $GeSe_4$ tetrahedra) and the mean number of Se between two polyhedra depends on the initial composition: the richer the composition in Se, the longer are the chains. In the second model, the so-called "outrigger raft model" proposed by Griffiths et al. [75], clusters of corner- or edge-sharing polyhedra are considered as being distributed in a chalcogenide matrix constituted by Se chains or rings [76, 77]. In order to favor one of these theories, the same strategy was applied as in the case of the ^{19}F experiments: by reference to the known structures of the parent crystalline phases, each type of selenium environment could be reliably characterized and quantified, leading to a precise description of the glassy network.

First, glasses belonging to the Se/As_2Se_3 system, situated just in between a 1D and 2D network, were investigated and their spectra compared with those of the crystalline phases c-Se and c-As_2Se_3 [78, 79]. Fig. 19.5 depicts the ^{77}Se NMR spectra of the five materials: pure Se in the vitreous and crystalline forms, the intermediate glass $AsSe_{4.5}$, and the 2D As_2Se_3 in the crystalline and glassy varieties. It is clear that the chemical shift at $\delta = 850$ ppm can be attributed to an Se atom coordinated by two others Se(a), i.e. Se–Se–Se, and that the line at $\delta = 350$ ppm is due to an Se atom connected to two As atoms, i.e. As–Se–As, as described in the orpiment crystalline structure of the arsenic selenide. Note that the spectrum of c-As_2Se_3 exhibits three sharp lines, in full agreement with the crystallographic data, showing that ^{77}Se NMR can be very accurate, even in solid-state materials. In Fig. 19.3, the intermediate glasses are represented by $AsSe_{4.5}$. Its vitreous network has to be considered as a chain-ramified glass due

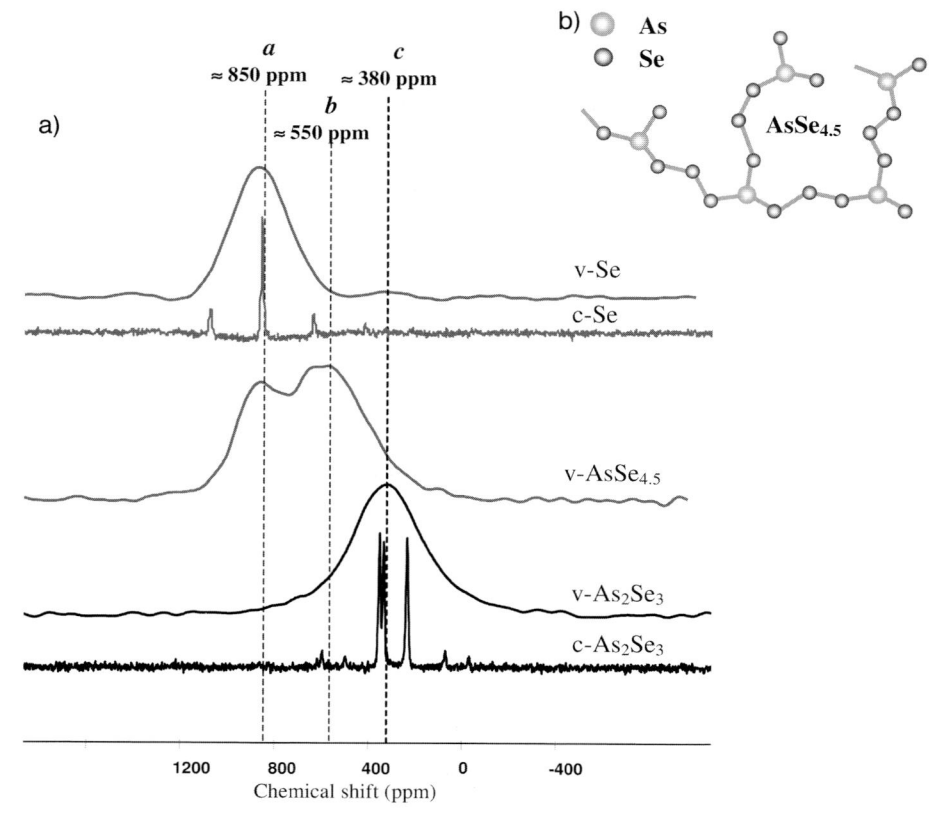

Fig. 19.5 a) Examples of ^{77}Se NMR spectra measured in the As$_x$Se$_{1-x}$ glass family, evidencing the presence of three types of line, labeled a, b, and c, assigned by comparison with parent crystalline phases. b) AsSe$_{4.5}$ representation as indicated by NMR experiments.

to the trivalency of As atoms, which allows for cross-linking of the chains. This situation leads to a new type of coordination for the Se atoms, Se–Se–As, in which the Se are coordinated by one Se and one As: the corresponding chemical shift Se(b) is located at $\delta = 550$ ppm. Measurements of the relative intensities of the three lines were carried out using Massiot's software [80]. Comparison with the expected relative proportions of selenium of each type proved the validity of the intuitive model, the so-called "chains crossing model", which predicts a homogeneous distribution of the AsSe$_3$ pyramids. It is particularly noteworthy that no signature of any clustering process was evidenced in these glasses.

Second, glasses belonging to the Se/GeSe$_2$ system have also been investigated [79, 81]. Fig. 19.6 displays three relevant spectra. Only two types of lines are exhibited, labeled 'a' and 'b', and respectively attributed to Se–Se–Se, as in pure Se, and Ge–Se–Ge, as expected in GeSe$_2$, in which all the selenium atoms are expected to be shared between two GeSe$_{4/2}$ tetrahedra. Note that no line appears

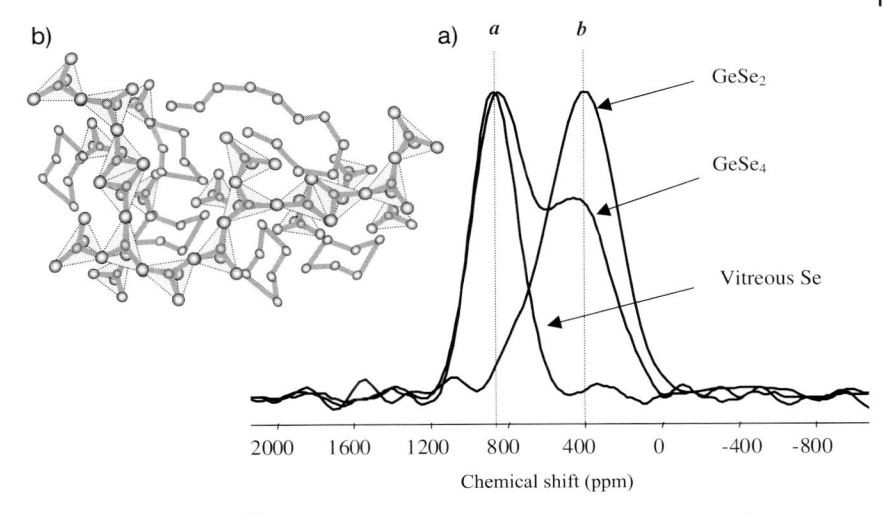

Fig. 19.6 a) Example of ^{77}Se NMR spectra measured in the Ge$_x$Se$_{1-x}$ glass family, evidencing the presence of three types of line, labeled a, b, and c, assigned by comparison with parent crystalline phases. b) GeSe$_4$ representation as indicated by NMR experiments, showing the existence of clusters.

at an intermediate position as in As$_x$Se$_{1-x}$ glasses, which would have been assigned to Se atoms linking Ge tetrahedra to Se strings, Se–Se–Ge. The 'a' and 'b' line intensities were then compared with the relative proportions of selenium expected with the two models. It appears that a scenario involving a selenium-clustering process has to be assumed to describe the Ge$_x$Se$_{1-x}$ medium-range order. This behavior has been corroborated by mass spectrometry experiments.

Third, ongoing works are being conducted on the Te$_x$Se$_{1-x}$ binary system and will be published soon. Then, the three interesting binary systems amenable to study by ^{77}Se NMR will have been investigated. Besides their fundamental interest in helping to understand the behavior of selenium in the presence of different neighbors, these preliminary studies also have to be considered as a first step towards a better comprehension of more complicated and technologically useful glassy networks, such as TeAsSe and GeAsSe systems, for example.

19.4
Optical Properties and Applications

19.4.1
Fiber Lasers

While very few fluoride glass lasers exist in bulk configuration [82, 83], more than 30 lasers have been demonstrated in fiber geometry. The optical confinement inherent in optical fibers results in high pump light density in the core of the fiber and, consequently, in a large population inversion. In addition, the fiber geometry is adequate for releasing the heat generated by possible non-radiative processes concomitant to lasing. The rare-earth concentration in the core of the single-mode fiber is typically 100 to 1000 ppm, possibly 1%. Usually, the fiber length ranges from a few centimeters to several meters. A numerical aperture as large as 0.4 can be obtained with ZBLAN fluoride glass by the addition of lead fluoride to the core and of hafnium to the cladding.

The lasing wavelengths that have been observed with ZBLAN fluoride glass fibers range from the UV to the mid-IR. This includes spectral domains in which silica does not show any significant lasing action, most notably in the visible with up-conversion lasers and at wavelengths greater than 2 μm. The characteristics are shown in Tab. 19.3 in terms of highest output power reported to date and pumping conditions.

Room temperature lasing has been demonstrated in the ultraviolet at 381 nm and in the violet at 412 nm with Nd^{3+}-doped ZBLAN fibers [84, 85]. The transitions involved are $^4D_{3/2} \rightarrow {}^4I_{11/2}$ and $^2P_{3/2} \rightarrow {}^4I_{11/2}$, as depicted in Fig. 19.7. Excitation of the $^4D_{3/2}$ and $^2P_{3/2}$ levels is obtained by a two-590 nm-photon up-conversion process. Output powers are equal to 74 μW and 500 μW, with 320 mW and 275 mW pumping powers, respectively. This UV up-conversion system serves applications for which high photon energy and compactness of the laser are of primary importance.

The generation of coherent blue light is a crucial requirement for applications in high density optical data storage, color displays, laser printing, and medical diagnostics. Blue lasing in fluoride glass fibers was demonstrated at 455 nm and 481 nm from the $^1D_2$ and $^1G_4$ levels of Tm^{3+} ions. The transitions involved are shown in Fig. 19.7. The highest output of 230 mW is obtained with Tm^{3+} at 481 nm with 1.6 W pump power at 1123 nm from a mini-YAG:Nd [86]. However, at these high pump powers, the laser performances degrade rapidly because of the presence of color centers. Sanders et al. have demonstrated up-conversion lasing with a Tm^{3+}-doped ZBLAN fiber pumped directly by two high-power laser diodes [87]. With a blue output of 106 mW at 482 nm and a 12% overall conversion efficiency from IR to blue radiation, this system shows promise for a powerful, small-size, all-solid-state blue laser source.

Green coherent light has also been obtained with Er- and Ho-doped fibers. Thus, up-conversion of 643 nm light leads to 38 mW of 549 nm light at the output of a Ho^{3+}-doped ZBLAN fiber [88].

ZBLAN microspheres are a very special configuration for lasing. Green lasing has been demonstrated in such Er^{3+}-doped structures [89]. Besides the small size, typically 100 μm in diameter or less, the interest in microspheres stems from a low laser threshold, 30 μW in the present case with up-conversion pumping at 801 nm.

Being resonant with the O–H absorption band, the 2.71 μm laser with Er^{3+} ions and the 3 μm laser with Ho^{3+} can also be applied in laser surgery as well as in dermatology and dental drilling. High-quality cutting or ablation has been demonstrated in biological tissue, which explains the increasing interest in such

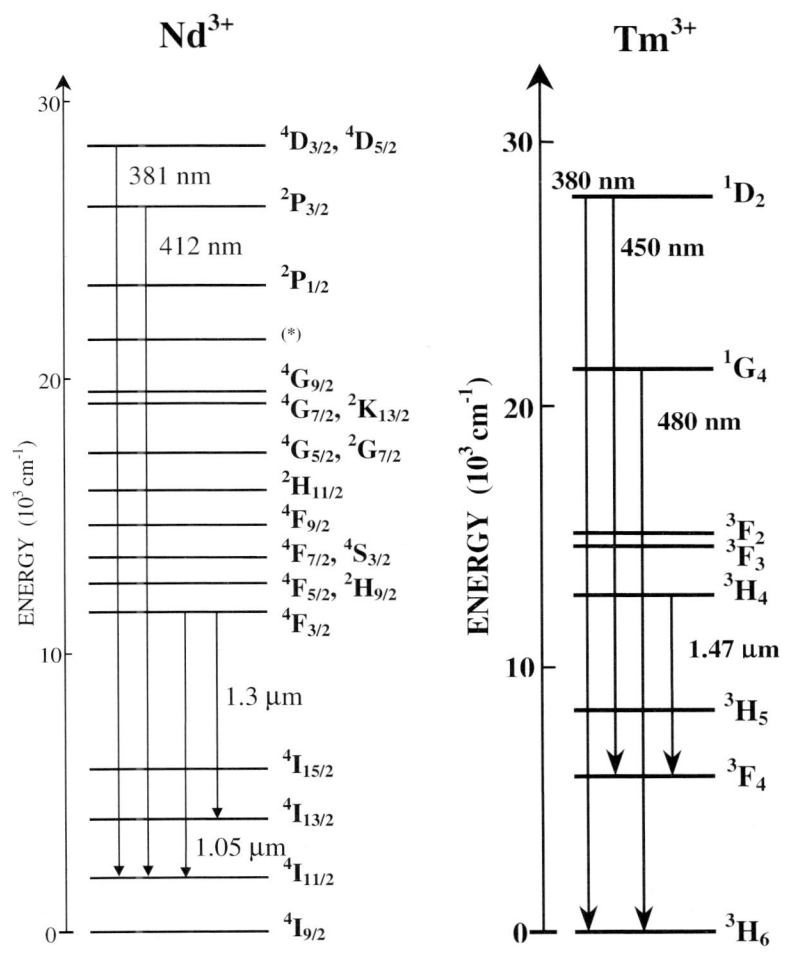

Fig. 19.7 Energy level diagram of Nd^{3+} and Tm^{3+} rare-earth ions.

lasers operating at around 3 μm. The transitions take place between the $^4I_{11/2}$ and $^4I_{13/2}$ levels of Er^{3+} and from $^5I_6$ to $^5I_7$ in Ho^{3+}. The highest output power is reported for the erbium fluoride glass laser, with a record value of 1.7 W, the slope efficiency being 17% [90]. Pumping is at 790 nm with a laser diode. Powerful lasing of 1.4 W has been demonstrated with the Ho^{3+} fluoride glass fiber laser as well. A fiber Raman laser was used for pumping at 1.15 μm [91].

The only glass lasers ever to generate wavelengths greater than 3 μm are a 3.45 μm Er^{3+}-doped fluoride fiber laser and a 3.9 μm Ho^{3+} laser [92, 93]. These correspond to the $^4F_{9/2} \rightarrow {}^4I_{9/2}$ and $^5I_5 \rightarrow {}^5I_6$ transitions, and are pumped at 650 nm and 885 nm, respectively. An output power of 2.5 mW is obtained in the CW mode at 3.45 μm. The position of the lasing wavelength is found to be temperature-dependent. Thus, tuning can be achieved from 3.456 μm at 5 °C up to 3.478 μm at

Tab. 19.3 Characteristics of some rare-earth-doped ZBLAN fluoride glass fiber lasers; (i) incident, (l) launched, and (a) absorbed pump powers; (uc) up-conversion pumping

λ (µm)	Rare-earth	Output power	Pump characteristics	Refs.
0.381	Nd	74 µW	590 nm (uc); 275 mW (i)	[84]
0.412	Nd	500 µW	590 nm (uc); 320 mW (i)	[85]
0.481	Tm	230 mW	1123 nm (uc); 1.6 W (i)	[86]
0.549	Ho	38 mW	643 nm (uc); 280 mW (a)	[88]
2.71	Er	1.7 W	790 nm; 10 W (l)	[90]
3	Ho	1.4 W	1150 nm; 4.8 W (i)	[91]
3.45	Er	2.5 mW	650 nm; 300 mW (a)	[92]
3.9	Ho	11 mW (77 K)	885 nm; 900 mW (l)	[93]

29 °C. This laser line is in the absorption range of various hydrocarbon groups, the C–H absorption typically being located at around 3.5 µm. This opens possibilities for the monitoring of trace gases and common pollutants.

In contrast to all the laser lines reported in Tab. 19.3, lasing at 3.9 µm with a ZBLAN:Ho^{3+} fiber is achieved at liquid-nitrogen temperature. A CW output power of 11 mW is obtained with 900 mW launched pump power at 885 nm. It should be noted that such lasers are of importance for military and space applications because they emit within an atmospheric window that is transparent from 3 µm to 5 µm. Few lasers emit in this spectral region.

All of the results mentioned above were obtained with optical cavities of the Fabry-Perot type with two external dielectric mirrors butted against the fiber ends. More sophisticated systems take advantage of Bragg gratings directly inscribed inside the fluoride glass optical fibers. Permanent Bragg gratings can be photo-induced in cerium-doped fluorozirconate glass plates and fibers by means of conventional holographic interferometry of 246 nm light from a pulsed UV laser [94]. Refractive index changes up to 4×10^{-4} and reflectivities of nearly 100% are achieved. Similar experiments have been conducted with Eu^{2+}- and Ce^{3+}-doped PZG thin films illuminated by a CW laser at 244 nm. Photoinduced modifications of the refractive index were as high as 1.3×10^{-2} and 3.9×10^{-3}, respectively [95].

With Nd^{3+}-doped GLS chalcogenide fibers, laser oscillation was observed at 1.08 µm. An output power of 1.2 mW was achieved with 550 mW incident pump power at 1075 nm [96].

19.4.2
Optical Amplifiers

Optical amplifiers are essential components for the development of high capacity telecommunications networks based on silica optical fibers. Silica fibers are characterized by a low-loss optical window from 1.2 µm to 1.7 µm, which is divided into ultrashort (XS), short (S), conventional (C), and long (L) bands. The XS-

Tab. 19.4 Performances of optical amplifiers based on lanthanide-containing non-oxide glass fibers

λ (μm)	Lanthanide	Glass	Max. Gain (dB)	Pump characteristics	Refs.
1.3	Pr	ZBLAN	30.5	1.017 μm 500 mW (MOPA)	98
1.34	Pr	GNS	32	1.017 μm 90 mW (Ti:sapph.)	99
1.475–1.502	Tm	ZBLAN	>25	1.4+1.502 μm (LD)	100
1.54	Er	ZBLAN	33	1.48 μm 50 mW (LD)	101

and S-bands are centered at 1.3 and 1.47 μm, while the C- and L-bands are equal to 1530–1565 nm and 1565–1625 nm, respectively. Amplification in the C- and L-bands is commonly achieved with Er^{3+}-doped silica fibers. On the other hand, amplification at 1.47 μm and 1.3 μm is achieved with Tm^{3+} and Pr^{3+} ions and requires the use of materials with phonon energies lower than that of silica. High-gain optical amplification has been demonstrated in the XS-, S-, and C-telecommunication windows with rare-earth doped fluoride glass fibers. *Real-world* amplifiers have been constructed and commercialized. With an estimated lifetime of more than 25 years under normal operating conditions, state-of-the-art fluoride fibers are compatible with long-term applications such as telecommunications [97]. The performances of some non-oxide glass amplifiers are listed in Tab. 19.4, in terms of maximum gain and pumping conditions [98–101].

A praseodymium-doped ZBLAN glass fiber appears to be the best compromise for 1.3 μm amplification with lanthanide ions. High-gain, low-noise Pr^{3+}-doped fluoride fiber amplifier modules (PDFA) have been developed by several laboratories around the world. For example, Yamada et al. have constructed a PDFA module with a Pr^{3+} fiber of only 10 meters in length, pumped by a MOPA (master oscillator power amplifier). A gain of 30 dB with 500 mW launched pump power was achieved in this configuration [98]. PDFAs have been successfully implemented in test systems for local communications, such as multichannel CATV, for example [102].

The low quantum efficiency of the 1.3 μm transition of Pr^{3+} ions accounts for the relatively small gain coefficients and high pump powers of PDFAs. A high gain coefficient of 0.81 dB mW^{-1} and an efficiency close to 30% were achieved with a praseodymium-doped GNS sulfide glass fiber amplifier [99]. However, refractive index compatibility between sulfide glass and silica remains to be solved for real devices.

In preliminary experiments, Tm^{3+}-doped fluoride glass fiber amplifiers (TDFA) have been recognized as having a high potential for the S-band, around 1.45 μm. A gain-shifted amplifier operating between 1475 and 1502 nm and delivering more than 25 dB of gain was demonstrated [100]. Pumping was achieved by means of two laser diodes at 1.4 and 1.502 μm. A major drawback of the $^3H_4 \rightarrow {}^3F_4$ transition at 1.47 μm is that the $^3F_4$ lifetime is greater than the $^3H_4$ lifetime (τ_{rad}=8.7 ms and 1.7 ms, respectively). In other words, the transition is self-terminating. Therefore, it is necessary to force the depopulation of the terminating level. This can be achieved by means of an up-conversion pumping mechanism at 1047 nm, the intermediate state of which is $^3F_4$.

An important result obtained some years ago by Alcatel Research was the world record of power conversion efficiency, 50%, in a gain-shifted thulium-doped fluoride fiber amplifier pumped at 1238 and 1400 nm [103]. The pump wavelengths were provided by a single Yb^{3+}-fiber laser emitting at 1117 nm coupled to a Raman resonator.

In the C-band, a net optical gain of 33 dB was achieved with Er^{3+}-doped ZBLAN optical amplifiers [101]. This gain was obtained with only 50 mW pump power at 1.48 μm, which resulted in a gain efficiency of 0.86 dB mW^{-1}. These fundamental characteristics are not as good as those reported for silica erbium-doped fiber amplifiers (EDFA). However, compared with silica amplifiers, fluoride EDFAs remain of practical interest because they possess a flat and broad gain bandwidth as a function of signal wavelength. This property is fundamental for wavelength division multiplexing (WDM).

In this context, ultra-broadband amplifiers were investigated over the entire 1.2 μm to 1.7 μm spectral range. Thus, a record cumulated bandwidth of 17.7 THz over the 1297–1604 nm range was achieved by combining C- and L-band silica-based EDFAs, fluoride-based TDFAs for the S-band, and Raman amplifiers for the XS-band [104].

Some research has also been devoted to integrated optics in fluoride glasses. Er^{3+}-doped fluoride glass channel waveguides can be obtained by ion exchange and photolithography techniques [105]. The experiment is carried out with doped ZBLA fluoride glass samples, which are treated with HCl gas at a temperature below the glass transition temperature T_g ($T_g = 307\,°C$ for ZBLA). The chemical reaction at the glass surface is:

$$MF_n + xHCl \rightleftarrows MF_{n-x}Cl_x + xHF$$

where M represents all the cations in the glass.

The chloride ions, which substitute part of the fluoride ions, locally modify the composition at the glass surface. The polarizability of chloride ions is greater than that of fluoride ions, and their introduction therefore increases the refractive index. Then, ionic diffusion of Cl^- ions within the glass generates a gradient-index fluorochloride guiding structure [106]. Typical results are "on/off" gains of 2.7 dB cm^{-1} at 1.047 μm for a neodymium waveguide and 3.9 dB cm^{-1} at 1.53 μm for an erbium waveguide [107].

19.4.3
Lenses for Infrared Cameras

Chalcogenide glasses would appear to be unrivalled as materials for infrared applications by virtue of their exceptional transparency in the mid-infrared range and their large nonlinear effects. Indeed, these low-phonon glasses exhibit

$\chi^{(3)}$ values of about 10^{-12} e.s.u, which are about one order of magnitude larger than those of semiconductor-doped glasses and two orders of magnitude larger than that of silica. Such high performance materials are able to meet the requirements of modern technology, such as fast switching and signal regeneration devices for telecommunications. On the other hand, the linear optical properties related to the low optical losses of such materials, in a broad range from about 1 to 14–22 µm, make them suited to applications associated with IR light transport. Thus, chalcogenide glasses allow the elaboration of IR lenses for night-vision systems as well as optical fibers for remote IR sensors.

Usually, lenses for infrared thermal imaging systems are made with germanium (Ge). This material satisfies an essential physical requirement, that is, low optical losses in the two spectral windows, 3 to 5 µm and 8 to 12 µm, in which the atmosphere is transparent enough to transmit the energy emitted by a thermal object or body. However, Ge presents several disadvantages that justify the search for alternative materials. Thus, Ge is intrinsically expensive and, moreover, costly diamond-turning operations are needed to shape Ge into sophisticated optical components such as aspherical and diffractive lenses. Consequently, the price of Ge-based infrared cameras is high and the evolution of night-vision technology towards mass production depends on the fabrication of low-cost optics, which can be obtained by molding chalcogenide infrared-transmitting glasses. Suitable glass compositions need to meet the following criteria. First, an excellent transparency from about 2 to 12 µm, which is easily obtained by selecting heavy anion Se-rich compositions. Second, the material has to present a complete resistance to crystallization: the presence of small crystallites in the heart of the glass would destroy the optical properties of the lenses. Third, and finally, the viscosity/temperature curve needs to be as flat as possible for ease of control of the molding process. Such a glass composition, a mixture of germanium (Ge), arsenic (As), antimony (Sb), and selenium (Se), has been optimized, and lenses with complicated designs such as aspherical or diffractive have been obtained in single molding operations, as shown in Fig. 19.8. This

Fig. 19.8 a) Photograph taken with a night-vision camera equipped with a chalcogenide glass molded lens (b).

work was conducted in the framework of contracts with the French Defense Department and in close collaboration with the Umicore IR Glass company. By using asphero-diffractive lenses, the design of the optical system is simplified, which consequently reduces the cost and size of the cameras. Note that, besides military outlets, there is a wide market for these goods, with various potential end users including fire-fighters and the police, and applications in heat conservation or driving assistance, for example.

19.4.4
Optical Sensors

19.4.4.1 Optical Fibers

Some selected compositions of chalcogenide glasses are very resistant to devitrification and can be drawn into optical fibers that offer exceptional spectral windows, typically extending from 3 to 12 µm. Although a theoretical loss of about 10^{-2} dB km^{-1} at 4.5 µm has been predicted for glassy $GeSe_3$ [108], actual experimental attenuations for chalcogenide fibers are considerably higher ($10-10^3$ at about 3 µm) [109–111]. Therefore, at present, chalcogenide glasses are clearly not suitable for long-distance trunk communication applications. Nonetheless, for smaller scale applications, their high infrared transparency makes them unrivalled. The only alternative materials are polycrystalline mixed silver halides, such as AgCl/AgBr solid solution, which exhibit a large optical window extending up to 20 µm and can be extruded into fibers [112, 113]. Nevertheless, pores, dislocations, and grain boundaries, as well as ageing, impose limitations for their propagation of light. Moreover, at present, core-clad structures are very difficult to realize.

In the last decade, chalcogenide glass fibers have been tested and have proved suitable for applications in several domains. These original waveguides can be used for low-power CO_2 laser energy transport for surgery as well as for cutting and/or welding [114], and for remote temperature measurements by guiding the IR light towards a detector [115]. The most exciting results, however, have been obtained by using these IR glass fibers for fiber evanescent wave spectroscopy (FEWS). This technique enables us to direct the signal from a Fourier-transform infrared (FTIR) spectrometer to a remote location, thereby allowing the in situ collection of the fingerprints of molecules that are involved in such chemical processes as fermentation [116], for pollutant analysis in waste water [117, 118], and in metabolic imaging in biology leading to early medical diagnosis [119–121].

19.4.4.2 Principle and Design

With the advent of FTIR spectrometers, infrared spectroscopy has become an indispensable and efficient tool for analytical studies. Nevertheless, classical data recording, by ATR (attenuated total reflection) or by transmission, necessitates the collection of samples and taking them to the spectrometer. An alternative method, which avoids the sampling steps, involves the use of optical fibers. For this, the fiber is employed, on the one hand, to transmit the IR beam from the spectrometer to the sample. On the other hand, the fiber is used as a probe by bringing a part of it, called the sensing zone, in contact with the environment to be studied. The experimental set-up is depicted in Fig. 19.9. This technique is called FEWS (fiber evanescent wave spectroscopy) because it is considered that the principle of the measurement is based on the presence of an evanescent wave around the fiber during the propagation of light into the fiber. Indeed, the light propagates within the fiber by total internal reflection (TIR). TIR occurs at the interface, for those rays having an angle of incidence above the critical angle, as long as the medium surrounding the fiber is weakly absorbent. In the case of a TAS glass fiber ($n_1 = 2.8$), the index difference is high, and the condition for TIR is fulfilled for all optical rays entering the fiber. When chemical species come into contact with the fiber, the IR optical rays are partially absorbed at the interface following the rules of ATR (attenuated total reflection). For optical fiber sensors, the FEWS acronym is preferable to avoid confusion with ATR crystal probe devices [122]. Due to the high index of the glass and the broad diameter of the fibers, the number of TIR inside the fiber is very high and the light propagation in such a multi-mode TAS glass fiber is complex. To cope with this complexity, a background spectrum is collected before each experiment. In this way, many effects can be neglected, i.e. the entrance and exit conditions of the infrared beam, the interaction and attenuation along the optical signal transportation section, the transition of the modes during the taper to the sensing zone, the absorption due to the fiber, and any effect related to fiber bending or surface roughness. Note that the evanescent wave penetration depth allows probing of only the very first microns of the sample [123].

The fibers used in these studies were fabricated from glass cylinders by means of an in-house drawing tower [124, 125]. Glass cylinders are heated up to softening temperatures and flowed to the appropriate diameter by selecting the best parameter combination of viscosity and drawing speed. The fibers are generally coated on-line with a polymer to improve their mechanical behavior and make them easier to handle. The composition of the chalcogenide glass used for the optical sensors is $Te_2As_3Se_5$, so-called TAS glass. This stoichiometry was chosen in the Te, As, Se ternary diagram according to the following specifications:

- Large optical window covering the spectral region 3 to 12 µm.
- Excellent resistance to devitrification during the fiber manufacturing process, thereby avoiding optical scattering losses.
- Good durability towards water and solvent corrosion.
- Suitable thermomechanical properties (vitreous transition point $T_g = 137\,°C$) that enable tapering of the fiber on-line during the drawing process.

Indeed, to improve the sensitivity of the sensor, the diameter of the fiber is locally reduced to create a tapered sensing zone that is brought into contact with the sample to be analyzed [114, 122]. Typically, the transportation section has a diameter of 400 µm, while the sensing zone has a diameter of 100 µm or less. The 400 µm diameter of the input and output parts of the guide is a compromise between imparting the system with sufficient flexibility and allowing correct light injection into the fiber as well as a good interface with the IR detector.

The procedure used for etching the glass surface allows for a controlled homogeneous decrease of the diameter of the sensing part of the fiber along a few tens of centimeters. In a first step, the fiber is mechanically tapered by modifying the temperature and fiber-forming speed during the preform drawing. Under given experimental conditions, the longitudinal profile of the fiber may correspond to a variation of the diameter of for instance 400/200/400 µm. The portion with diameter 200 µm is about a few tens of centimeters in length. The second step of the procedure involves a chemical etching treatment, performed on the pre-tapered portion of the fiber in order to reduce the diameter to ∼ 100 µm [116, 126, 127]. Another great benefit of this etching process is that it results in a chemical polishing of the glass, leading to high optical quality, shiny surfaces, thereby reducing the optical propagation losses.

19.4.4.3 Application in the Environment, Biology, and Medicine

For the analysis of pollutants in waste water, glass fibers play an essential role in that the infrared beam has to be guided to a remote location, e.g. groundwater [117, 118]. The experimental set-up consists of a Vector 22 FTIR spectrometer (Bruker) coupled with a tapered fiber and a Hg/Cd/Te (MCT) detector. A single fiber is used both as a waveguide and as a sensing element, as depicted in Fig. 19.9.

Pilot-scale tests were conducted in an artificial aquifer system consisting of a tank containing 1 m^3 of simulated polluted water and filled with gravel

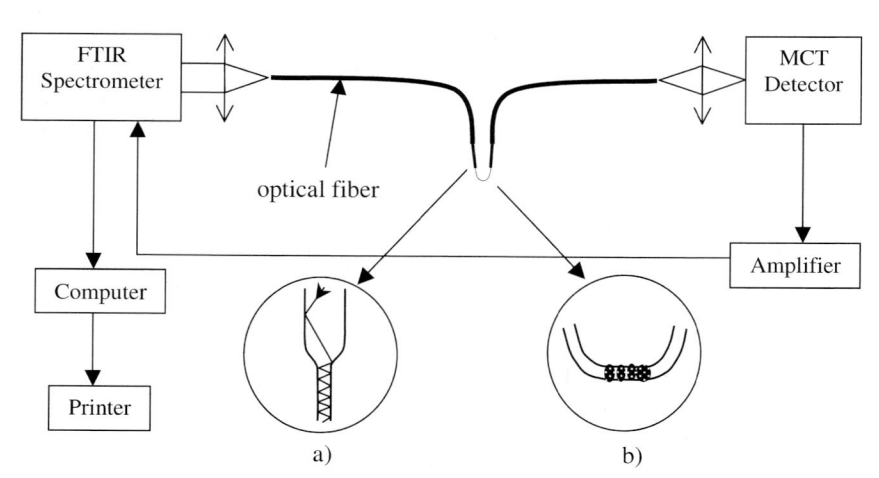

Fig. 19.9 Experimental set-up used for fiber evanescent wave spectroscopy measurements: (a) is an enlarged schematic of the fiber tapering zone in which the optical beam path is shown; the number of reflections within the fiber is greater in the tapered zone; (b) shows the mice liver tissues, or anything else, in contact with the optical fiber.

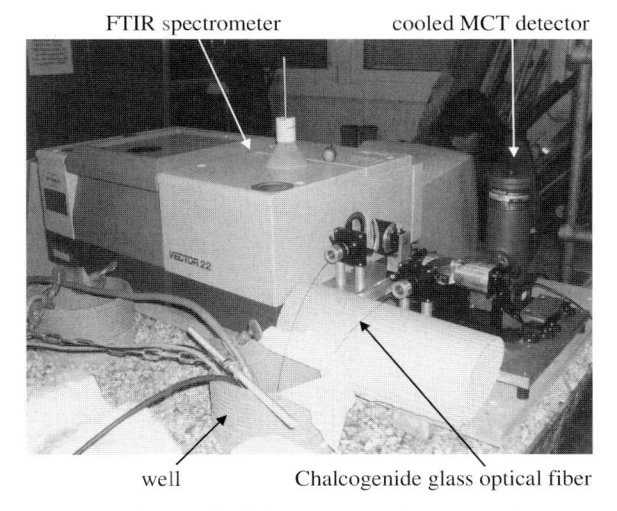

Fig. 19.10 Photograph of the experimental set-up used to carry out the pilot-scale tests. The chalcogenide glass fiber is connected to the FTIR spectrometer to direct the IR beam into the wells, and is dipped in the polluted water of the tank.

(Fig. 19.10). The measurement series was initiated using tetrachloroethylene (C_2Cl_4) as the pollutant. During the migration of the contaminants through the tank, the signals from the sensor systems were logged to a PC and, simultaneously, samples were withdrawn from wells for chemical analysis. The main relevant absorption band identified in the wide-range spectra was located at 910 cm^{-1} and could be attributed to the C–Cl stretching vibration of C_2Cl_4. It was checked that the absorbance of this absorption line increases linearly with the concentration of the pollutant in agreement with the pseudo Beer-Lambert law. Overall, it was shown that the original design of the tapered fiber enabled detection of low concentrations down to about 1 mg dm^{-3}, although no selective polymer was applied on the fiber.

A second investigation took place outdoors in a natural aquifer system. These real-world conditions enabled in situ studies of the sensor response to the spreading of pollutants injected into the system with controlled ground water flow. The sensor was immersed in a monitoring well, and dichlorobenzene ($C_6H_4Cl_2$) and tetrachloroethylene (C_2Cl_4) as pollutant analytes were injected into the inlet of the aquifer. The analytes were introduced according to a well-defined profile of concentration versus time and the sensor measurements proved that it was possible to discriminate between these two pollutants under real-world conditions and to retrieve the general evolution of their concentrations versus time.

In view of the fact that chalcogenide glasses are very resistant to chemical degradation by air or water, these results are promising with regard to the installation of permanent checking devices in wells to monitor the levels of pollutants in waste water from landfill sites.

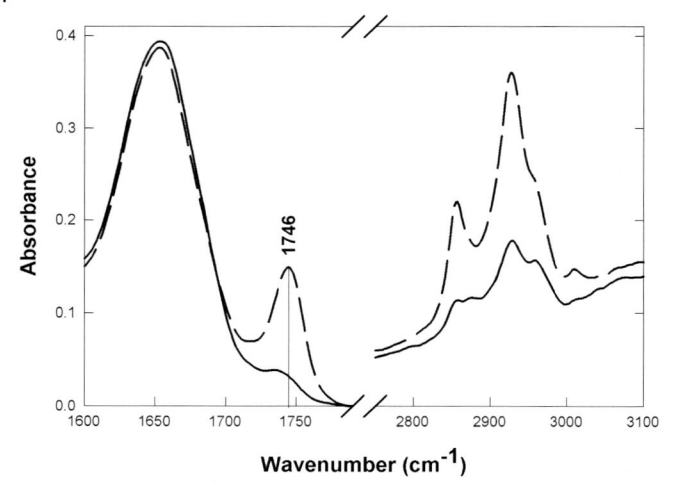

Fig. 19.11 FEWS IR spectra of tumoral (dashed line) and non-tumoral (solid line) mice liver tissues; the accumulation of triglycerides within tumoral hepatocytes gives rise to strong absorbances at 1746, 2850, and 2950 cm^{-1} [121].

The most promising domains of application, however, lie in biology and medicine. Indeed, infrared fingerprints of biomolecules contain a lot of information on cell metabolism, allowing to distinguish between healthy and altered tissues. Thus, the experimental set-up described above was used to record mouse liver biopsy infrared spectra. For example, the metabolic modifications related to starvation were identified and supported by histological studies conducted simultaneously [119]. Also, modifications of IR signatures in liver tumors were studied [121]. For this purpose, mice received an injection of both diethylnitrosamine, a liver tumor inductor, and iron-dextran, which induces liver iron overload. After a few months, all animals developed macroscopic liver tumors, and tumoral and non-tumoral areas were examined with the fiber sensor. Fig. 19.11 displays the mid-IR spectra of non-tumoral (solid line) and tumoral (dashed line) zones in the 1750–1600 cm^{-1} and 3000–2800 cm^{-1} frequency domains. The most prominent effects observed in the cells that have undergone the tumoral process are: i) a dramatic increase in the CH$_3$ and CH$_2$ vibrational bands of fatty acids, ii) a 10 cm^{-1} up-shift and a large increase in intensity of the C=O ester band, which now appears at 1746 cm^{-1}, corresponding to the C=O vibration observed in triglycerides. Taken together, these two spectral alterations are strongly suggestive of an accumulation of hepatic triglycerides during the tumoral process, reflecting the associated metabolic changes. Thus, simply by depositing the mouse liver biopsies on the fiber, this device has enabled us to differentiate between tumorous and healthy tissues.

Ongoing promising studies are being carried out with serum, plasma, and blood from human patients with a view to characterizing their disease states.

The IR fiber sensor is also being used to follow in real time and in situ the spatio-temporal spreading of biofilms of bacterial cultures [120]. Biofilms are the predominant "way of life" of bacteria, which can thereby colonize all kinds of surfaces. Proteus Mirabilis is an opportunist pathogen agent of the human urinary tract that presents two types of phenotype, namely swarming and vegetative. Remote, in situ, and real-time analysis using chalcogenide glass fibers has been implemented directly in Petri dishes. Then, unsupervised methods of analysis (principle component analysis) could be successfully used to differentiate the two phenotypes and to provide a view of their distribution in space and time.

The aforementioned examples of applications demonstrate that accurate local spectroscopic information can be obtained by the FEWS technique, provided that the design of the tapered IR fiber sensor permits an enhancement of the signal-to-noise ratio. Moreover, chalcogenide glass fibers possess an intrinsic hydrophobic behavior that enhances the infrared signal of the molecule in solution at the expense of that of water. This property makes chalcogenide glass sensors particularly suitable for applications in biology. More generally, such sensors may allow the physician to obtain in vivo, on-line, local spectroscopic diagnosis at the early stage of a disease.

References

1 G. Rault, J.L. Adam, F. Smektala, J. Lucas, *J. Fluorine Chem.* **2001**, *110*, 165.
2 N. Rigout, J.L. Adam, J. Lucas, *Eur. J. Solid State Inorg. Chem.* **1993**, *30*, 997.
3 K. Fujiura, T. Kanamori, Y. Ohishi, Y. Terunuma, K. Nakagawa, S. Sudo, K. Sugii, *Appl. Phys. Lett.* **1995**, *67*, 3063.
4 C. Henriel-Ricordel, J.-L. Adam, B. Boulard, C. Sourisseau, *Eur. J. Solid State Inorg. Chem.* **1997**, *34*, 125.
5 C. Charron, G. Fonteneau, J. Lucas, *J. Non-Cryst. Solids* **1993**, *122*, 213.
6 D.R. MacFarlane, P.J. Newman, Z. Zhou, J. Javorniczky, *J. Non-Cryst. Solids* **1993**, *161*, 182.
7 J. Qiu, K. Maeda, R. Terai, *Phys. Chem. Glasses* **1995**, *36*, 70.
8 D.C. Tran, G.H. Sigel, B. Bendow, *J. Lightw. Technol.* **1984**, *LT-2*, 566.
9 Y.L. Page, G. Fonteneau, J. Lucas, *Rev. Chim. Min.* **1984**, *21*, 589.
10 J.L. Adam, *Properties of glasses and rare-earth doped glasses for optical fibres* (Ed.: D. Hewak), IEE, Herts, UK, **1998**, p. 161, chapter 1.1.
11 N. Rigout, J.L. Adam, J. Lucas, *J. Non-Cryst. Solids* **1993**, *161*, 161.
12 I. Chiaruttini, G. Fonteneau, J. Lucas, P.S. Christensen, S. Mitachi, *Mater. Sci. Forum* **1991**, *67/68*, 245.
13 Y. Messaddeq, A. Delben, M.A. Aegerter, *J. Mater. Res.* **1993**, *8*, 885.
14 T. Iqbal, M.R. Shahriari, G. Merberg, G.H. Sigel, *J. Mater. Res.* **1991**, *6*, 401.
15 J.M. Jewell, E.J. Friebele, I. Aggarwal, *J. Non-Cryst. Solids* **1995**, *188*, 285.
16 J. Lucas, J.-L. Adam, *Glass Techn. Ber.* **1989**, *62*, 422.
17 H. Ohno, N. Igawa, Y. Ishii, N. Umesaki, M. Tokita, *J. Nucl. Mater.* **1992**, *191–194*, 525.
18 P. Nemec, B. Frumarova, M. Frumar, J. Oswald, *J. Phys. Chem. Solids* **2000**, *61*, 1583.
19 B. Cole, L.B. Shaw, P.C. Pureza, R. Miklos, J.S. Sanghera, I.D. Aggarwal, *J. Mater. Sci. Lett.* **2000**, *20*, 465.
20 X.H. Zhang, H.L. Ma, J. Lucas, *Opt. Mater.* **2004**, *25*, 85–89.

21 C. Blanchetière, K. LeFoulgoc, M.L. Ma, X.H. Zhang, J. Lucas, *J. Non-Cryst. Solids* **1995**, *184*, 200–203.

22 V.S. Shyriaev, J.-L. Adam, X.-H. Zhang, C. Boussard-Plédel, J. Lucas, M.F. Churbanov, *J. Non-Cryst. Solids* **2004**, *336*, 113.

23 J.-M. Bobe, J.-M. Reau, J. Senegas, M. Poulain, *J. Non-Cryst. Solids* **1997**, *209*, 122.

24 J.-M. Reau, Y.Y. Xu, J. Senegas, Ch. Le Deit, M. Poulain, *Solid State Ionics* **1997**, *95*, 191.

25 J.-M. Bobe, J.-M. Reau, J. Senegas, M. Poulain, *Solid State Ionics* **1995**, *82*, 39.

26 J.-M. Reau, J. Senegas, J.-M. Rojo, M. Poulin, M. Poulain, *J. Non-Cryst. Solids* **1990**, *116*, 175.

27 J.-M. Reau, H. Kahnt, M. Poulain, *J. Non-Cryst. Solids* **1990**, *119*, 347.

28 W.C. Hasz, J.H. Whang, C.T. Moynihan, *J. Non-Cryst. Solids* **1990**, *161*, 127.

29 C.A. Angell, *Chem. Rev.* **1990**, *90*, 523.

30 S. Aasland, M.A. Einarsrud, T. Grande, P.F. McMillan, *J. Phys. Chem.* **1996**, *100*, 5457.

31 S. Aasland, M.A. Einarsrud, T. Grande, A. Grzechnik, P.F. McMillan, *J. Non-Cryst. Solids* **1997**, *213/214*, 341.

32 C.C. Phifer, D.J. Gostola, J. Kieffer, C.A. Angell, *J. Chem. Phys.* **1991**, *94*, 3440.

33 R.M. Almeida, M.I. de Barros Marques, M.C. Gonçalves, *J. Non-Cryst. Solids* **1994**, *168*, 144.

34 J.H. Simmons, C.J. Simmons, R. Ochoa, A.C. Wright, in *Fluoride Glass Fiber Optics* (Eds.: I.D. Aggarwal, G. Lu), Academic Press, New York, 1991.

35 S. Sen, R.E. Youngman, *Phys. Rev. B* **2002**, *66*, 134209, 1.

36 H. Eckert, *Prog. NMR Spectrosc.* **1992**, *24*, 159.

37 K.J.D. MacKenzie, M.E. Smith, *Multinuclear Solid-state NMR of Inorganic Materials* (ed. R.W. Cahn), Pergamon Materials Series (2002).

38 B. Bureau, G. Silly, J.Y. Buzaré, J. Emery, *J. Phys.: Condens. Matter* **1997**, *9*, 6719–6736.

39 B. Bureau, G. Silly, J.Y. Buzaré, C. Legein, D. Massiot, *Solid State NMR* **1999**, *15*, 129–138.

40 B. Bureau, G. Silly, J.Y. Buzaré, C. Jacoboni, *J. Non-Cryst. Solids* **1999**, *15*, 129–138.

41 B. Bureau, G. Silly, J.Y. Buzaré, *Solid State NMR* **1999**, *15*, 79–89.

42 B. Bureau, G. Silly, J.Y. Buzaré, J. Emery, *Chem. Phys.* **1999**, *249*, 89–104.

43 B. Bureau, H. Guérault, G. Silly, J.Y. Buzaré, J.M. Grenèche, *J. Phys.: Condens. Matter Lett.* **1999**, *11*, L423–L431.

44 B. Bureau, G. Silly, J.Y. Buzaré, B. Boulard, C. Legein, *J. Phys.: Condens. Matter.* **2000**, *12*, 5775–5788.

45 H. Guérault, B. Bureau, G. Silly, J.Y. Buzaré, J.M. Grenèche, *J. Non-Cryst. Solids* **2001**, *287*, 65–69.

46 J.P. Miranday, C. Jacoboni, R. De Pape, *J. Non-Cryst. Solids* **1981**, *43*, 393.

47 A. Lebail, C. Jacoboni, R. De Pape, *J. Solid State Chem.* **1984**, *52*, 32.

48 A. Lebail, C. Jacoboni, R. De Pape, *J. Solid. State Chem.* **1983**, *48*, 168.

49 C. Dupas, K. Le Dang, J.-P. Renard, P. Veillet, J.P. Miranday, C.J. Jacoboni, *J. Physique* **1981**, *42*, 1345.

50 B. Boulard, C. Jacoboni, M. Rousseau, *J. Solid State Chem.* **1989**, *80*, 17.

51 A. Lebail, C. Jacoboni, R. De Pape, *J. Physique* **1985**, *46*, 163.

52 C. Legein, J.Y. Buzaré, J. Emery, C. Jacoboni, *J. Phys.: Condens. Matter* **1995**, *7*, 3853.

53 C. Legein, J.Y. Buzaré, B. Boulard, C. Jacoboni, *J. Phys.: Condens. Matter* **1995**, *7*, 4829.

54 C. Legein, J.Y. Buzaré, C. Jacoboni, *J. Solid State Chem.* **1996**, *121*, 149.

55 C. Legein, J.Y. Buzaré, G. Silly, C. Jacoboni, *J. Phys.: Condens. Matter* **1996**, *8*, 4339.

56 J.C.C. Chan, H. Eckert, *J. Non-Cryst. Solids* **2001**, *284*, 16.

57 D. Massiot, I. Farnan, N. Gautier, D. Trumeau, A. Trokiner, J.P. Coutures, *Solid State NMR* **1995**, *4*, 241.

58 G. Czjzek, J. Fink, F. Gotz, H. Schmidt, J.M.D. Cooey, J.P. Rebouillat, A. Liénard, *Phys. Rev. B* **1981**, *23*, 2513.

59 G. Czjzek, *Phys. Rev. B* **1982**, *25*, 4908.

60 Y. Iwadate, T. Hattori, S. Nishiyama, K. Fukushima, Y. Mochizuki, M. Misawa, T. Fukunaga, *J. Phys. Chem. Solids* **1999**, *60*, 1447.

61 A. J. Leadbetter, A. J. Apling, *J. Non-Cryst. Solids* **1974**, *15*, 250.
62 V. Mastelaro, H. Dexpert, S. Benazeth, R. Ollitrault-Fichet, *J. Solid State Chem.* **1992**, *96*, 301.
63 A. L. Renninger, B. L. Averbach, *Phys. Rev. B* **1973**, *8*, 1507.
64 P. Boolchand, W. J. Bresser, P. Suranyi, *Hyperfine Interact.* **1986**, *27*, 385.
65 J. C. Phillips, C. A. Beevers, S. E. B. Gould, *Phys. Rev. B* **1980**, *21*, 5724.
66 P. Armand, A. Ibanez, Q. Ma, D. Raoux, E. Philippot, *J. Non-Cryst. Solids* **1993**, *167*, 37.
67 P. Armand, A. Ibanez, E. Philippot, NIM B **1995**, *97*, 176.
68 A. Ibanez, M. Bionducci, E. Philippot, L. Descôtes, R. Bellissent, *J. Non-Cryst. Solids* **1996**, *202*, 248.
69 N. Ramesh Rao, K. S. Sangunni, E. S. R. Gopal, P. S. R. Krishna, R. Chakravarthy, B. A. Dasannacharya, *Physica B* **1995**, *213*, 561.
70 P. Armand, A. Ibanez, H. Dexpert, E. Philippot, *J. Non-Cryst. Solids* **1992**, *139*, 137.
71 G. Lucovsky, F. L. Galeener, R. C. Keezer, R. H. Geils, H. A. Six, *Phys. Rev. B* **1974**, *10*, 5134.
72 P. Tronc, M. Bensoussan, A. Brenac, C. Sebenne, *Phys. Rev. B* **1973**, *8*, 5947.
73 R. Maxwell, D. Lathrop, H. Eckert, *J. Non-Cryst. Solids* **1995**, *180*, 244.
74 C. Rosenhahn, S. E. Hayes, B. Rosenhahn, H. Eckert, *J. Non-Cryst. Solids* **2001**, *284*, 1.
75 J. E. Griffiths, G. P. Espinosa, J. C. Phillips, J. P. Remeika, *Phys. Rev. B* **1983**, *28*, 4444.
76 J. C. Phillips, *Phys. Rev. B* **1985**, *31*, 8157.
77 M. F. Thorpe, *Physics of Disordered Materials*, Plenum, New York, 1985, p 55.
78 B. Bureau, J. Troles, M. Le Floch, F. Smektala, G. Silly, J. Lucas, *Solid State Sciences* **2003**, *5*(1), 219–224.
79 B. Bureau, J. Troles, M. Le Floch, F. Smektala, J. Lucas, *J. Non-Cryst. Solids* **2003**, *326*, 58–63.
80 D. Massiot, F. Fayon, M. Capron, I. King, S. Le Calvé, B. Alonso, J.-O. Durand, B. Bujoli, Z. Gan, G. Hoatson, *Magn. Reson. Chem.* **2002**, *40*, 70.
81 B. Bureau, J. Troles, M. Le Floch, P. Guénot, F. Smektala, J. Lucas, *J. Non-Cryst. Solids* **2003**, *319*, 145–153.
82 T. Sandrock, A. Diening, G. Huber, *Opt. Lett.* **1999**, *24*, 382.
83 J.-L. Adam, *Advanced Inorganic Fluorides: Synthesis, Characterization and Applications*, Elsevier, Amsterdam, 2000, p 235.
84 D. S. Funk, J. W. Carlson, J. G. Eden, *Electron. Lett.* **1994**, *30*, 1859.
85 D. S. Funk, J. W. Carlson, J. G. Eden, *Opt. Lett.* **1995**, *20*, 1474.
86 R. Paschotta, N. Moore, W. A. Clarkson, A. C. Tropper, D. C. Hanna, G. Maze, *IEEE J. Select. Topics Quant. Electron.* **1997**, *3*, 1100.
87 S. Sanders, R. G. Waarts, D. G. Mehuys, D. F. Welch, *Appl. Phys. Lett.* **1995**, *67*, 1815.
88 D. S. Funk, S. B. Stevens, S. S. Wu, J. G. Eden, *IEEE J. Quant. Electron.* **1996**, *32*, 638.
89 W. von Klitzing, E. Jahier, R. Long, F. Lissillour, V. Lefèvre-Seguin, J. Hare, J. M. Raimond, S. Haroche, *J. Optics B: Quantum and Semiclassical Optics* **2000**, *2*, 204.
90 S. D. Jackson, T. A. King, M. Pollnau, *Opt. Lett.* **1999**, *24*, 1133.
91 T. Sumiyoshi, H. Sekita, T. Arai, S. Sato, M. Ishihara, *IEEE J. Select. Topics Quant. Electron.* **1999**, *5*, 936.
92 H. Többen, *Electron. Lett.* **1993**, *29*, 667.
93 J. Schneider, C. Carbonnier, U. B. Unrau, *Appl. Opt.* **1997**, *36*, 8595.
94 H. Poignant, S. Boj, E. Delevaque, M. Monerie, T. Taunay, P. Niay, P. Bernage, W. X. Xie, *Electron. Lett.* **1994**, *30*, 1339.
95 W. X. Xie, P. Bernage, D. Ramecourt, M. Douay, T. Taunay, P. Niay, B. Boulard, Y. Gao, C. Jacoboni, A. Da Costa, H. Poignant, M. Monerie, *Opt. Commun.* **1997**, *134*, 36.
96 T. Schweizer, B. N. Samson, R. C. Moore, D. W. Hewak, D. N. Payne, *Electron. Lett.* **1997**, *33*, 414.
97 K. Fujiura, Y. Nishida, T. Kanamori, Y. Terunuma, K. Hoshino, K. Nakagawa, Y. Ohishi, Y. Sudo, *IEEE Photon. Technol. Lett.* **1998**, *10*, 946.
98 M. Yamada, T. Kanamori, Y. Ohishi, M. Shimizu, Y. Terunuma, S. Sato, S.

Sudo, *IEEE Photon. Technol. Lett.* **1997**, *9*, 321.

99 H. Tawarayama, *Properties, Processing and Applications of Glass and Rare-Earth Doped Glasses for Optical Fibres*, INSPEC, London, UK, 1998, p 355.

100 T. Kasamatsu, Y. Yano, T. Ono, *Electron. Lett.* **2000**, *36*, 1607.

101 Y. Miyajima, T. Komukai, T. Sugawa, T. Yamamoto, *Opt. Fiber Technol.* **1994**, *1*, 35.

102 H. Yoshinaga, M. Yamada, M. Shimizu, T. Kanamori, *Electron. Lett.* **1994**, *30*, 2042.

103 F. Roy, F. Leplingard, L. Lorcy, A. Le Sauze, P. Baniel, D. Bayart, *Electron. Lett.* **2001**, *37*, 943.

104 D. Bayart, P. Baniel, A. Bergonzo, J.Y. Boniort, P. Bousselet, L. Gasca, D. Hamoir, F. Leplingard, A. Le Sauze, P. Nouchi, F. Roy, P. Sillard, *Electron. Lett.* **2000**, *36*, 1569.

105 J. L. Adam, E. Lebrasseur, B. Boulard, B. Jacquier, G. Fonteneau, Y. Gao, R. Sramek, C. Legein, S. Guy, *SPIE* **2000**, *3942*, 130.

106 H. Haquin, G. Fonteneau, J.-L. Adam, M. Couchaud, M. Rabarot, *J. Non-Cryst. Solids* **2004**, *326/327*, 460.

107 H. Haquin, I. Vasilief, G. Fonteneau, V. Nazabal, J.-L. Adam, S. Guy, B. Jacquier, M. Couchaud, M. Rabarot, L. Fulbert, *SPIE* **2004**, vol. 23, 5350.

108 A. M. Andriesh, *J. Non-Cryst. Solids* **1985**, *77*, 1219.

109 T. Kanamori, Y. Terunuma, S. Takahashi, T. Miyashita, *J. Non-Cryst. Solids* **1985**, *69*, 231.

110 S. Shibata, M. Horiguchi, K. Jinguji, S. Mitashi, T. Kanamori, T. Manabe, *Electron. Lett.* **1981**, *17*, 775.

111 J. Nishi, S. Morimoto, R. Yokota, T. Yamagishi, *J. Non-Cryst. Solids* **1987**, *95*, 641.

112 K. Takahashi, N. Yoshida, K. Yamauchi, *Elect. Tech. Review* **1987**, *26*, 102.

113 N. Barkay, A. Levite, F. Moser, D. Kowal, A. Katzir, *Proc. SPIE* **1989**, *1048*, 9.

114 S. Hocdé, C. Boussard-Plédel, G. Fonteneau, D. Lecoq, H. L. Ma, J. Lucas, *J. Non-Cryst. Solids* **2000**, *274*, 17.

115 F. Gilbert, F. F. Ardouin, P. Morillon, K. Le Foulgoc, X. H. Zhang, C. Blanchetière, H. L. Ma, J. Lucas, *Proc. SPIE* **1996**, *2839*, 239.

116 D. Le Coq, K. Michel, J. Keirsse, C. Boussard-Plédel, G. Fonteneau, B. Bureau, J.M. Le Quéré, O. Sire, J. Lucas, *C. R. Chim.* **2003**, *5*, 1–7.

117 K. Michel, B. Bureau, C. Boussard-Plédel, T. Jouan, C. Pouvreau, J.C. Sangleboeuf, J.L. Adam, *J. Non-Cryst. Solids* **2003**, *327*, 434–438.

118 H. Steiner, M. Jakush, M. Kraft, B. Mizakov, M. Kerlowatz, B. Mizakov, T. Baumann, R. Niessner, Y. Reichlin, A. Katzir, W. Konz, A. Brandenburg, N. Fleischmann, K. Staubmann, R. Allabashi, K. Michel, C. Boussard-Plédel, B. Bureau, J. Lucas, J.M. Bayona, *Applied Spect.* **2003**, *57*, 607–613.

119 J. Keirsse, C. Boussard-Plédel, O. Loréal, O. Sire, B. Bureau, B. Turlin, P. Leroyer, J. Lucas, *J. Non-Cryst. Solids* **2003**, *327*, 430–433.

120 J. Keirsse, C. Boussard-Plédel, O. Loréal, O. Sire, B. Bureau, P. Leroyer, B. Turlin, J. Lucas, *Vibrational Spectroscopy* **2003**, *32*, 23–32.

121 S. Hocdé, O. Loréal, O. Sire, C. Boussard-Plédel, B. Bureau, B. Turlin, J. Keirsse, P. Leroyer, J. Lucas, *J. Biomed. Optic* **2004**, *9*, 1–4.

122 S. MacDonald, K. Michel, D. Le Coq, C. Boussard-Plédel, B. Bureau, *Optical Material*, to appear in 2004.

123 N. J. Harrick, *Internal Reflection Spectroscopy*, Ossining, New York, 1979.

124 X. H. Zhang, H. L. Ma, C. Blanchetière, J. Lucas, *J. Non-Cryst. Solids* **1993**, *161*, 327.

125 S. Hocdé, C. Boussard-Plédel, G. Fonteneau, J. Lucas, *Solid State Sciences* **2001**, *3*, 279.

126 D. Le Coq, K. Michel, C. Boussard-Plédel, G. Fonteneau, J. Lucas, *Proc. SPIE* **2001**, *4253*, 19.

127 D. Le Coq, K. Michel, G. Fonteneau, C. Boussard-Plédel, S. Hocdé, J. Lucas, *Int. J. Inorg. Materials* **2001**, *3*, 233.

Index

a

alkoxyaluminates 65 ff
  perfluorinated 65
  polyfluorinated 65
aluminum(I)
  chemistry 89 ff
  halides 89
  reactions 90 ff
aluminosilicates 319
analysis
  Franck-Condon 4
  LeRoy-Bernstein-Lam 4
anion stabilities 82
antimonides 167 ff
argon compounds 16, 26
aragonite-type orthoborates 296, 304

b

band structure
  of Li_2Sb 176
  of $Li_{13}Si_4$ 148
  of $Na_{0.5}ScI_3$ 117
  of Zr_7Sb_2 (schematic) 186
BIG glasses 366
bismuth
  homopolyatomic cations 42, 43
  "monochloride" 35
bond(ing)
  distances
    in H-Ng-Y compounds 20
    in H-Xe-Y compounds 18
  nonclassical Sb-Sb 167
  order (Pauling) 170
  properties 1, 16
Born-Haber cycles
  for Ag–ethene complexes 82
  for polycationic clusters 36
boron phosphide 57

c

cage clusters 208
cage orbitals 144
calcite-type orthoborates 296, 304
carbenium ions 78
carbon-carbon bond-forming reactions 221
catalysis
  catalytic processes 1
  of cyclotrimerization of alkynes 223
  metal-catalyzed dehydrocoupling 53 ff
cations
  binary phosphorus–halogen 74
  eight-atomic $E_8^{2+/4+}$ 41
  highly electrophilic 74
  homopolyatomic 39
  prism-shaped Te_6^{2+4+} 40
  silver in complexes 67
  square planar E_4^{2+} 39
cell volume vs. temperature of
  $Y_{0.92}Er_{0.08}BO_3$ 301
chemistry of $(AlCp^*)_4$ 95 ff
complexes
  Ag–C_2H_4–WCA 72
  Ag–P_4–WCA 70
  Ag–P_4S_3–WCA 69
  Ag–S_8–WCA 68
  with HNgY 27
conductivity
  vs. temperature of Ag(SPy) 198
  vs. temperature of $Ni_2(C_4H_3N_2S)_4$ 202
coordination polymers 193 ff
  chain-like 194
  lamellar 197
  tubular 202
clusters
  acid stabilization 36
  bare / naked 36
  polycationic 35
  Zintl type 35

Index

crystal orbital Hamiltonian population (COHP)
  for "CsScI$_3$" and Cs$_4$Sc$_3$□I$_{12}$ 116
  for MoSb$_2$S 179
  for Na$_{0.5}$ScI$_3$ and "NaScI$_3$" 118
  for "ScI$_2$" 109
crystal orbital overlap population (COOP)
  for Dy$_2$Te 126
  for Sc$_6$AgTe$_2$ 131
cyclotrimerisation of alkynes 221
  mediated by the Cp'RuCl fragment 223
  with unsaturated organic molecules 227

d
dehydrocoupling 53 ff
  application 62
  catalytic, of primary amine-borane adducts 59
  catalytic, of primary phosphine-borane adducts 56
  catalytic, of secondary amine-borane adducts 58
  catalytic, of secondary phosphine-borane adducts 54
  mechanism 60 ff
  thermal, of amine-borane adducts 57
  thermal, of phosphine-borane adducts 54
dehydrogenative coupling 53
density functional theory (DFT) 1, 224, 261
density of states (DOS)
  for "CsScI$_3$" and Cs$_4$Sc$_3$□I$_{12}$ 116
  for Dy$_2$Te 126
  for HfMoS$_4$ 182
  for Li$_{12}$Si$_7$ 152
  for Li$_{13}$Si$_4$ 148
  for Na$_{0.5}$ScI$_3$ 117
  for NbS$_2$ 182
  for Sc$_6$AgTe$_2$ 131
  for "ScI$_2$" 109
diffuse scattering 347
disorder in YbSi$_{1.4}$ 160
dissociation energy 3, 4

e
effective atom charges 135
electron density 19
electron diffraction
  patterns (EDP) 350
  temperature dependent 349
electron localization function (ELF) 19
  for Li$_{12}$Si$_7$ 156
  for Si$_2$ in Li$_{13}$Si$_4$ 146
energy level diagram of Nd^{3+} and Tm^{3+}
energy profile of B3LYP potential energy surface 230, 234, 237

f
fiber evanescent wave spectroscopy (FEWS) 385, 388
fiber lasers 377
folded sheet mechanism (FSM) 321
fluoride ion affinities 80
framework structures 347 ff
fresnoites 354

g
gallium
  dimers 5
  vapor 3
glasses
  applications 386
  chalcogenide 368, 374
  compositions 365
  fluoride glasses 365
  non-oxide optical 365 ff

h
helium compounds 26
heterodehydrocoupling 53
highest occupied molecular orbital (HOMO) of PBr$_2^+$ and P(NH$_2$)$_2^+$ 75
homodehydrocoupling 53
hypervalence 48, 49, 167
  trans influence 259

k
krypton compounds 16, 26

l
lanthanide
  bis(benzimidazolyl)pyridines 277
  crown ether complexes 283
  β-diketonates 275
  disilicides 157
  liquid-crystalline complexes 267 ff
  phthalocyanines 280
  porphyrins 281
lattice dynamic program (CRUSH) 349
lattice energies of polycationic clusters 36
lenses for infrared cameras 382
Lewis acid-base
  complexes 67
  concept 36
Lewis base adducts of OsO$_4$ 247

Lewis octet rule 167
ligand affinity 83, 84
liquid crystal display (LCD) 267
liquid crystals 267ff
    alignment in a magnetic field 284
    calamitic mesogens 269
    clearing point 269
    discotic mesogens 269
    enantiotropic 269
    hexagonal columnar mesophase 270
    luminescence 287
    lyotropic 269
    magnetic anisotropy 284
    monotropic 269
    nematic phase (N) 269
    smectic phase (SmA, SmC) 269
    thermotropic 269
local crystal chemistry 347
Lux-Flood concept 293

m

magnetic behavior of $Li_{12}Si_7$ 154
matrix isolation 1, 2, 21
MCM-41 322
MCM-48 323
mechanism of formation of mesoporous structures 325, 327
mesophase sequence 332
mesophases 268ff
mesoporous materials 319ff
mesostructured materials 319ff
    ordered siliceous 319
metal atom
    clusters 1
    dimers 1
metallacyclopentatriene intermediates 221ff
metallicity (one-dimensional) 144
metallomesogens 267
    mixed f-d 279
metal-metal bond 3
metal-salt interface 121
metal semiconductor compounds 143
mobil composition of matter (MCM) 321
molar volume
    vs. temperature for $Sc_{0.9}I_2$ 109
molecular orbital (MO) diagram
    of P_4 71
    for SF_6 168
molecular sieve materials 352
morphology control 319
Mulliken charges of osmium 261
Mulliken population 169

n

noble gas
    chemical reactivity 15
    hydrides 15
neon compounds 26
nitridoosmates 249

o

optical amplifiers 380
optical sensors 384
osmium(VIII)
    dioxide tetrafluoride 255
    nitridoosmates 249
    organo-imido osmates 249
    osmium tetroxide 243
    osmium tetroxide adducts 245
    osmium trioxide difluoride 250
    oxides 243ff
    oxide fluorides 243ff
    perosmates 243

p

Peierls distortion 177
perosmates 243ff
photoconversion 9
photodissociation 22
polyborazylene 58
polycationic clusters
    heteronuclear 44ff
    homonuclear 39ff
polymers
    one-dimensional 46, 208
    polyphosphinoborane films 57
polymerization
    ring-opening (ROP) 53
    of Sc_2Te by transition metals 130
polyphosphinoboranes 57
polytellurides 48, 49
precursor to boron nitride ceramics 58
pseudo gas-phase conditions 81
PZG glasses 366, 370

q

quantum chemical calculations 1
    coupled cluster 17
    extended Hückel (EH) 146
    linear muffin tin orbitals (LMTO) 146
    MRCI method 5, 21
    perturbation theory 17
quantum size effects 1

r

rare-earth
  borates 293 ff
    anhydrous polyborates 304
    hydrated polyborates 311
    metaborates 307
    orthoborates 296
    oxyborates 294
    pentaborates 307
  metals 121
rationally designed syntheses 207
reactions
  of $Al[HC(CMeCNAr)_2]$ 98
  between Ga_2 and H_2 7
  between HNgY and H 29
  intermediates 1
  isodesmic 11
  mechanisms 1
  between Ti_2 and N_2 10
reactive building blocks 207
reactivity 1, 208
  of unsaturated organic compounds 221 ff
replacement of Cl in Cp′RuCl 228
resistivity
  vs. temperature of $Li_{12}Si_7$ 154
  vs. temperature of Mo_3Sb_7 and $Mo_3Sb_5Te_2$ 183
resonance structures of XeF_2 170
rigid unit mode (RUM) approach 349
ring-opening polymerization (ROP) 53
ruthenium(II) centers 221

s

Santa Barbara acid (SBA) 321
SBA-15 322
Sb-Sb interactions 170 ff
scandium
  "diiodide" $Sc_{0.9}I_2$ 106 ff
  divalent 105 ff
  ternary iodides ASc_xX_3 110 ff
scanning electron microscopy (SEM) 336
Schiff's base complexes 267 ff
  salicylaldimine 271
  two-ring 271
semiconductivity
  of transition metal thiolates 214
silaheterocycles 54
silica polymorphs 357
Si-O-Si bond angle distribution 358
spectroscopy
  absorption 2, 4, 114
  electron spin resonance (ESR) 2, 108, 114
  fluorescence 2
  infrared (IR) 8, 23
    transmission edges of fluoride glasses 367
  mass spectroscopy 94
  nuclear magnetic resonance (NMR)
    ^{27}Al 93
    ^{11}B 60
    ^{19}F 252, 256, 262, 371, 372
    $^{69,71}Ga$ 371
    ^{14}N 325
    ^{187}Os 252, 262
    ^{31}P 375
    ^{207}Pb 371
    ^{77}Se 375, 376, 377
    ^{89}Y 295
  resonance Raman 3, 4, 5
  ultraviolet (UV) 24
  vibrational 2, 24
stability
  of HNgY molecules 21
  of polycationic clusters 36
  of weakly bound complexes 81
stereochemistry of heptacoordination 258
structures
  of $Ag_2(bpsb)_3(ClO_4)_2$ 199
  of $[Ag(C_2H_4)_3][Al\{OC(CF_3)_3\}_4]$ 72
  of $[Ag(P_4)][Al\{OC(CH_3)(CF_3)_2\}_4]$ 71
  of $[Ag(P_4)_2][Al\{OC(CF_3)_3\}_4]$ 71
  of Ag(SPy) 198
  of $Ag_2(i\text{-mnt})_2(SPyH)_4$ 196
  of $Ag_4(Et_2NCS_2)_2(SPy)_2$ 195
  of $Ag_5(SPy)_4(SPyH)(BF_4)$ 197
  of $Ag_6(SPy)_6$ 195
  of $Ag_7(tpst)_4(ClO_4)_2(NO_3)_5$ 203
  of $Al_4Br_4(NEt_3)_4$ 92
  of $(AlCp^*)_4$ 91, 93
  of $Al[HC(CMeCNAr)_2]$ 95
  of AlPO$_4$-5 348
  of $Ba_2TiGe_2O_8$ (fresnoite-type) 355
  of $[CI_3]^+[Al\{OC(CF_3)_3\}_4]^-$ 79
  of $[CS_2Br_3]^+[Al\{OC(CF_3)_3\}_4]^-$ 78
  of Cs_5Sb_8 173
  of cyclic dimer $[Me_2NH \cdot BH_3]_2$ 59
  of $DyBO_3$ 302
  of $Dy_2B_4O_9$ 305
  of $Dy_4B_6O_{15}$ 305
  of $EuBO_3$ 303
  of $FeSb_2$ 178
  of GdB_5O_9 308
  of Gd_3MnI_3 139
  of heteropolycations 45
  of homopolycations 40

of KSb   173
of KSb$_2$ 172
of LaBO$_3$ 297
of LaB$_5$O$_9$ 310
of La$_{26}$(BO$_3$)$_8$O$_{27}$ 295
of [La(dbm)$_3$L$_2$]   276
of Li$_2$Sb   174
of Li$_{12}$Si$_7$ 150
of Li$_{13}$Si$_4$ 145, 150
of Li$_{13}$Ge$_4$ 150
of LuBO$_3$ 297
of Lu$_8$Te and Lu$_7$Te   129
of MoSb$_2$S   178
of [N(CH$_3$)$_4$][*fac*-OsO$_3$F$_3$]   253
of NaSc$_2$I$_6$ 117
of NbSb$_2$ 182
of [Nd(LH)$_3$(NO$_3$)$_3$]   272, 274
of Ni(spcp)$_2$ 200
of Ni$_2$(C$_4$H$_3$N$_2$S)$_4$ 201
of one-dimensional polycations   47, 48
of [(OsO$_2$F$_3$)$_2$F][Sb$_2$F$_{11}$]   257
of [OsO$_3$F][HF]$_2$[AsF$_6$]   253
of [OsO$_3$F][SbF$_6$]   254
of [OsO$_3$F][Sb$_3$F$_{16}$]   255
of OsO$_2$F$_4$ 256
of OsO$_3$F$_2$ 251
of OsO$_4$ · quinuclidine   246
of (OsO$_4$)$_2$ · hexamethylenetetramine   246
of (OsO$_4$) · (*R,R*)-*trans*-1,2-bis(*N*)pyrrolidino)cyclohexane   246
of OsO$_4$F$^-$ 246
of phosphine-borane adducts   55
of [P$_2$Br$_5$]$^+$][Al{OC(CF$_3$)$_3$}$_4$]$^-$ 76
of [P$_5$Br$_2$]$^+$][Al{OC(CF$_3$)$_3$}$_4$]$^-$ 77
of [PI$_4$]$^+$][Al{OC(CF$_3$)$_3$}$_4$]$^-$ 76
of [P$_3$I$_6$]$^+$][{(CF$_3$)CO$_3$}$_3$Al-F-Al{OC(CF$_3$)$_3$}$_3$]$^-$ 77
of [P$_5$S$_2$I$_2$]$^+$][Al{OC(CF$_3$)$_3$}$_4$]$^-$ 78
of scandium diiodide   106
of Sc$_5$Ni$_2$Te$_2$ 138
of Sc$_6$PdTe$_2$ 131
of Sc$_3$TnTe$_2$ 132
of Sc$_2$Te   124
of Sc$_8$Te$_3$ 128
of ternary alkali metal scandium halides ASc$_x$X$_3$ 111 ff
of ThSi$_2$ 160
of V$_{8-s}$Sb$_9$ 184
of YBO$_3$ 299
of Y$_5$Cu$_2$Te$_2$ 138
of Y$_5$Fe$_2$Te$_2$ 137
of YbSi$_{1.4}$ 161 ff
of Zn(spcp)(OH)   204

of ZrSb   184
of (Zr,Ti)Sb   185
of ZrSb$_2$ 181
of Zr$_7$Sb$_2$ 186
of (Zr,V)$_{11}$Sb$_8$ 185
structure-directing agents   330
superacidic media   37
surfactants   330
synthesis
   of polycationic clusters   37
   of W(Mo)/Ag/S polymers   210
   of mesoporous materials   323, 329
   rationally designed   207

t

tetrachloroaluminates   38, 45
tellurides
   binary   124
   rare-earth metal-rich   121 ff
tethered phosphine ligand   235
thiolates   207 ff
thiometalate polymers
   cage-like clusters   215
   chains with square units   212
   double chains   213
   hanging ladder chains   213
   helical chains   212
   linear chains   209
   wave-like chains   211
   zig-zag chains   211
titanium
   dimers (Ti$_2$)   5
   vapor   3
trans influence   259
transition metal antimonides
   valence-electron poor   180 ff
   valence-electron rich   177 ff
transition metal organosulfur compounds   193 ff
transition metal thiolates   207 ff
transmission electron microscopy (TEM)
   dehydrocoupling   60
   of MCM-41, -48 (HRTEM)   338
   of Y$_{17.33}$(BO$_3$)$_4$(B$_2$O$_5$)$_2$O$_{16}$ (HRTEM)   295
   of YbSi$_{1.4}$ (HRTEM)   161

v

valence bond (VB) theory   167
valence shell electron pair repulsion (VSEPR) theory   167
valence electron concentration (VEC)   145
vapor phase transport, chemical   38
vaterite-type orthoborates   297, 304

W

weakly coordinating anions (WCA)   36, 65
wonder metal   43

X

xenon compounds   16, 25

Z

ZBLAN glasses   365, 370, 378
zeolites   319 ff
Zintl-Klemm concept   143
Zintl Phases   143 ff, 170 ff
   electronic structures   143
   extended antimony units   172
   molecular antimony units   171
   of tetrelides   143 ff